Lecture Notes in Computer Science   6776

Commenced Publication in 1973
Founding and Former Series Editors:
Gerhard Goos, Juris Hartmanis, and Jan van Le

Masaaki Kurosu (Ed.)

Human Centered Design

Second International Conference, HCD 2011
Held as Part of HCI International 2011
Orlando, FL, USA, July 9-14, 2011
Proceedings

 Springer

Volume Editor

Masaaki Kurosu
The Open University of Japan
2-11 Wakaba, Mihama-ku, Chiba-shi, Chiba 261-8586 Japan
E-mail: masaakikurosu@spa.nifty.com

ISSN 0302-9743   e-ISSN 1611-3349
ISBN 978-3-642-21752-4   e-ISBN 978-3-642-21753-1
DOI 10.1007/978-3-642-21753-1
Springer Heidelberg Dordrecht London New York

Library of Congress Control Number: 2011929164

CR Subject Classification (1998): H.5.2, H.5, H.3-4, C.2, K.4, D.2, K.6

LNCS Sublibrary: SL 3 – Information Systems and Application, incl. Internet/Web and HCI

Typesetting: Camera-ready by author, data conversion by Scientific Publishing Services, Chennai, India

Printed on acid-free paper

Springer is part of Springer Science+Business Media (www.springer.com)

Foreword

The 14th International Conference on Human–Computer Interaction, HCI International 2011, was held in Orlando, Florida, USA, July 9–14, 2011, jointly with the Symposium on Human Interface (Japan) 2011, the 9th International Conference on Engineering Psychology and Cognitive Ergonomics, the 6th International Conference on Universal Access in Human–Computer Interaction, the 4th International Conference on Virtual and Mixed Reality, the 4th International Conference on Internationalization, Design and Global Development, the 4th International Conference on Online Communities and Social Computing, the 6th International Conference on Augmented Cognition, the Third International Conference on Digital Human Modeling, the Second International Conference on Human-Centered Design, and the First International Conference on Design, User Experience, and Usability.

A total of 4,039 individuals from academia, research institutes, industry and governmental agencies from 67 countries submitted contributions, and 1,318 papers that were judged to be of high scientific quality were included in the program. These papers address the latest research and development efforts and highlight the human aspects of design and use of computing systems. The papers accepted for presentation thoroughly cover the entire field of human–computer interaction, addressing major advances in knowledge and effective use of computers in a variety of application areas.

This volume, edited by Masaaki Kurosu, contains papers in the thematic area of human centered design (HCD), addressing the following major topics:

- Human centered design methods and tools
- Mobile and ubiquitous interaction
- Human centered design in health and rehabilitation applications
- Human centered design in work, business and education
- Applications of human centered design

The remaining volumes of the HCI International 2011 Proceedings are:

- Volume 1, LNCS 6761, Human–Computer Interaction—Design and Development Approaches (Part I), edited by Julie A. Jacko
- Volume 2, LNCS 6762, Human–Computer Interaction—Interaction Techniques and Environments (Part II), edited by Julie A. Jacko
- Volume 3, LNCS 6763, Human–Computer Interaction—Towards Mobile and Intelligent Interaction Environments (Part III), edited by Julie A. Jacko
- Volume 4, LNCS 6764, Human–Computer Interaction—Users and Applications (Part IV), edited by Julie A. Jacko
- Volume 5, LNCS 6765, Universal Access in Human–Computer Interaction—Design for All and eInclusion (Part I), edited by Constantine Stephanidis
- Volume 6, LNCS 6766, Universal Access in Human–Computer Interaction—Users Diversity (Part II), edited by Constantine Stephanidis

- Volume 7, LNCS 6767, Universal Access in Human–Computer Interaction—Context Diversity (Part III), edited by Constantine Stephanidis
- Volume 8, LNCS 6768, Universal Access in Human–Computer Interaction—Applications and Services (Part IV), edited by Constantine Stephanidis
- Volume 9, LNCS 6769, Design, User Experience, and Usability—Theory, Methods, Tools and Practice (Part I), edited by Aaron Marcus
- Volume 10, LNCS 6770, Design, User Experience, and Usability—Understanding the User Experience (Part II), edited by Aaron Marcus
- Volume 11, LNCS 6771, Human Interface and the Management of Information—Design and Interaction (Part I), edited by Michael J. Smith and Gavriel Salvendy
- Volume 12, LNCS 6772, Human Interface and the Management of Information—Interacting with Information (Part II), edited by Gavriel Salvendy and Michael J. Smith
- Volume 13, LNCS 6773, Virtual and Mixed Reality—New Trends (Part I), edited by Randall Shumaker
- Volume 14, LNCS 6774, Virtual and Mixed Reality—Systems and Applications (Part II), edited by Randall Shumaker
- Volume 15, LNCS 6775, Internationalization, Design and Global Development, edited by P.L. Patrick Rau
- Volume 17, LNCS 6777, Digital Human Modeling, edited by Vincent G. Duffy
- Volume 18, LNCS 6778, Online Communities and Social Computing, edited by A. Ant Ozok and Panayiotis Zaphiris
- Volume 19, LNCS 6779, Ergonomics and Health Aspects of Work with Computers, edited by Michelle M. Robertson
- Volume 20, LNAI 6780, Foundations of Augmented Cognition: Directing the Future of Adaptive Systems, edited by Dylan D. Schmorrow and Cali M. Fidopiastis
- Volume 21, LNAI 6781, Engineering Psychology and Cognitive Ergonomics, edited by Don Harris
- Volume 22, CCIS 173, HCI International 2011 Posters Proceedings (Part I), edited by Constantine Stephanidis
- Volume 23, CCIS 174, HCI International 2011 Posters Proceedings (Part II), edited by Constantine Stephanidis

I would like to thank the Program Chairs and the members of the Program Boards of all Thematic Areas, listed herein, for their contribution to the highest scientific quality and the overall success of the HCI International 2011 Conference.

In addition to the members of the Program Boards, I also wish to thank the following volunteer external reviewers: Roman Vilimek from Germany, Ramalingam Ponnusamy from India, Si Jung "Jun" Kim from the USA, and Ilia Adami, Iosif Klironomos, Vassilis Kouroumalis, George Margetis, and Stavroula Ntoa from Greece.

This conference would not have been possible without the continuous support and advice of the Conference Scientific Advisor, Gavriel Salvendy, as well as the dedicated work and outstanding efforts of the Communications and Exhibition Chair and Editor of HCI International News, Abbas Moallem.

I would also like to thank for their contribution toward the organization of the HCI International 2011 Conference the members of the Human–Computer Interaction Laboratory of ICS-FORTH, and in particular Margherita Antona, George Paparoulis, Maria Pitsoulaki, Stavroula Ntoa, Maria Bouhli and George Kapnas.

July 2011   Constantine Stephanidis

Organization

Ergonomics and Health Aspects of Work with Computers

Program Chair: Michelle M. Robertson

Arne Aarås, Norway
Pascale Carayon, USA
Jason Devereux, UK
Wolfgang Friesdorf, Germany
Martin Helander, Singapore
Ed Israelski, USA
Ben-Tzion Karsh, USA
Waldemar Karwowski, USA
Peter Kern, Germany
Danuta Koradecka, Poland
Nancy Larson, USA
Kari Lindström, Finland

Brenda Lobb, New Zealand
Holger Luczak, Germany
William S. Marras, USA
Aura C. Matias, Philippines
Matthias Rötting, Germany
Michelle L. Rogers, USA
Dominique L. Scapin, France
Lawrence M. Schleifer, USA
Michael J. Smith, USA
Naomi Swanson, USA
Peter Vink, The Netherlands
John Wilson, UK

Human Interface and the Management of Information

Program Chair: Michael J. Smith

Hans-Jörg Bullinger, Germany
Alan Chan, Hong Kong
Shin'ichi Fukuzumi, Japan
Jon R. Gunderson, USA
Michitaka Hirose, Japan
Jhilmil Jain, USA
Yasufumi Kume, Japan
Mark Lehto, USA
Hirohiko Mori, Japan
Fiona Fui-Hoon Nah, USA
Shogo Nishida, Japan
Robert Proctor, USA

Youngho Rhee, Korea
Anxo Cereijo Roibás, UK
Katsunori Shimohara, Japan
Dieter Spath, Germany
Tsutomu Tabe, Japan
Alvaro D. Taveira, USA
Kim-Phuong L. Vu, USA
Tomio Watanabe, Japan
Sakae Yamamoto, Japan
Hidekazu Yoshikawa, Japan
Li Zheng, P. R. China

Human–Computer Interaction

Program Chair: Julie A. Jacko

Sebastiano Bagnara, Italy
Sherry Y. Chen, UK
Marvin J. Dainoff, USA
Jianming Dong, USA
John Eklund, Australia
Xiaowen Fang, USA
Ayse Gurses, USA
Vicki L. Hanson, UK
Sheue-Ling Hwang, Taiwan
Wonil Hwang, Korea
Yong Gu Ji, Korea
Steven A. Landry, USA

Gitte Lindgaard, Canada
Chen Ling, USA
Yan Liu, USA
Chang S. Nam, USA
Celestine A. Ntuen, USA
Philippe Palanque, France
P.L. Patrick Rau, P.R. China
Ling Rothrock, USA
Guangfeng Song, USA
Steffen Staab, Germany
Wan Chul Yoon, Korea
Wenli Zhu, P.R. China

Engineering Psychology and Cognitive Ergonomics

Program Chair: Don Harris

Guy A. Boy, USA
Pietro Carlo Cacciabue, Italy
John Huddlestone, UK
Kenji Itoh, Japan
Hung-Sying Jing, Taiwan
Wen-Chin Li, Taiwan
James T. Luxhøj, USA
Nicolas Marmaras, Greece
Sundaram Narayanan, USA
Mark A. Neerincx, The Netherlands

Jan M. Noyes, UK
Kjell Ohlsson, Sweden
Axel Schulte, Germany
Sarah C. Sharples, UK
Neville A. Stanton, UK
Xianghong Sun, P.R. China
Andrew Thatcher, South Africa
Matthew J.W. Thomas, Australia
Mark Young, UK
Rolf Zon, The Netherlands

Universal Access in Human–Computer Interaction

Program Chair: Constantine Stephanidis

Julio Abascal, Spain
Ray Adams, UK
Elisabeth André, Germany
Margherita Antona, Greece
Chieko Asakawa, Japan
Christian Bühler, Germany
Jerzy Charytonowicz, Poland
Pier Luigi Emiliani, Italy

Michael Fairhurst, UK
Dimitris Grammenos, Greece
Andreas Holzinger, Austria
Simeon Keates, Denmark
Georgios Kouroupetroglou, Greece
Sri Kurniawan, USA
Patrick M. Langdon, UK
Seongil Lee, Korea

Zhengjie Liu, P.R. China
Klaus Miesenberger, Austria
Helen Petrie, UK
Michael Pieper, Germany
Anthony Savidis, Greece
Andrew Sears, USA
Christian Stary, Austria

Hirotada Ueda, Japan
Jean Vanderdonckt, Belgium
Gregg C. Vanderheiden, USA
Gerhard Weber, Germany
Harald Weber, Germany
Panayiotis Zaphiris, Cyprus

Virtual and Mixed Reality

Program Chair: Randall Shumaker

Pat Banerjee, USA
Mark Billinghurst, New Zealand
Charles E. Hughes, USA
Simon Julier, UK
David Kaber, USA
Hirokazu Kato, Japan
Robert S. Kennedy, USA
Young J. Kim, Korea
Ben Lawson, USA
Gordon McK Mair, UK

David Pratt, UK
Albert "Skip" Rizzo, USA
Lawrence Rosenblum, USA
Jose San Martin, Spain
Dieter Schmalstieg, Austria
Dylan Schmorrow, USA
Kay Stanney, USA
Janet Weisenford, USA
Mark Wiederhold, USA

Internationalization, Design and Global Development

Program Chair: P.L. Patrick Rau

Michael L. Best, USA
Alan Chan, Hong Kong
Lin-Lin Chen, Taiwan
Andy M. Dearden, UK
Susan M. Dray, USA
Henry Been-Lirn Duh, Singapore
Vanessa Evers, The Netherlands
Paul Fu, USA
Emilie Gould, USA
Sung H. Han, Korea
Veikko Ikonen, Finland
Toshikazu Kato, Japan
Esin Kiris, USA
Apala Lahiri Chavan, India

James R. Lewis, USA
James J.W. Lin, USA
Rungtai Lin, Taiwan
Zhengjie Liu, P.R. China
Aaron Marcus, USA
Allen E. Milewski, USA
Katsuhiko Ogawa, Japan
Oguzhan Ozcan, Turkey
Girish Prabhu, India
Kerstin Röse, Germany
Supriya Singh, Australia
Alvin W. Yeo, Malaysia
Hsiu-Ping Yueh, Taiwan

Online Communities and Social Computing

Program Chairs: A. Ant Ozok, Panayiotis Zaphiris

Chadia N. Abras, USA
Chee Siang Ang, UK
Peter Day, UK
Fiorella De Cindio, Italy
Heidi Feng, USA
Anita Komlodi, USA
Piet A.M. Kommers, The Netherlands
Andrew Laghos, Cyprus
Stefanie Lindstaedt, Austria
Gabriele Meiselwitz, USA
Hideyuki Nakanishi, Japan

Anthony F. Norcio, USA
Ulrike Pfeil, UK
Elaine M. Raybourn, USA
Douglas Schuler, USA
Gilson Schwartz, Brazil
Laura Slaughter, Norway
Sergei Stafeev, Russia
Asimina Vasalou, UK
June Wei, USA
Haibin Zhu, Canada

Augmented Cognition

Program Chairs: Dylan D. Schmorrow, Cali M. Fidopiastis

Monique Beaudoin, USA
Chris Berka, USA
Joseph Cohn, USA
Martha E. Crosby, USA
Julie Drexler, USA
Ivy Estabrooke, USA
Chris Forsythe, USA
Wai Tat Fu, USA
Marc Grootjen, The Netherlands
Jefferson Grubb, USA
Santosh Mathan, USA

Rob Matthews, Australia
Dennis McBride, USA
Eric Muth, USA
Mark A. Neerincx, The Netherlands
Denise Nicholson, USA
Banu Onaral, USA
Kay Stanney, USA
Roy Stripling, USA
Rob Taylor, UK
Karl van Orden, USA

Digital Human Modeling

Program Chair: Vincent G. Duffy

Karim Abdel-Malek, USA
Giuseppe Andreoni, Italy
Thomas J. Armstrong, USA
Norman I. Badler, USA
Fethi Calisir, Turkey
Daniel Carruth, USA
Keith Case, UK
Julie Charland, Canada

Yaobin Chen, USA
Kathryn Cormican, Ireland
Daniel A. DeLaurentis, USA
Yingzi Du, USA
Okan Ersoy, USA
Enda Fallon, Ireland
Yan Fu, P.R. China
Afzal Godil, USA

Ravindra Goonetilleke, Hong Kong
Anand Gramopadhye, USA
Lars Hanson, Sweden
Pheng Ann Heng, Hong Kong
Bo Hoege, Germany
Hongwei Hsiao, USA
Tianzi Jiang, P.R. China
Nan Kong, USA
Steven A. Landry, USA
Kang Li, USA
Zhizhong Li, P.R. China
Tim Marler, USA

Ahmet F. Ozok, Turkey
Srinivas Peeta, USA
Sudhakar Rajulu, USA
Matthias Rötting, Germany
Matthew Reed, USA
Johan Stahre, Sweden
Mao-Jiun Wang, Taiwan
Xuguang Wang, France
Jingzhou (James) Yang, USA
Gulcin Yucel, Turkey
Tingshao Zhu, P.R. China

Human-Centered Design

Program Chair: Masaaki Kurosu

Julio Abascal, Spain
Simone Barbosa, Brazil
Tomas Berns, Sweden
Nigel Bevan, UK
Torkil Clemmensen, Denmark
Susan M. Dray, USA
Vanessa Evers, The Netherlands
Xiaolan Fu, P.R. China
Yasuhiro Horibe, Japan
Jason Huang, P.R. China
Minna Isomursu, Finland
Timo Jokela, Finland
Mitsuhiko Karashima, Japan
Tadashi Kobayashi, Japan
Seongil Lee, Korea
Kee Yong Lim, Singapore

Zhengjie Liu, P.R. China
Loïc Martínez-Normand, Spain
Monique Noirhomme-Fraiture,
    Belgium
Philippe Palanque, France
Annelise Mark Pejtersen, Denmark
Kerstin Röse, Germany
Dominique L. Scapin, France
Haruhiko Urokohara, Japan
Gerrit C. van der Veer,
    The Netherlands
Janet Wesson, South Africa
Toshiki Yamaoka, Japan
Kazuhiko Yamazaki, Japan
Silvia Zimmermann, Switzerland

Design, User Experience, and Usability

Program Chair: Aaron Marcus

Ronald Baecker, Canada
Barbara Ballard, USA
Konrad Baumann, Austria
Arne Berger, Germany
Randolph Bias, USA
Jamie Blustein, Canada

Ana Boa-Ventura, USA
Lorenzo Cantoni, Switzerland
Sameer Chavan, Korea
Wei Ding, USA
Maximilian Eibl, Germany
Zelda Harrison, USA

HCI International 2013

The 15th International Conference on Human–Computer Interaction, HCI International 2013, will be held jointly with the affiliated conferences in the summer of 2013. It will cover a broad spectrum of themes related to human–computer interaction (HCI), including theoretical issues, methods, tools, processes and case studies in HCI design, as well as novel interaction techniques, interfaces and applications. The proceedings will be published by Springer. More information about the topics, as well as the venue and dates of the conference, will be announced through the HCI International Conference series website: http://www.hci-international.org/

General Chair
Professor Constantine Stephanidis
University of Crete and ICS-FORTH
Heraklion, Crete, Greece
Email: cs@ics.forth.gr

Table of Contents

Part I: Human Centered Design Methods and Tools

Part II: Mobile and Ubiquitous Interaction

Part III: Human Centered Design in Health and Rehabilitation Applications

Part IV: Human Centered Design in Work, Business and Education

Part V: Applications of Human Centered Design

Part I

Human Centered Design Methods and Tools

Investigating Users' Interaction with Physical Products Applying Qualitative and Quantitative Methods

Chun-Juei Chou[1] and Chris Conley[2]

[1] Department of Industrial Design, College of Planning and Design,
National Cheng Kung University, No. 1, University Road, Tainan City 70101, Taiwan (R.O.C.)
cjchou@mail.ncku.edu.tw
[2] gravitytank, inc. 114 West Illinois Street, Floor 3, Chicago, IL 60654, USA
chris.conley@gravitytank.com

Abstract. When using products, people are sometimes involved in activities other than the products' primary use. Some of these activities are peripheral, while others may reinforce people's experiences with the products. The latter is related to the focus of this research – user engagement. User engagement is defined as a situation in which a product provides one or more additional features related to its primary function, so the user engages more senses through the product experience. This research investigates how six product samples engage subjects. The result shows that the six product samples can engage users and therefore result in an interesting user–product relationship. Based on the subjects' reactions, user engagement can be categorized into at least three types: sensory, physical, and emotional engagement. In addition, products can enable user engagement because they possess particular properties that represent mimicking, inspiring, or staging a function.

Keywords: user engagement, engaging products, user-product interaction.

1 Introduction

When using products, people are sometimes involved in activities other than the products' primary use. Some of these activities are peripheral, while others may reinforce people's experiences with the products. The latter is related to the focus of this research–*user engagement*. User engagement is defined as a situation in which a product provides one or more additional features related to its primary function, so the user engages more senses through the product experience. For example, if a toaster is transparent, the user can see the bread darkening; the transparent sides stage the toasting process as a visually engaging performance. The user has additional interaction or meaning with the toaster. Another example is a tape dispenser with a simple odometer. In addition to acquiring tape, the person who uses such a dispenser can observe how much tape has been dispensed in terms of distance. This kind of product satisfies people who expect participation, contribution, or involvement when using products. Interestingly, people use "engaging products" in a way that is frequent, intense, active, vivid, or complete, etc.

M. Kurosu (Ed.): Human Centered Design, HCII 2011, LNCS 6776, pp. 3–12, 2011.
© Springer-Verlag Berlin Heidelberg 2011

2   User Engagement - A Special Case of User-Product Interaction

Several studies summarized in Table 1 discuss specific user behavior related to user engagement. Engagement could be a synonym for participation, involvement, and immersion. It refers to a person taking part in an activity in order to affect the activity or to be affected by the activity. In fact, studying on user engagement is valuable in understanding how people participate in activities. It is common that memorable experience can be occasioned by a process, prop, souvenir, activity, and/or interactivity. Designers can apply these elements to enhance user engagement. This research follows this concept and user engagement is defined thusly for this research:

When a person uses a product, he or she also participates in another activity enabled by the product other than the primary use. The person:

. is entertained by the activity through his or her senses and participation, or
. creates something as a part of the activity and enjoys the result, or
. escapes temporarily from his or her normal daily reality, or
. is attracted by an interesting aspect of the activity.

That is, the person has more senses, actions and/or feelings engaged by the product. As a result, his or her product experience is reinforced.

Table 1. Studies related to user engagement

Domain	Researchers	Characteristics
Design	Wright et al.[1]	Four aspects to describe experience: (1) the compositional aspect, (2) the sensual aspect, (3) the emotional aspect, and (4) the spatio-temporal aspect.
Marketing	Pine and Gilmore[2]	Three elements to engage consumers: (1) shopping process as a performance, (2) services as a stage, and (3) goods as props.
	Pine and Gilmore[2] Gupta and Vajic[3]	Three levels of consumer engagement: (1) theme, (2) central activity, and (3) supporting activity.
Social interaction	Kearsley and Shneiderman[4]	Three components of engagement in learning: (1) relate, (2) create, and (3) donate.
	Chung[5]	Four phases of social engagement: (1) initiation, (2) participation, (3) cooperation, and (4) solidarity.
	Falk[6]	Four elements for live role-playing games: (1) character, (2) costume, (3) props, and (4) stage.
Interaction design	Blackwell et al.[7] Hoven et al.[8]	TUIs makes user-product interaction more engaging.
	Nack[9]	Four aspects of experiences: (1) attributes, (2) presence, (3) temporality, and (4) interactivity.
	Hoven[10]	Interactive devices function as souvenirs
Product design	Overbeeke et al.[11]	How to make products engaging: (1) beauty in interaction, (2) rich actions, and (3) irresistibles.

3 Selecting Product Samples

Based on the definition of user engagement, six primary criteria for assessing product candidates are determined. On the contrary, five secondary criteria are considered in order to narrow the scope and control the quality and categories of product candidates. Next, to verify the six primary criteria, six professionals in product design are involved, including three males and three females. They are practicing product designers or product managers who have more than 10 years of professional experience. Analytic methods applied in this verification include a well-structured interview and a digital questionnaire, both of which take for around one hour. Besides, a web-based video editing tool for processing interview data was used for protocol analysis. This step helps to rate, refine the primary criteria and add new ones for assessing product candidates. However, the five secondary criteria are not examined.

Then, the refined criteria are clear enough for determining if one product is engaging or not. Any product that meets any of the criteria is a potential product candidate. To collect product candidates, hundreds of products were examined online. For variety, only one candidate was selected among products that provided the same functions and that met the same set of criteria. Subsequently, 16 product candidates are gathered, each of which meets two to five criteria, respectively.

Last, to select product samples, an online survey is applied to examine the 16 product candidates. This survey consists of 16 product pairs, including the 16 product candidates and 16 typical products that provide similar functions. Thus, it is easy for online respondents to rate each product candidate against to the six primary criteria in a five-point rating scale. For statistical significance, there are at least 30 respondents who rates 4 of the 16 product pairs. For randomization, the questionnaire was presented in four different sequences; therefore, the resulting data was less susceptible to respondent fatigue. In this way, whether each of the product candidates is a qualified product sample can be determined based on respondents' examination. Regarding methods, Independent-Samples T Test is applied to examine qualified product samples. Factor analysis and mean rating are used to examine the effectiveness of the six survey questions (criteria). Finally, the six product samples are selected as shown in Fig. 1.

4 Qualitative User Investigation

This research argues that the six product samples possess particular properties that foster specific types of user engagement. The product properties and user engagement can be identified through product trials and subject interview, that is, the qualitative investigation of user engagement.

4.1 Investigation Design

The purpose of this investigation is to uncover ordinary users' experience with the six product samples, focusing on the type(s) of user engagement that each product sample enables and the distinctive product properties that foster user engagement. Thus, this research requests each subject to use one product sample for one week at home or at

Dinnersaurs™, a Fork and Spoon Set for Kids

Still Life Fruit Bowl

Sunbeam®/Oster® Toaster

Viceversa® Weight Scale

Pop Art Toaster™

FlexibleLove™ Chair

Fig. 1. Six Product Samples

work. The reasons are three-fold: (1) The subjects needed time to become familiar with the product sample. The one-week trial allowed them to use the product sample in daily life. (2) The one-week trial minimized the sample's novelty and the subject's curiosity. (3) Unlike in a laboratory, the subjects felt comfortable using the product at home or at work. Next, an observational interview was conducted to investigate each subject's actual experience with the product sample. Each interview was videotaped.

This investigation recruits 24 subjects, four for each product sample. There are three requirements for selection: (1) Each subject must have no expert knowledge of design-related domains in order to represent ordinary users. (2) Each subject must be willing to use the product sample for one week at home or at work. (3) The four subjects who try out the same product sample must comprise two males and two females of different backgrounds. In addition, one subject, a graduate student in design, was recruited for a pilot test in order to improve the investigation.

The investigation contains seven steps:

1.   Introduction – Each subject was given a product sample, a one-time-use camera, and instruction guidelines. The disposable camera was used to capture the subject's experiences with the product sample. The instruction guidelines presented tasks for the subjects to perform.
2.   Product trial – Each subject used a product sample for one week at home or at work. During this period, he or she took pictures to document significant experience with the product sample. All pictures were developed before the follow-up interview.
3.   Meeting with each subject – At the beginning of the observational interview, three questions were asked:
   .   Where did you use the product, and why did you use it there?
   .   How many days have you used the product for?
   .   How many times total was the product used?

4. Observation – The researcher observed how each subject typically used the product sample at home or at work. Thus, the subject's situational experiences with the product sample were recorded.
5. Interview Pictures – Each subject explained the pictures that he or she took. The purpose was to enable the subject to actively discuss user engagement and product properties without biased prompting.
6. Interview Question 1 – Please tell me about your experience with this product sample in comparison to an everyday product. Each subject was encouraged to talk about his or her experience with the product sample and with a corresponding typical product.
7. Interview Question 2 – What specific functions, features, or aspects of this product enable the experience that you just described?

Throughout the observational interview, the researcher asked follow-up questions depending on the subject's responses in order to specify experiences, especially user engagement, with the product sample.

To analyze the 24 subjects' verbal reports, a web-based video editing tool was applied for protocol analysis. Each subject's verbal report related to user engagement is paraphrased for better comprehension. Therefore, user engagement enabled by each product sample is determined.

4.2 How the Six Product Samples Engage Subjects- One Example

The Pop Art Toaster™ engaged subjects in four ways (Table 2). Fig. 2 shows three subjects' experiences with the branded toast. First, the toaster elicited sensory engagement. When the toast popped up, four subjects liked to check the images made

Table 2. How the Pop Art Toaster™ engages subjects

Types of Engagement	Description
. Sensory engagement	. When the toast pops up, four subjects like to check the images made on the toast.
. Emotional engagement	. The toast amuses three subjects and provokes an emotional response.
. Physical engagement (1)	. One subject likes to play or make a joke with the toast in front of his family or friends.
. Physical engagement (2)	. When using the toaster, one subject is inclined to toast an extra slice of bread.

Fig. 2. Subjects' reactions with the branded toast

on the toast. This engagement fulfilled their curiosity, amused them, and reminded them of what to do next time in order to create better images on the toast.

The branded toast also provoked an emotional response. For example, the smiley face image plate gave the three subjects a funny feeling, because they imagined that the toast was smiling at them. Three subjects reported that branding images on toast enabled a wonderful breakfast experience in the morning. Regarding physical engagement, one subject liked to make a joke with the toast in front of family or friends, for example, by placing the smiley face toast next to her own smiling face. With a slice of "BITE ME" bread in hand, one female user pretended to bite a chunk of bread and laugh wildly. Another female user held up the "i'm hot!" bread to indicate that the toast was complimenting her. This type of user engagement enabled social interaction and allowed subjects to share amusing experiences with others.

One subject was inclined to toast an extra slice of bread in order to create different images or better images. This engagement allowed for a test through which the subject found out how to correctly brand images on bread. In addition, the toaster engaged one subject by relating to a larger activity. This female subject hosted a toasting party and invited her female friends to make pop art toast, taking turns or using the toaster together. This type of engagement turned breakfast into a social occasion.

Regarding product properties, the subjects reported that the Pop Art Toaster™ differs from a typical toaster in appearance and use. The six removable image plates were key components that inspired the subjects to brand images on the toast. However, because the image plates are not fastened components, they make the toaster's appearance less attractive. These accessories also affect toaster use. Two image plates must be selected and correctly installed in order to brand pleasing images on the toast. This is not necessary with a typical toaster. Thus, the Pop Art Toaster™ engages subjects, but its appearance and ease of use are compromised.

4.3 Types of User Engagement and Product Properties

The qualitative user investigation shows that the six product samples can engage users and therefore result in an interesting user–product relationship. Based on the resulting reactions, user engagement can be categorized into at least three types, sensory, physical, and emotional engagement, each of which represents an additional interaction and product meaning. To articulate, *sensory* engagement takes place when products engage the user's senses. Sensory engagement is similar to what tourists do from the top of a skyscraper; they comprehend the distant scenery by seeing and listening. For example, the Sunbeam®/Oster® toaster can visually engage users. *Physical* engagement takes place when products cause users to act. Physical engagement is similar to how a person acts if given a stick to hold. He may become a baseball player, an orchestra conductor, or even a Jedi knight. With imagination and enjoyment, physical engagement can be triggered by many objects, including products. For example, the Still Life Fruit Bowl can engage users in playful activities. *Emotional* engagement takes place when products evoke specific ideas and/or feelings that affect users. Emotional engagement is similar to the way in which people are inspired by reading an old diary or when given a precious souvenir. For example, the Viceversa® Weight Scale can cause users to imagine they are lighter (or heavier).

In addition, the qualitative user investigation reveals that the six product samples can enable user engagement because they possess particular properties that represent mimicking, inspiring, or staging a function. *Mimicking* means that the appearance of a product is designed to simulate a character, object, or circumstance. For example, the Dinnersaurs[TM], a fork and spoon set for kids, simulates dinosaurs. *Inspiring* means that a product is designed to be associated with another interesting object, activity, or circumstance. For example, the Still Life Fruit Bowl associates a pile of fruit with a still life painting. *Staging a function* means that a product is designed to present a compelling view of an invisible mechanism or to display its attractiveness. For example, the Sunbeam®/Oster® toaster with viewing window allows users to see the internal wires. These three attributes potentially foster user engagement.

5 Quantitative User Investigation (Online Survey)

In the previous section, how the six product samples engaged subjects is investigated. The finding is confirmed through a quantitative user investigation- online survey.

5.1 Investigation Design

The purpose of the online survey is to confirm (1) the types of user engagement that each product sample enables and (2) the product properties that foster these types of user engagement. The online survey includes six product samples. Each product sample was compared to a corresponding typical product (similar to the product pairs in section 3). The questions were developed according to the user engagement and product properties investigated in the previous section. For randomization, the questions were presented in four different sequences. Therefore, the resulting data were less susceptible to participant fatigue. In this way, how the six product samples engage users can be confirmed. For statistical significance, this online survey required at least 30 respondents for each product sample. Respondents must have no expert knowledge in design-related domains.

The procedure for taking the digital questionnaire is summarized as follows.

1. Obtaining respondent information – Each respondent provides their name, gender, specialty, nationality, and age.
2. Presenting the product – Pictures and description of each product appear. After the respondent has reviewed the product, he or she proceeds to the next page.
3. Rating the product – The respondent rates the product by answering questions.
4. Rating the corresponding typical product – The respondent rates the corresponding typical product by answering the same set of questions.
5. Rating other product pairs – The respondent rates other product pairs by answering related questions.

The online respondents' answers indicated whether they agree with the user engagement and product properties investigated in the previous section. The author assumes that if a product sample is rated significantly higher than its corresponding typical product, then the question that specifies the user engagement or product property is confirmed. In order to calculate statistical significance, Independent-Samples T Test was applied.

5.2 Result of Online Survey- One Example

For the Pop Art Toaster[TM], 32 online respondents are involved in this quantitative user investigation. Table 3 shows that all types of user engagement are supported by the online respondents. The author assumes two reasons for this result. First, branding images on bread makes sense to online respondents, and they are excited about it. This feature would significantly affect the experience of using the toaster. Second, online respondents are familiar with a typical toaster in appearance, use, and function. Compared to a typical toaster, it is easier for them to assess what the Pop Art Toaster[TM] could prompt them to do, as the survey questions describe.

Before the online survey, the author doubted that Question 4 would be confirmed. The reason is that people usually toast slices of bread for breakfast or snack; extra slices are only toasted when the user prepares breakfast for others or during the first few uses of the Pop Art Toaster[TM]. Thus, the online respondents likely agree with Question 4, because they have never used the Pop Art Toaster[TM]. They answered the question based on what they would do at the first time if they use it.

The questions verifying product properties are also supported. That is, the online respondents confirm that the Pop Art Toaster[TM] would inspire them to make special pieces of toast. The image plates and the different ways of use are two factors that enable user engagement.

The quantitative user investigation presents the validation of this research. The identified product properties and user engagement were confirmed by an online survey. The results show that (1) the six product samples provide additional interaction and meaning to engage users; (2) there are at least three types of product attributes that can foster user engagement (mimicking, inspiring, and staging a function); and (3) user engagement can be sensory, physical, and/or emotional. The causal relationship between product properties and user engagement would reveal

Table 3. Results for the Pop Art Toaster[TM]

Type of Engagement	Description supported by online respondents
. Sensory engagement	Q1: When the toast pops up, I would like to check the pattern made on it. (Four subjects)
. Emotional engagement	Q2: The toast from this toaster would amuse me and provoke an emotional response. (Three subjects)
. Physical engagement (1)	Q3: I would like to play or make a joke with the toast in front of my family or friends. (One subject)
. Physical engagement (2)	Q4: When using this toaster, I would be inclined to toast an extra slice of bread. (One subject)
Product Property	**Description**
. Appearance	Q5: This toaster's accessories enable an interesting experience. (Four subjects)
. Use	Q6: Using this toaster would result in an interesting experience. (Four subjects)
. Attribute: inspiring	Q7: This toaster would inspire me to make a special piece of toast. (Four subjects)

patterns among the product samples. By relating product properties to the elements of engagement, guidelines for enabling user engagement can be discussed. These guidelines must lead to a preliminary framework for generating product ideas that enable user engagement.

6 Conclusion

To conclude, this research claims that an engaging activity is comprised of an original activity and at least one reinforcing activity, both of which attract people to participate. As a result, each person involved in the engaging activity has a better experience. In terms of user-product relationship, the original activity is use of the primary function of a product. The reinforcing activity is a user-product interaction in which the person senses more, acts more, and/or feels more during the product experience.

This research shows that the six product samples can engage users and therefore result in an interesting user–product relationship. The results show that products that enable user engagement are currently on the market. Thus, user engagement is an existing product value that certain users desire. Based on the subjects' reactions, user engagement can be categorized into at least three types, sensory, physical, and emotional engagement, each of which represents an additional interaction and product meaning. This research shows that products can enable user engagement, because they possess particular properties that represent mimicking, inspiring, or staging a function. Thus, it is clear that typical products can become engaging through the addition of mimicking, inspiring, or staging a function.

Not surprisingly, people use engaging products in a way that is frequent, intense, active, vivid, or complete, etc. User engagement makes the user-product relationship more interesting. It is different from, but as significant as, domains such as functionality, usability, aesthetics, interaction, pleasure, and emotion, all of which are important to satisfy user needs. It satisfies users who expect participation, contribution, or involvement when using products. The author expects that this research will lead to a better understanding of how products can be designed to enable user engagement, providing additional interaction and meaning with products and bringing another dimension of value to product design.

References

1. Wright, P., Mccarthy, J., Meekison, L.: Making Sense of Experience. In: Blythe, M.A., Overbeeke, K., Monk, A.F., Wright, P.C. (eds.) Funology: from Usability to Enjoyment, pp. 43–53. Kluwer Academic, Dordrecht (2005)
2. Pine II, B.J., Gilmore, J.H.: The Experience Economy- Work Is Theatre & Every Business a Stage. Harvard Business School Press, Massachusetts (1999)
3. Gupta, S., Vajic, M.: The Contextual and Dialectical Nature of Experiences. In: Fitzsimmons, J., Fitzsimmons, M. (eds.) New Service Development: Creating Memorable Experiences, pp. 33–51. Sage, London (1999)
4. Kearsley, G., Shneiderman, B.: Engagement Theory: A framework for technology-based teaching and learning. Educational Technology 38(5), 20–23 (1998)

5. Chung, Y.-C.: How Can Design Support Collaborative Experience in Human-Product Interaction? Master's Thesis, The School of Design. Carnegie Mellon University, Pittsburgh, PA, USA (2005)
6. Falk, J.: Interfacing the Narrative Experience. In: Blythe, M.A., Overbeeke, K., Monk, A.F., Wright, P.C. (eds.) Funology: from Usability to Enjoyment, pp. 249–256. Kluwer Academic, Dordrecht (2005)
7. Blackwell, A.F., Fitzmaurice, G., Holmquist, L.E., Ishii, H., Ullmer, B.: Tangible user interfaces in context and theory. In: Conference on Human Factors in Computing Systems, CHI 2007 Extended Abstracts on Human Factors in Computing Systems, pp. 2817–2820 (2007)
8. Hoven, E., van den Frens, J., Aliakseyeu, D., Martens, J.-B., Overbeeke, K., Peters, P.: Design Research & Tangible Interaction. In: Proceedings of the 1st International Conference on Tangible and Embedded Interaction, pp. 109–115 (2007)
9. Nack, F.: Capturing Experience- a matter of contextualising events. In: Proceedings of the 1st ACM MM WS on Experiential Telepresence (ETP 2003), pp. 53–64 (2003)
10. van den Hoven, E.A.W.H.: Graspable Cues for Everyday Recollecting. Doctoral dissertation, the J. F. Schouten School for User-System Interaction Research, Technische Universiteit Eindhoven (2004)
11. Overbeeke, K., Djajadiningrat, T., Hummels, C., Wensveen, S., Frens, J.: Let's Make Things Engaging. In: Blythe, M.A., Overbeeke, K., Monk, A.F., Wright, P.C. (eds.) Funology: from Usability to Enjoyment, pp. 7–17. Kluwer Academic, Dordrecht (2005)

Human Interaction and Collaborative Innovation

Kevin A. Clark

Content Evolution LLC,
510 Meadowmont Village Circle, Suite 320
Chapel Hill, North Carolina 27517 USA
ce@contentevolution.net

Abstract. Collaborative innovation is on the rise. The tools, techniques and technologies to foster human interaction in the service of collaborative innovation are increasing every year. Interactions that lead to win-win outcomes are also on the rise. Examples in this paper include the IBM advisory council program and process, the design of a global cross-company and cross-culture derivative in Content Evolution Labs, and the emergence of EduPresence to drive a global education network using telepresence and other technology-enabled forms of interactive learning. Samples of collaborative innovation techniques are explored, including the team use of Post-Its™, journey mapping, and voting and group-commitment.

Keywords: Advisory, board, collaboration, collaborative, commitment, connection, continuous, coping, council, diversity, education, human, innovation, interaction, interactor, members, membership, outsourcing, progress, team, technology, telepresence.

1 Introduction

Human interaction is gravitating to collaboration – and technology-enabled collaboration is speeding things up. Collaborative behaviors unheard of just a few years ago, or just months or weeks ago are becoming commonplace. The speed of interaction and connection increases each day – as is the ability to create and participate in non-zero-sum activities, where multi-party winning is maximized and losing is minimized.

Creating the basis for trust amongst "interactors" is required for breakthrough collaborations. *Interactor* is a term borrowed from biology, referring to the individual evolutionary paths of organisms. In this case, the interactors are <u>the participants that interact with each other in a collaborative innovation process</u>. This paper focuses on both how interactors and their worldviews evolve, along with the groups they choose to join. More background about the originals of the term interactor here:

- **"Interactor** is a term used to describe a part of an organism with evolution selection acts upon.[1] Interactors are the individual evolutionary paths which are subject to real-life interactions, such as phenotype and the outward traits most affected by natural selection. In this way the interactor interacts with the environment in a way which creates differential reproduction." - http://en.wikipedia.org/wiki/Interactor
- "Memes are also Interactors" by H.C.A.M. Speel, Delft University of Technology - http://www.hanscees.com/memesym.htm

M. Kurosu (Ed.): Human Centered Design, HCII 2011, LNCS 6776, pp. 13–21, 2011.
© Springer-Verlag Berlin Heidelberg 2011

2 IBM Advisory Councils

One of the ways the IBM management team stays abreast of customer wants and needs is the use of global advisory councils. Major programs both started and ended in many of these gatherings. It is a powerful way to make forecasting research, technology roadmaps, and business plans come alive in the minds and hearts of the leadership team.

IBM ThinkPad advisory councils sponsored by the Personal Systems Group (pre-Lenovo transaction) represent a good example of human interaction and collaborative innovation. During a particularly frothy era of inventiveness in the personal computer business over a ten year period of time, these councils harness collective intelligence and professional diversity to drive collaborative success. They were formed around the world to advise the management team about crucial decisions – and at the high-water-mark there were 17 of these membership councils in operation around the world by the year 2000:

- Industry Advisory Councils (6) – composed of industry analysts, magazine editors, and selected industry consultants:
    - o Asia-Pacific Industry Advisory Council
    - o China Industry Advisory Council
    - o European Industry Advisory Council
    - o Japan Industry Advisory Council
    - o Latin American Industry Advisory Council
    - o North American Industry Advisory Council
- Customer Advisory Councils (7) – composed of information technology (IT) executives from some of IBM's most largest companies:
    - o Asia-Pacific Customer Advisory Council
    - o China Customer Advisory Council
    - o Japan Customer Advisory Council
    - o Latin American Customer Advisory Council
    - o North American Customer Advisory Council
    - o Northern European Customer Advisory Council
    - o Southern European Customer Advisory Council
- Road Warrior Advisory Council – entrepreneurs and small business owners that run their businesses on the road using notebook computers and mobile technology.
- Marketing Advisory Board
- Scientific Advisory Board
- Human-Centered Innovation Advisory Board

These councils and boards are composed of culturally cohesive, and language-common groups, with 20-25 interactors each, and an expectation of 15 to 18 of them showing up for any given gathering. They form "human salt marshes" – where the diversity of the expertise in the room helps create broader worldviews and wider avenues of potential action.

In a physical salt marsh there is a great deal of biological diversity. The place where the ocean touches the land is a point of rich interaction with brackish water and lots of material moving back and forth between ocean, fresh water, and ground materials and life forms.

The same diversity is required for the selection of participant interactors - it is pivotal to the success of the groups over time. They must possess:

- Exceptional industry knowledge and insights
- An open, collaborative spirit when behind closed doors
- Trustworthiness in keeping the secrets of and maintaining the intellectual property rights of other interactors.

These are physical meetings – in-person gatherings that helped people bond to each other and create the basis for trust. We structure rooms so all participants can see each other and have the same relative status to each other – either in a circle or in a U-shape pattern.

A major innovation spawned in these gatherings is "Conjoint Live" – a method pioneered by several members of the management team, yet none more vibrant than those sessions hosted by IBM PC company executive Leo Suarez.

Similar to the work done to segment the global customer marketplace, this is a live feature/function tradeoff exercise based on real product roadmaps with industry experts and some of IBM's most valuable customers. "Would you rather have *this* or *this*? – a standard part of the dialogue in these sessions. Later, after getting to a smaller set of features and functions, the third dimension of price was added – "Do you still want this if it adds US$XX.xx to the price of the ThinkPad?"

The tradeoffs were documented visually in the room on easel charts or whiteboards to make the results known and easily accessible during discussions.

What make these sessions work is an underlying sense of trust amongst the participants. They had been working with each other for years in this setting. Outside the context of these advisory groups, especially for the industry advisors (consultants, industry analysts, magazine editors), these professionals were natural competitors. IBM's regular schedule and format for these meetings made collaboration both intense and candid.

As a result of these in-person meetings, interactors as trusted cohorts can also be called upon between meetings to be briefed by teleconference and offer advice, or answer brief questions by e-mail.   The willingness to participate and the candor is similar to the physical meetings due to imprinting and bonding that takes place over time in-person.

3   Content Evolution Labs

As described in the IBM advisory council case, trusted cohorts can have powerful influences over time and distance. To bring these methods into a new and contemporary multi-organization context, **Content Evolution Labs** (or **[c]e Labs**) is a designed membership organization that will draw on these experiences and is dedicated to cross-organization *collaborative innovation*.

[c]e Labs is designed to co-create a continuous global innovation conversation in the fields of human engagement and interaction. **[c]e Labs** provides a surrogate research and development (R&D) function for professions lacking internal capacity for this research in the past.   Discoveries will have implications for a wide range of leadership and management applications, including general management, marketing, human resources, human-centered design, and supply-chain and stakeholder communications.

Content Evolution Labs is designed to be cross-cultural, cross-discipline, and cross-industry. It has an approach centered on creating mutual trust inside the membership network.

[c]e Labs functions in a way similar to 24/7 global open-source programming for software development, where code is constantly improving each hour through the contributions of many programmers. In the case of Content Evolution Labs, the focus is not on software – it is on co-creating continuous innovation by invited members. Ideas are held private for a brief period of time for relevant pre-market differentiation and action by members, before all co-created concepts are published on an annual basis. Content Evolution Labs functions for the mutual benefit of its members and for the greater advancement of represented member disciplines.

[c]e Labs exists to benefit its members with a process that drives a global and continuous innovation conversation. It provides the basis for innovation across multi-organization boundaries. The rapid embrace and progress of nanotechnology is a good example of this cross-discipline and cross-organization catalyst common denominator. Breakthrough innovation requires exposure to thinking that is not captive to any single organization, profession, or culture. Competitive advantage is delivered in the differentiated marketplace expressions of what members collaboratively discover together.

Game-changers come from new business models; new alliances; cross-pollination interests that before were considered proprietary and competitive, vs. new collaborative and cooperative pre-competitive idea space.

Content Evolution Labs is both a pre-competitive collaborative conversation and a form of *innovation outsourcing.*

[c]e Labs is designed to be a series of intimate interactor groups consisting of approximately 15 - 20 members each – and a planned 14 boards around the world. Membership is limited and by invitation only. These groups are chartered geographically – and are administered by trusted curators to ensure cultural and professional diversity.

The curator function is designed to provide both a "home base" for each interactor cohort, and connectivity back to the global **[c]e Labs** conversation and mission. The curators also provide translation as needed from native language into English – the base linguistic currency of the labs worldwide. As with the IBM advisory councils, careful selection of interactor members and curators for the labs is essential to the success of the global organization.

Curated groups are meant to meet twice yearly – once in a smaller recurring base cultural cohort – and once with all members annually providing an opportunity for broad professional cross-pollination.

Monthly cycle design:

- 15th hold two cohort meetings
- 1st teleconference summary of last council proceedings for all members; agenda for next council meetings
  Yearly cycle design:

- May to November – member board meetings
- December to February – idea consolidation and writing
- March – annual all member Spring Strategy Summit
- April – previous year proceedings report to members
- August – public summary of previous year proceedings

Content Evolution Labs is designed to create its own self-evolving agenda and direction. It is an adaptive and resilient organization designed to draw on the collaborative intelligence of its interactor members.

4 Telepresence: The Network Yet to Be Created

One of the great untapped potential interactive technologies of our time is telepresence. It holds so much untapped energy – yet it functions as an inert bundle of technology instead of a vibrant gateway to personal learning over vast distances.

Telepresence was designed and deployed on the metaphor of the executive boardroom, now distributed to connect leaders over time and distance. It functions in the same role as an executive jet – exclusive and generally unobtainable by the masses.

Now imagine if there were a public network of addressable telepresence facilities in hotels and community centers where conferences have never taken place...where no college or university has ever been in local driving distance for those eager to learn. Where experts who have retired to golf resorts can connect to institutions of learning conveniently and offer their expertise to campuses small, large, and virtual.

What if you could have dinner at a conference with an internationally recognized expert in an intimate setting with several other professionals after a stimulating day of interaction – and it was within a two hour drive from your home? What if deep conversations in rooms were connected with other conversations being held in different places and countries? What if it were possible to converse with a rich diversity of thoughts and experiences building insights to change your world?

What if you were on an ocean cruise and decided to learn something each day from an expert on land from a noted university?

This is the potential of telepresence as yet untapped. The elitist beginnings of the telepresence movement need to reach mainstream locations, and drive a rapidly expanding network effect.

Further, telepresence as hardware needs network applications drive greater purpose. The applications we have in mind drive the telepresence medium to greater outcomes. If you ask someone for a referencefor "telepresence training" – there is no such education available. You can learn to be a great public speaker, a riveting classroom professor, an on-camera television personality, an on-stage actor – yet nothing is currently available to fully exploit the capabilities of the telepresence medium.

What telepresence as a pure technology lacks in processes, training, and local hosting skills, is now being developed in an initiative called **EduPresence**.

4.1 Harnenessing Telepresence for Education

EduPresence is a global conversation that sits at the intersection of continuous learning, delivery diversity, and executive education. Technology advances continue to enable and enhance the opportunities for learning with diverse forms of delivery, both in-person, distance, and virtual. We see new locations taking on educational roles over time, such as hotels and places where people vacation and work. We see classrooms and conferences taking place where the creative classes are gathering, yet where no classroom has ever been or conference has ever taken place.

EduPresence is committed to delivering replicable wisdom, innovative relevance, and practical solutions to bring about a step-function evolution in the desire for and benefits of lifetime learning at both at a personal and societal level. It is a reflection of a new personal learning system available to all in the not too distant future.

New information is being created faster than ever – and being shared at the speed of light. The coping skills required navigate and make sense of this tidal wave of knowledge can be found in the accumulated wisdom and frameworks of a well-rounded liberal arts education. Refracted through the lens of practical concerns such as business, commerce, governance, and the good society – fostering a new yearning for lifetime learning and other ways of thinking tied to a commitment for useful applications moves from a utopian ideal to a practical necessity.

EduPresence is a catalyst for a new educational conversation.

EduPresence is committed to new forms of continuous learning tied to practical action.

EduPresence is convening this catalytic conversation for collaboration amongst liberal arts colleges, business schools, schools of government, school systems, government systems, and technology innovators, executive leaders, thought leaders, and global enablement and delivery organizations.

We don't have time to wait to create a global movement of leaders and professionals committed to the continuous learning needed to bring about a thrivable world. This is personal level change coupled with new and diverse education delivery methods and processes to drive human success and significance.

4.2   Interactors in Haiti and the Connective

After the earthquake in Haiti, many aid organizations dispatched relief teams and supplies to the ravaged city of Port-au-Prince and surrounding region. They acted independently and in many cases at cross-purposes for transportation access, resource distribution, and overall coordination for effectiveness.

Work by David Hodgson online created critically needed connections here in the U.S. so these organizations could know each other. Aid organizations that had never interacted with each other were connected by Hodgson in the U.S. so they could more effectively coordinate operations on the ground in Haiti.

Similar roles have been present at the time of this writing are roiling countries and cultures in Northern Africa and the Middle East. The era of spontaneous collaboration enabled by internet-mediated connections is gaining traction. This has implications for governance at both the country and company level.

Hodgson is taking his work and learning to a new initiative called The Connective that holds the promise of matching and connecting professionals and resources to critical needs in the marketplace for social good. He is taking social networking to a new level of principle and purpose.

4.3 Techniques for In-Person Interaction and Collaborative Innovation

Post-Its. The humble Post-It™ invented by 3M and now widely available under a variety of brands, is a uniquely useful tool for collaborative work sessions and strategy workshops. While there are a number of electronic versions of group input and collaborative work environments – highly useful when people are collaborating remotely, nothing beats Post-Its for in-person collaboration.

Getting people to express ideas simultaneously to a single question is the power of the Post-It. It eliminates groupthink and gives everyone in the room an equal voice. Ask a question, have the group write down their answers all at once, and then reveal the answers one-by-one. Put them up on a wall or chart paper and group them. You now have a permanent record of what every person said in the room – in their own handwriting.

Take pictures as the session progresses. Once taken, you can rearrange and re-cluster the Post-Its to drive new conversations and insights.

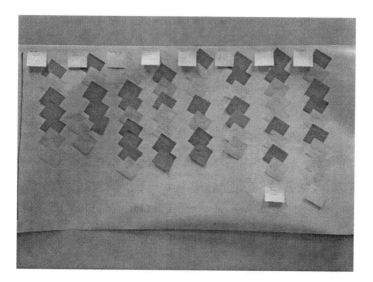

Fig. 1. Collaborative ideas collected with Post-Its

Journey Mapping. You can also take the ideas created on Post-Its and arrange them into a journey over time. This can be for customers or any constituency you wish to understand and serve better. Name the critical points in time and put key ideas underneath – both ideas that will help make the experience being studied better, and ideas that hold back satisfaction and delight.

Multi-track journey mapping is used to look at touch-points across customer and organization boundaries – and how the different people and points in time affect each other. Multi-track mapping reveals insights and dependencies not found in single-constituency journey depictions.

Let the map stay in a space where new ideas can be added over time – this way it functions as continued stimulus for innovation, not a static artifact from a single work session.

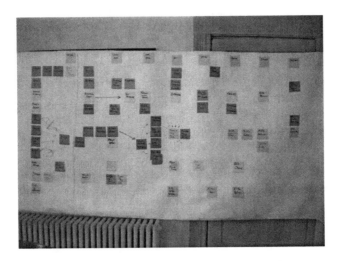

Fig. 2. Journey mapping by idea category over time

Voting and Group Commitment. Great planning sessions that have robust interaction and create new innovative concepts will inevitably exceed resources to act on all of them. This is where the group can work to get a handful of ideas "nominated" and then "elected to office."

This concept has been regularly delivered for the past seven years as a case study about customer and market segmentation at Duke University Fuqua School of Business – and now taught at the University of Colorado at Boulder Leeds School of Business, University of Rochester Simon Graduate School of Business, and Yale University School of Management.

If you want people to act on the results of a collaborative strategy – invite them to participate and interact in its development. Find out what they need to know to act effectively; find out what they would need to embrace the outcomes and discover commitment.

In-person exercises can use simple colored sticky circles to vote for ideas participants find most valuable. These can then be prioritized into a set of actions the group can develop a preliminary plan about and a rough timeline. Assign time to come back with actions and assessments – get names assigned to tasks – leave with a sense of shared commitment and progress.

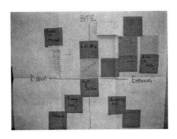

Fig. 3. A sample group commitment prioritization chart

5 Conclusion

The tools, techniques and technologies for interactors to collaborate and create meaning are on the rise. Desirable win-win outcomes for multiple economic actors become increasingly likely every day. IBM advisory council programs, the emergence of Content Evolution Labs, EduPresence, and The Connective all point to global collaboration and technology-enabled forms of interactive learning. A co-created and more meaningful global conversation is underway.

A Pattern Approach for Designing Application

Kohei Daimaru and Buntaro Kaji

IBM Japan
{daimaruru,cazy06}@gmail.com

Abstract. We propose to include a pattern approach in design-process for application. Pattern approach is typically used as reference of best practice. In this case, we use pattern of current operation as input to inspect requirement of next-new application. We tried case study and studied about those results.

Keywords: Design, Pattern Language, Application, Requirement.

1 Introduction

1.1 About Pattern Language

Originally, Pattern Language is theories about architectural design proposed by Christopher Alexander (born October 4, 1936 in Vienna, Austria).

Patterns are ways to describe best practices, explain good designs, and capture experience so that other people can reuse these solutions. When a designer is designing something, they must make many decisions about how to solve problems. A single problem, documented with its most common and recognized good solution seen in the wild, is a single design pattern. Each pattern has a name, a descriptive entry, and some cross-references, much like a dictionary entry. A documented pattern must also explain why that solution may be considered a good one for that problem, in the given context. [1],[2]

1.2 Patterns in Software

The concept of Pattern Language has been inherited in Software engineering. For example, "wiki" was made with Pattern Language and citation of architectural pattern.

Ward Cunningham developed first "wiki" (WikiWikiWeb in Portland, Oregon in 1994, and installed it on the Internet domain c2.com on March 25, 1995). He have adopted some pattern from Patterns in Urban Architecture. [3]

1.3 Leading Research about Patterns in Software Design

In papers, there is advanced research about Patterns in software design. [A Pattern Approach to Interaction Design] [4] shows outline of present of Patterns in perspective and example usage of patterns for interaction design.

M. Kurosu (Ed.): Human Centered Design, HCII 2011, LNCS 6776, pp. 22–27, 2011.

2 Case Study

In this chapter, we present about case study. First, making Patterns of current operation about "Claim expenses" in 2.1. Second, extracting Application requirement from that patterns in 2.2. In the end, Showing proposal of Application form those requirement in 2.3.

2.1 Patterns of Current Operation

We have conducted hearing to current users who are using application system to claim expenses about that know-how, how to use it more efficiently. And we have made patterns of current usage from result of hearing.

■ Name:
No1. Submit all at once

■ Context
They have needs to claim action in short time, without interrupting their major work.

■ Problem
If they have many things to claim, they should have use many time to do against that. It is waste of time.

■ Solution
They do it once in every duration (it depends on that person)

■ Example
Stocking evidence or having memo, and in the end of month, they checked them and claim expenses all at once.

■ References
This pattern matches in case of having many things to claim but it is also valid for submitting several things at once.

Fig. 1. An Example of current operation's Pattern (No1. Submit all at once)

And we classified Patterns into Category and User Roll. Category is property of Pattern. In this case, Category is composed of "Action" is know-how about operation and "Remind" is things to keep in mind. And There are two main User Roll those are "Applicant" and "Approver" as basic work-flow system.

Table 1. Mapping of current operation's Patterns

Category＼User Roll	Applicant	Approver
Action:	1 Submit all at once	6 Review for certain
	2 Input same as evidence	-
	3 Alert to rejection	-
	4 Submit in fast	-
Remind:	5 Alert to forget submitting	7 Take care of pending things

2.2   Outcome of Application Requirement from Current Requirement

We had an insight into Application Requirement from current operation's Patterns as a hypothesis, and listed them corresponding current requirement.

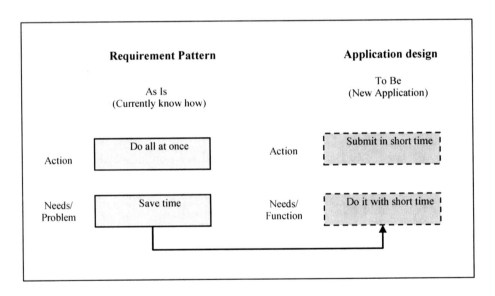

Fig. 2. Image of flow of deriving Current Requirement to Application Requirement

Table 2. Mapping of Current Requirement×Application Requirement

Current Pattern	Application Requirement	
1 Submit all at once	1.	Submit plural things at a time
2 Input same as evidence	2.	Automatic Validate function to check difference of Input form and evidence.
	3.	Automatic Input function same as evidence.
3 Alert to rejection	4.	Display to alert about rejection
4 Submit in fast	5.	Remainder or Alert for due date
5 Alert to forget submitting	6.	Cross function with other relational system (For example, relation to)
	7.	Access by various platform and add and stock contents at once
6 Review for certain	8.	Automatic Validate function to check difference of Input form and evidence and assist reviewing
7 Take care of pending things	9.	Display to alert for pending things

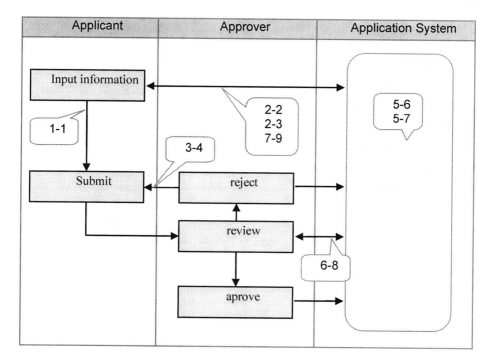

Fig. 3. Operation flow of Application with compatible Application Requirement

2.3 Designing of Application from Requirements

We designed application from Application Requirement. It is simple application that can access from web browser and local application.

Applicants can use the application with smart phone. And they can take a picture of till receipt and input automatically with OCR function. Approver can user widget of the Application and can operate of check things easily.

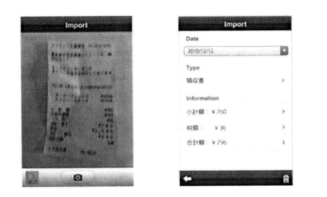

Fig. 4. Screen Image of Application for Applicant

3   Conclusions

3.1 Lookout in Result of Case Study

In Application system development project, we define Application requirement in first step (generally in requirement definition phase) by hearing or interview for clients or end users. We expect that we can discover more numerous requirements efficiently with pattern approach. Especially, potential user requirements are hard to discover from general hearing. Business analyst is usually familiar with current operation. If all project members share Pattern approach, it has possible to share knowledge of business side and development side and improve Application architecture and design.

3.2 Further Research

What scene can we adopt Pattern Approach efficiently? We expect it is effective when we can get voice of expert using current operation, lead user often add function or customize it for usage, often represents of potential needs. And typically in Business Application or own Application as services provided by themselves is adopted for that, we may have a tendency survey about users because we can figure out actual users. It also has potential for adopting to the scene when we plan new application by referring analog operation.

To improve Pattern approach for Applications, we should define details of how to extract patterns, how to derive Application requirements from patterns efficiently in short times. And in actual development project, it is essential to consider about cost and technical aspect so that we need to set priority of requirement from Pattern Approach and consider it with other entity. Application work flow is depends on type of operation. There is room to analyze operations of affairs and make framework of requirement by Application type.

References

1. Alexander, C.: A Pattern Language. Kajima Institute Publishing Co., Ltd. (1977)
2. Alexander, C.: Timeless Way of Building. Kajima Institute Publishing Co., Ltd. (1979)
3. Eto, K.: Pattern, Wiki, XP ~ Timeless principles of creation (2009)
4. Borchers J.O.: A Pattern Approach to Interaction Design (2000)
5. Walkman 2.1 Development Background Wikipedia, (as of January 30, 2011),
   http://ja.wikipedia.org/wiki/%E3%82%A6%E3%82%A9%E3%83%BC%E3%
   82%AF%E3%83%9E%E3%83%B3#.E9.96.8B.E7.99.BA.E7.B5.8C.E7.B7.AF
6. Matsuo, H., Ogawa, S.: Innovating Innovation: The Case of Seven-Eleven Japan (2000)

A Holistic Model for Integrating
Usability Engineering and Software Engineering
Enriched with Marketing Activities

Holger Fischer, Karsten Nebe, and Florian Klompmaker

University of Paderborn, C-LAB, Fürstenallee 11,
33102 Paderborn, Germany
{holger.fischer,karsten.nebe,florian.klompmaker}@c-lab.de

Abstract. To support the integration of usability engineering and software engineering this paper analyses corresponding international standards and introduces a model that consists of activities and artifacts highlighting dependencies, similarities and possible points for integration. In addition the model presents activities that serve as potential integration points for the third discipline of marketing. By using this model processes can be aligned easier on a common information base (e.g. activities, artifacts). Innovative thinking will be forced by considering the business perspective of marketing activities likewise.

Keywords: Integration, Usability Engineering, Software Engineering, Marketing, Marketing Research, Standards ISO 9241-210, ISO/TS 18152, ISO/IEC 12207 and ISO/IEC 15504.

1 Introduction

In todays industry usability has already been recognized as an important quality aspect of the software development process. However, the integration of usability engineering and software engineering still remains to be a challenge in practice [1], even though implications of good usability are obvious: End-users will be able to work more effectively and efficiently, but it is also beneficial for the developing organization in different aspects, e.g. in monetary form, such as the reduction of support- and training-costs. Thus, usability is not an exclusive attribute of the generated product; it is also a fundamental attribute for the development process itself. Furthermore the usability of the product will have major impact on its proposition on the market. According to the increasing competition and the variety of alternative solutions a product will only survive on the market if it is easy to use and if it supports the user's needs. A good example is Apple Inc. that provides products that are easy to use and well accepted by costumers and the market. Consequently it is assumed by the users that even 'any' future product will have the same benefits in terms of usability and therewith this will increase the market relevance of company's products. It can be said that usability is an important marketing aspect [2].

In this paper the authors present an approach that integrates usability engineering with software engineering and that identifies potential links for marketing, too. This

M. Kurosu (Ed.): Human Centered Design, HCII 2011, LNCS 6776, pp. 28–37, 2011.
© Springer-Verlag Berlin Heidelberg 2011

includes an equal consideration of interests and aims for all these three disciplines. In order to do so, the authors analyzed existing integration approaches and identified the abstraction level of standards as a basis for the integration. Common activities have been identified; artifacts and their dependencies are worked out and extended by marketing perspective. As a result a holistic model for integration was created. This was evaluated based on the results of interviews with experts from usability.

2 Background

In order to create a common integration model of the three addressed disciplines usability engineering, software engineering and marketing it is important to create a common understanding.

Usability engineering (UE) is a discipline that involves user participation during the development of software and systems and ensures the effectiveness, efficiency and satisfaction of the product through the use of a usability specification and metrics [3]. Software engineering (SE) is a discipline that addresses the whole software lifecycle (from the phase of requirements specification up to the maintenance of a released product) in a systematic and predictable way adopting several engineering approaches [4, 5].

Both disciplines use process models to plan and systematically structure the activities and tasks to be performed during software creation. In general these models detail activities, the sequence in which these activities have to be performed as well as the resulting deliverables (in various levels of abstraction). The goal is to define a process where the project achievement does not depend on individual efforts of particular people or fortunate circumstances and aims to create good software. The essential attributes of good software are maintainability, dependability, security, efficiency and acceptability [5]. Even like in UE and SE especially the attribute of acceptability is an important marketing characteristic.

Marketing (and Marketing Research) (MM) is a discipline that offers market-oriented business strategies to satisfy customer needs and expectations [6, 7]. One important issue that needs to be considered is that the customer is not necessarily the end user. This discrepancy exists in current practice not only in marketing but also in software development: The customer is not (always) the user. However, the end user is a promising stakeholder for MM even like the customer, as the users needs will have a major impact on the success of the product. Benefits for SE appear in less maintenance and those for UE are obvious – the satisfaction of the users. Thus, this seems to be one common dominator of all three disciplines.

While looking on integration approaches a considerable amount can be found - at least for UE and SE - in theory and practice. The integration with MM is still an open issue, which will be addressed later on in this paper.

2.1 Existing Integration Approaches

While looking on existing approaches the authors investigated in those for UE and SE primarily. The reason is because SE is usually the driver in software development and the amount of existing integration activities with UE is remarkable. An investigation on these two disciplines should therefore serve as a basis to build on.

Most of the UE and SE integration approaches can be organized along four general categories [8, 9]: Approaches that a) concern the concrete implementation; b) present a common specification; c) address the definition of processes and process models; and d) focuses on abstract or generic approaches.

Those *concerning the concrete implementation* define activities and artifacts as well as links to existing SE activities. For example, Ferre [10] specified a set of 51 generic UE techniques that had been reviewed by experts following some criteria, e.g. how adaptable the techniques are to software development processes, how the applicability of the techniques is in general, how much it costs to perform the techniques and how the acceptance is in the field of human computer interaction. To ensure the relation of the techniques to SE activities the concepts and the terminology of SE have been adapted. A second group of integration approaches presents a *common specification*. For example, Juristo et al. [11] developed an approach concerning the measuring and evaluation of usability topics using architectural patterns embedded in the system architecture design. Therefore they identified several patterns of UE and adopted them to SE. Other integration approaches address the *definition of processes and process models distinguishing* between independent UE models and SE models with integrated UE activities. As an example, Düchting et al. [12] discuss existing agile process models (e.g. Scrum or eXtreme Programming [13]) for the implementation of UE activities and derived practise-oriented recommendations. A fourth group of integration approaches *focuses on abstract or generic approaches* specifying general conditions to be considered for the integration. As an example, Metzker and Reiterer [14] present an 'Evidence-Based Computer-Aided Usability Engineering Environment (CAUSE)' for the organizational level of an organization. They use the paradigm of a situation-based decision-making and developed a process meta-model supporting the selection of UE methods.

As the variety of UE and SE integration approaches show there is a lot of research going on. However, the discipline of marketing and its integration appears to be very sparse. Only few approaches can be found such as [15, 16]. They focus on user centered design processes and try to implement activities or results in the product innovation process. These can be mapped to the third category of defining processes and process models.

It can be said that most of the approaches for integration were applied in practice appear to be on an operational level, which results in very specific activities or customized methods. The challenge is that these approaches cannot be easily transferred to any other situation in practice. Therefore it seems to be promising to investigate in a more abstract view of integration.

2.2 Hierarchy of Integration Approaches

While looking on the different types and categories of integration approaches for UE and SE, Nebe [17] differentiates into three levels of abstractions: 'standards that define the overarching framework, process models that describe systematic and traceable approaches and the operational level in which the models are tailored to fit the specifics of an organization' (see Figure 1). This hierarchy exists in both disciplines software engineering and usability engineering and can be exploited for integration. At the level of standards general integration strategies can be defined that are applicable to a large number of existing development processes. For the purpose of integration Nebe has focused on the SE standard ISO/IEC 12207 [18] and the UE standard ISO 13407 [19].

ISO/IEC 12207 addresses essential processes in the software development lifecycle (from acquisition to maintenance) and supports the development and management of a software product. ISO/IEC 15504 [20] aims at the evaluation of the maturity of a software development process in a specific organization and is also known as SPICE ('Software Process Improvement and Capability dEtermination').

ISO 13407 defines four activities of human-centered design that should take place during system development. In the meantime a revised version of the UE standard Nebe used has been published (ISO 9241-210 [21]). ISO 9241-210 focuses on human-centered design for interactive systems and is mainly structured in four parts: 'understand and specify the context of use', 'specify the user requirements', 'produce design solutions', 'evaluate the design against requirements' [21]. ISO/TS 18152 [22] is based on the ISO 9241-210 and specifies issues for an assessment of human-systems.

Nebe compared standards and defined compliancy and key requirements for the integration of the disciplines. These requirements can be used in addition to existing base practices (based on ISO/TS 18152) to access existing processes in an organization and to specify concrete activities to be performed. This work shows that even when looking on the abstract level of standards concrete strategies on an operational process level can be perceived. Thus, as mentioned before the level of standards seems also to be promising for integration of UE, SE and MM. The idea in behind is, that no process-specific or organizational-details would being an obstacle and the common information exchange would succeed.

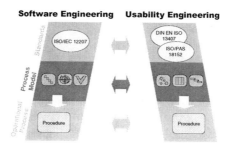

Fig. 1. Similar hierarchies in the two disciplines SE and UE: standards, process models and operational processes [13]

3 Proceedings

Looking forward to create a common integration model the authors have analyzed specific standards for UE and SE. The aim was to create a concrete basis of activities and artifacts for further investigations. In addition, dependencies of those activities and artifacts have been examined to show the flow of information between the activities of each discipline. After that, interdisciplinary dependencies between UE and SE have been analyzed to show the information exchange and the correlation between activities of both perspectives. Then, the authors verified the activities and identified potential gaps according to assessment standards.

Then, the authors analyzed literature to search MM activities, which show commonalities in the outcomes regarding to the artifacts of UE and SE. By doing this, potential integration points are being identified, where MM can enrich the overall development process. Having analyzed and listed the activities and artifacts of all three disciplines the authors visualized them in a holistic model highlighting while showing all the commonalities and interconnections. The goal was to create a clearly laid out and comprehensible picture of the complexity. Finally, the authors evaluated the model in two ways (with the focus on UE). First they compared the activities with the compliancy and key requirements from Nebe [8] to verify the list and to incorporate changes according to revised UE standard ISO 9241-210 (formerly ISO 13407). Second, the authors proved the completeness, the relevance to practice, as well as the correlations between UE and SE artifacts of the UE activities- and artifacts-list by interviewing UE experts.

As a result, the model was partially approved and seems to have potential for the integration of UE, SE and MM. The detailed proceedings will be described as follows.

3.1 Common Activities and Artifacts

As mentioned previously, standards seemed to be a reasonable basis for the development of a holistic model. Starting with the integration of UE and SE, the two standards ISO 9241-210 and ISO/IEC 12207 have been analyzed in detail. Therefore the specification documents have been examined. The goal was to operationalize these standards in terms of defining lists of activities and artifacts that can be used as a basis for an integrated model. While having this in mind, the authors scanned the documentations and looked for 'verbs' that represent potential activities and added them to a list. In addition, potential inputs and outputs as well as dependencies of those activities are identified and added to the list, too. This was done for the UE as well as for the SE standard likewise. As a result, two independent lists were created highlighting all activities and artifacts for the corresponding standard and discipline.

Next, the flow of information was analyzed, as communication is crucial for different disciplines to work together. By doing this, textual correlations of activities and artifacts between the two lists are identified and documented.

Having listed activities from the UE and SE the authors verified them by comparing with Base Practices of the corresponding assessment standards (ISO/TR 18152 for UE; and ISO 15504 for SE). Each base practice has been compared to the list of activities while seeking for semantic relationships between the base practices and an activity. By doing this, the activities- and artifacts-lists' completeness could be verified. By doing this, gaps in the list of base practices have also been noticed but these appeared to be irrelevant for the model. However, as a result of the verification gaps in the fundamental standards (ISO 9241-210 and ISO/IEC 12207) have also been identified. In the process, this has not been further investigated until now. However, it is expected that this originates either from the incompleteness to their corresponding assessment standard or the missing activities have not been identified during the author's analysis of the standards.

An excerpt of the resulting list is shown in Table 1. The activities are listed with a number for identification (e.g. 'AKT.UE.12' – means: activity from UE + increasing identifier), the reference to their textual position in the origin standard (chapter/paragraph) and the reference to their correlated base practices (existing reference number).

Table 1. Examples of identified activities for the perspectives of UE and SE based on the analysis of standards

Reference	Activity	Origin	Base Practice
AKT.UE.12	Identify the characteristics of the users or group of users.	9241-210:6.2.2.b	HS.3.1.BP3
AKT.UE.18	Identify user and other stakeholder needs.	9241-210:6.3.2	HS.1.1.BP5 HS.1.2.BP5 HS.2.3.BP1
AKT.UE.33	Allocate tasks and sub-tasks to the user and to other parts of the system.	9241-210:6.4.2.2	HS.2.7.BP2 HS.3.3.BP1
AKT.SE.8	Specify functions and capabilities of the system.	12207:5.3.2	ENG.1.1.BP1 ENG.1.1.BP2
AKT.SE.36	Develop and document a top-level design for the interfaces external to the software item and between the software components of the software item.	12207:5.3.5.2	ENG.1.3.BP2
AKT.SE.51	Test each software unit and database ensuring that it satisfies its requirements.	12207:5.3.7.2	ENG.1.4.BP3 ENG.1.5.BP4 ENG.1.5.BP6

3.2 Integration of Marketing

In order to add MM details to the approach, basic literature of the discipline has been analyzed [23] in order to identify activities, which have commonalities to the content of the artifacts-lists of UE and SE. Therefore the literature was scanned for 'verbs' that gave hints to similar activities or outcomes. The overall aim was to highlight potential points for integration where results from marketing can enrich the overall development process. The perspective how marketing can obtain was recently not in the focus of investigation. However, there are potential benefits for marketing as considering the users' needs will certainly improve the quality of innovative products.

Having the focus in mind, the authors looked for conductive integration points to an overall process. Examples are the tasks of synchronizing 'the identification of user characteristics' (UE) with the 'analysis of the consumer behavior' (MM) or 'the identification of the system environment' (UE) with the 'analysis of the macro-economic environment' (MM), which both can influence and increase the quality of the context of use description. A complete list of MM activities can be found in [24].

Those similarities have been visualized in the overall diagram of the model, which is exemplarily shown in the next section.

3.3 Visual Model of Integration

In order to present the complex results of investigation the authors created a concrete visualization: the holistic model. This includes the lists of activities, artifacts and their correlations as well as the disciplinary and interdisciplinary dependencies of UE, SE and MM.

Therefore different visual representations were chosen. Activities are presented using squares and artifacts are presented using pentagons. To highlight dependencies between activities and artifacts the authors use arrows. The distance between two or more objects represents interdisciplinary correlations – the closer they are the higher the correlation is. Objects were colorized according to the corresponding discipline (performing the activity or creating the artifact): green = UE, blue = SE, and magenta = MM. An excerpt of the visualized model is shown in Figure 2.

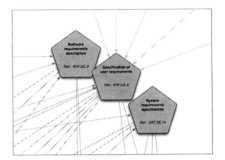

Fig. 2. Examples of the holistic model focusing on integration points for MM (left)

Fig. 3. The requirements specifications as a central artifact (right)

The illustrated model gives an overview of all activities, which have to be performed, and all artifacts, which have to be delivered. Similarities and dependencies are highlighted according to their visual closeness. The flow of information can easily be understood following the directed arrows (shown in Figure 3). The range of the activities to be carried out can be perceived and an approximately timeline of the activities is visualized. In addition the model enables organizations to easily compare and discuss their process along the model and as a result helps to identify integration aspects between UE and SE. The model can also be used to specify new and to enhance or evaluate existing process models.

3.4 Evaluation from UE Perspective

In the first step of evaluation the authors compared the lists of activities with the earlier mentioned compliancy and key requirements according to Nebe [8]. Therefore the textual and semantic existence of each requirement had been checked against the lists of activities. Any not confirmed requirement has been formulated as an activity and has been included in the list. By doing this, the authors wanted to verify the conformance of the list to the integration approach of Nebe and to incorporate changes considering the revised UE standard (ISO 9241-210). The result was an extended list of activities.

In addition the authors performed an evaluation focused on the UE perspective. Therefore they carried out interviews with experts in the field of UE who work in ISO committees as well. At first a guideline for the interviews had been created to operationalize the topics to be questioned. Then the questions had been developed and validated according to their precision, uniqueness and comprehensibility. Further on the lists of UE activities and artifacts had been extended with a five point Likert scale. The aim was to get answers considering the relevance of the model in the current work practice of the experts, the completeness of the lists according to the experts and the prioritization of the activities in regard to their importance in the development process. An additional aim was to review the textual dependencies of artifacts between UE and SE. Finally the experts were interviewed.

As a result the experts confirmed the completeness of the lists of activities and artifacts and prioritized them. They acknowledged that an analysis of the context of use at the beginning of the development process can create an overview of the expected users, their characteristics and the tasks they want to solve. All participating disciplines in a software developing process have specific requirements that can be merged and reviewed in a shared requirements specification (compare Figure 3). In addition design solutions of a UE process should be compared with technical solutions of the SE perspective to discuss possibilities and constraints. The results are primary related to the UE perspective. With regard to the SE perspective no further statements can be made and with regard to the MM perspective no statements can be made. Therefore additional evaluations have to be realized, which the authors plan to do.

4 Summary and Outlook

This paper presents a holistic model for the integration of Usability Engineering (UE) and Software Engineering (SE) as well as Marketing and Marketing Research (MM), which is created, based on standards in UE and SE. Therefore it basically consists of fundamental activities and artifacts from UE and SE, flavored with contributions from MM. Activities are linked with artifacts showing dependencies and relationships. Similarities between the disciplines are highlighted.

The beneficial use of the model is manifold. Hence, the model enables organizations to easily compare and discuss their process along the model and as a result helps to identify integration aspects between UE and SE. The model can also be used to specify new and to enhance or evaluate existing process models. It also shows activities that may serve as potential integration points for a third discipline: marketing and marketing research. Thus, UE and SE can profit from MM activities even as MM can, based on UE. An early involvement of users in the process as well as the extensive analysis of their environment can create innovative ideas with business perspective, too. Existing processes of development may be extended with a MM perspective and also MM may be enriched with innovative ideas resulting from a user centered design process.

Currently the authors plan further validations with experts in the field of SE and MM in order to reflect the integration of all perspectives likewise. Additional perspectives are currently being analyzed to increase the quality of the final product. For example accessibility should be an issue considering that existing software products often lack of access for disabled people.

Further on the authors identified the need of a common understanding of requirements from different perspectives. When creating software, various participating parties specify their own requirements from their individual perspective and speak different domain specific languages. Extended requirements and redundant requirements are formulated as well as contrary requirements. Thus in future the authors want to analyze how requirements are specified and communicated in an understandable manner reducing the addressed problems. Therefore a framework will be developed to share, understand, resolve conflicting and prioritize requirements in a common way.

References

1. Seffah, A., Desmarais, M.C., Metzker, E.: HCI, Usability and Software Engineering Integration: Present and Future. In: Seffah, A., Gulliksen, J., Desmarais, M.C. (eds.) Human-Centered Software Engineering – Integrating Usability in the Software Development Lifecycle. Springer, Heidelberg (2005)
2. Haghirian, P., Madlberger, M., Tanuskova, A.: Increasing Advertising Value of Mobile Marketing – An Empirical Sutdy of Antecedents. In: Proceedings of the 38th Hawaii International Conference on System Sciences, HICSS. IEEE Xplore, Big Island (2005)
3. Faulkner, X.: Usability Engineering. Palgrave Macillan, New York (2000)
4. Balzert, H.: Lehrbuch der Software-Technik – Software-Entwicklung, 2nd edn. Spektrum Akademischer Verlag, Heidelberg (2000)
5. Sommerville, I.: Software Engineering, 9th edn. Pearson Education, Essex (2010)
6. American Marketing Association: Definition of Marketing and Marketing Research, http://www.marketingpower.com/_layouts/Dictionary.aspx?dLetter=M
7. Drucker, P.: Managing in Turbulent Times. Harper & Row Publishers, New York (1980)
8. Nebe, K.: Integration von Usability Engineering und Software Engineering: Konformitäts- und Rahmenanforderungen zur Bewertung und Definition von Softwareentwicklungs prozessen, Doctoral Thesis. Shaker Verlag, Aachen (2009)
9. Nebe, K., Paelke, V.: Key Requirements for Integrating Usability Engineering and Software Engineering. In: Kurosu, M. (ed.) HCII 2011. LNCS, vol. 6776. Springer, Heidelberg (2011)
10. Ferre, X.: Integration of Usability Techniques into the Software Development Process. In: Proceedings of the 25th International Conference on Software Engineering ICSE 2003, Portland, Oregon, pp. 28–35 (2003)
11. Juristo, N., Lopez, M., Moreno, A.M., Sánchez, M.I.: Improving software usability through architectural patterns. In: Proceedings of the 25th International Conference on Software Engineering, ICSE 2003, Portland, Oregon, pp. 12–19 (2003)
12. Düchting, M., Zimmermann, D., Nebe, K.: Incorporating User Centered Requirement Engineering into Agile Software Development. In: Jacko, J.A. (ed.) HCI 2007. LNCS, vol. 4550, pp. 58–67. Springer, Heidelberg (2007)
13. Kniberg, H.: Scrum and XP from the Trenches. Lulu Enterprises, Raleigh (2007)
14. Metzker, E., Reiterer, H.: Evidence-Based Usability Engineering. In: Kolski, C., Vanderdonckt, J. (eds.) Computer-Aided Design of User Interfaces III: Proceedings of the 4th International Conference on Computer-Aided Design of User Interfaces, Valenciennes, pp. 323–336. Academics Publishing, Dordrecht (2002)
15. Mahlke, S., Götzfried, S., Hofer, R.: User Centered Product Innovation – Methoden zur Bewertung von innovativen Produktideen aus Nutzersicht. In: Brau, H., Diefenbach, S., Göring, K., Peissner, M., Petrovic, K. (eds.) Proceedings of the Usability Professionals, German UPA (2010)

16. Sangwan, S., Koh, H.C.: User-centered Design – Marketing Implications from Initial Experience in Technology Supported Products. In: Proceedings of the Engineering Management Conference (2004)
17. Nebe, K., Zimmermann, D., Paelke, V.: Integrating Software Engineering and Usability Engineering. In: Pinder, S. (ed.) Advances in Human-Computer Interaction, pp. 331–350. I-Tech Education and Publishing KG, Vienna (2008)
18. ISO/IEC 12207: Information technology - Software life cycle processes, Amendment 1. ISO/IEC, Genf (2002)
19. ISO 13407: Human-centered design processes for interactive systems. ISO, Genf (1999)
20. ISO/IEC 15504: Information technology - Process Assessment - Part 1-5. ISO/IEC, Genf (2003-2006)
21. ISO 9241-210: Human-centered design processes for interactive systems. ISO, Genf (2010)
22. ISO/TS 18152: Ergonomics of human-system interaction - Specification for the process assessment of human-system issues. ISO, Genf (2010)
23. Kotler, P., Armstrong, G., Saunders, J., Wong, V.: Grundlagen des Marketing. Pearson Studium, Munich (2010)
24. Fischer, H.: Integration von Usability Engineering und Software Engineering: Evaluation und Optimierung eines ganzheitlichen Modells anhand von Konformitäts- und Rahmenanforderungen, Master Thesis (2010), http://nbn-resolving.de/urn:nbn:de:101:1-201101142473

Investigation of Indirect Oral Operation Method for Think Aloud Usability Testing

Masahiro Hori, Yasunori Kihara, and Takashi Kato

Graduate School of Informatics, Kansai University,
2-1-1 Ryozenji-cho, Takatsuki, Osaka 569-1095 Japan
{horim,tkato}@res.kutc.kansai-u.ac.jp

Abstract. Usability testing with prototypes is typically conducted with a concurrent think-aloud protocol. Due to the simultaneous process of prototype operation and verbalization, participants of the think-aloud testing sometimes say very little and are likely to become silent when they are required to think abstractly or complete complex tasks. In this paper, we propose a method of user operation with oral instruction, which facilitates thinking aloud because oral operation would help participants to keep a continuous flow of verbalization. To investigate the quantity and quality of utterances made during think aloud protocols, we conducted a comparative study between oral and conventional manual operation methods. The study was carried out with two test objects: an interactive prototype of a touch-screen digital camera and photo album software with standard mouse/keyboard user interface. Our results demonstrated that the oral operation method was more effective in drawing more utterances for explanation and observation that would be an important source of discovering usability problems although the effect was dependent on the user interface of test objects.

Keywords: Concurrent think-aloud protocols, keep talking, usability testing, prototype evaluation.

1 Introduction

Prototyping is widely used for the evaluation of design ideas, in particular, aspects of usability and user experiences [1]. Prototypes are becoming richer and more interactive [2][3], and that would be helpful for testing usability of emerging products enhanced their capabilities by input components such as touch screen and sensors. However, interactive prototypes may not necessarily achieve operational performance expected in the final products, and that would expose slow or inaccurate responses to test users. The limitation of an interactive prototype in its early development stage is that the users cannot distinguish if flaws they experienced are intrinsic problem of the artifact or come from insufficient operational performance of the prototype.

Usability testing with prototypes is typically conducted with a concurrent think-aloud protocol, which is the process of having the potential users verbalize what they are thinking while completing tasks with an artifact to be tested [4][5][6][7]. Think-aloud protocols have been widely used to collect qualitative data through the actual

M. Kurosu (Ed.): Human Centered Design, HCII 2011, LNCS 6776, pp. 38–46, 2011.
© Springer-Verlag Berlin Heidelberg 2011

use of an artifact [8][9]. The assumptions underlying the think-aloud method is to provide insights from the information resides in the short-term memory in parallel with the users' thought process [10]. To put it another way, the think-aloud method allows to reveal cognitive activities that may not be visible without users' verbal reports.

In usability testing with think-aloud protocols, users are instructed to continuously verbalize their thoughts. However, if users keep silent for a long time, the verbalization will not be usable because significant information in the short-term memory may not be tracked down [11]. The difficulty thus stems from the simultaneous process of verbalization and cognitive task to be made relying on the user's short-term memory. Actually, users say very little and are likely to become silent during concurrent think aloud protocols when they are required to think abstractly or complete complex tasks [12].

It is suggested in the original think-aloud method [13] that with careful instructions for the participants at the beginning of a test session, the session should be conducted with minimum interruption except for reminding the participants to keep talking when they fall silent. However, it is pointed out [8] that actual practice of think aloud protocols for usability testing diverges from the theoretical foundation [13] in the manner of reminding participants to think aloud.

In pursuit of the constant verbalization, there has been proposed a dialogue approach where a test session is conducted interactively between participant and facilitator [14]. Although the dialogue approach may be helpful when the participant is stuck or the prototype system crashes, it would be possible for the test facilitator to detect such situations and handle as extraneous intervention without relying on the dialog approach. Boren and Ramey [8] proposed an interactive think-aloud method based on a speech communication framework, where thinking aloud is regarded as a shared task for participant and facilitator rather than just a participant task. In particular, it is argued in the method that careful use of acknowledged tokens (*e.g.,* "OK", "yeah", or "mm hm") will keep continuous flow of verbalization. However, such an interactive approach may reflect the facilitators' judgments on the participant thoughts, and cause validity problems in usability studies [11]. For instance, acknowledgements from the facilitator might be given earlier than the response of a test object, and hinder participant's opportunities to think about actual feedback from the test object.

In this paper, we propose a method of user operation with oral instruction for think aloud usability testing with interactive prototypes. The proposed operation method allows participants to concentrate on the evaluation of a test object regardless of the insufficient operational performance of prototypes. Moreover, the proposed method facilitates thinking aloud because oral operation would help participants to keep a continuous flow of verbalization, which is not necessarily easy in the conventional thinking aloud method [9].

The oral operation approach does not entail dialogue between the participant and facilitator, but include verbal interaction with a test object through the facilitator. The idea behind this approach is to involve the facilitator into one side of the interaction, namely, from the participant to a test object, and minimize intervention into another side of from the test object to the participant. The oral or verbalized operation does not interrupt the participant's flow of thinking process and contributes to giving more chances for the participant to think aloud because the oral instruction is the action the participant is going to take to work with the test object with reference to information in the short-term memory.

In order to figure out situations that allow participants to concentrate on evaluation of a test object regardless of its insufficient operational performance of its prototype, this paper describes an experiment to compare conventional manual operation and indirect oral operation methods for think-aloud protocols. Following research questions will be addressed in the experiment presented in the remainder of this paper:

1. Do the two methods differ in terms of the easiness of operation?
2. Do the two methods differ in terms of the quantity/quality of utterances?

2 Method

2.1 Test Object

In pursuit of the above research questions, we conducted a study to compare results of usability testing for an existing touch-screen digital camera and its accompanying photo album software. In particular, we used two test objects: a working product of the photo album software, and an interactive prototype of a digital camera created with presentation software.

Although it is assumed in this study that operational performance of prototypes is insufficient, such insufficiency may not occur depending on type of test objects. For instance, if indirect oral operation was forced for use of ordinarily-familiar product such as desktop PCs with mouse/keyboard interface, it may hinder users to make spontaneous verbalization due to cumbersome interaction with the test object. Therefore, we adopted two test objects with different user interface. One is a prototype of a touch-screen digital camera created with presentation software on a notebook tablet PC (Figure 1). The prototype of a digital camera provides single-touch screen interface that allows operations such tapping for selection, flick for scrolling the screen, and drag-and-drop for moving an object. Another test object is a product version of photo album software operated with ordinary mouse/keyboard user interface.

Fig. 1. Prototype of touch-screen digital camera on a tablet PC

2.2 Design

The experiment had a nested design with mixed within and between-participant factors. The main within-participant factor was operation method (direct manual vs. indirect oral), and the between-participant factor was user interface (single-touch screen vs. standard mouse/keyboard). Participants were thus divided into two test groups. Dependent variables were: task-completion time, quantity/quality of utterances, subjective measures from a questionnaire given to the participants.

2.3 Participants

The experiment was conducted with a sample of 32 participants, all of whom were third or fourth year students of Faculty of Informatics at the Kansai University. All the participants had no prior experience in think-aloud protocols, and were paid to participate in the experiment. The participants were recruited by means of e-mail and printed announcements, and selected on a first-come, first-serve basis.

2.4 Tasks

To consider differences between the manual and oral operation methods, four tasks listed in Table 1 were given to all the participants. Tasks 1 and 2 were designed for the prototype of a digital camera, and tasks 3 and 4 for the photo album software. Given on the right of each task description is the number of operation steps to complete the task. For example, the task 2 requires 7 steps, which consist of four tapping, two flick, and one drag-and-drop operations on the test object.

Table 1. List of tasks

Test object		Task description	Task steps
Digital camera	(1)	Delete a photo of strawberry	8
	(2)	Move a photo of muscat into the folder '1'	7
Photo album software	(3)	Put letters 'apple' on a photo of apple, and change the photo file name as 'apple.jpg'	9
	(4)	Change image quality of a banana photo to standard, and move the photo into the folder '1'	11

2.5 Questionnaires

Two forms of questionnaire were prepared to obtain participants' evaluations on a seven-point Likert scale (1 being very negative and 7 being very positive), and handed out to all the participants at the end of a session for each test object. Q1 asked participants rate the ease of finding target objects (*e.g.*, menu, button, and icon) to interact with, and Q2 to rate the ease of applying actions (*e.g.*, clicking, tapping, and flicking) to the objects found as targets.

2.6 Procedure

The experiment was conducted individually in a meeting room (7 x 8 m) at the Research Center of Advanced Informatics, Kansai University, and each session lasted about 30 minutes. One facilitator was present in the room throughout the sessions to observe participants. Before starting a session, participants were informed that all the sessions would be videotaped from behind. However, participants were not told that their verbalizations would be analyzed for comparing operation methods because that may affect participants' verbalization quantity and/or content. Instead, participants were told that the session was held for the usability evaluation of target objects.

When starting a session with each test object, participants were provided with a printed user manual. All the participants were asked to complete a set of tasks, and to constantly verbalize their thoughts while working on the tasks. In particular, one half of the participants were asked to operate prototypes directly by themselves [Figure 2 (a)] whereas another half was asked to operate indirectly, giving oral instructions to an operator [Figure 2 (b)].

After a practice task on making crane of paper folding (*Origami*) to rehearse oral or manual operation method as well as thinking aloud, participants attempted the four tasks. The two tasks were given for the prototype of a digital camera with touch screen interface, and the other two for the photo album software with a standard mouse/keyboard interface. In each group, the task order was counterbalanced across participants to prevent order effect. After completion of the two tasks for each test object, participants were asked to rate on a seven-point scale the ease of operation with the test objects.

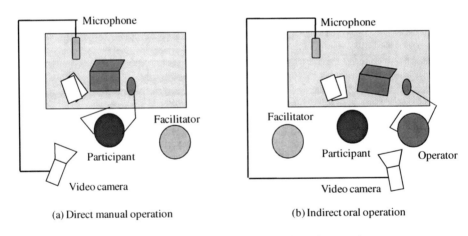

(a) Direct manual operation   (b) Indirect oral operation

Fig. 2. Experimental setting for participant's operation

3   Results and Discussion

3.1   Easiness of Operation

Since the tasks in this study consist of different number of steps (Figure 1), the task completion time was normalized and compared as operation time per task step. A two-way ANOVA for the single-step operation time (Table 2) revealed that there was a signifiant interaction [$F(1,30)=12.76$, $p<0.005$] between the operation method and user interface. The simple main effect of operation method was significant when the touch-panel interface was used [$F(1,30)=23.05$, $p<0.0001$]. For the user interface, the simple main effect was significant when the oral operation method was used [$F(1,30)=38.43$, $p<0.0001$]. These results show that tasks with touch-panel interface took more time (46.24 sec. per task step) to complete when the oral operation method was used. In contrast, when the manual operation method was used, there was no significant difference between the two types of user interface.

The participants' rating for the easiness of operation is given in Table 3. A two-way ANOVA for the easiness of finding target objects (Q1 in the quesionnaire) revealed a significant interaction [$F(1,30)=4.95$, $p<0.05$] between the operation method and user interface. The simple main effect of user interface was significant when the oral operation method was used [$F(1,30)=21.67$, $p<0.001$]. In addition, for the easiness of applying actions to target objects (Q2 in the quesionnaire), there was a significant interaction [$F(1,30)=4.48$, $p<0.05$] between the operation method and user interface. The simple main effect of user interface was significant when the manual operation was used [$F(1,30)=7.09$, $p<0.05$], and was not significant under the condition of the oral operation.

Table 2. Operation time per task step (in seconds)

User interface	Manual (n=16)		Oral (n=16)	
	Mean	SD	Mean	SD
Touch screen	25.17	14.22	46.24	16.52
Mouse/keyboard	20.73	7.76	22.24	6.51

Table 3. Participants' rating for the easiness of operation. (Higher means indicate more positive assessment).

	User interface	Manual (n=16)		Oral (n=16)	
		Mean	SD	Mean	SD
Q1: Finding objects	Touch screen	3.88	1.05	3.31	1.67
	Mouse/keyboard	4.63	1.41	5.63	1.05
Q2: Applying actions	Touch screen	4.81	1.88	4.94	1.44
	Mouse/keyboard	6.31	1.04	4.75	1.56

The above results of the participants' rating particularly for the oral operation method indicate that finding target objects is easier for the software with mouse/keyboard (5.63) than the interactive prototype with a touch-panel screen (3.31). However, there is no difference in the perceived easiness of applying actions with the oral operation. When an action is being applied to an object, the target object is already identified on the screen by the user. Therefore, it is probable that the intrinsic difficulty of the oral operation will be in the process of identifying a target object.

3.2   Quantity/Quality of Utterances

In addition to the easiness of operation, the utterances made during the experiment were investigated with regard to the quantity and quality. Verbalization categories assumed in this experiment are given in Table 4, and overall numbers of utterances made during the four tasks are given in Figure 3 with classification into the categories.

Table 4. Category of transcribed unit of verbalization

Verbalization category	Description
Explanation	Describe the reason for participant's behavior; make a prediction about a result of participant's activity
Procedure	Describe participant's activities
Observation	Describe the observation of what occurred as a result of participant's activity
Other	Utterances that do not fit into one of the other categories

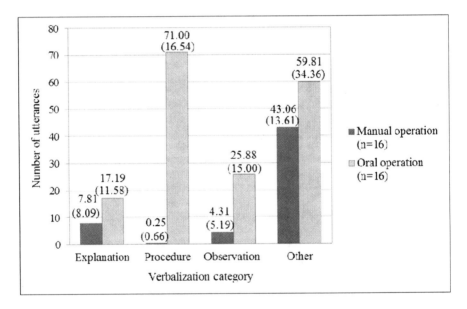

Fig. 3. Number of utterances made during the four tasks (SD in parentheses)

Since the participants in the group of manual operation performed tasks without verbalizing their operations, the average number of procedure utterances with manual operation 0.25 was much fewer than those with oral operation 71.00 as shown in Figure 3. More importantly, the numbers of explanation (17.19) and observation (25.88) utterances with the oral operation were both significantly higher than the corresponding utterances with manual operation ($p < 0.05$ and $p < 0.01$ respectively, Welch's t-test). In addition, for the number of 'other' category utterances, there was no significant difference between the oral and manual operations (p > 0.05, Welch's t-test). These results indicate that the number of utterances for the explanation and observation increased substantially when the oral operation method was used. Note here that participants' utterances for explanation and observation are important source of discovering usability problems and crucial for the improvement of target artifact.

Figure 4 shows the sum of explanation and observation utterances in each experimental condition, which is normalized as a value per single task step. A two-way ANOVA for the utterances revealed a significant interaction [F (1, 30) = 10.12, p < 0.005] between the operation method and user interface. The simple main effect of the operation method was significant when the touch-panel interface was tested [F (1, 30) = 27.58, p < 0.0001]. For the user interface, the simple main effects were significant in both oral [F (1, 30) = 45.24, p < 0.0001] and manual [F (1, 30) = 4.96, p < 0.05] operation methods, and more utterances were made when the touch-panel was tested.

These results suggest that the oral operation method could contribute to the increase of utterances for explanation and observation, while the prototype of a touch-pane interface might have an influence on the facilitation of verbalization. The oral operation method avoids bringing users' attention overly to insufficient performance of an interactive prototype, and that will help users concentrate more on the other aspects of a target object to be tested. Consequently, the effects of the oral operation method in think aloud usability test still depends on types of interface. However, it would be worthy of consideration when it is crucial to draw more utterances in usability testing.

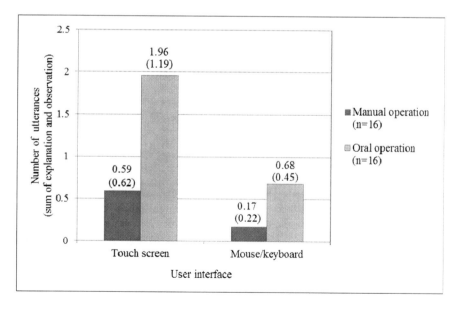

Fig. 4. Number of utterances for explanation and observation as values normalized per task step (SD in parentheses)

References

1. Vredenburg, K., Mao, J.-Y., Simith, P.W., Cary, T.: A survey of user-centered design practice. In: Proceedings of CHI 2002, Minneapolis, Minnesota, pp. 471–478 (2002)
2. Arnowitz, J., Arent, M., Berger, N.: Effective Prototyping for Software Makers. Morgan Kaufmann, San Francisco (2007)
3. Berger, N., Arent, M., Arnowitz, J., Sampson, F.: Effective Prototyping with Excel: A practical Handbook for Developers and Designers. Morgan Kaufmann, San Francisco (2009)
4. Hennipman, E.-J., Oppelaar, E.-J.R.G., van der Veer, G.C., Bongers, B.: Rapid and rich prototyping: Proof of concepts for experience. In: Proceedings of the 15th European Conference on Cognitive Ergonomics, Madeira, Portugal (2008)
5. Lim, Y.-K., Pangam, A., Periyasami, S., Aneja, S.: Comparative analysis of high- and low-fidelity prototypes for more valid usability evaluations of mobile devices. In: Proceedings of NordiCHI 2006, pp. 291–300 (2006)
6. Snyder, C.: Paper Prototyping: The Fast and Easy Way to Define and Refine User Interfaces. Morgan Kaufmann Publishers, San Francisco (2003)
7. Virzi, R.A., Sokolov, J.L., Karis, D.: Usability Problem Identification Using Both Low- and High-Fidelity Prototypes. In: Proceedings of CHI 1996, Vancouver, BC Canada, pp. 236–243 (1996)
8. Boren, M.T., Ramey, J.: Thinking aloud: Reconciling theory and practice. IEEE Trans. on Professional Communication 43(3), 261–278 (2000)
9. Van den Haak, M.J., de Jong, M.D.T., Schellens, P.J.: Retrospective versus concurrent think-aloud protocols: Testing the usability of an online library catalogue. Behaviour & Information Technology 22(5), 339–351 (2003)
10. Nielsen, J., Clemmensen, T., Yssing, C.: Getting access to what goes on in people's heads? Reflections on the think-aloud technique. In: Proceedings of NordiCHI, pp. 101–110 (2002)
11. Krahmer, E., Ummelen, N.: Thinking about thinking aloud: A comparison of two verbal protocols for usability testing. IEEE Trans. on Professional Communication 47(2), 105–117 (2004)
12. Cooke, L.: Assessing concurrent think-aloud protocol as a usability test method: A technical communication approach. IEEE Trans. on Professional Communication 53(3), 202–215 (2010)
13. Ericcson, K.A., Simon, H.A.: Protocol Analysis: Verbal Reports as Data, revised ed. MIT Press, Cambridge (1993)
14. Buur, J., Bagger, K.: Replacing usability testing with user dialogue - How a Danish manufacturing company enhanced its product design process by supporting user participation. Communications of the ACM 42(5), 63–66 (1999)

Exploring the Impact of Cultures on Web Usability Test

Hsiu Ching Laura Hsieh

Graduate School and Department of Creative Design,
National Yunlin University of Science and Technology,
123 Section3, University Rd, Douliou,
Yunlin, Taiwan 64002
laurarun@gmail.com

Abstract. Many previous studies reveal that cultures not only affect web design, but also impact web usability test. Most previous cross cultural web usability research adopted a western usability method to collect data from users, however some researchers have recognized that there is a need to define usability by considering in cultural context [7] [12] [15]. The traditional usability method is not a properly way to investigate people from different cultures. New usability methods are required to be developed to completely understand users from different cultures. Much more research is required to be developed to reach at guidelines for possible differences which web developers could integrate in methodology when web usability evaluating in a target culture market. In this paper, the literature review begins in illustrating the constant transformation and evolution of the web usability concept, then come to the discussion of the problems in the previous cross cultural usability methodologies. Furthermore an usability evaluation is constructed. Finally the ways for improving cultural usability methodology are recommended based on the initial results of the experiment. It is hoped that this research would contribute to an increased awareness of how culture may impact usability evaluation and the implications can help and ensure more efficient usability evaluation.

Keywords: Cross Cultural Web Usability.

1 Introduction

According to previous research, different cultures not only affect web design, but also influence web usability. Many types of usability evaluation method have been identified, such as interviews, think-aloud protocols, moderated tests, group walkthrough, card sorts. Most cross cultural web studies applied Western usability method to collect data from target users and the most popular usability is defined as a general quality of a product that can be added at a certain stage in the design process [11] [22]. However some researchers have recognized that there is a need to define usability by considering in cultural context. For examples, Nielsen [22] and Fernandes [13] revealed that the results of usability tests which were established by web developers may not work efficiently across different cultures. The web products have to be tested in the target culture to make sure that the product is accepted by the target market and would not offend a specific culture. One of the core issues that the web

M. Kurosu (Ed.): Human Centered Design, HCII 2011, LNCS 6776, pp. 47–54, 2011.
© Springer-Verlag Berlin Heidelberg 2011

developers are required to pay attention is cultural aspect may impact on cross cultural usability evaluation. The popular usability test method in Western culture may not work properly in other cultures. In 1998 Barber and Badre [4] suggest "culturability" which combines the two words "usability" and "culture". They constructed a cultural marker approach which is a kind of systematic usability method to examine hundreds of websites, and then define cultural markers such as colors, fonts, icons, metaphors, language, preferences for text and graphics, help features and navigation tools to facilitate user performance. In 2002, Ever also suggested that different data collection methods need to be developed for the different cultures involved while investigating the same topic for cross cultural web research. The issue "Are there ways we can adapt methods to suit different cultures" is raise by some researchers [3]. Also Gould [15] echoed that culture affects usability tests and interviews, and doubted that if the participants always express themselves honestly in cross cultural usability survey.

Indeed, new usability methods and techniques are required to be developed to fit users from different cultural contexts. The social - cultural context surrounding the product is often ignored in research and practice. As usability is one of the core terms in human-computer interaction research, it should represent the interdisciplinary attributes of this area. More research is required to be developed to reach at guidelines for possible differences that web developers could integrate in methodology when web usability evaluating in a target culture market.

2   Web Usability and Web Usability Evaluation Method

Initially web usability is derived from engineering ways, the ISO Standard 9241 defines usability as the extent to which a product can be used by a specific group of users to accomplish their goals with efficiency, effectiveness, efficiency and satisfaction in a specified context to use. This definition implies that a usable web site design should be intuitive and transparent. It supports users in carrying out the intended task efficiently, easily, enjoyably, functionally, and quickly. Nielsen [22] defines five attributes of web usability as learnability, efficiency, memorability, errors, satisfaction. Based on Human-computer interaction (HCI) studies, Preece [25] defines usability as follows. The design of computer systems that are safe, efficient, easy, functional and enjoyable to use is the main concern in HCI.

In the past, many types of usability method have been identified, such as interviews, think-aloud protocols, moderated tests, group walkthrough. Group walkthrough is frequently applied to help orient designers in the process of creation of the websites, providing feedback on the progress of the work. This method is properly for testing the interaction of the website with completely novice and it helps the developing of guidelines for improving the usability of the website. Questionnaires can be used to obtain information of the choices, desires, expectations and satisfaction of the users of the websites. Questionnaires are useful and informative in all phases of the development and design of the website. Interviews and Focus groups. Two techniques differ from questionnaires. The evaluator interacts directly with the participants, inducing opinions and comments on the web product. The participants in this type of investigation response to the questions according to their experiences and preference regarding interaction with the website. The interviews are often structures formally, the focus group is less formal and allow participants to discuss their experiences together. The participants were asked to express their opinions to acquire suggestions for adding new services or improving the website. These investigation techniques can be applied in all phases of the development and design of websites. Think Aloud method involves a participant expressing about what they are thinking while they interact with a web product. The method can be applied by assigning the users a specific task or allowing users navigation freely, and it is applied particularly for the evaluation of prototypes or already existing websites. The experimenter play a "leader" role during the investigation process and they have to stimulate the participant to keep on thinking aloud, eliciting them to depict what is happening. This technique is especially useful, as it can help the experimenter to capture a wide range of cognitive activities of the participants.

3   Culture and Web Usability Evaluation

The relationship between culture and usability has been brought by Barber and Badre [4] and they suggested culturability, which combines two words, usability and culture. They built up a approach "cultural markers" which are systematic usability methods to evaluate hundreds of websites, and then define cultural design elements such as colors, fonts, icons, metaphors, geography, sounds, motions, flags, language, preferences for text vs. graphics, directionality of how language is written, help features and navigation tools to facilitate user performance. The merging of culture and usability - Culturability has implications for web design. Usability must be re-defined based on cultural context.

Sun [28] applied "cultural markers" approach to explore how cultural markers affect web usability by interviewing target users about their interaction experiences of websites. Some important implications are revealed from her study, "Culture is moving from borders of web usability to the forefront. Cultural marker should be one metric in usability matrix as learnability, efficiency, satisfaction, and so on." [28]. Later, Sun [28] stated that the traditional usability method which is derived from the engineering way, user and task analysis actually could not provide thorough discoveries of cultural factors in the observed site. When the traditional usability method is applied, culture is approached statically, and researchers seek universal patterns for different cultures. Most researchers have not paid attentions on the ever-changing cultural context and some related studies of cross cultural interface design elements usually stay at the state of ethnic cultural preferences and ignore to discuss the dynamic relationship between the cultural preferences and digressive power.

According to Mantovani [18], "The meeting place becomes the Internet, the World Wide Web, which is increasingly considered no longer as a pure physical structure, but as a cultural space in which new forms of social relations and identity are experimented". Also, Gamberini and Valentini [14] revealed that the role of usability keeps on expanding and transforming. It is recommended that web usability is required to take cultural and social context into account in the investigation of web products, as they are not to be regarded as only tools, unrelated to the concrete environment in which they are used. Thus, usability evaluation has to be carried out within cultural context, from which actions take their meaning.

4   Previous Cross Cultural Web Usability Evaluation Research

According to Yeo's study [31], participants who own higher rank than the experimenter comments the software negatively; those with equal or lower status than the experimenter were more positive and the way to criticize the software is more polite and subtle in his study he had done to localize the software in Malaysia. This kind of self-censorship was based on the relationship of the participants to the test moderator. Based on Hofstede's five cultural dimension [16], Malaysia society is categorized as a high power-distance, moderately high collectivism. Malaysia people sought to keep relationship in harmony and save face for each own by refusing to be harsher and negative, whilst the participants whose status is higher sensed their problems had made them look clumsy so they picked up the drawbacks of the software. The implication of Yeo's study is that to acquire honest response of a users' experiences, an experimenter of the equal status or lower might be needed when establish a think-aloud experiment. Yeo [32] suggested that usability assessment techniques from Western should only be applied with participants who were already experienced users, familiar with the experimenter, and own higher status than the experimenter. Apparently, some problems for our traditional focus on inexperienced users were revealed in Yeo's suggestions, meanwhile those problems unbalanced the empirical design of the previous usability measurement.

An Indian researcher Chavan [9] found the problems of relationship between participants and moderators in her research. The willingness of participants to comments on products is influenced by gender, youth, and class. Usually women would talk only with women; younger researchers had more success than older, more senior people. The difference to stronger social relationship based on liking for people similar to one's self was featured by Chavan [9]. Also, Chavan [9] recommended that India users usually do not want to comments about the software products under any condition and suggests using an approach which is derived from Indian Bollywood drama theories.

In the usability study of Clemmensen et al. [10], it is found support for both the above explanations in a study comparing the role of test moderators in usability measurements in India, China, and Denmark. According to gender and age among conventional end users in India, Indian experimenter is supposed to deal with self-censorship. While a male research needed to interview with a rural woman, this woman's male relative is supposed to accompany with her. In Denmark, usability measurements operate most smoothly when the researchers and users were the similar age, gender, and have the similar level of job experience. Clemmensen et al. [10] recommended that cross cultural usability assessments need to be consistent with suggestions as follows, including consider those users who are less comfortable with foreigner or more traditional, assess the evaluator effect and choose researchers properly to those users (who are more traditional) and modify the test protocols to localize scenarios, ask different questions and apply more direct approach. Clemmansen et al. [10] found that moderators operate measurements in China have to apply lots of more straight investigations since users would not identify their actions unless prompted. However, many often provided a backward think- aloud analysis of their choices, after a period of silence.

Shi [26] also found that Chinese people needed steady instigating when he reviewed usability assessments in Bejing. Chinese peoples' silence is imputed to the holistic considering mode and interpersonal needs of East Asians. According to

Nisbett's description [24], Asian thought concentrate on relations among people and events, social concordance, and the acceptance of instinctive operation change; Western thought is more heedful to a tangible and visible entity, formal logical system, categories, manipulate, and stable theories of explanation when confront rightness and the true.

Based on the above literature review, many studies imply that the current cross cultural usability methods are not properly for different culture. Cultures impact moderated tests, interviews, and think - aloud protocols. The existing cross cultural usability approaches are not methodical as we think [15].

5   Construction of a Web Usability Evaluation

During the author's earlier study in 2008, a cross cultural web usability evaluation is constructed. Taiwanese and British users were recruited in the interviews. It was found that the way young Taiwanese users (who were graduate students in the UK) to criticize the web product and their explicit and straightforward attitudes are quite different as Hofstede's cultural measure. Therefore, the inconsistency formulate the motivation for this study. The aim of this research is to identify the cultural factors that influence usability evaluation, then examine how these cultural factors influence participants' attitude in usability evaluation, further identify ways to improve cultural usability testing. Finally, the author wants to gain awareness of how Taiwanese users use websites.

5.1   Method

Eight Taiwanese users were recruited in this experiment and all of them have 10 years experiences in using internet and navigating in websites. Four of them are undergraduates and graduates students and their average age is twenty five years old. The other four participants have many years experiences in work and their average age is forty two years old. "Facebook" social web is used as the instrument to test the feedback of Taiwanese users. All of the participants have two years experiences in using "Facebook". SUS method is applied to gauge the users' response. During this experiment, users are asked to fill in a survey sheet which is called SUS (System Usability Scale), based on the users' experiences using "Facebook". According to the related international usability evaluation research, SUS method is an efficient method to understand the target users [2] [5] [30]. System Usability Scale is developed at Digital Equipment Corp in1986. It consists of ten items. One of the items is adapted by replacing "system" with "website". The ten items are translated into traditional Chinese as Taiwanese use traditional Chinese. After finishing the SUS, the participants were interviewed and asked three questions which is related to their experiences in using "Facebook".

5.2   Initial Results and Discussions

According to Hofstede's cultural dimensional model, Taiwan is ranked tenth among seventy four countries in collectivism dimension and is ranked third among thirty nine countries in long term time orientation dimension. Collectivist culture tends to value group welfare more than the individual's target, where the achievement of an individual is not regarded as important as the accomplishment of the group, and believes in group relationship, cares about saving face for others, value harmony more than truth. Long-term time orientation plays a crucial role in Asian countries (e.g., Taiwan, China, Hong Kong and Singapore) that have been influenced by Confucianism. People in these countries believe strongly that an unequal state of connection is required to keep a society stable, a clear hierarchical relationship is needed to keep family and society in harmony, aged people own more authority, younger people have to be filial to older people, students have to obey teacher's order, and virtuous behavior is identified as hard-working and perseverant.

Based on the initial observation, four of young Taiwanese participants are willing to express their opinions about the website and present the advantage and disadvantage very directly, actually they are not afraid to criticize the website, even the experimenter who is ranked higher that the users (the experimenter is participants' teacher). Whist older Taiwanese participants have different attitude to answer the questions from the experimenter, they do not response the question directly and immediately, and the experimenter needs to encourage them to express their opinions more, even they criticize the website, usually they use an indirect and polite way to show their feedback. Two of Older Taiwanese users does not feel comfortable in the process of this experiment. The presence of the evaluator makes this user uncomfortable and it is considered as in an examination circumstance. Actually, according to the observation in the interview, young Taiwanese users' attitude reveal the reverse of Hofstede's theories. Based on SUS scores, it indicated a neutral or favorable response.

5.3 The Implication

While young Taiwanese users joint an usability experiment, the status, age, or authority of the experimenter would not influence their attitude to response to the questions in interview. While older Taiwanese users participate the experiment, the status, age, or authority of the experimenter is a matter to the participants. It would influence them the way to express themselves. Based on Hofstede's measure of collectivism and long term time orientation, Taiwanese were expected to be less critical of the technology. However, the initial results of this experiment reveal that young Taiwanese are willing to criticize the website and not afraid to express their ideas, but the attitudes of older Taiwanese are still consistent with Hofstede's dimension theories. The implication is that western usability evaluation can work efficiently, while the usability participants are young generation Taiwanese. While the usability participants are older Taiwanese, the usability evaluation method needs to be adapted. According to the initial result, it is revealed that cultural dimension theories could be too stereotype. Actually culture is dynamic, it is why the attitudes and responses of young Taiwanese are reverse to Hofstede's cultural dimension, but older Taiwanese users' response is still quite consistent with Hofstede's theory.

6 Conclusions and Future Work

Results from this Taiwanese study might be revelatory of results in other Asian cultures which have been influenced by Confucianism a lot. It is recommended that if web developers want to comprehend the accurate response of target culture users, the properly means is to recruit users from target cultures to get the direct response. Even evaluate the users in the same culture, the method needs to be adapted based on different age. Those cultural dimensional model are categorized in a particular time. Those models could be reference but needs to be applied cautiously as Hofstede's cultural model treats culture as a stable phenomenon, but culture is always changing and developing in reality.

Alternatively applying an ethnographic study seems to be a properly solution. It will allow the researcher to comprehend the real users from a different cultural perspective, learn to understand the thought processes of another culture and look at it from the indigenes' viewpoint. Applying the 'participant observation' method of an ethnographic study, the behavior of the different cultures should be observed to reveal ordinary activities of actual interaction with cultural website. This approach leads to consideration the context in which these behaviors take place. A good rapid ethnography - time deepening strategies for HCI field research developed by Millen [19] is recommended and explained in the following. Step first, narrow the field research focus and scope, such as zoom in on the important activities and use the core informants. Step second, use multiple approaches of interactive observation to gain the likeliness of finding eccentric and useful user behaviors. Step third, apply cooperative and computerized iterative data analyze methods.

In the forthcoming paper, a more in-depth analysis of eight participants data will be executed. Experiments using other culture users are managed and more rigorous evaluation method would be applied. It is hoped that outcome of the research will gain increased awareness of how culture may affect usability testing and also ensure more efficient usability testing with minimal cultural influences.

References

1. Anderson, R.J.: Representations and requirements: The value of ethnography in system design. Human-Computer Interaction 9(2), 151–182 (1994)
2. Bangor, A., Kortum, P.T., Miller, J.T.: An Empirical Evaluation of the System Usability Scale. International Journal of Human-Computer Interaction 24, 574–594 (2008)
3. Barab, S.A., Thomas, M.K., Dodge, T., Newell, M., Squire, K.: Design ethnography: Building a collaborative agenda for change. Anthropology & Education Quarterly (2003)
4. Barber, W., Bardre, A.: Proceedings of the 4th Conference on Human Factors and Usability (1998),
   http://zing.ncsl.nist.gov/hfweb/att4/proceedings/barber
5. Brooke, J.: SUS: A "quick and dirty" usability scale. In: Jordan, P.W., Thomas, B., Weerdmeester, B.A., McClelland (eds.) Usability Evaluation in Industry, pp. 189–194. Taylor & Francis, London (1996)
6. Chavan, A.: The Bollywood method. In: Schaffer, E. (ed.) Institutionalization of Usability: A Step-By-Step Guide, pp. 129–130. Addison-Wesley, New York (2004)
7. Chavan, A.L.: Another culture, another method. In: Proceedings of the 11th International Conference on Human Computer Interaction [CD-ROM]. Lawrence Erlbaum Associates, Hillsdale (2005)
8. Chavan, A.: Usability in India, is it different? In: HCII 2005. Lawrence Erlbaum, Mahwah (2005)
9. Chavan, A.: What about a "local" wrapper around a "unuversal" core. In: CHI 2008, pp. 2605–2607. ACM Press, New York (2006)
10. Clemmensen, T., Shi, Q., Kumar, J., Li, H., Sun, X., Yammiyavar, P.: Cultural usability tests – how usability tests are not the same all over the world. In: Aykin, N. (ed.) HCII 2007. LNCS, vol. 4559, pp. 281–290. Springer, Heidelberg (2007)

11. Dumas, J.S., Redish, J.: A practical guide to usability testing. Ablex Publishing Corp., NJ (1993)
12. Evers, V.: Cultural aspects of user interface understanding: an empirical evaluation of an E-learning Website by international user group. Ph.D thesis, Open University, pp. 52–580 (2002)
13. Fernandes, T.: Global Interface Design. Academic Press, Chestnut Hill (1995)
14. Gamberini, L., Valentini, E.: Web usability today: Theories approach and methods (2001)
15. Gould, E.: Intercultural usability surveys: Do people always tell "The truth"? In: Aykin, N. (ed.) IDGD 2009. LNCS, vol. 5623, pp. 254–258. Springer, Heidelberg (2009)
16. Hofstede, G.: Cultures and Organizations: Software of the Mind. McGraw-Hill, London (2005)
17. Lewis, J.R., Sauro, J.: The Factor Structure of the System Usability Scale. In: Kurosu, M. (ed.) HCD 2009. LNCS, vol. 5619, pp. 94–103. Springer, Heidelberg (2009)
18. Mantovani, G.: Network. Reti elettroniche e reti di significato, in Ergonomia. Lavoro, sicurezza enuove technologie. In: Mantovani, G. (ed.) Mulino, Bologna 2000, pp. 153–177 (2000)
19. Millen, D.R.: Rapid ethnography: time deepening strategies for HCI field research. In: Proceedings of the 3rd Conference on Designing Interactive Systems: Processes, Practices, Methods, and Techniques, DIS 2000. ACM, Brooklyn (2000)
20. Nawaz, A., Plocher, T., Clemmensen, T., Qu, W., Sun, X.: Cultural differences in the structure of categories in denmark and China. Working paper, Department of Informatics, Copenhagen Business School (March 2007)
21. Nielsen, J.: Usability Testing of International Interfaces. In: Nielsen, J. (ed.) Designing User Interfaces for International Use. Elsevier, Amsterdam (1990)
22. Nielsen, J.: Usability Engineering. Academic Press, San Francisco (1993)
23. Del Galdo, E.: Culture and Design. In: Del Galdo, E., Nielsen, J. (eds.) International User Interfaces, pp. 74–87. John Wiley and Sons, Inc., Chichester (1996)
24. Nisbett, R.: The geography of thought: How Asiana and Westerns think differently and why. Free Press, New York (2003)
25. Preece, J.: A guide to usability: human factors in computing. Addison-Wesley, New York (1993)
26. Riva, G., Galimberti, C.: Mind, Cognitions and Society in the Internet Age. IOS Press, Amsterdam (2003)
27. Shi, Q.: A field study of the relationship and communication between Chinese evaluators in thinking aloud usability test. In: NordiCHI 2008, pp. 344–352. ACM Press, New York (2008)
28. Sun, H.: Building a culturally-competentcorporate web site: an exploratory study of cultural markers in multilingual web design. In: Proceedings of the 19th Annual International Conference on Computer Documentation, pp. 95–102 (2001)
29. Sun, H.: Exploring Cultural Usability. In: Proceedings of IEEE International Professional Communication Conference, Portland OR, pp. 319–330 (September 2002)
30. Tullis, T.S., Stetson, J.N.: A Comparison of Questionnaires for Assessing Website Usability. unpublished presentation Given at the UPA Annual Conference (2004), http://home.comcast.net/~tomtullis/publications/UPA2004Tulli sStetson.pdf
31. Yeo, A.: Cultural effects in usability assessment. In: CHI 1998, pp. 74–75. ACM Press, New York (1998)
32. Yeo, A.: Are usability assessment techniques reliable in non-Western countries? Elec. J. on Info. Sys. in Dev. Countries 3(1), 1–21 (2000)
33. Willis, P.: The ethnographic imagination. Polity Books UK (2003)

A Three-Fold Integration Framework to Incorporate User–Centered Design into Agile Software Development

Shah Rukh Humayoun[1], Yael Dubinsky[2], and Tiziana Catarci[1]

[1] Dipartimento di Informatica e Sistemistica "A. Ruberti",
SAPIENZA - Università di Roma, Via Ariosto – 25, 00185, Roma, Italy
[2] IBM Research – Haifa, Mount Carmel, Haifa 31905, Israel
{humayoun,catarci}@dis.uniroma1.it,
dubinsky@il.ibm.com

Abstract. We present a framework that incorporates user-centered design (UCD) philosophy into agile software development through a three-fold integration approach: at the process life-cycle level for the selection and application of appropriate UCD methods and techniques in the right places at the right times; at the iteration level for integrating UCD concepts, roles, and activities during each agile development iteration planning; and at the development-environment level for managing and automating the sets of UCD activities through automated tools support. We also present two automated tools—UEMan and TaMUlator, which provide the realization of the development-environment level integration.

Keywords: User-centered design (UCD), agile software development, usability evaluation, integrated development environment (IDE), UEMan, TaMUlator.

1 Introduction

One of the challenges in software development is to involve end users in the design and development stages so as to collect and analyze their behavior and feedback in an effective and efficient manner and then to manage the ensuing development accordingly. One way to achieve this is by applying user-centered design (UCD) [4] philosophy. This philosophy puts the end users of the system at the centre of the design and evaluation activities, through a number of methods and techniques. UCD is applied in software projects with the aims of increasing product usability, reducing the risks of failure, decreasing long-term costs, and increasing overall quality. Integrating UCD activities into software development processes fuses the user experience with the development process, attaining a high level of usability in the resulting product. One description of UCD that we find particularly motivating is: "*User-centered system design (UCSD) is a process focusing on usability throughout the entire development process and further throughout the system life-cycle*" [10] (p. 401).

The agile approach [1] is one software development approach that has emerged over the last decade. This approach is used for constructing software products in an iterative and incremental manner; in which each iteration produces working artifacts that are valuable to the customers and to the project. This is performed in a highly-collaborative fashion to produce quality products that meet the requirements in a cost-effective and timely manner.

M. Kurosu (Ed.): Human Centered Design, HCII 2011, LNCS 6776, pp. 55–64, 2011.

Generally, in software development practice, software teams hesitate to imply UCD activities due to their time-consuming and effort-intense nature. Using the agile approach, in which customers and product owners lead the prioritization of the development, helps developers overcome these hesitations. By emphasizing the benefits common to both the end users and the developers, UCD and the agile approach can be dynamically integrated to get benefits from both, resulting in the development of high-quality and usable software products.

We have been working with agile and non-agile software development teams in industry and in academia [5, 11, 17] for several years, and we have successfully integrated parts of UCD philosophy into agile development environment [6, 7, 12]. Based on this experience, we identified that agile development teams were often lacking a properly-integrated approach that utilizes the UCD philosophy from end-to-end at all levels. To overcome this gap, we propose a three-fold integration framework that gives suggestions and recommendations for involving UCD in agile software development at different levels. Our approach identifies ways to apply appropriate UCD activities alongside agile development activities, with the aim of developing high-quality and usable products.

The remainder of this paper is structured as follows: In Section 2, we describe other approaches that also integrate UCD into software development processes. In Section 3, we describe our three-fold integration approach in detail. In Section 4, we present two automated tools—UEMan and TaMUlator, which provide realization of parts of our framework at the integrated development environment (IDE) level. In Section 5 we briefly describe two case studies in which development teams used our development-environment level integration approach for evaluating their software projects along development. Finally, we present our conclusion and suggest future directions in Section 6.

2   Related Work

Integrating UCD philosophy into software development processes, and specifically into the agile development approach, is not a new idea. Some past works focus on applying several UCD techniques to agile development, while others focus on the benefits a particular technique can bring to agile development. However, the literature lacks a consolidated approach to cover the integration from all aspects.

Göransson et al. [9] defined a usability design process that integrates their previously defined user-centered system design (UCSD) [10] approach with software development processes. Their defined process is iterative-in-nature and works with well-planned iterative and incremental development approaches, such as one provided for Rational Unified Process (RUP). Their process is divided into three phases: *requirements analysis, growing software with iterative design,* and *deployment.* Their approach is not very well suited for agile development, in which emphasis is given to the working artifacts and small iterations.

Chamberlain et al. [3] described a framework for integrating UCD techniques into agile development and provided a field study. They identified a set of five principles as being significant in integrating the two approaches—user involvement, collaboration and culture, prototyping, project lifecycle, and project management.

Sy [16] described the use of cycle-zero in his integrating approach. Pattern [15] proposed using of interaction design for agile development through Constantine and Lockwood's usage centered design approach. Hussain et al. in [13] proposed an approach of integrating UCD and XP development based on evaluating the usability of user interfaces of developing application in small iterative steps. Fox et al. provided a study [8] that researched participants experienced in combining these two approaches and concluded that there can be a common model from the existing models.

3   The Three-Fold Integration Framework

The UCD and agile development processes differ in nature. The two approaches were developed in different environments and disciplines, but can be integrated if care is properly taken, as their basic concepts and philosophies have no fundamental contradictions. Our approach emphasizes a tight integration from top-to-bottom, in which shared ideas are combined at every level, from the process life cycle to the development environment, gaining the benefits of both approaches. Our three-fold integration framework incorporates UCD into agile development at three levels: the process life-cycle level, the iteration level, and the development-environment level. The following sub-sections describe each of these levels.

3.1   The Life-Cycle Level Integration—Selection and Application of UCD Methods and Techniques in the Agile Process Life Cycle

We define life-cycle level integration as performing appropriate UCD methods and techniques in the right places at the right times, alongside the other development tasks. We distinguished the different types of UCD methods into two groups, elicitation and evaluation, based both on the way the methods perform and on their impact on software project development.

Elicitation Methods: These UCD methods are used for eliciting requirements and design of the software project. Normally, the end users or UCD experts are involved during the initial phases of development lifecycle. We suggest using these in early activities of agile development iterations to give more attention to eliciting requirements and design. Among the different elicitation methods, those that take less time, efforts, and give high feedbacks, such as focus groups and card sort methods, are more suitable as they fit perfectly in agile development.

Evaluation Methods: These UCD methods involve end users, UCD experts, and automated tools and are used to evaluate developing/developed products by identifying usability issues. Each type of evaluation method highlights only parts of usability issues, and only to a certain extent, so using more than one method is recommended [14] for covering a higher rate of usability issues. The evaluation methods performed by UCD experts are useful in early design activities, in which only paper prototypes or only parts of working prototypes are available, as these methods take less time and effort, both of which are at a premium during these early activities of development iteration. On the other hand, evaluation methods performed by end users give better results when they are used on working prototypes, as these prototypes help end users understand the system, taking better advantage of the methods. Heuristic evaluation, question-asking protocol, and performance measurement [4, 14] are all examples of evaluation methods.

Our framework provides a set of attributes for selecting appropriate elicitation and evaluation methods to apply during agile development iteration activities. Table 1 shows these attributes and describes each one from the agile development perspective, i.e., short-time iterations, high-level of collaboration with customers, focus on working artifacts, and dynamic processes.

Table 1. Set of attributes for determining the selection of appropriate UCD methods

Attribute	Description
Automation	The automated tools support and the level of automation (e.g., *None, Capture, Analysis, Critique* (taxonomy by Balbo [2])
Effectiveness	Feedback effects (*Low, Medium, High*) on design or development
Dynamicity	The ability of the method to be changed according to the target environment (*Low, Medium, High*)
Time-cost	How much minimum time is needed to complete this method
Effort-cost	How much efforts are needed to perform this method (e.g., man power, equipments, experiment place, other resources)
Ease of Learning	How easy it is to learn the method, both for responsible persons on the development team and/or for the end users who will perform it
Results Accuracy	Accuracy of results
Covering Area	The usability issues covered by a method

Along these attributes, selecting an appropriate UCD method also depends on other factors, such as life-cycle stage, availability of evaluators, etc. For early design activities, we recommend an emphasis on paper-based or simple UI prototype-based evaluation methods to improve design early. While in later iteration activities, we recommend using formal evaluation methods to get formal results. We also recommend using a mixture of evaluation methods, preferably supported by automated tools and performed by end users and UCD experts for maximum results. Automation tools support gives more accurate results and save time and costs—all factors that complement agile development.

On the basis of the above recommendations, we suggest a life cycle of four UCD activities for involving UCD philosophy alongside the agile development iteration. Figure 1 shows the UCD activities (solid ovals), in which agile activities that are done per each user story or development task are represented by dashed lines ovals. Sometimes a UCD activity overlaps one or more agile activities.

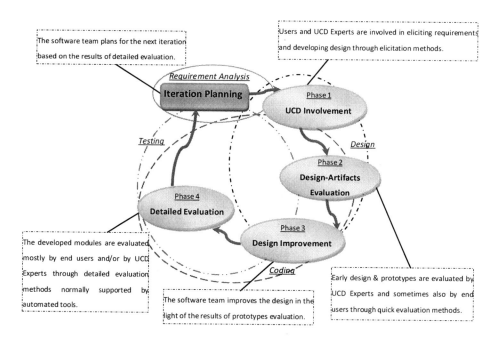

The software team plans for the next iteration based on the results of detailed evaluation.

Users and UCD Experts are involved in eliciting requirements and developing design through elicitation methods.

Requirement Analysis

Iteration Planning

Phase 1
UCD Involvement

Testing

Design

Phase 2
Design-Artifacts Evaluation

Phase 4
Detailed Evaluation

Phase 3
Design Improvement

Coding

The developed modules are evaluated mostly by end users and/or by UCD Experts through detailed evaluation methods normally supported by automated tools.

The software team improves the design in the light of the results of prototypes evaluation.

Early design & prototypes are evaluated by UCD Experts and sometimes also by end users through quick evaluation methods.

Fig. 1. Life cycle for involving UCD philosophy into agile development iteration

UCD Involvement: The software team involves end users and UCD experts when appropriate, mostly through the use of elicitation methods during work on requirements, design, and early prototypes.

Design-Artifacts Evaluation: The early design and prototypes are then quickly evaluated through short-time consuming evaluation methods normally by UCD experts (i.e., system analysts, usability evaluators, UI designers) and sometimes by a small group of end users. The results of this phase become input for the third phase.

Design Improvement: The software team corrects and improves the design according to the feedback and performs implementation of the target modules.

Detailed Evaluation: The developed modules are evaluated in detail by end users and/or by UCD experts, normally through rigid evaluation methods with automated tools support. The results, feedback, and suggestions serve as input for making plans for improvements in the design and for the implementation of developing products in the upcoming iterations.

3.2 The Iteration Level Integration—Integrating UCD Concepts, Roles, and Activities in Agile Development Iteration

This level of integration helps to align UCD concepts, roles, and activities within the development iteration activities for maximum benefit. For example, our framework suggests that UCD tasks exist in addition to the development tasks in every iteration planning. The results of these tasks are presented in the iteration presentation. The level integration resulting effects into agile development are performing iterative design activities, taking measurements, and defining UCD roles.

Iterative Design Activities: In many cases, when UCD techniques are used (if at all), the design of the system is refined according to the users' evaluations mainly during the design phase. In the agile development approach, the design is updated regularly as the product evolves. When combining the UCD approach with agile development, the user evaluation is fostered by performing UCD tasks in each iteration of two to four weeks, and the design is updated according to the evaluations' ongoing outcomes.
Measures: Taking measurements is a basic activity in software development processes. When combining the agile and UCD approaches, a set of evaluation tools is built and refined during the development process and is used iteratively to complement the process and the product measures.
Roles: Different roles are defined to support software development environments. The agile approach adds roles for better management of the project [17]. Combining the agile and UCD approaches adds UCD roles, such as the design evaluator or the usability expert, to support and carry on UCD activities in the development.

3.3 The Development-Environment Level Integration—Managing and Automating User and Usability Evaluation in IDE

UCD guides integrating user experience into the software development process. One of the challenges of this integration is to automate the management of UCD activities during development. When we analyzed current software design practices, we identified a lack of *UCD management*, which we define as the ability to steer and control the UCD activities within the development environment of the project [6, 7, 12]. Defining evaluation experiments and running them from within the IDE equips the software team with the mechanism to monitor and control a continuous evaluation process, tightly coupled with the development process, thus receiving ongoing user feedback while continuing development.
Experiment Entity: Automating the evaluation methods means that we can add a new kind of an object in the development area of a software project. These objects, known as experiments because of the controlled environment in which they are performed, can be created and executed to provide evaluation data. Furthermore, an experiment's results can be associated with future development tasks as they emerge. We further divide these experiments into three categories: expert-based methods, user-based methods, and system-based methods. Expert-based methods (e.g., heuristic evaluation, cognitive walkthrough) are performed by UCD experts/system experts to evaluate the system usability with respect to standards or guidelines. User-based methods (e.g., question-asking protocol) involve end users performing different tasks on target systems or evaluating the system based on any given criteria. System-based methods (e.g., log file *or* task-environment analysis) use automated evaluation tools to record users' and system behavior and produce analysis results of the recorded data.
Derived development tasks: Each kind of evaluation experiment has its own criteria for judging the usability level of the product. Support for the analysis of the experiments' results enables the comparison of these results against the targeted usability criteria. If the results show a failure to achieve the target usability level, then new development tasks can be defined accordingly. Each development task is associated with the relevant data, thus providing its rationale.

Code Traceability: Automating the process of backward and forward traceability among different evolving parts, at the development-environment level provides a better traceability of the refinement carried out in the design to improve the product. Such parts could include code parts, experiments, and derived development tasks.

Developing Evaluation Aspects: Automating UCD activities in the development environment can enable developers to add automatic evaluation hooks to the software under development. For example, an aspect could be created to control the use of a specific button or key that is part of the developing software. These system-based methods that include such measures provide insights about the users' behavior.

4   Automated Tools Support

Our research approach, which involved working with software teams to understand how UCD can be embedded in the software development process, helped us to shape a set of requirements for tooling as well as guidelines and techniques to accompany it [6, 7, 12, 17]. We developed two tools that automate and manage user and usability evaluation at IDE level to provide realization of development environment-level integration.

UEMan: A Tool for UCD Management in Integrated Development Environment

UEMan (User Evaluation Manager): [7, 12] is an Eclipse plug-in that supports the automation and management of UCD activities as part of the Eclipse IDE. UEMan includes a Java library of evaluation aspects that enables software developers to integrate automatic measures as part of the developing project. In a nutshell, UEMan evaluation life cycles consist of four phases that occur iteratively.

Evaluation Definition: The evaluation is defined through the following: different types of experiments (such as questionnaires, heuristic evaluations, etc); experiment tasks (which end users/UCD experts need to perform during experiments); role holders who are involved in experiments (e.g., end user, UI expert, evaluator); and other management considerations (such as experiment details, time to execute, etc).

Evaluation Execution: The evaluation is executed by running the experiments either locally at the evaluation site or remotely at the participant's site; thus, tasks in the experiments can be performed by participants and the results can be stored in the system.

Evaluation Analysis: The results are analyzed by software teams as per evaluation experiment and any cross experiments to identify users' and system behavior as well as usability issues.

Feedback to Development: New development tasks for upcoming iterations are derived according to the evaluation analysis for improving the product, whereas connectivity is kept between the evaluation results and the relevant code parts during development.

TaMUlator: A Tool for Managing and Automating Task Model-based Usability Evalaution at IDE level

TaMUlator[1] (**Ta**sk **M**odel-based **U**sability Eval**uator**) is a Java-based tool that provides a set of APIs and interfaces for managing and automating task model-based usability evaluation at the IDE level.

TaMUlator allows the development team to tag tasks and variables of interest at code-level to be used in task models. TaMUlator provides an easy and dynamic way to define different usability scenarios for evaluation. This is achieved by compiling TaMoGolog-based task models that can be aggregated into evaluation experiments, which can then be evaluated at any time by the built-in analyzer using the recorded data of these experiments, or manually evaluated by exporting the recorded data into a CSV format for analysis.

5   Case Studies

We present two case studies in which development teams[2] in academia worked with agile approach for developing their software projects while applying our UCD management approach at IDE level.

In the first case study, six development teams developed an application, named FTSp (Follow the Sun plug-in), to support synchronization between distributed teams that have no synchronous communication between them due to large time zone differences. The development teams (each developed their version of the same application) conducted the evaluation (by other development teams) of the developed application using UEMan by defining and executing the heuristics evaluation experiment and logging-aspect experiment. A total of twelve experiments were conducted to evaluate FTSp. Different teams provided feedback on the contribution of the evaluation process using UEMan. Among other comments, teams mentioned the good collaboration between the team and the participants, the benefit of recognizing the new issues raised that had not been seen before, and the ability of UEMan to automate the results summary, enabling them to identify significant problems and define development tasks accordingly. As part of the evaluation study, we found that software team members engage in UCD management activities in a natural and intuitive manner. They can easily analyze experiments' results for their project and derive significant development tasks accordingly.

In the second case study, six development teams developed their software project through agile development approach while using TaMUlator to evaluate their developed project. The usability evaluation of the product that is being developed was implemented as part of this project. Following are the main practices we used: 1) Iterative design activities that include cycles of development that contain development tasks that were derived from usability evaluation. 2) Role holders in the

---

[1] TaMUlator demo video:
  http://www.dis.uniroma1.it/~humayoun/tamu/tamulator.html
[2] The team members were, both times, 4th year CS-major students participating in the 'annual project in software engineering' course of the Computer Science Department at Technion, IIT.

subject of usability evaluation and using TaMUlator. 3) Measurements that were taken by the role holders as part of fulfilling their responsibilities.

In addition to the developing their project, named 'Brain Fitness-Room' aims to develop a system for maintaining and strengthening memory and brain capabilities as well as identifying any decline in these capabilities, the subject of usability evaluation using task models was presented to the teams. In the first development iteration, all teams had to develop a tool with a Java library to enable writing basic task models in a formal way. Based on the teams' work, one tool was selected (TaMUlator) and all teams used this tool during two iterations of evaluation.

In the final retrospective on the course, team members were asked to grade their satisfaction between 1 to 5 (very satisfied) with respect to the project topic, course methodology (agile, time management and early detection of problems, emphasis on testing and usability), tools that were used (Trac and Moodle), and the services in the physical lab they worked in. 32 team members answered and the average grade for the methodology was high (4.09) (for project topic 4.36, tools 3.98/3.28 respectively, and for lab services 2.66). Specifically, regarding the roles that concerned with usability, team members referred to the importance of learning and dealing with usability while developing. Following are some of their comments on this matter: "It is important to get feedback from the users...", "It does not matter how good the product is, [people] will use it only if it is simple and user friendly. A lot of things that seem clear to developers are not clear to the end users," and "The role of being in charge of the evaluation experiment was an important role with which we specified the usage of our system by the user."

6 Conclusion

We have presented a framework that integrates UCD philosophy into agile development practice at three levels. Our framework provides a way to get equal benefits from both approaches, thus enabling the development of high-quality and usable software products. We also presented two automated tools for the realization of the development and that provide environment-level integration and two case studies where they used these tools to evaluate their software projects.

In the future, we intend to apply our framework in medium- to large-scale software projects to properly check its effectiveness. We also intend to work on the presented tools to support better management and automation.

Acknowledgments. Our thanks to Eli Nazarov and Assaf Israel from Technion IIT for their work in developing TaMUlator.

References

1. Agile Alliance. 2001: Manifesto for Agile Software Development. Technical Report by Agile Alliance (2001), http://www.agilealliance.org/
2. Balbo, S.: Automatic evaluation of user interface usability: Dream or reality. In: Balbo, S. (ed.) Proceedings of the Queensland Computer-Human Interaction Symposium. Bond University, Queensland (1995)

3. Chamberlain, S., Sharp, H., Maiden, N.A.M.: Towards a Framework for Integrating Agile Development and User-Centred Design. In: Abrahamsson, P., Marchesi, M., Succi, G. (eds.) XP 2006. LNCS, vol. 4044, pp. 143–153. Springer, Heidelberg (2006)

4. Dix, A., Finlay, J.E., Abowd, G.D., Beale, R.: Human Computer Interaction, 3rd edn. Prentice-Hall, Englewood Cliffs (2003)

5. Dubinsky, Y., Hazzan, O.: The construction process of a framework for teaching software development methods. Computer Science Education 15(4), 275–296 (2005)

6. Dubinsky, Y., Catarci, T., Humayoun, S.R., Kimani, S.: Integrating user evaluation into software development environments. In: 2nd DELOS Conference on Digital Libraries, Pisa, Italy (December 5-7, 2007)

7. Dubinsky, Y., Humayoun, S.R., Catarci, T.: Eclipse Plug-in to Manage User Centered Design. In: Workshop on the Interplay between Usability Evaluation and Software Development (I-USED), Pisa, Italy (2008)

8. Fox, D., Sillito, J., Maurer, F.: Agile Methods and User-Centered Design: How These Two Methodologies are Being Successfully Integrated in Industry. In: AGILE 2008, pp. 63–72 (2008)

9. Göransson, B., Gulliksen, J., Boivie, I.: The Usability Design Process - Integrating User-Centered Systems Design in the Software Development Process. Software Process: Improvement and Practice 8, 111–131 (2003)

10. Gulliksen, J., Goransson, B., Boivie, I., Blomkvist, S., Persson, J., Cajander, A.: Key principles for user-centered systems design. Behaviour & Information Technology 22(6), 397–409 (2003)

11. Hazzan, O., Dubinsky, Y.: Agile Software Engineering. In: Undergraduate Topics in Computer Science Series. Springer-Verlag London Ltd, Heidelberg (2008)

12. Humayoun, S.R., Dubinsky, Y., Catarci, T.: UEMan: A tool to manage user evaluation in development environments. In: ICSE, pp. 551–554. IEEE press, Vancouver (2009)

13. Hussain, Z., Milchrahm, H., Shahzad, S., Slany, W., Tscheligi, M., Wolkerstorfer, P.: Integration of Extreme Programming and User-Centered Design: Lessons Learned. In: Abrahamsson, P., Marchesi, M., Maurer, F. (eds.) Agile Processes in Software Engineering and Extreme Programming. LNBIP, vol. 31, Part 3, Part 5, pp. 174–179. Springer, Heidelberg (2009)

14. Ivory, M., Hearst, M.: The State of the Art in Automating Usability Evaluation of User Interfaces. ACM Computing Surveys 33(4), 470–516 (2001)

15. Patton, J.: Hitting the target: adding interaction design to agile software development. In: OOPSLA 2002. ACM, New York (2002)

16. Sy, D.: Adapting Usability Investigations for Agile User-Centered Design. Journal of Usability Studies 2(3), 112–130 (2000)

17. Talby, D., Hazzan, O., Dubinsky, Y., Keren, A.: Agile software testing in a large-scale project. IEE Software, Special Issue on Software Testing, 30–37 (2006)

Development of Web-Based Participatory Trend Forecasting System: urtrend.net

Eui-Chul Jung[1], SoonJong Lee[2], HeeYun Chung[2], BoSup Kim[2],
HyangEun Lee[2], YoungHak Oh[2], YounWoo Cho[2], WoongBae Ra[2],
HyeJin Kwon[2], and June-Young Lee[2]

[1] Yonsei University, 50 Yonsei-Ro, SeoDaeMun-Gu,
Seoul 120-749, Korea
[2] Seoul National University, 599 GwanAk-Ro, GwanAk-Gu
Seoul 151-742, Korea
jech@yonsei.ac.kr, leesj1@snu.ac.kr, heeyj2@hanafos.com,
equququ@empas.com, xmozil@nate.com,
{onebuttwo,dusn,rnskdl,bluyoung}@naver.com,
hyejin.kwon80@gmail.com

Abstract. The goal of this research is to develop a participatory system that can capture live trend issues and people's latent needs in the issues. Web 2.0 technology is adopted because open and sharable information platform is important for this development. The urtrend.net is developed with three sub systems: issue monitoring & generation system, imagination & creation system, and value finding system. This paper focuses on the development of the first and second sub systems. Using the System 1, trend related data are gathered and analyzed to extract emerging trend issues in our lives. Using the System 2, people can join freely the public discussion on the issues from the System 1. System 3 will be developed to analyze people's discussion to provide deep insights for designers. The urtrend.net enables designers and planners to be more creative and innovative because the system will produce more sophisticated trend information with rich and informative resources.

Keywords: Participatory System Design, Web 2.0, Trend Forecasting System.

1 Introduction

Many trend issues and values come into being continuously in daily life environments and shape our new experience. It is difficult but important to read and manage the trend information because forecasting people in the future is one of competitive factors in many fields such as product design, software design, urban design, architecture, planning or even medicine. By this reason, many institutes and individuals have published trend information regularly. However, many designers and planners must invest a lot of time in order to understand background of the trend information and explore people needs hidden in the trend information. Some trend information providers on the left side in the Figure 1 only deliver trend information one-sidedly with some keywords, descriptions, and design resources. They do not provide channels for people's participation to make the trend information more live.

M. Kurosu (Ed.): Human Centered Design, HCII 2011, LNCS 6776, pp. 65–73, 2011.
© Springer-Verlag Berlin Heidelberg 2011

Due to the recent development of social networking systems and Web 2.0 technology, many people can share their voice on several issues with others [1]. The items on the lower right side in the Figure 1 are good examples. Among people's voices on the example sites, there are many clues for extracting trend issues and solutions for the issues. People can express their own ideas as one of stakeholders in design activities [2]. However, the data from people are too raw to be used directly, so they must be tuned and interpreted for specific purposes.

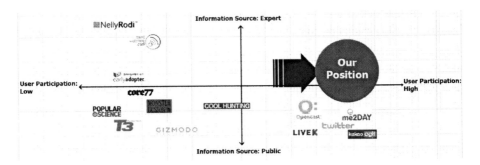

Fig. 1. Positioning of this research

Figure 1 shows how this research is positioned against other trend related institutes and systems. The aim of this research is to develop a people participatory system that can not only make people joined in capturing and discussing trend issues, but also provides insights with experts' opinions on trend issues. The name of the system, 'urtrend,' means that 'you yourself are trend' and/or 'trend is created by you.' The 'urtrend.net' will produce more practical and tangible trend information through people's participation and experts' involvement.

2   System Overview

In order to understand such diverse aspects of hidden and latent trend information in our lives and to bridge product and system design processes, it is critical to introduce a mechanism for sharing and integrating trend information.

2.1   System Structure

Figure 2 shows the whole mechanism of the 'urtrend.net,' a participatory trend forecasting system. It consists of three systems: issue monitoring & generation, imagination & creation, and value finding systems. A design processes can be divided with two axes: 1) defining issues and 2) generating solutions [3]. The 'urtrend.net' is developed by considering this general design process.

The System 1, issue monitoring & generation system has five sub-systems: the issue note composer, the issue diary composer, the issue diary-tree analyzer, the matrix analyzer, and the issue composer. Many trend resources are gathered and composed in the issue note composer by trend collectors and the collected resources are used to compose issue diary which is more refined and valid trend information edited by trained researchers. In order to find hidden pattern and meanings amongst each issue note and issue diary, analyses using issue trees and matrix are conducted. The issue trees are used to find the flow of current phenomena and the matrixes are used to find opportunities of issues hidden in people lives.

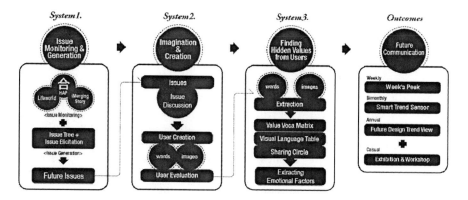

Fig. 2. Structure of 'urtrend.net'

Fig. 3. Expected outcomes from 'urtrend.net'

The System 2 is for imaging and creating the future by referring to the issues with issue notes and diary from the System 1. This system is open in the public, so people who have an interest in each issue can join freely in the issue discussion. The participants can liberally post their ideas and imaginations of the future regarding the issue and each posting can trigger others' participation spontaneously.

The System 3 is developed for designers to concentrate on creative design thinking. The basic concept of the part is that participants leave their latent needs in their writings and images as replies on the System 2. This system can extract the valuable vocabularies from participants for providing insights for designers and system developers.

The expected outcomes from these systems are weekly, bimonthly, and annual trend reports for stakeholders in each issue and/or project as shown in Figure 3. The reports from these systems can provide richer and more informative contents because each reports contains well-grounded resources, participants' discussions, and the analyses of participants' latent needs.

2.2 System Architecture

This paper introduces the development of the 'urtrend.net,' the implemented web-based system. Figure 4 shows the system architecture of 'urtrend.net.' The 'urtrend.net' is implemented on three layered architecture: database server, foundation package, and application package. The database server handles the meta-structure of issue-based project and consists of seven key tables and their sub-tables. Its primary roles are to manage system information, to generate and store project data structure and to handle reports, resources, issues, issue discussion, and users' profiles. The foundation package consists of several classes for implementing shared functions of the system. Modules in the application packages such as report composer, report analyzer, issue composer, issue discussion, people activity analyzer, and stamp manager are developed on the shared platform, so the whole encoded data can be integrated and deployed as the ways people want to utilize.

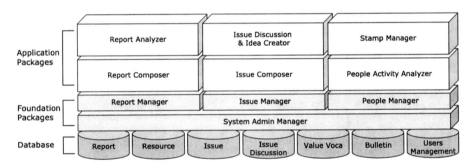

Fig. 4. System architecture of 'urtrend.net'

The implemented key features as shown in Figure 4 are introduced as follows:

1. Issue notes and diaries can be edited and composed with period, primary region and target users, multimedia data, and bibliographic data.
2. Issues can be suggested by analyzing the flow and patterns in issue notes and diaries by plotting them in an analytic frame with axes of time, region, keywords, user group, and so on. The format of issue is important because the issue should work as triggering statements for stimulating participants' imaginations not as questionnaires to be answered.
3. Public open discussion boards provide features to express participants' thoughts in diverse ways.
4. Different user levels such as system administrators, expert panels, researchers, trend collectors, and normal users can be managed to perform their tasks in the 'urtrend.net.'

3   Process of Using 'urtrend.net'

The whole process of using 'urtrend.net' consists of four steps as shown in Figure 5. For better explanation how to utilize 'urtrend.net' for practical purposes, a case study of what and how will you do if you can afford to design your home will be used as an example.

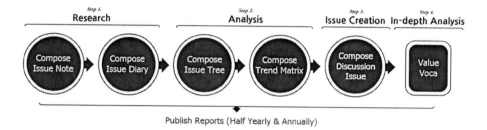

Fig. 5. Process of using 'urtrend.net'

Using this case study, performance of the following four features can be evaluated and refined: 1) methods for framing trend information by integrating multiple resources, 2) methods for managing, encoding, and analyzing trend information using the System 1, 3) representation and visualization of trend information to suggest new issues, and 4) usability of the System 2 to make participants engaged in the issue discussion voluntarily and seamlessly. This case study can demonstrates that urtrend.net achieved primary goals to provide usable and applicable trend information platform.

3.1   Step 1: Research

Collecting diverse data is important for innovative and creative design because they will inspire designers' and planners' insights. Moreover, the data work as criteria at critical decision points for better design results [4].

Fig. 6. Issue Note (left) and Issue Diary (right)

Issue Note. An issue note is collected from many external trend resources such as web sites, magazines, and books as shown in the left of Figure 6. The note contains live and current facts, so it will be used for composing issue diary which shows emerging recent issues. The trend hunters hired and trained by the 'urtrend.net' collect data with the origin of each resource to prevent copyright issues.

Issue Diary. An issue diary is composed by merging several issue notes that contain the shared issues as shown on the right of Figure 6. The diary has links to relevant issue notes, so researchers can refer to the origin if they want to. The diary shows an emerging issues in a certain region, era, and/or people group depending on which aspects are considered when the diary is composed.

3.2 Step 2: Analysis

Once issue notes and diary has been collected and composed, analysis will be followed. In order to reveal patterns and meaning hidden in collected trend data, this step is very important. Issue tree and issue matrix methods are used in the step 2.

Issue Tree. Issue trees are methods to find patterns amongst issue diaries. Each issue diary can be connected with others in some aspects. For example, a trend of 'people's pursuing their own characteristic' has some relations with a trend of 'tremendous popularity of house that enables customers to design in their own ways. As shown in Figure 9, there are four types of issue trees. Each picture in the Figure 9 represents issue diary. Time-based tress can be composed by deploying composed issue diaries in a timeline and connect each other by causal relation. Venn Diagram-based tree can be created by nesting one diary into another. Hierarchy-based tree is a kind of bottom up approaches. By categorizing diaries with similarities and naming the each category, the meaning of collected diaries can be revealed. Card sorting method can be used to create hierarchy-based tree. Composite trees can be used to analyze issue diaries by considering several aspects. This method enables researchers to analyze issue diaries in an integrated way.

Trend Matrix. Issue matrix is an issue generating tool by setting industry fields at the X axis and themes extracted from issue trees at the Y axis as shown in Figure 8. When selecting themes for the Y axis, five criteria are used: durability, extensity, reliability, originality, and potentiality. The high rated themes are selected from issue tree analysis. For example, the social soft mentoring, the third theme at the Y axis, is selected through these processes. It represents the recent phenomenon that people tend to reply on the opinions from popular people on the network who exercise their influence over others due to the development of social network. Many people try to imitate their styles and want to hear their comments. If this trend is applied to housing industry field, the issue of online house life consulting will be suggested as shown inside the bold box at the third row in housing column as shown in Figure 8. By doing this, many hidden and emerging issues in each industry can be discussed. If a company and/or institutes wants to adopt the 'uretrend.net' for a certain research purpose, the relevant field should come into the X axis for customized analysis.

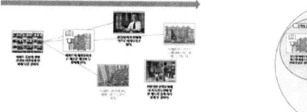

01: Time-based tree

02: Venndiagram-based tree

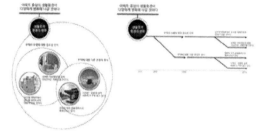

03: Hierarchy-based tree

04: Composite trees

Fig. 7. Four types of Issue Trees

Theme	Industry Field	Housing	Fashion / Beauty	Vehicle	Outdoor Goods	Health	IT / Electronic
Personal Governance	Positive Mind						
Polydentity	Social Sexual, etc.						
Social Soft Mentoring	Star standard Manual clipper						
Eternal value	Humanism						
Niche Design	Design as a solution						
Extended UX	Reel & Cyber						
Extraordina-riness	Explore newness						

Fig. 8. Example of Issue Matrix

3.3 Step 3: Issue Creation

Once an issue matrix is generated, the issue is developed by selecting related issues in each cell to be discussed. This issue selecting step is very important in design process because well defined design opportunities and problems lead to successful design results in many cases. Figure 9 show how the issue, what and how will you do if you can afford to redesign your home in your own ways. Related generated issues in each cell in Figure 8 are selected and filtered in some ways to set the main emerging issue in housing industry field.

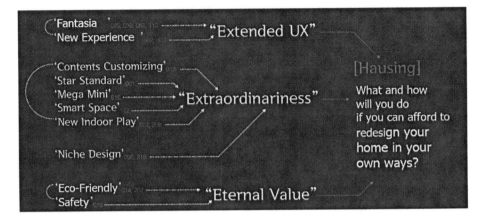

Fig. 9. Example of Issue Matrix

The describing format of an issue should not ask people directly. The issue should provide plentiful insights and background information that make people think in their own ways. That is, it should be a start point of the issue discussion.

Fig. 10. Issue discussion (left) and people (right)

The left in Figure 10 shows the interface design for issue discussion. People freely join the discussion and upload their ideas and opinions in their own ways. For the rewards for people's volunteer activities, various stamps are designed to represent the level of people in 'urtrend.net.' as shown in the right of Figure 10. This will encourage people's participation. For copyright issues of people's posting, the 'urtrend.net' will pay their ideas by analyzing their activities and stamps, if the postings will be used commercially.

3.4 Step 4: In-Depth Analysis

In this step, valuable vocabularies and images from participants are analyzed to provide insightful information to designers and planner from the users' viewpoints. As people discuss the issue with other, they show their latent thoughts unconsciously in text and images. Those are very important and valuable resource to be analyzed. This analysis processes will be conducted manually, but this is one of benefits of using the 'urtrend.net.'

4 Conclusion

The benefits of developing 'urtrend.net' can be summarized as follows: 1) it can manipulate trend forecasting in an integrated way, 2) it can produce trend information from and by crowds, and 3) it can produce practical and concrete trend information.

First of all, the trend forecasting method in the 'urtrend.net' is based on intuition and reasoning from collected facts. Especially, the method can work effectively for the design fields in an integrated way because the method includes not only defining issues and problems but also exploring design ideas and solutions about the issues. It will lead designers and planner to focus on innovative and creative concept development because they can understand the rationales behind the discussion. Secondly, all stakeholders related to a trend can participate in the 'urtrend.net.' People always produce a new trend in their lives, so this system is designed to lead people's participation by adopting Web 2.0 technology. Finally, the trend information from the 'urtrend.net' is not abstract but concrete because many imaginations about the future life and products come up as issues are composed and discussed. By this reason, the trend information and ideation in this system will be more persuasive and informative.

For the further evaluation of using the 'urtrend.net,' more applicable case studies should be conducted. The studies will demonstrate its effectiveness to frame trend information reflecting different resources and to support composing various analytical, representational and synthetic tools for product and system design in general.

References

1. O'Reilly, T.: What is Web 2.0 (2005),
   http://oreilly.com/web2/archive/what-is-web-20.html
2. Tapscott, D., Williams, A.D.: Wikinomics: how mass collaboration changes everything, p. 324. Portfolio, New York (2006)
3. Ulrich, K.T., Eppinger, S.D.: Product design and development, 4th edn. McGraw-Hill Higher Education, Boston (2008)
4. Hippel, E.V.: Democratizing innovation. MIT Press, Cambridge (2005)

Consideration of HCD Methods for
Service Innovation Design

Akira Kondo[1] and Naoko Kondo[2]

[1] Hitachi Intermedix Co., Ltd.
2-1-5 Kandanishikicho, Chiyodaku, Tokyo, 101-0054 Japan
kondo@hipri.com
[2] OpensourceCRM, Inc.
UtokuTamachi Bldg., 2-13-7 Shibaura, Minato-ku, Tokyo 108-0023, Japan
kondo@osscrm.com

Abstract. In modern society the service industry has took main role in advanced countries and the service innovation, how to design to improve productivity of the service, become a major issue in the business world. The service has four features as intangible, concurrency, heterogeneity, and extinction, then the service design process and perspective are considered to be different from the product design in the industrial age. When we build the service business, we should think about service elements such as a service receiver, a service provider, field of services, and time axis, in comprehensive viewpoint. In the service industry, we have to provide the service to satisfy customers, but it is necessary to understand the varying needs of different customers. Traditionally this process is relied on the ability of the person providing the service. The improvement of the service productivity was depends on the individual's tacit knowledge in the large part, there are also limits of human ability, then it is difficult to generalize. On the other hand, modern Web services that are provided through the internet, information processing technology could be speculated the information needs of users through the human computer interaction, it has become possible to improve the service productivity. In this paper, in order to achieve improved the service productivity by information technology services, and we considered how to embody changeable user desires as explicit knowledge using the human-centered design techniques. As a concrete methodology, in order to systematically understand the varying needs of users, is considered to be a ethnography and contextual inquiry method, as output in order to incorporate the inference engine need to be written as a structured form. As for the psychological needs of users, I think it is appropriate to consider developing a persona, the issue is a how to build a appropriate emotional model. What may be modeled using the technique of human-centered design to the desire for services that change these users, such as shops electrons on the current Internet, analyze the user's preferences, select the information that may be of interest for each user show to take a case recommendation service system, and consider.

Keywords: Service Innovation, Human Centered Design.

M. Kurosu (Ed.): Human Centered Design, HCII 2011, LNCS 6776, pp. 74–80, 2011.
© Springer-Verlag Berlin Heidelberg 2011

1 Introduction

In industrialized society, the main economic activity was trading products. That time, the service was associated with products. But now the customer wants, not just get good products, comfortable process of trading, and favorable experience using products. In retail trade service, as well as the actual value of products, customer service attitude, store image, product line, and appropriate products information are affecting customer satisfaction. Usui advocates that the following four perspectives are important for retail trade service.

- Providing information and prior expectations
- Provision of products and services themselves
- The quality of providing process
- Duration of follow-up and provide satisfaction

Then we should think about not a provider perspective, but it is important to the customer's perspective in service design.

The other hand, implementing ICT in service process are getting that aiming at improving the productivity in the service market. Retail business market continues to expand ecommerce. As a result, many companies try to refine their service to get good position in the market. Then such companies start to take in HCD (Human-Centered Design) process.

In Japan, the government launch e-government usability guideline to satisfy most user. This guideline introduces the useful e-government website and system building process by HCD methods.

Thus, the expanding Internet services market has been focused on the need for HCD process.

2 Purpose of Research

With the expansion of the Internet services market it is necessary to provide services to various users via the Internet. Traditional service contact point was person but such a current service, PC or tablet, and mobile phone interface become a new contact point. Service Business on the Internet, not just simply replace the traditional service delivery models into internet fields, service provider should think about such as service innovation to improve productivity and customer satisfaction.

In this paper, first we survey the concept of service innovation to improve the productivity of services.

Then, as a concrete case of service innovation, we investigate in ecommerce recommendation system that aim at consider about ICT based service delivery issues. At the last part, we consider how we can use a HCD approach in service design and discuss its effectiveness.

3 Survey of Service Innovation

3.1 Definition of Innovation from Economic View Point

In Japan, many people think that "Innovation" means "Technological Revolution." However famous economist Joseph Schumpeter defined economic innovation as follows;

- The introduction of a new good — that is one with which consumers are not yet familiar — or of a new quality of a good.
- The introduction of a new method of production, which need by no means be founded upon a discovery scientifically new, and can also exist in a new way of handling a commodity commercially.
- The opening of a new market, that is a market into which the particular branch of manufacture of the country in question has not previously entered, whether or not this market has existed before.
- The conquest of a new source of supply of raw materials or half-manufactured goods, again irrespective of whether this source already exists or whether it has first to be created.
- The carrying out of the new organization of any industry, like the creation of a monopoly position (for example through trustification) or the breaking up of a monopoly position

We have this definition in mind we survey the service innovation as "the service deliver new value."

3.2 Activity of "Service Industry Productivity Council" in Japan

"Service Industry Productivity Council" is Japanese cooperative research groups consisting of experts from universities and companies. They announced the award to companies and organizations doing innovative work to help improve innovation and productivity. They evaluate companies and organizations efforts from the following perspective.

- Scientific and engineering approaches
- Improvement Services Process
- High value-added services
- Human resource development
- International Expansion
- Regional contribution

I think these perspectives are almost close to Schumpeter's Innovation definition. This award also intend to introduce the importance of ICT and HCD process to build good service value chain.

3.3 Direction of Service Innovation

From the previous literature survey, it is necessary to design a new value procreation service value chain to realize service innovation. On its way, Takada and Koike analyzed that following directions are important, "Building the Web service value" for including creating a value chain for service, and "Module Development Services ", "Improvement of service interaction, "" development, utilization, service platform "for having both way diversity and standardization. The creating a value chain for service is dependent on the part of basic service delivery, so in this study we focus on having both way diversity and standardization, particularly "Improvement of service interaction."

As cases of "Improvement of service interaction," the report introduced "financial aggregation" that means the office handles a variety of financial products, and "tailor-made service" that is fabricated a personal computer using a direct order from user. This direction of service innovation to achieve this service the needs of various customers, as embodied by the appropriate interaction mechanism. Present in ecommerce "Improvement of service interaction" realized as a system, the recommendation system to be an example case.

4   Case Study

4.1   Recommendation System Theories and Issues

Ecommerce market is expanding, many EC site provide the recommendation system for user to find products from various products. However, there are findings that user feel disgusting recommendations because of unknown system data processing. The current recommendation systems have been constructed by following methods "rules based," "content based," "collaborative filtering," and "Bayesian network." Most methods are showing items those are selected from the user's action history. These recommendation theories, the following problems were identified was conducted user interviews.

- Not meet the user's intention to buy the relevant product recommendations
- Feel uncomfortable used historical data without permission
- Items appear that user have purchased products or not want

We consider about this issue from "Model of service quality gaps (Parasuraman et al., 1985; Curry, 1999; Luk and Layton, 2002)." As a result, service gaps have become obvious in GAP1 and GAP5 in the current ecommerce. So we analyze that there are no inconsistent within the service provider side but in reality service are not meet user expectations and service provider attempt. In addition, we consider that a service provider have not investigated about management system to correct these service gaps.

4.2   Improvement Proposal of Recommendation System by HCD

We consider how to use HCD method in development process of recommendation system in Fig 2.

We presented the scenario of improved recommendation service by HCD methods to subjects then they express positive response. So we believe that HCD process is effective for service innovation powered by ICT such as recommendation system.

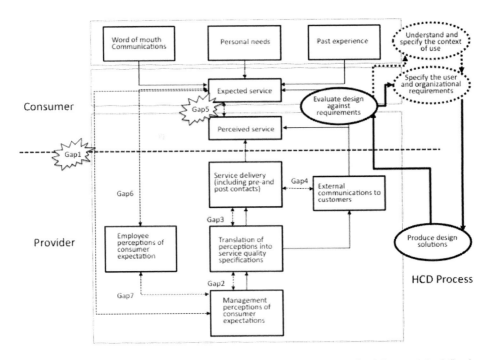

Fig. 1. Model of service quality gaps and HCD methods (Kondo revised from original fig. by Parasuraman et al., 1985; Curry, 1999; Luk and Layton, 2002)

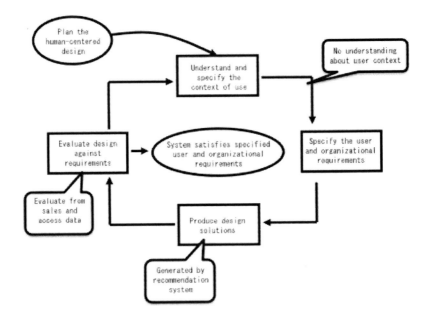

Fig. 2. HCD development process for the recommendation system

5 Future Agenda

Service innovation will not be brought by HCD methods but collaboration work with system developers, UX designers and other specialists. Then we would like to survey various cases to find appropriate team and methods for service innovation.

The other hand, we need to consider about service business sustainability. In this business perspective, service design process should establish a cooperative mechanism with business management methods. So we proposed this service business methods flame and investigate efficiency.

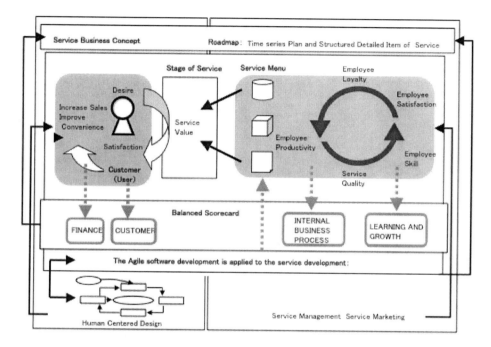

Fig. 3. Service design methods flame

References

[1] Makoto Usui, Promote Service Innovation, Nikkei IT Pro Website, Japanese (2009),
     http://itpro.nikkeibp.co.jp/article/COLUMN/20090529/330954/
[2] Cabinet Secretariat, Usability e-Government Guideline, Japanese (2009),
     http://www.kantei.go.jp/jp/singi/it2/guide/security/kaisai_
     h21/dai37/h210701gl.pdf
[3] Schumpeter, J.: The Theory of Economic Development. Harvard University Press, Boston (1934)
[4] Service Industry Productivity Council, High Service Best 300 in Japan Website (2007-2010) (Japanese), http://www.service-js.jp/cms/page0600.php

[5] Takada, T., Koike, K.: Service Innovation of Japanese Companies. Creation of intellectual property (CHITEKISHISANSOUZOU in Japanese). Nomura Research Institute Ltd., Japan (2002)

[6] KDDI R&D Laboratories, Successful elucidation of receptor structure recommended services using personal information, News Release Website (2010),
http://www.kddilabs.jp/press/img/154_1.pdf

[7] Carroll, J.M.: Scenario-based Design envisioning work and technology in system development. Wiley, US (1996)

[8] Carroll, J.M.: Making Use of Scenario-based design of human-computer interactions. The MIT Press, US (2000)

[9] Kondo, A., Kondo, N.: Research of Design Method for Service Business Development. In: 2nd International Service Innovation Design Conference, Hakodate Japan (2010)

[10] Yamazaki, K., Furuta, K.: Proposal for design method considering user experience. In: 11th International Conference on Human-Computer Interaction, Las Vegas (2005)

[11] Kameoka, A.: Service Science (Japanese), NTS, Inc., Japan (2007)

Descriptive Words for Expressing the User Experience

Masaaki Kurosu

Center of ICT and Distance Education, Open University of Japan, 2-11 Wakaba,
Mihama-ku, Chiba-shi, Chiba 261-8586, Japan
masaakikurosu@spa.nifty.com

Abstract. User experience is a function of various traits of the artifact including the usability. In the first part of this article, various traits of the artifacts were examined before the purchase, during the purchase and after the purchase (usage) on how values of each trait vary depending on the phase. In the second part, the direct examination on the descriptive words in terms of the user experience was examined based on the proposed concept of GOB, POB and SOB.

Keywords: user experience, usability, satisfaction, pleasure, happiness.

1 Introduction

The User Experience (UX) (Roto et al. 2011) can be regarded as a function of various traits of the artifact including the usability. In other words, the UX is a dependent variable and various traits are independent variables, hence the UX and the usability is not the same. In the first part of this study, the analysis of 24 artifact's traits was conducted in terms of three different phases to see what kind of traits are regarded to be important in each phase. In the second part, what kind of subjective impression regarding the hedonic quality will matter depending on the situation. In other words, the first part concerns the dependent variables and the second part the independent variable.

2 Method

The same informants participated in two studies and the same method (questionnaire) was used to specify the nature of dependent and independent variables. Informants included 15 undergraduate students and 17 graduate school students. But due to an unexpected accident, 4 data was missed and only 28 data was used in the analysis. The research was conducted on December, 2010.

3 Study 1: Dependent Variables of UX

Informants were asked to rate the importance of the 24 quality traits including 1. brand image, 2. initial cost, 3. reputation, 4. familiarity, 5. novelty, 6. beauty, 7. self-expression, 8. fuctionality, 9. performance, 10. Infrastructure, 11. running cost,

M. Kurosu (Ed.): Human Centered Design, HCII 2011, LNCS 6776, pp. 81–90, 2011.

12. ease of operation, 13. compatibility, 14. environmental adaptability, 15. durability, 16. maintenance, 17. reliability, 18. safety, 19. disposability, 20. physical fitness, 21. ease of memorizing, 22. ease of learning, 23. efficiency and 24. effectiveness. The evaluation was conducted on the 4 point scale (1: unrelated, 2: do not matter, 3: mind a bit, 4: mind very much) in terms of 3 phases including a. before the purchase, b. at the purchase, and c. usage (after the purchase). The third phase c. means the real usage of the artifact in the real situation for a certain long time. Targeted artifacts were the cell phone and the automobile.

3.1 Results for the Cell Phone

Following graphes show the average value and the standard deviation of ratings for each of 24 independent variables (quality traits) for the cell phone and the the graphes are combined for three periods including a.before purchase, b. purchase and c.usage. Wherever there was found the significant difference based on the multiple comparison (Turkey method), the asterisk is shown (* for 5 %) on the graph.

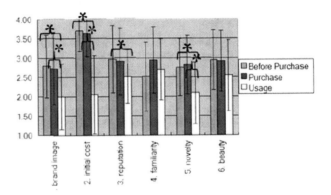

Fig. 1. Scale values for quality trait 1 through 6 (cell phone)

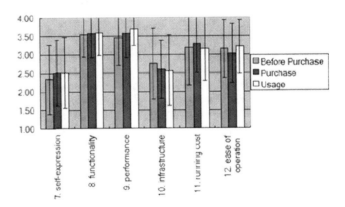

Fig. 2. Scale values for quality trait 7 through 12 (cell phone)

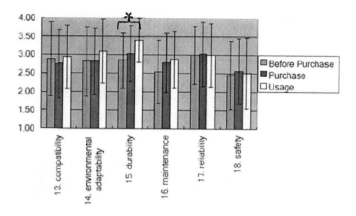

Fig. 3. Scale values for quality trait 13 through 18 (cell phone)

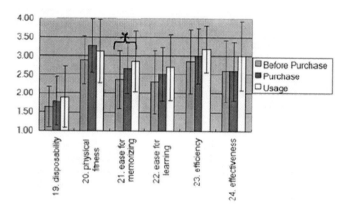

Fig. 4. Scale values for quality trait 19 through 24 (cell phone)

As can be seen in Fig 1-4, almost all quality traits had similar significance except 1, 2, 3, 5, 15, 21, there were significant differences especially between before the purchase and usage (after the purchase). Quality traits 1,2,3 and 5 become less important after the purchase, where quality traits 15 and 21 become more important after the purchase.

3.2   Results for the Automobile

For the automobile, only scale 1, 2 and 3 were significantly different between before the purchase and the usage (after the purchase). All other scales were judged to have equal importance regardless of the phase.

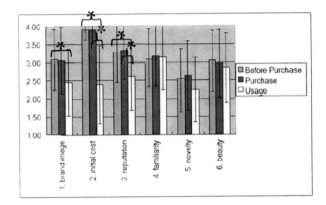

Fig. 5. Scale values for quality trait 1 through 6 (automobile)

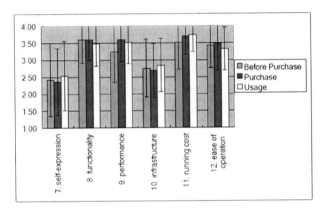

Fig. 6. Scale values for quality trait 7 through 12 (automobile)

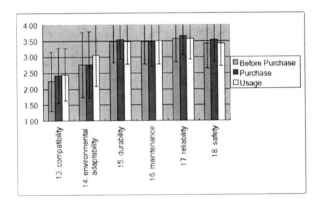

Fig. 7. Scale values for quality trait 13 through 18 (automobile)

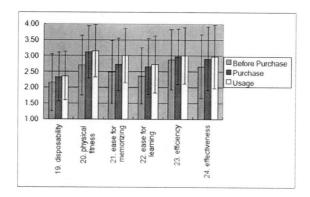

Fig. 8. Scale values for quality trait 19 through 24 (automobile)

4   Study 2: Independent Variables of UX

The UX can be expressed in various ways. 10 different situations were evaluated by 6 different scales, namely 1. Comfort, b. pleasure, c. delight, d. pleasantness, e. happiness and f. satisfaction.

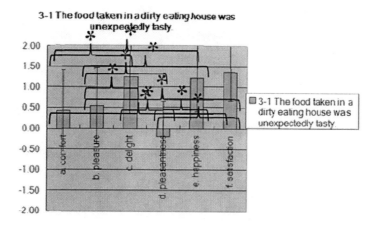

Fig. 9. The food taken in a dirty eating house was unexpectedly tasty

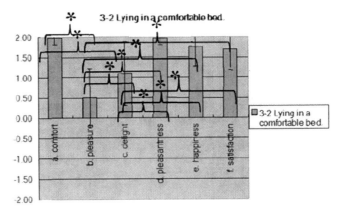

Fig. 10. Lying in a comfortable bed

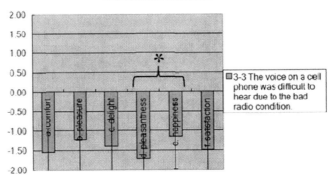

Fig. 11. The cellphone was difficult to hear due to the bad condition

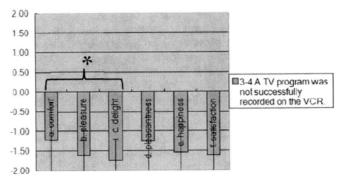

Fig. 12. A TV program was not successfully recorded on the VCR

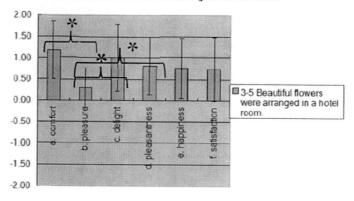

Fig. 13. Beautiful flowers were arranged in a hotel room

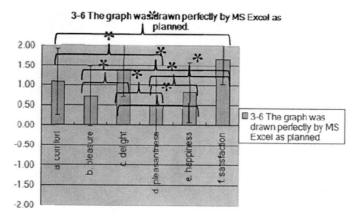

Fig. 14. The graph was drawn perfectly by MS Excel as planned

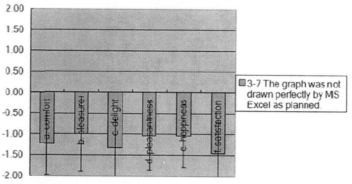

Fig. 15. The graph was not drawn perfectly by MS Excel as planned

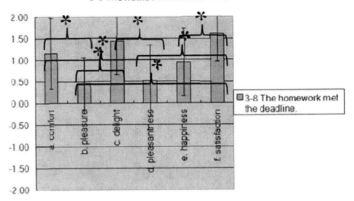

Fig. 16. The homework met the deadline

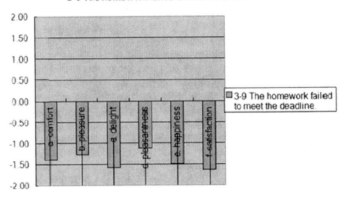

Fig. 17. The homework failed to meet the deadline

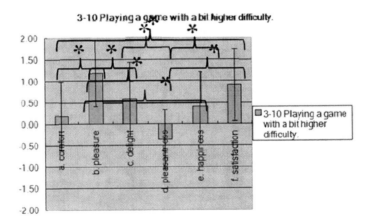

Fig. 18. Playing a game with a bit higher difficulty

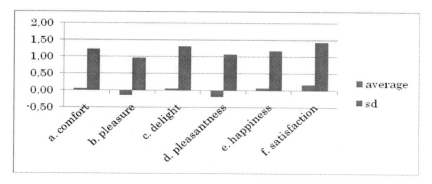

Fig. 19. The standard deviation (and the average) of each scale

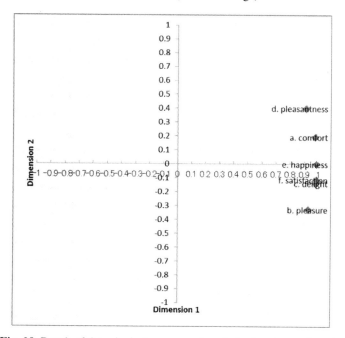

Fig. 20. Result of the principal component analysis showing each scale

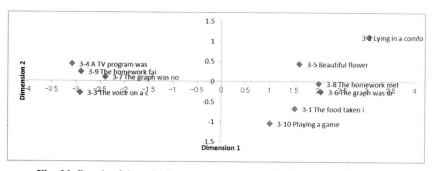

Fig. 21. Result of the principal component analysis showing each situation

5 Conclusion

Regarding the quality traits, just some of them differed in evaluating before / after the purchase. Regarding the description of UX, each scale used in experiment a bit of differences among them.

Reference

Roto, V., et al.: User Experience White Paper (2011)

Expert-Sided Workflow Data Acquisition by Means of an Interactive Interview System

Daniel Ley

Fraunhofer Institute for Communication, Information Processing and Ergonomics FKIE
Neuenahrer Straße 20, 53343 Wachtberg, Germany
daniel.ley@fkie.fraunhofer.de

Abstract. This paper outlines an approach for gathering workflow data via an interactive interview system. By means of this approach, data acquisition for a subsequent task and/or process analysis is conducted by a process expert instead of a process analyst as customary in application of conventional data acquisition methods. Beside other problems concerning existing techniques, this may solve the dilemma of a lacking common basis between expert and analyst in terms of process knowledge and process thinking.

A classification method is described which allows a definition of processes acquirable by the system. Furthermore, a procedure for decomposing processes is used to gather workflow data in a systematic way. During system application, feedback by sub-process models directs experts to process thinking while system records impart process knowledge for the analyst. The applicability of this approach is shown by results of a first system evaluation. Advantages and disadvantages in relation to common data acquisition methods are stated.

Keywords: Data Acquisition Method, Process Analysis, Task Analysis, Interview System, Process Thinking, Process Knowledge.

1 Introduction and Purpose

Process and task analyses are important and essential ways for improving workflows in various domains. Therefore, different semi-standardized procedures are described in literature, especially for the domain of business processes [1], [2]. These procedures mostly include a generation of workflow models and their analyses via simulations. Consequent optimization recommendations are developed among numerous aspects like processing time, costs or/and workload of involved operators. An essential requirement for building valid process models is the acquisition of complete relevant process data on the basis of expert process knowledge. Methods like questionnaires, interviews or observations are regularly used by process analysts for data acquisition.

There are two main challenges in application of conventional methods, which are considered in the current approach. First, analyses of workflows are mostly conducted by external analysts, which do not have any experience or specific knowledge about the work domain. Their expertise is limited to their knowledge about model notations and the analysis of process models. In contrast, operators are experts with respect to their field of work, but mostly do not know anything about process model notations.

M. Kurosu (Ed.): Human Centered Design, HCII 2011, LNCS 6776, pp. 91–100, 2011.
© Springer-Verlag Berlin Heidelberg 2011

Consequently, a dilemma exists between analyst and expert: an analyst possesses process thinking and an expert possesses process knowledge – without any or only limited overlap.

Secondly, analysts have to prepare questionnaires and/or interviews and probably have to apply them repetitively in order to capture all relevant data. In case of interview situations it is improbable that analysts adapt their questions during an interrogation in order to capture any unexpected and unconsidered relevant data at the first attempt. Hence, conventional data acquisition methods mostly require time-consuming preparations and have to be iteratively applied. Furthermore, after data acquisition an enormous effort must be made to edit the data and to transfer them into process models for further analyses.

This contribution describes the development of a system called "Process Interviewer" (hereinafter "PI") which allows workflow data acquisition by process experts by a dynamic interview procedure. In this way, no effort of a process analyst is required. Data acquisition can be conducted by experts autonomously by the use of intuitive graphical user interfaces. Periodically depicted graphical sub-process models enable the expert to reflect own workflows and to learn thinking in processes. Apart from that, data and models serve the analyst as information input to understand the conduction of workflows and thus impart process knowledge. Recorded data can be used in process modelling and simulation tools – so there is a minimal effort for data transformation.

Initially, methodical aspects concerning system design are described in the paper. For this purpose, common data acquisition methods are delimited, processes are classified and systematized as well as a method for process decomposition is introduced. Subsequently, the generation of interview steps and an exemplary representation in a user-friendly graphical interface are described. Furthermore, considerations in terms of initiating process thinking via graphical models are appointed. Finally, results of a system evaluation are discussed and the paper is accomplished by a summary and a description of future work.

2   Methodical Aspects for Acquiring Workflow Data by Experts

2.1   Dissociation of Common Data Acquisition Methods

There are several commonly deployed data acquisition methods described in literature [3], [4], [5], [6], usually varying in kind of execution in a wide range, whereby a similar approach like here described has not been found. Summarized, there are six main methods to gather data for task and process analyses: interviews, questionnaires, observations, workshops, document analyses and protocols. In the current work advantages and disadvantages have been collected and compared with expected advantages and disadvantages of the new data acquisition approach.

The feedback by process models is an expected advantage of the system PI in comparison to common methods, directing the expert to process thinking and the analyst to process knowledge with minimal effort. Beside this, analysts do not need previous knowledge concerning the considered work domain: a system application can be conducted without any preparatory work – on condition that an expert with appropriate verbalization skills is using the system. A possible acquisition of cognitive processes, as opposed to e.g. common observations, provides a further advantage. Data acquisition can be interrupted and continued at any point of time via a system application. In addition, the dynamic interview procedure ensures a capture of complete workflows which is problematic in a human expert/analyst situation, e.g. in case of conducting an interview. Due to autonomous system usage, no social-psychological effects like social desirability and investigator bias are expected, too. This involves the problem of no immediate requests during data acquisition. Disadvantageously, data are not acquired in real working situations but only on the basis of the expert's memory. A possible remedy is a system application at the expert's workplace – so she/he can put oneself in the own labor situation. A further disadvantage may that a trained human analyst cannot intervene in order to increase the expert's motivation and to react to his/her needs.

2.2 Classification and Systematization of Processes

Processes are analyzed in a wide range of work domains. Therefore, numerous kinds of process denotations exist, e.g. business processes, cognitive processes, knowledge processes or security processes. A classification scheme is needed for determining process classes ascertainable by PI. A literature and online research on properties of different process entities resulted in the following eight differentiating process characteristics:

- structuredness (structured – unstructured),
- state transition (stochastic – deterministic),
- flow (continuous – discrete),
- action (inter-individual – intra-individual),
- object of transformation (physical – immaterial),
- specificity (specific – unspecific),
- level of abstraction (leading – operational) and
- regularity (regular – irregular).

According to this classification scheme, PI gathers data of discrete workflows with deterministic state transitions of single experts (intra-individual), which are either structured or unstructured. Workflows in question are conducted regular on an operational abstraction level and can be specific or unspecific. They can transform physical as well as immaterial objects. The eight classification characteristics enable an analyst to decide, if data of a specific process or class of processes are ascertainable by PI. Thereby, acquisition is basically domain independent.

Figure 1 shows the placement of several ascertainable processes in a structure model of a three-dimensional working system. It is assumed that there are domain experts E_i (i = 1..n) working in an organization. These experts conduct several activities A_j (j = 1..n) to fulfill their tasks in processes P_k (k = 1..n) within a specific duration $D_{i,l}$ (l = 1..n). It is supposed that an expert E conducts several activities A in different processes P and that she/he is working together with other experts E, which is shown by the overlapping control flows of experts in process P_1.

According to the process classification scheme above we are considering only intra-individual actions for now. We acquire sub-processes of one expert with interfaces (information or material input/output) to other experts. By means of individual interviews of all experts involved in the whole process, hereinafter the data should be connectable to this whole process.

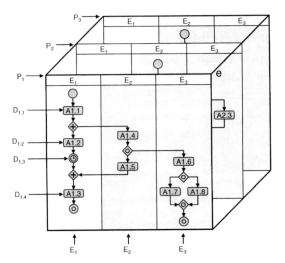

Fig. 1. Structure model of a working system

Therefore, we extract the tasks and actions of only one expert from the working system in one PI application. In Figure 1, these process fragments of expert E_1 within the three processes P_1 to P_3 are marked. Purpose of the interview system for acquiring workflow data of an expert E is to capture these process fragments up to a particular granularity dependent on the purpose of a subsequent analysis. The single process steps (activities A) and their relationships among each other – the process flow (e.g. activity order, alternative activities, random structures and iterations) – are required data. In addition it is necessary to gather activity data like activity durations, interfaces to other experts in a superior process scenery, means of labor and the place of accomplishment. It might not be easy for an expert to communicate the process flow and therewith the activities on the required level of detail, immediately. Therefore, it is essential to choose a step by step procedure, which is outlined in the following section.

2.3 Decomposition of Processes

A particular challenge consists in the generation of a systematic procedure to gather process data by an expert who has no experience in terms of process thinking. For this purpose, a dynamic interrogation procedure with appropriate graphical user interfaces is developed. On this basis a process is captured in user-friendly single interview steps. At first the procedure inquires data of superior process structures, which are then refined step by step until the final level of single activities is reached, similar to common task analyses [4], [6], [7]. This approach is based on a procedure mentioned in investigations to process model abstraction [8], [9], [10]. The authors use 'single entry single exit fragments' (SESE fragments) to abstract an existing process model step by step. Thereby, SESE fragments are process elements which have exactly one incoming and one outgoing edge. Thus, all activities contained in a SESE fragment can be synthesized into one superordinated activity by which means the process is abstracted gradually.

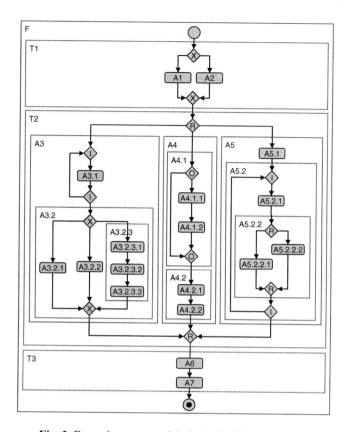

Fig. 2. Exemplary process labeled with SESE fragments

In Figure 2 an exemplary process model is shown in which SESE fragments are labeled on different abstraction levels. The process notation consists of activity elements (small rounded rectangles), gateways (diamonds) and their connections for labeling chronological orders (arrows). Gateways are distinguished in X-Gateways (XOR – labeling alternative flows), I-Gateways (ITERATION – labeling the repetition of activities or fragments), A-Gateways (ARBITRARY – labeling activities or fragments conduct in arbitrary order) and O-Gateways (OPTIONAL – labeling activities or fragments which are not necessarily conducted). This gateway-based process notation supports an easy identification of SESE fragments, since gateways join incoming or outgoing edges and therefore determine boundaries of SESE fragments. SESE fragments on the lowest level are single activities, e.g. A4.1.2.

As already mentioned above, the present work exploits this concept of SESE fragments. However, it is a process decomposition rather than a process abstraction. So we choose the opposite way and begin with inquiring the function F of an expert, e.g. his profession or job, which encompasses the sum of her/his workflows (note 'F' in Figure 2). Next step is the acquisition of her/his work tasks T, which represent SESE fragments on the highest level (note the single entry edge and the single exit edge at the border lines in each case – enabled through the gateway concept). These tasks match the process oriented activity sequences suggested in Figure 1. All further steps deal with the decomposition of activity sequences on different levels. In doing so, it is important that every fragment satisfies the SESE requirement: only one ingoing and one outgoing edge may occur. Principally, every process transformable in SESE notation can be acquired by PI.

3   Realization and Evaluation of the Interview System PI

3.1   Questions and Appropriate User Interfaces

Additional data are necessary for a further analysis of workflows, affecting properties of activities beside the acquisition of single process steps (activities) and their relationships. In order to determine relevant properties, different modeling languages like the Business Process Modeling Notation (BPMN, [11]) were taken into account. Finally, properties of activities relevant for standard process and task analyses were fixed, like activity duration, interfaces to other experts in a superior process scenery (resource input, output) and means of labor. Contrary to most other approaches, the place of accomplishment of activities is additionally considered – by this means a subsequent analysis of spatial conditions is possible.

All necessary data for a workflow analysis are allocated to different interview steps. In each step only limited data can be acquired in order to prevent an overload of experts. Another requirement for system design concerns the unique application of PI by an expert without the help of an analyst. This circumstance requires an easy respectively intuitive usage of the system by inexperienced users. Anyway, the allocation of questions to a single step took place dependent on related content and structure of workflow components. After allocation the development of a dynamic interview procedure has been the next step, whereas steps concerning the structure of workflow determined the fundamental flow. For each interview step a human-machine interface has been designed. As an example, the acquisition of structured and unstructured activity orders is presented below.

Figure 3 shows the user interface for acquiring the order of activities. At the beginning activities of a partial process structure are presented in the upper area of the interface. According to the number of activities fields are displayed, among which the activities can be moved. In Figure 3 there are three fields and the corresponding three activities are already allocated to these fields – two activities to the left field and one activity to the middle field. Allocating activities to different fields constitute a specified order of activity execution (from left to right – indicated through arrows between the fields), whereas the allocation of more than one activity to one field represents an arbitrary execution order of these activities. Thus, the two activities in the left field in Figure 3 are conducted in any order during the workflow and only after their completion the activity in the middle field will be executed. This example shows a 'drag and drop' question. There are other question types like multiple choice or fill-in-the-blank text. The example shows an interview step repeated for every partial structure of a workflow. An interlacing of structured and unstructured partial flows is possible by the iterative and dynamic interview procedure. According to this, structured, unstructured as well as semi-structured processes can be gathered.

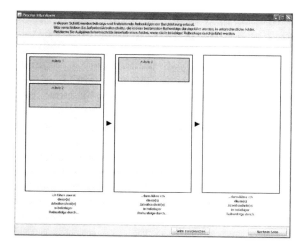

Fig. 3. Interview step of PI for the acquisition of the order of activities

3.2 Initiating Process Thinking: Feedback via Graphical Sub-process Models

The dilemma of a missing common knowledge base of analysts and experts has already been mentioned above: in the majority of cases an analyst does not know anything about the process she/he has to analyze. But by applying PI she/he receives information by means of data logging. By contrast, the expert mostly does not know anything about process modeling, thereby a subsequent participative validation of analyst's process models is quite difficult. As a countermeasure, the expert should be introduced to a simple process model notation in the course of an interrogation. For this purpose the system provides a feedback for every acquired SESE fragment in form of a partial graphical model representation. This allows the expert to check her/his entered data whereby a thinking in processes is initiated. In order to restrict the complexity of process models, activity generation is limited up to ten for each partial workflow.

In Figure 4 a simple process model is represents the order of activities determined in Figure 3. It shows the activities order in control flows and the arbitrary activity order by the concept of blobs [12]. In this case, the concept of blobs does not additionally represent simultaneous actions of experts. Reason for that is the assumption that simultaneous actions are only possible on the level of rules [13], [14]. However, in the current approach we are acquiring activities on rule-based level. By proofing the shown model the expert decides if it represents her/his mental concept of the partial process or not – and gets the opportunity to adjust it. In the next interrogation steps the system asks if activities can be divided in further sub-activities.

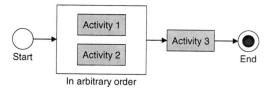

Fig. 4. Expert feedback through a simple partial process model

The knowledge gap between expert and analyst is reduced by such simple partial process model representations: Experts get a simple model-based representation of their workflows and learn process thinking, whereas the analyst can understand the content of experts' workflows by tracing back the survey records. This should build the missing common basis between expert and analyst in terms of process knowledge and process thinking and constitutes a good condition for further process analysis.

3.3 Results of a First Evaluation

Twelve scientists (average age=26.2, x_{median}=28, x_{min}=24, x_{max}=32); six females and six males) out of the field of human-systems engineering have been asked for the workflows of their everyday job for the first expert evaluation of PI. Beside the analysis of logged data (application time, number of activities entered…), the testing of the independent applicability by experts, the identification of improvement suggestions concerning the system usability and the examination of a possible initiation respective improvement of a thinking in processes have been the aims of the evaluation.

The analysis of log files shows an average application time of μ=70.2 minutes (x_{median}=68, x_{min}=34, x_{max}=114), whereby instructions of the investigator provide an application termination in case of a lack of concentration or motivation. During application time every expert generated an average of 20.6 nodes (x_{median}=21.5, x_{min}=10, x_{max}=30) including function, tasks and activities. Hence, the acquisition of one node (including all information like input, output, duration, tools needed, order concerning other nodes, child nodes…) amounts 3.7 minutes (x_{median}=3.1, x_{min}=2.1, x_{max}=7.9). These data have to be compared with applications of standard acquisition methods (like interviews, questionnaires and observations), which will be conducted in a future, more extensive evaluation. Out of this can be proved, if an application of PI is less time-consuming than the application of a standard method. During the development of this approach it has been considered in which way a restriction of interview granularity can be ensured. A feasible solution seemed to be the limitation of an activity duration. According to the following process or task analysis the analyst might define the interview granularity by determining a minimum activity duration. This would enable the system to terminate an interview by averting a splitting of appropriate activities into sub-activities. However, the first evaluation showed that experts defined at most x_{max}=6 hierarchical levels (excluding function and tasks), whereas the mean has been 3.2 levels (x_{median}=3, x_{min}=1). This indicates that the need for a termination criterion does not exist due to an already limited granularity. Otherwise it must be examined, if the existing granularity is sufficient for the purpose of the subsequent analysis – perhaps an activity division must be enforced by the system.

The evaluation revealed a possible independent applicability of the system by experts. Experts have been allowed to talk to the investigator and to ask questions. However, those concerned almost exclusively improvement suggestions instead of system handling. A further structured identification of improvement suggestions related to the system usability has been conducted by surveys after applying the system (NASA-TLX [15] and ISONORM 9241/10 [16]).

In order to test an initiation or improvement of thinking in processes, the test persons have been asked to transform a continuous text into a process model before the application of PI and afterwards. The evaluation of these process models showed that there has been no relevant improvement. This circumstance is ascribed to the already existing experience concerning process and task analysis, inquired in the course of investigation.

Altogether, the evaluation reveals positive assessments by the usability experts concerning the applicability of the system and numerous suggestions for improvement regarding the usability. A potential to use the system in projects for an acquisition of workflows has been confirmed. However, the discussion in terms of the motivation of users was divided. Some experts assessed the system handling as exciting, others expressed that it is boring. So there is a demand for the identification of opportunities to increase the motivation of users while interacting with the system. Finally, it should be mentioned, that consequent process models indeed showed weakly structured processes (i.e. not many activity orders) which probably corresponds to the overall creative work of scientists.

4 Summary and Future Work

In this paper a new approach for workflow data acquisition by the interactive interview system PI has been outlined. Acquirable processes have been structured by SESE fragments for the development of a dynamic interview structure with corresponding questions and user interfaces. During system application, feedback through partial process models should initiate an expert's thinking in processes. A first evaluation showed the functioning of the approach and the favored independent applicability of the system by experts. It gave information concerning application durations, quantity of obtained data and suggestions for improvement in terms of usability.

Investigation issues like a comparison with conventional data acquisition methods and initiation of process thinking will be considered in a future evaluation. Prior to that, the system will be adapted in terms of usability, inclusive the integration of a structure tree indicating the actual interview position for a better orientation. Furthermore, since the current system enables interviews on operational level (see the process classification in section '2.2'), a future version of the system will contain a configuration for data acquisition on management level.

Overall, feedback by simple partial process models intent to initiate an expert's process thinking. System records impart an analyst's process knowledge. This provides a common knowledge basis and therefore a simplification of further analyses of workflows.

References

1. Havey, M.: Essential Business Process Modeling. O'Reilly Media, Inc., Sebastopol (2005)
2. White, S.A., Miers, D.: BPMN Modeling and Reference Guide: Understanding and Using BPMN. Future Strategies Inc. (2008)
3. Burge, J.E.: Knowledge Elicitation for Design Task Sequencing Knowledge, Thesis (1998)

4. Kirwan, B., Ainsworth, L.K.: A Guide to Task Analysis. Taylor & Francis, London (1992)
5. Lindgaard, G.: Usability Testing and System Evaluation. A guide for designing useful computer systems. Chapman & Hall, London (1994)
6. Stanton, N.A., Salmon, P.M., Walker, G.H., Baber, C., Jenkins, D.P.: Human Factors Methods. A Practical Guide for Engineering and Design. Ashgate (2005)
7. Annett, J.: Hierarchical Task Analysis. In: Diaper, D., Stanton, N. (eds.) The Handbook of Task Analysis for Human-Computer Interaction, pp. 67–82. Lawrence Erlbaum Associates, Mahwah (2004)
8. Polyvyanyy, A., Smirnov, S., Weske, M.: Process Model Abstraction: A Slider Approach. In: Proc. International IEEE EDOC 2008. IEEE Computer Society, Los Alamitos (2008)
9. Polyvyanyy, A., Smirnov, S., Weske, M.: On Application of Structural Decomposition for Process Model Abstraction. In: Proc. 2nd International Conference on Business Process and Services Computing (2009)
10. Smirnov, S.: Structural Aspects of Business Process Diagram Abstraction. In: Proc. IEEE Conference on Commerce and Enterprise Computing (2009)
11. OMG Business Process Modeling Notation (BPMN): Specification, Version 1.2, 2009-01-03. OMG (2009)
12. Harel, D.: Statecharts: A visual formalism for complex systems. Science of Computer Programming 3(8), 231–274 (1987)
13. Rasmussen, J.: Skills, Rules, Knowledge, Signals, Signs and Symbols and other Distinctions. Human Performance Models. IEEE Transactions on Man, Systems and Cybernetics SMC-13(3), 257–266 (1983)
14. Rasmussen, J., Pejtersen, A.M., Goodstein, L.P.: Cognitive Systems Engineering. John Wiley & Sons, Inc., U.S (1994)
15. Hart, S.G., Staveland, L.E.: Development of NASA-TLX (Task Load Index): Results of empirical and theoretical research. In: Hancock, P.A., Meshkati, N. (eds.) Human Mental Workload, pp. 139–183. Elsevier, Amsterdam (1988)
16. Prümper, J., Anft, M.: ISONORM 9241/10, Beurteilungsbogen auf Grundlage der Internationalen Ergonomie-Norm ISO 9241/10. Büro für Arbeits- und Organisations psychologie, Berlin (1997)

Human Systems Integration Design: Which Generalized Rationale?

Romain Lieber[2,*] and Didier Fass[1,2]

[1] ICN Business School, Artem Augmented Human project
[2] Mosel team, LORIA, University of Lorraine,
Campus Scientifique - BP 239, France
didier.fass@loria.fr

Abstract. In this paper, we present a synthesis of our fundamental and theoretical research on human system integration and human *in-the-loop* system for enhancing human performance - especially for technical gestures, in safety critical systems operations such as surgery, astronauts' extra-vehicular activities and aeronautics. Grounding humans-systems integration engineering and design (modelling and simulation) on a formally and experimentally verified theoretical framework, is a necessity to make sure of human in-the-loop system security, safety and reliability. The rise issues concerned with scientific principles of human systems integration and rationale for human in-the-loop systems technical engineering and managerial specific rules.

Keywords: human systems integration, human in-the-loop system, performance, security, safety, reliability, theoretical principles, generalized rationale.

1   Introduction

A human being, by its biological nature, cannot be reduced to properties of mathematical or physical automaton. Thus, connecting up humans and artefacts is not only a question of technical interaction and interface; it is also a question of integration.

Human systems integration is an umbrella term for several areas of "human factors" research and engineering that include human performance, technology design, and human-interactive systems interaction on six levels, from socio-technical systems to human devices interaction [1]. These are concerned with the integration of human capabilities and performances into the design of complex human-machine systems supporting safe, efficient operations; there is also the question of reliability. Human systems integration is traditionally based on technical and managerial principles [2] [3].

Human systems integration involves augmented human design with the objectives of increasing human capabilities and improving human performance using behavioural technologies [4]. By using wearable interactive systems, made up of virtual environments technologies-like or wearable robotics, many applications offer technical gesture assistance e.g. in aeronautics, human space activities or surgery.

---

*
Convention CIFRE N°393/2008.

M. Kurosu (Ed.): Human Centered Design, HCII 2011, LNCS 6776, pp. 101–109, 2011.

Gesture is a highly integrated neuro-cognitive behaviour, based on the dynamical organization of multiple physiological functions [5] [6]. Assisting gestures and enhancing human skill and performances requires coupling sensorimotor functions and organs with technical systems through artificially generated multimodal interactions. Thus, augmented human design has to integrate human factors - anatomy, neurophysiology, behaviour - and assistive cognitive and interactive technologies in a safe and coherent way for extending and enhancing the ecological domain of life and behaviour.

1.1 Human In-the-Loop System

The goal of this type of human *in-the-loop* system design is to create entities that can achieve goals and actions (predetermined) beyond natural human behavioural, physical and intellectual abilities and skills – force, perception action, awareness, decision...

Augmenting cognition and sensorimotor loops with automation and interactive artefacts enhances human capabilities and performance. It is extending both the anatomy of the body and the physiology of human behaviour. Designing augmented human beings by using virtual environment technologies requires integrating both artificial and structural elements and their structural interactions with the anatomy, and artificial multimodal functional interactions with the physiological functions.

Therefore, the scientific and pragmatic questions are: how to best couple and integrate in a coherent way, a biological system with physical and artifactual systems, the less or more immersive interactive and invasive or not artefact, in a behaviourally coherent way by design? How augmented human engineering can anticipate and validate a technical and organizational design? How modelling and assessing such a design?

This paper focuses on one of the main issues for augmented human engineering: integrating the *biological user's needs* in its methodology for designing human-artefact systems integration requirements and specifications. To take into account biological, anatomical and physiological requirements we need a validated theoretical framework. We propose to ground augmented human engineering on the Chauvet mathematical theory of integrative as a fundamental framework for human system integration and augmented human design. We propose to validate and assess augmented human domain engineering models and prototypes by experimental neurophysiology.

2 Augmented Human Domain Engineering

Human-Artefact systems are a special kind of systems of systems. They are made up of two main categories of systems. These two kinds of systems differ in their nature: their fundamental organization, complexity and behaviour. The first category, the traditional one, includes technical or artifactual systems that could be engineered. The second category includes biological systems: the human that could not be engineered. Thus, integrating human and complex technical systems in design is to couple and integrate in a behaviourally coherent way, a biological system (the human) with a technical and artifactual system. Augmented human engineering needs to model the human body and its behaviour to test and validate augmented human reliability and human systems integration (HSI).

2.1 Domain Engineering

According to system engineering, taking into account user needs in the world of activities and tasks, designing system requirements is to find the system model, its three dimensional organizational dimensions of requirements - structural, functional and dynamical - and its three view plans of system specifications -architecture, behavior and evolution. (Fig.1).

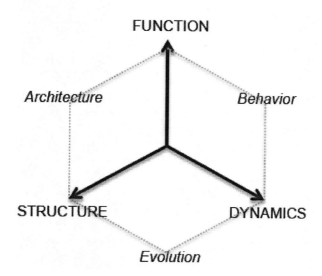

Fig. 1. Grounding human systems integration design is integrating in a total system model, its three dimensional organizational dimensions of requirements – structure, function and dynamics, and its three plans of system specifications – architecture, behavior and evolution, according to fundamental scientific principles. These basic classes of elements of *"integrative design"* are defined by generalization at the more common levels and domains of hierarchical structural and functional organization of a biological system or an artefactual system.

Thus, system engineering requires both expert skills and validated formal modelling methodologies. To some extent, the main difficulty is to build a system model from a collection of informal and sometimes imprecise, redundant and unstructured descriptions to the domain of expertise. A formal model could be relevant to highlight a hidden structure according to an intended function and its dynamics, or to apply operations or transformation on the system itself.

From domain engineering to requirements, our approach is situated inside Dines Bjoemer's framework [7] [8] [9] based on the triptych: D,S -> R, where D is the problem domain and where requirements R are satisfied by the relation ->, which intends to mean entailment ; so, S is some model of our system built or expressed from D. If that triptych is able to express, in a synthetic manner, a situation related to the problem domain, a system model and the requirements, it remains at a global level and can thus be applied in different problem spaces and instances.

The domain provides a way to express properties and facts of the environment of the system under construction. The system model S is intended to summarize actions and properties of the system and it is a link between the requirements and the final resulting system. The relation -> is conceptualized as a deduction-based relation which can be defined in a formal logical system, and which helps to derive requirements from domain and model. This relation is sometimes called entailment and is used to ground the global framework. When one considers an application, one should define the application domain from the analysis and this may integrate elements of the world. The triptych helps for defining a global framework and offers the possibility to use tools which are useful for assessing the consistent relation between D, S and R; because we aim to use proof techniques for ensuring the soundness of the relation.

2.2   Human System Integration

The major benefits of using augmented human modelling in design include reducing the need for physical development; reducing design costs by enabling the design team to more rapidly prototype and test a design; avoiding costly design 'fixes' later in the program by considering human factors requirements early in the design process; and improving customer communications at every step of product development by using compelling models and simulations. Thus, designing an artefact consists of organizing a coherent relation between structures and functions in a culture and context of usage [design=structure/function]. Modelling human beings consists of taking into account anatomical and physiological elements in the same model. It is to design functions by organizing a hierarchy of structural elements and their functions [human modelling=physiology (functions)/anatomy (structures)]. Such models should be used to create models of individuals rather than using aggregated summaries of isolated functional or anthropometric variables that are more difficult for designers to use. Therefore augmented human modelling in design requires an integrative approach according to the three necessities we defined for human systems integration [10].

2.3   Human Systems Integration Domain

Since technical systems are mathematically grounded and based on physical principles, HITLS needs to be considered in mathematical terms. There are several necessities to make HIS and augmented human reliable [11].

Necessity 1 – Designing a HITLS is to couple two systems from different domains organized and grounded on different principles theory and framework: biological, physical, numerical.

Necessity 2 – HITLS design is a global and integrative model based method ground on Chauvet's Mathematical Theory of Integrative Physiology and domain system engineering.

Necessity 3 – Modelling augmented human and HSI is to organize the required hierarchically structures and functions and their functional interactions related to dynamics

Consequently, designing an augmented human following human systems integration rationale is to organize hierarchically and dynamically human and artifact coupling. This requires a new domain engineering approach for requirements and specification based on biological user's needs.

2.4   Augmented Human Engineering

Dealing with augmented human engineering is being able to situate and limit its domain for specifying the whole system – biological and artefactual integrated system- in accordance with the high-level and global requirements:

D: The ecology of the augmented human: scientific validated principals of augmented human needs;

R: Augmented human teleonomy, augmented human economy and ethics;

S: Biological, Technical and organizational specification of the human-system integration architecture, behavioural performance, stability and reliability.

3   Augmented Human's Need

3.1   Epistemological Needs

Converging technologies for improving human performances [12], *augmented human*, need a *new epistemological and theoretical* approach to the nature of knowledge and cognition considered as an integrated biological, anatomical, and physiological process, based on a hierarchical structural and functional organization. Current models for human-machine interaction or human-machine integration are based on symbolic or computational cognitive sciences and related disciplines. Even though they use experimental and clinical data, they are not yet based on logical, linguistic and computational interpretative conceptual frameworks of human nature, where postulate or axiomatics replace predictive theory. It is essential for the robust modeling and the design of future rules of engineering for HIS, to enhance human capabilities and performance. *Augmented human* design needs an integrative theory that takes into account the specificity of the biological organization of living systems, according to the principles of physics, and a coherent way to organize and integrate structural and functional artificial elements. Consequently, virtual environments design for *augmented human* involves a shift from a metaphorical, and scenario based design, grounded on *metaphysical* models and rules of interaction and cognition, to a predictive science and engineering of interaction and integration. We propose to ground HSI and *augmented human* design on an integrative theory of the human being and its principles.

3.2   CHAUVET's Mathematical Theory of Mathematical Physiolgy (MTIP) Needs

The mathematical theory of integrative physiology, developed by Gilbert Chauvet [14] [15] [16], examines the hierarchical organization of structures (i.e., anatomy) and functions (i.e., physiology) of a living system as well as its behaviour. MTIP introduces the principles of a functional hierarchy based on structural organization within spaces limits, functional organization within time limits and structural units that are the anatomical elements in the physical space. It copes with the problem of structural discontinuity by introducing functional interaction, for physiological function coupling, and structural interaction y from structure-source s into structure-sink S, as a coupling between the physiological functions supported by these structures.

Unlike interaction in physics, at each level of organization functional interactions are non-symmetrical, leading to directed graph, non local, leading to non local fields, and increase the functional stability of a living system by coupling two hierarchical structural elements. As G. Chauvet said: "we have chosen a possible representation related to hierarchical structural constraints, and which involves specific biological concepts. We also made the important hypothesis that a biological system may be mathematically represented as a set of functional interactions of the type: $s \xrightarrow{\psi} S$. However, the main issue now is to determine whether there exists a cause to the existence of functional interactions, i.e. to the set of triplets $s \xrightarrow{\psi} S$? What is the origin of the existence (the identification) of s, S and y that together make a component $s \xrightarrow{\psi} S$ of the system? The answer to this issue is the existence of a mathematical principle, the stabilizing auto-association principle or PAAS, a principle that makes of a framework, the MTIP, a veritable theory. *The PAAS may be stated as follows*: For any triple (syS), denoted as $s \xrightarrow{\psi} S$, where s is the system-source, S the system-sink, and y the functional interaction, the area of stability of the system $s \xrightarrow{\psi} S$ is larger than the areas of stability of s and S considered separately. In other words, increasing in complexity the system $s \xrightarrow{\psi} S$, corresponds to increase in stability. MTIP consists in a representation (set of non-local interactions $s \xrightarrow{\psi} S$), an organizing principle (the PAAS), and a hypothesis (any biological system may be described as a set of functional interactions) that gives rise to two faces of the biological system, the (O-FBS) and the (D-FBS). The first one may be studied using the potential of organization, the second one using the S-Propagator formalism, which describes the dynamics in the structural organization, making an n-level field theory. Both are based on geometrical/topological parameters, and coupled via geometry/topology that may vary with time and space (state variables of the system) during development and adult phases. The structures are defined by the space scale k, hence the structural hierarchy, the functions are defined by the time scale T, hence the functional hierarchy. Any model built in this theoretical framework will use the same representation, the same basic principle and hypothesis, and consequently will be comparable and able to be coupled with any other on".

4   Rationale for a Model of *Human In-the-Loop System*

Who would even think about separating a living goldfish from its water and its fishbowl!

As claims by Fass [4], since artifactual systems are mathematically founded and based on physical principles, HSI needs to be thought of in mathematical terms. In addition, there are several main requirements categories to make HIS and human in-the-loop system design, modelling and simulation, safe and efficient. They address the technology - virtual environment-, sensorimotor integration and coherency.

Requirement 1: Virtual Environment is an Artifactual Knowledge based Environment. As an environment, which is partially or totally based on computer-generated sensory inputs, a virtual environment is an artificial multimodal knowledge-based environment. Virtual reality and augmented reality, which are the most well known technologies of virtual environments, are obviously the tools for the augmented human design and the development of human in-the-loop systems. Knowledge is gathered from interactions and dynamics of the individual-environment complex. It is an evolutionary, adaptive and integrative physiological process, which is fundamentally linked to the physiological functions with respect to emotions, memory, perception and action. Thus, designing an artifactual or a virtual environment, a sensorimotor knowledge based environment, consists of making biological individual and artifactual physical system consistent. This requires a neurophysiological approach, both for knowledge modeling and human in-the-loop design.

Requirement 2: Sensori-motor Integration and Motor Control. Humans use multimodal sensori-motor stimuli and synergies for interacting with their environment, either natural or artificial (vision, vestibular stimulus, proprioception, hearing, touch, taste…) [17]. When an individual is in a situation of immersive interaction, wearing head-mounted display and looking at a three-dimensional computer-generated environment, his or her sensorial system is submitted to an unusual pattern of stimuli. This dynamical pattern may largely influence the balance, the posture control, the spatial cognition and the spatial motor control of the individual. Moreover, the coherence between artificial stimulation and natural perceptual input is essential for the perception of the space and the action within. Only when artificial interaction affords physiological processes is coherence achieved.

Requirement 3: Coherence and HIS. If this coherence is absent, perceptual and motor disturbances appear, as well as balance troubles, illusions, vection or vagal reflex. These illusions are solutions built by the brain in response to the inconsistency between outer sensorial stimuli and physiological processes. Therefore, the cognitive and sensorimotor abilities of the person may be disturbed if the design of the artifactual environment does not take into account the constraints imposed by human sensory and motor integrative physiology. The complexity of physiological phenomena arises from the fact that, unlike ordinary physiological systems, the functioning of a biological system depends on the coordinated action of each of the constitutive elements. This is why the designing of a virtual environment as an augmented biotic system, calls for an integrative approach.

Integrative design strictly assumes that each function is a part of a continuum of integrated hierarchical levels of structural organization and functional organization as described above within mathematical theory of integrative physiology (MTIP). Thus, the geometrical organization of the artifactual structure, the physical structure of interfaces and the generated patterns of artificial stimulations, condition the dynamics of hierarchical and functional integration. Functional interactions, which are products or signals emanating from a structural unit acting at a distance on another structural unit, are the fundamental elements of this dynamic.

5 Conclusion and Perspective

By designing a human-artifact system consists in organizing the linkage of multimodal biological structures, sensorimotor elements at the hierarchical level of the living body, with the artificial mechanical or interactive elements of the system, devices and patterns of stimulation. There exists a "transport" of functional interaction in the augmented space of both physiological and artefactual units, and thus a *function* may be viewed as the final result of a set of functional interactions that are hierarchically and functionally organized between the artificial and biological systems.

Architecture: spatial organization of the structural elements, natural and artificial, coupled by non-local and non-symmetric functional interactions according to PAAS. It is specifying the function(s) of the integrated system. Different organizations specify different architecture and their specific functions:

Behavior: temporal organisation of the patterns of artificial functional interactions, condition and specify the dynamics of augmented sensorimotor loops. It is determining human in-the-loop system behaviour.

Evolution: the spatio-temporal organization of the structural elements and the functional interactions they produce and process specify functional stability of human-artefact system according to an optimum principle -the Chauvet's orgatropy principle, during the *life of human in-the-loop system*.

Contingent on ecology and economy, architecture, behaviour and evolution as specified, define and limit the *life domain of human in-the-loop system*.

MTIP is thus applicable to different space and time level of integration in the physical space of the body and the natural or artificial behavioural environment; from molecular level to socio-technical level; from drug design to wearable robotics, and to life and safety critical systems design.

Future works should address questions related to the development of formal models [18], [19] or co-simulation [20] related to human systems integration engineering and design. New questions arise when dealing with deontic or ethical questions that might be handled by human systems integration together with classical formal modelling languages based on deontic or modal languages.

Industrial scientific and ethical challenges rely on designing intelligent and interactive artefactual systems relating machines and human beings. This relationship must be aware of its human nature and its body: it is anatomy and functions. The man-machine interface becomes an integrated continuation of the body between perception-action and sensory and motion organs. By integrating human body and behaviours, the automaton is embodied but this embodiment grounds on the user's body; it enhances capabilities and performances. Safety and reliability rely on aspect of these fundamental necessities.

That is a generalized rationale for guaranteeing the effectiveness of the overall system.

References

1. Nasa Human System Integration Division, http://humanfactors.arc.nasa.gov/index.php
2. Booher, H.R.: Introduction: human systems intégration. In: Booher, H.R. (ed.) Handbook of Human Systems Integration, pp. 1–30 (2003)

3. Hobbs, A., Adelstein, B., O'Hara, J., Null, C.: Three principles of human-system integration. In: Proceedings of the 8th Australian Aviation Psychology Symposium, Sydney, Australia (April 8-11, 2008)
4. Fass, D.: Rationale for a model of human systems integration: The need of a theoretical framework. Journal of Integrative Neuroscience 5(3), 333–354 (2006)
5. Kelso, J.A.: An Essay on Understanding the Mind. Ecol. Psychol. 20(2), 180–208 (2008)
6. de'Sperati, C., Viviani, P.: The relationship between curvature and velocity in two-dimensional smooth pursuit eye movements. J. Neurosci. 17(10), 3932–3945 (1997)
7. Bjorner, D.: Software Engineering 1 Abstraction and Modelling. In: Texts in Theoretical Computer Science. An EATCS Series. Springer, Heidelberg (2006), ISBN: 978-3-540-21149-5
8. Bjorner, D.: Software Engineering 2 Specification of Systems and Languages. In: Texts in Theoretical Computer Science. An EATCS Series. Springer, Heidelberg (2006), ISBN: 978-3-540-21150-1
9. Bjorner, D.: Domain Engineering Technology Management, Research and Engineering. COE Research Monograph Series, Vol. 4, JAIST (2006)
10. Fass, D.: Integrative Physiological Design: A Theoretical and Experimental Approach of Human Systems Integration. In: Harris, D. (ed.) HCII 2007 and EPCE 2007. LNCS (LNAI), vol. 4562, pp. 52–61. Springer, Heidelberg (2007)
11. Fass, D., Lieber, R.: Rationale for human modelling in human in the loop systems design. In: 3rd Annual IEEE International Systems Conference, SysCon, Vancouver, pp. 27–30 (2009)
12. Roco, M.C., Brainbridge, W.S.: Converging technologies for improving human performance. National Science Foundation (2003)
13. Chauvet, G.A.: Hierarchical functional organization of formal biological systems: a dynamical approach. I. An increase of complexity by self-association increases the domain of stability of a biological system. Phil Trans Roy Soc London B 339, 425–444 (1993)
14. Chauvet, G.A.: Hierarchical functional organization of formal biological systems: a dynamical approach. II. The concept of non-symmetry leads to a criterion of evolution deduced from an optimum principle of the (O-FBS) sub-system. Phil Trans Roy Soc London B 339, 445–461 (1993)
15. Chauvet, G.A.: Hierarchical functional organization of formal biological systems: a dynamical approach. III. The concept of non-locality leads to a field theory describing the dynamics at each level of organization of the (D-FBS) sub-system. Phil Trans Roy Soc London B 339, 463–481 (1993)
16. Sporns, O., Edelman, G.: Bernstein's dynamic view of the brain: the current problems of modern neurophysiology. Motor Control 2, 283–305 (1988)
17. Cansell, D., Méry, D.: The Event-B Modelling Method: Concepts and Case Studies, in Logics of specification languages. In: Bjørner, D., Henson, M.C. (eds.) Monographs in Theoretical Computer Science, pp. 47–152. Springer, Heidelberg (2008)
18. Mermet, B., Méry, D.: Safe combinations of services using B. In: McDermid, J. (ed.) SAFECOMP 1997 The 16th International Conference on Computer Safety, Reliability and Security. Springer, New York (1997)
19. Leclerc, T., Siebert, J., Chevrier, V., Ciarletta, L., Festor, O.: Multi-Modeling and Co-Simulation-based Mobile Ubiquitous Protocols and Services Development and Assessment. In: 7th International ICST Conference on Mobile and Ubiquitous Systems (2010)

The Impact of Human-Centred Design Workshops in Strategic Design Projects

André Liem[1] and Elizabeth B.-N. Sanders[2]

[1] Norwegian University of Science and Technology, Department of Product Design, 7491 Trondheim, Norway
[2] MakeTools, LCC, Columbus, OH 43214, USA
andre.liem@ntnu.no, liz@maketools.com

Abstract. Implementation of Human-centred Design methods in the Fuzzy Front-End is not likely to lead to diversification in educational product planning exercises, where time lines are short and executors lack experience. Companies, interested to collaborate with M.Sc. students on strategic design projects, should have realistic ambitions with respect to innovation and value creation. Moreover, diversification is not the only generic growth strategy to gain competitive advantage. Value can also be created from developing new products for existing markets, or creating new markets for existing products. On the contrary, companies who aim for diversification in their generic growth strategies, may not always end up with a complementary "high valued" design outcome. From a learning perspective, the understanding of HCD methods created awareness among students and companies that respect and empathy for the end-user are important for enriching their design processes, as such increasing the chances for diversification in subsequent projects with clients.

Keywords: User-centred Design, Human-centred Design, Co-creation, Design-led Innovation, Front End of Innovation, Positioning Maps, Diversification.

1 Introduction

Design offers a potent way to position and differentiate products as competition intensifies, product complexity increases and technological differentiation becomes more difficult [1]. Within the context of integrated product development, formulating an effective product strategy and a design goal is one of the greatest challenges of the innovation process; however effective management of the Fuzzy Front End (FFE) may result in a sustainable competitive advantage [2].

A User-centred Design (UCD) approach, whereby the needs of potential end-users are assessed in the product development process, can then be important for achieving a company's strategic and innovation goals. However, the main problem is that too many projects suffer from 'insufficient market input, a failure to build in the voice of the customer, and a lack of understanding of the market place [3]. Furthermore, it has been noted that limited and inadequate market research, resulting in problematic translation of engineers' wishes into customers' needs, is a key factor of failure of innovations [4].

M. Kurosu (Ed.): Human Centered Design, HCII 2011, LNCS 6776, pp. 110–119, 2011.
© Springer-Verlag Berlin Heidelberg 2011

As a response, user involvement is seen as a way to obtain valuable input from end-users. According to Kujala [5], involving end-users in research and design activities can have diverse positive effects: on the quality or speed of the research and design process; on a better match between a product and end-users' needs or preferences; and on end-users' satisfaction.

2   Methods and Paradigms for Innovation

As the global environment is continuously changing, organizations and businesses are compelled to permanently seek the most efficient models to maximize their innovation management efforts through new methods and paradigms, which efficiently serve existing and new markets with new and/or modified products as well as services [6]. Hereby, Ansoff's Product-Market matrix is a frequently used model to position generic innovation strategies [7].

Considering the four generic growth strategies [7], this article argues that the implementation of UCD methods in the Fuzzy Front-End (FFE) is not likely to lead to diversification in product planning exercises conducted in an educational setting, where time lines are short and executors lack experience, as exemplified in this 4th year collaborative product strategy project. However from a "Design Strategy" and "Value Creation" perspective, end-user and other stakeholder's input can be valuable, if not decisive, in promoting a company's products and services to the "Upper Right Quadrant" of the 3-D "Style" versus "Technology" positioning map [8].

3   The Concept of Value Creation in Products and Services

In their investigation of what it takes to create breakthrough products, Cagan and Vogel concluded that one of the key attributes that distinguishes breakthrough products from their closest followers is the significant value they provide for users [8]. Taking it one step further, the more value in a product, the higher price people are willing to pay, with the price increasing more rapidly than the costs, resulting in a profit margin, significantly higher for higher valued products. After all, as Drucker has pointed out, *"customers pay only for what is of use to them and gives them value"* [9].

Boztepe has categorised user value according to utility, social significance, emotional and spiritual value [10]. Utility value refers to the utilitarian consequences of a product. Social significance value refers to the socially oriented benefits attained through ownership of and experience with a product. Emotional value refers to the affective benefits of a product for people who interact with it. Similarly, Sanders and Simons identified 3 types of values related to co-creation, which are inextricably linked. These values are monetary, use /experience and societal [11].

According to Dewey, experience is not something that is totally internal to the individual, but instead, "an experience is always what it is because of a transaction taking place between an individual and what, at the time, constitutes his environment" (p.43). [12]. Experiences are context- and situation-specific; which means they change from one set of immediate circumstances, time, and location to another. In a similar way, value changes as cultural values and norms, and external contextual factors, change [13].

In summary, consumers are willing to pay a higher price for product purchases that connect with their own personal values, although monetary value is important in determining market penetration strategies [8].

4 User-Centred versus Design-Driven Innovation

Significant efforts in this recent literature have been concentrated into investigating a specific approach to design, usually referred to as *user-centred design* [14, 15, 16]. This approach implies that product development should start from a deep analysis of user needs. In practice, researchers spend time in the field observing customers and their environment to acquire an in-depth understanding of customer's lifestyles and cultures as a basis for better understanding their needs and problems [17].

Design-driven innovation, which plays such a crucial role in the innovation strategy of design intensive firms, has still remained largely unexplored [18]. One of explanations for why design-driven innovation has largely remained unexplored is that its processes are hard to detect when one applies the typical methods of scientific investigation in product development, such as analyses of phases, organizational structures, or problem-solving tools [19, 20]. Unlike user-centred processes, design-driven innovation is hardly based on formal roles and methods such as ethnographic research.

Design-driven innovation may be considered as a manifestation of a *reconstructionist* [21] or *social-constructionist* [22] view of the market, where the market is not "given" a priori, but is the result of an interaction between consumers and firms. Hereby, users need to understand the radically new language and message, to find new connections to their socio-cultural context, and to explore new symbolic values and patterns of interaction with the product. In other words, radical innovations of meaning solicit profound changes in socio-cultural regimes in the same way as radical technological innovations, which solicit profound changes in technological regimes [23].

Currently, design-driven innovation is starting to be explored and discussed [24]. However, the industrial applications tend to be design-led innovation accomplished through user-centred design research methods. Besides this, design curricula are also in the midst of discussion and change. Although user-centred design methods are being taught, it is often difficult for students to bridge the gap between research and design. Students tend to take a design-driven innovation approach, because they find it difficult to extract and incorporate user involvement in the "later" designing stages.

5 Design-Driven Innovation vs. Innovation through Co-creation

A third perspective on non-technological push approaches to innovation is that of co-creation (sometimes referred to as co-designing). This perspective can also be considered to be co-design-led innovation [25].

The map of design research and practice as shown in Figure 1 (updated from the map in Sanders, 2008) [26] can serve as a framework on which to compare the three perspectives: User-centred, design-led and co-creation. The map is defined and described by two intersecting dimensions: approach and mind-set. Approaches to design research have come from research-led thinking (shown at the bottom of the map) and from design-led thinking (shown at the top of the map). The research-led perspective has the longest history and has been driven by applied psychologists, anthropologists, sociologists, and engineers. The design-led perspective, on the other hand, has come into view much more recently. There are also two opposing mind-sets evident in the practice of design research today. The left side of the map describes a culture characterized by an expert mind-set. Designers and researchers here are involved with designing *for* people. They consider themselves to be the experts, and they see and refer to people as "subjects," "users," "consumers," etc. The right side of the map describes a culture characterized by a participatory mind-set. Designers and researchers on this side design *with* people. They see the people as the true experts in domains of experience such as living, learning, working, etc. Designers and researchers who have a participatory mindset value people as co-creators in the design process. It is difficult for many people to move from the left to the right side of the map (or vice versa), as this shift entails a significant cultural change.

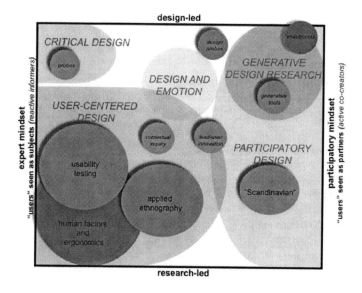

Fig. 1. A map of Design Research and Practice

If we strip the map of the design research tools and methods it serves well as a framework for positioning the three perspectives on non-technologically driven product development processes (Figure 2). The user-centred perspective uses research-led approaches coming primarily from marketing and the social sciences to make incremental improvements to existing products or product lines. The design-led perspective uses design thinking and has the potential for significant innovation but it does not value the input of potential end-users as being participants in the early front end of the process. The co-creation perspective puts the tools and methods of design thinking into the hands of the people who will be the future end-users (and the other stakeholders) early in the front end of the product development process.

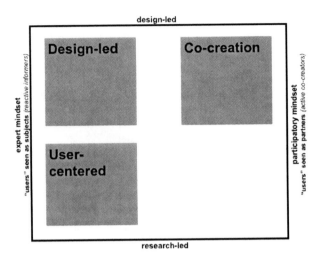

Fig. 2. Framework for positioning the three perspectives on non-technologically driven product development processes

6 Design and Research Approach

The strategic design reports were developed based on different collaborative projects with Industry, where 4th Year Industrial Design students acted as consultants and were required to formulate a design strategy as well as materialise the strategy into a product and / or service. Students were subjected to a short but intensive hands-on workshop on co-creation methods, tools and techniques early in the semester. The students worked in groups of 2 or 3 in a design studio setting. In the initial stages of the project, students planned a series of participatory design sessions with various groups of stakeholders to support their strategic and industrial design process. UCD as well as co-creation methods, which were suggested and later on implemented included, for example: Observations, Function Mapping [27], Bulls Eye Collage, Participatory Design through Making and Acting [28], Storytelling [29], What-If Scenario Building, etc. Students were free to choose what kind of approach and what kinds of tools and methods made sense based on their client and the challenge they were faced with.

Sources of evidence were mainly based on the analysis of nine strategic design reports followed by interviews. A case study research approach was used to compare how various methods were instrumental in determining the level and type of innovation [30]. The analysis of the strategic design reports was carried out through a procedure of "Explanation Building".

7 Results and Analysis of Workshops

A detailed description of how the participatory workshops were managed and executed within each of the projects will be shown below. The workshop results were analysed based on the following criteria:

- Client criteria and constraints: This covers: nature, size and business activities of the client company, etc.
- Involvement of internal / external stakeholders and end-users
- Approach: Processes and methods used in the workshops
- Results: This mainly elaborates upon the insights gained during the workshops and how these have been implemented in the follow up product planning and designing activities.

Appendix A provides a comparative overview of the analysis of the workshop sessions.

8 Discussion

Even with the broadening of the approach to design, a fundamental tension between design-driven and user-centred driven innovation is prevalent [16, 18]. In 5 of the 9 projects, a "New Product – Existing Market" strategy was targeted, whereas 2 projects aimed at creating a new market for the companies' based on existing products and technologies. In addition, two (2) companies adopted a "natural" diversification strategy, as they were contract manufacturers and do not have a history in developing their own products. The two (2) reports showed that end users were not very much involved in the product /service idea generation process with respect to these contract manufacturers. Establishment of design goals and generation of concepts mainly took place through discussions among company management and design students, based on a conjecture – analytical design approach [31].

As summarized and mapped onto Ansoff´s Product-Market matrix [7], overall results indicate that Human-centred Design (HCD) methods may not be directly applicable for establishing a diversification strategy in an educational setting, where 4th year design students were for the first time subjected to co-creation tools and methods.

New Market	• Sweets • Social Game Play "Lego"	• Multi- functional Outdoor Fire Place • Load-Crosser
Existing Market		• Heating Systems for the Future • Energy Control Systems for the Future • Monitoring Fish health • New Thinking in Bridge Design • Bridge and Identity
	Existing Product	New Product

Fig. 3. Mapping of 9 design projects according onto Ansoff's Product- Market matrix [7]

However, the design outcome of these industrial projects (see figures 3 and 4) suggest that students were capable of producing innovative design concepts by proposing products or services to be positioned in the "Upper Left and Right Quadrants" challenging new technologies and style (= ergonomics and form).

Companies who have the interest to collaborate with students on design / product innovation projects should have realistic ambitions with respect to value creation. Instead of being fixated or aiming too hard for diversification, they should also consider that value can be derived from developing new products for existing markets, or creating new markets for existing products.

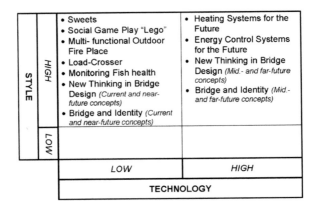

Fig. 4. Mapping of 9 design projects according to Cagan´s and Vogel´s Positioning Map [8]

In support of value creation, HCD can be considered a useful tool in educating companies and prospective design consultants about how end-users and other stakeholders are to be involved in certain aspects of the co-designing process. However, interviews with the students have surfaced the following limitations and opportunities for implementing HCD in search of a suitable generic growth and design strategy:

- Nature, history and pragmatic attitudes of some of the companies
- Most of the companies have unconsciously influenced the students to focus on the "new product / existing market" or "existing product / new market" strategies
- Although in some cases a radical product idea is *"in the making"*, very aggressive time frames for the projects as well as the lack of experience among students to frame and communicate, did not provide a convincing atmosphere for the company to pursue diversification

On the contrary, companies, who aim for diversification in their generic growth strategies may not always end up with a complementary "high valued" design outcome, as illustrated through the "Multi-functional Outdoor Fire Place" and "Load Crosser" projects.

From an educational perspective, this learning experience showed that Front-End of Innovation (FEI) processes, and HCD methods and tools should be taught to students in conjunction with Ansoff's Product-Market matrix [7], as well as the 3-D Positioning Maps [8]. This will then lead to the following thinking approaches:

- Diversification on its own is not the only generic growth strategy to gain significant competitive advantage
- Focusing on "Development of new products for existing markets" or "Creation of new markets for existing products" as generic growth strategies in combination with a design strategy targeted at the "Upper Right Quadrant" can also lead to a significant value creation for companies.

From a learning perspective, the understanding of HCD methods (both user-centred and co-creation) created an awareness among students and companies that respect and empathy for the end-user are important aspects to consider for enriching their design processes, as such increasing the chances for diversification in subsequent projects with clients.

References

1. Cova, B., Svanfeldt, C.: Societal Innovations and the Postmodern Aestheticization of Everyday Life. International Journal of Research in Marketing 10(3), 297–310 (1993)
2. Koen, P.A., Ajamian, G., Burkart, R., Clamen, A., Davidson, J., D'Amoe, R., Elkins, C., Herald, K., Incorvia, M., Johnson, A., Karol, R., Seibert, R., Slavejkov, A., Wagner, K.: New Concept Development Model: Providing Clarity and a Common Language to the 'Fuzzy Front End' of Innovation. Research Technology Management 44(2), 46–55 (2001)
3. Cooper, R.G.: The invisible success factors in product innovation. Journal of Product Innovation Management 16(2), 115–133 (1999)
4. v. d. Panne, G., Beers, C.V., Kleinknecht, A.: Success and failure of innovation: A literature review. International Journal of Innovation Management 7(3), 309–338 (2003)
5. Kujala, S.: User involvement: a review of the benefits and challenges. Behaviour and Information Technology 22(1), 1–17 (2003)
6. Christiansen, J.A.: Building the Innovative Organization. MacMillan Press, London (2000)
7. Ansoff, H.I.: Corporate Strategy: An Analytic Approach to Business Policy for Growth and Expansion. Penguin, Harmondsworth (1968)
8. Cagan, J., Vogel, C.M.: Creating breakthrough products: Innovation from product planning to program approval. Prentice Hall, Upper Saddle River (2002)
9. Drucker, P.F.: The essential Drucker: The best of sixty years of Peter Drucker's ideas on management. Harper Business, New York (2001)
10. Boztepe, S.: User Value: Competing Theories and Models. International Journal of Design 1(2) (2007)
11. Sanders, L., Simons, G.: A Social Vision for Value Co-creation in Design. Open Source Business Resource (December 2009),
    http://www.osbr.ca/ojs/index.php/osbr/article/view/1012/973
12. Dewey, J.: Experience and education. Free Press, New York (1938)
13. Overby, J.W., Woodruff, R.B., Gardial, S.F.: The influence of culture upon consumers' desired value perception: A research agenda. Marketing Theory 5(2), 139–163 (2005)
14. Chayutsahakij, P., Poggenpohl, S.: User-Centered Innovation: The Interplay between User-Research and Design Innovation. In: Proceedings of the European Academy of Management 2nd Annual Conference on Innovative Research in Management (EURAM), Stockholm, Sweden (2002)

15. Vredenburg, K., Isensee, S., Righi, C.: User-Centered Design: An Integrated Approach. Prentice Hall, Upper Saddle River (2002)
16. Veryzer, R.W., Borja de Mozota, B.: The Impact of User-Oriented Design on New Product Development: An Examination of Fundamental Relationships. Journal of Product Innovation Management 22, 128–143 (2005)
17. Belliveau, P., Griffin, A., Somermeyer, S.M.: The PDMA Toolbook for New Product Development. Wiley, Hoboken (2004)
18. Verganti, R.: Design Driven Innovation. Harvard Business School Press, Boston (2008)
19. Brown, S.L., Eisenhardt, K.M.: Product Development: Past Research, Present Findings, and Future Directions. Academy of Management Review 20(2), 343–378 (1995)
20. Shane, S.A., Ulrich, K.T.: Technological Innovation, Product Development, and Entrepreneurship in Management Science. Management Science 50(2), 133–144 (2004)
21. Kim, W.C., Mauborgne, R.: Blue Ocean Strategy: From Theory to Practice. California Management Review 47(3), 105–121 (2005)
22. Prahalad, C.K., Ramaswamy, V.: Co-opting Customer Competence. Harvard Business Review, 79–87 (January-February 2000)
23. Geels, F.W.: From Sectoral Systems of Innovation to Socio-Technical Systems. Insights about Dynamics and Change from Sociology and Institutional Theory. Research Policy 33, 897–920 (2004)
24. Bucolo, S., Matthews, J.: Design led innovation: Exploring the synthesis of needs, technologies and business models. In: Participatory Innovation Conf., pp. 354–357 (2011)
25. Sanders, E.B.-N.: Sustainable innovation through participatory prototyping. Formakademisk (2011) (in press)
26. Sanders, E.B.-N.: An Evolving Map of Design Practice and Design Research. Interactions (November/December 2008)
27. Moolenbeek, J.: Function Mapping: A Sound Practice for System Design. In: SysCon 2008 – IEEE International Systems Conference, Montreal, Canada (2008)
28. Sanders, E.B.-N., Stappers, P.J.: Co-creation and the new landscapes of design. Co-design 4(1), 5–18 (2008)
29. Sametz, R., Maydoney, A.: Storytelling through Design. Design Management Journal (Fall 2003) (issue)
30. Yin, R.K.: Case Study Research: Design and Methods, 3rd edn. Sage Publ., Inc., Thousand Oaks (2003)
31. Bamfort, G.: From analysis/synthesis to conjecture/analysis: A review of Karl Popper's influence on design methodology in architecture. Design Studies 23, 245–261 (2002)
32. Hekkert, P., Van Dijk, M.: Designing from context: Foundations and Applications of the ViP approach. In: Lloyd, P., Christiaans, H. (eds.) Designing in Context: Proceedings of Design Thinking Research Symposium 5. DUP Science, Delft (2003)

Appendix A: A Comparative Overview of the Analysis of the Workshop Sessions

	Client criteria and constraints	Involvement of internal stakeholders	Involvement of end-users	Involvement of external stakeholders	Approach taken	Results
Heating Systems for the Future	The client is open to various forms of innovation and design input within the context of their business activity, which is heat production	YES, different departments (finance, marketing, development, purchasing, etc).	NONE	NONE	Co-design led with internal stakeholders, comprising of various exercises, such as: • Bulls-eye method • Visualization of values • Future (vision) mapping, • Scenario-based interviews	Insights to develop visions and design concepts. Water-based heating system, Multi-purpose and portable heat pump and Modular, decorative heating panels Unclear how uses were addressed for the service providers, such as suppliers, contractors, etc. across the three concepts
Energy Control Systems for the Future		NONE	NONE			In conjunction with the development of personas and interviews, the design brief and problem definition were reformulated. Focus towards user-centeredness combined with a lack of design directives from the company led to weak physical 3-D concepts, but an interesting interface concept.
New Thinking in Bridge Design	The client is interested in a stage-wise future development of bridges. However, many constraints were communicated concerning production, assembly and management of suppliers.	NONE	YES	NONE	Co-design led with end users: • Functional mapping • Collages • Future mapping • Designing from context [32].	Input from the workshop provided mainly insight for the near future development of bridges. A design-driven innovation approach has been adopted, as it was difficult for users to concretely comment on how the system and elements should be designed and developed over time.
Bridge and Identity		NONE	YES	NONE	User-centered with end users • Stakeholder Analysis • Observations • Informal Interviews	An emphasis was placed on consistent identity development as well as a design driven approach in the ergonomic development of the "overall" bridge with various stakeholders in mind
Sweets	15 year-old adolescents were defined as a target group	NONE	YES	NONE	Co-design led with end-users • Informal Interviews • Co-designing	Interesting spread of ideas and design cues. However there was a miss-match between workshop results (cues) and development of concrete design concepts
Monitoring Fish Health	The client is open to innovative concepts concerning all forms of remote operations and monitoring systems for the aqua-culture industry	YES	YES	YES	Co-design led with all stakeholders	Insights for concept development of a fish health surveillance interface from various stakeholders. Two main developments are achievable: • The creation of an interface to improve usability and accessibility among end-users and external stakeholders based upon accessibility of complementary services • The development of technology, just to facilitate information flow between various programs and the service provider's services. However, an integrated service solution with improved usability, accessibility and compatibility has not been achieved yet.
Multi-functional Outdoor Fireplace	NONE. The client's aim is to optimise unused manpower and machine capacity	YES	NONE	NONE	Research led and client-centered	A standard design driven "Product Planning and Goal Finding" exercise
Load Crosser	The client's aim is to optimise unused manpower and machine capacity related to sheet metal construction	YES	NONE	NONE	Research led and client-centered	A design driven exercise, based upon iterative rounds of functional prototyping and testing
Social Game Play - LEGO	Development of design directions and concepts around the theme "Social Game Play"	NONE	YES	YES	Co-design led with end-users • Positioning robot images on a spectrum • Creating a favorite robot • Developing a Storyline around a favorite robot • Focus group discussions with teachers about social play	Workshops provided a good foundation for the development of design ideas and concepts. The following objectives were met: • To identify product attributes for social play with robots • To find out how boys visualise and adapt robots • To gain insight how boys interact and stimulate social play. Adaptation and customisation of robots were found to be essential in stimulating social play.

Quantitative Evaluation of the Effectiveness of
Idea Generation in the Wild

Lassi A. Liikkanen[1,2], Matti M. Hämäläinen[2,3], Anders Häggman[4],
Tua Björklund[2], and Mikko P. Koskinen[2,3]

[1] Helsinki Institute for Information Technology HIIT, P.O. Box 19215, FI-00076 AALTO,
Finland, Aalto University and University of Helsinki
[2] Aalto Design Factory, P.O. Box 17700, FI-00076 Aalto University, Finland
[3] Engineering Design and Production, P.O. Box 14100, FI-00076 Aalto University, Finland
[4] Dept. of Mechanical Engineering, Massachusetts Institute of Technology,
77 Massachusetts Ave, 3-449B, Cambridge, MA 02139, USA
{Lassi.Liikkanen,Matti.M.Hamalainen,Tua.Bjorklund,
Mikko.P.Koskinen}@Aalto.fi, haggman@mit.edu

Abstract. New ideas are the primary building blocks in attempts to produce novel
interactive technology. Numerous idea generation methods such as Brainstorming
have been introduced to support this process, but there is mixed evidence
regarding their effectiveness. In this paper we describe an experimental,
quantitative methodology from the domain of product design research for
evaluating different idea generation methods. We present prominent results from
relevant literature and new data from a study of idea generation in the wild. The
study focused on the effects of the physical environment, or in other words, the
physical context, on designers' capacity to produce ideas. 25 students working in
small groups took part in an experiment with two design tasks. Moving from an
office environment to the actual surroundings of the intended use, we discovered
that the change in resulting ideas was surprisingly small. Of the measured
dimensions, the real-world context influenced only the feasibility of ideas, leaving
quantity, novelty, utility and level of detail unaffected. This finding questions the
value of diving into the context as a design idea generation practice.

Keywords: Design methods, idea generation, creativity, psychology.

1 Introduction

Finding the right direction, that is, defining what to design is the evident challenge in
most development projects. This also holds for the design of new interactive
technologies. In attempts to organize creative efforts effectively and maximize their
impact, many research and development (R&D) organizations employ innovation
management. A variety of methods have been introduced to ensure quantitatively and
qualitatively sufficient results during the fuzzy front-end of design. These methods,
called idea generation (IG) techniques, and comparable methods for requirements
elicitation have been available for some time. In the tradition of this craft, formal IG
methods such as synectics, six thinking hats, morphological analysis, lateral thinking,

M. Kurosu (Ed.): Human Centered Design, HCII 2011, LNCS 6776, pp. 120–129, 2011.
© Springer-Verlag Berlin Heidelberg 2011

TRIZ, and naturally Brainstorming [1], can be found. The variety of available methods is nowadays considerable if we also include a category called creativity support tools [2] to the collection of ideation techniques.

Despite the extent of idea generation methodology, we and many others find something essential missing from the debate. The presentation of new ideation methods commonly lacks an objective assessment of the impact that the tools and practices are supposed to yield. They lack proof regarding their effectiveness and suitability for different types of projects. This has led design researchers and psychologists to question the effectiveness of these procedures. Indeed, Brainstrorming alone has received attention from psychologists since its introduction in the 1950's [see, for example, 3-5]. There are many approaches to proving the effectiveness of ideation. One could target social interaction and communication [6], provide self-evaluation instruments for the designers [7], or analyze the output of the process as will be done in this study. The historical source of inspiration for the latter approach is the creativity measurement literature [8] from which most of the methods are derived from.

In this paper, we discuss the quantitative *idea analysis methods*. The aim of this methodology is to focus on the qualities and quantities of ideas produced in experimental settings. With this approach, we can investigate how different controlled and manipulated variables influence the outcomes of the IG process. As an example of applying this methodology, we include an experiment studying the effects of design surroundings in generating new interactive product-service concepts.

2 Studying Idea Generation through Produced Ideas

Idea analysis has so far been utilized within different fields of psychology, namely cognitive [9] and social [10], but also in management science [11]. These studies have dealt with IG as a generic topic, assessing ideas for large-scale social and environmental problems. Nowadays there exists a wealth of studies that have particularly addressed IG in the context of design. They most often concern product design and have been carried out in the engineering domain [12-14]. However, they are relevant for interaction design as well, given that interaction design is a design discipline too [15]. For instance, the research findings repeatedly challenged the notion of group work's efficiency in idea production in comparison to nominal groups consisting of independent individuals. These findings on the illusion of group productivity are paralleled by the discovery of design fixation, a phenomenon caused by exposure to inspirational material.

For successful empirical research it is important to have a theoretical basis to help make sense of the experimental data. The majority of studies share a psychological perspective in which IG is considered a predominantly memory-based activity. There are several theoretical models on how new ideas can be produced based on re-using knowledge that designers have acquired previously [10, 16, 17]. These models describe ideation as knowledge reuse and refer to psychological concepts such as memory recall and conceptual combination. With these models, we have just started to understand the dynamics of idea generation as it occurs in reality. This means abandoning the romantic or mystic accounts of ideas just emerging in one's mind.

2.1 Research in Practice

Design IG studies that apply idea analysis methods [see, for instance 13, 18] are typically experimental psychological studies. The goal is to determine causal relations; i.e. what consequences do changes in the fixed factors (*independent variables*) have in terms of certain outcomes (*dependent variables*). The effects should be explained by referring to the properties of the background theory, i.e. human cognition. IG studies might be more accurately described as quasi-experimental research. This refers to the fact that the assignment of individuals to groups may not be truly random, and to external variables that cannot be controlled which may affect the data collection. A major reason for this is that the participating designers are never completely identical to each other in terms of all the relevant factors in their background experience.

The idea analysis method has a few requirements. First of all, there must be a lot of data. An interesting repeated finding in design IG research [17], as well as from creativity tests [8], is that independent minds will repeatedly generate similar ideas. The empirical evidence is commonly acquired in one or more successive design sessions. In each session, a design brief advices the participants to generate ideas for a given situation. In design studies, a visual representation of the idea is usually preferred. The sessions are commonly managed and facilitated by the experimenter(s). An example of an experimental IG study setup would be to investigate how the group size (independent variable) affects the amount of ideas produced by one person (dependent variable). This has been the prevalent choice in studies of Brainstorming, which have often compared real, interactive groups with nominal groups, equal in size, but consisting of individuals working in isolation [see, for example, 4, 5]

Experimental research on design IG also requires limiting the freedom of designers in order to maintain control over the experiment. The participants need to follow quite strict rules and adapt to working in ways new to them. This can be problematic due to the fact that it induces some learning effects. There are also some concerns regarding subjects who have been traditionally selected by convenience. The majority of published design studies compare groups of (design) students rather than professionals. This is problematic to some extent, as there are known effects of gaining skills and knowledge, i.e. expertise in design [19]. However, for IG research, a uniform education and lack of design experience may be beneficial if it results in greater homogeneity of background knowledge from which the ideas emerge, and consequentially create more pronounced experimental effects.

After the design experiment is carried out, the analysis proceeds by measuring the dependent variables. The analysis always requires some domain expertise to understand and interpret the ideas. This is usually carried out by select domain experts, but recently interesting proposals about crowd-sourcing have also been suggested [20]. Typical dependent variables are *Quantity*, *Creativity*, *Feasibility*, *Elaboration*, and *Categorization*. The most commonly used variable is the number of ideas produced, labeled quantity or productivity. Typically only unique, non-repetitive ideas are counted for this measure. This simple figure is often complemented by some measure of creativity. Although challenging to define [21], creativity is usually assessed through *novelty* and *quality* of the idea. Novelty, diversity, or commonality all represent how common the idea is. This can be referenced either to the pool of ideas collected or to solutions known to a domain expert (the evaluator). Quality can be operationalized as functionality, feasibility, or even usability. Feasibility and elaboration are quite straightforward variables describing the readiness for implementation and the detail of description. Elaboration is generally negatively associated with quantity. Finally, categorization refers to assigning ideas to clusters by similarity. This is theoretically justified and provides a measure of novelty [for more information, see 12, 14, 22].

An important question regarding the dependent variables is that are they valid? Do they measure things that we think are important regarding the success of ideation? This is a difficult question because the application of any of these measures is a compromise. Ultimately, we are interested in the ideas that can lead to breakthrough innovations. But that is nearly impossible to measure. Thus far, many researchers have adopted the thinking that quantity breeds quality – having many ideas guarantees finding a few good ones [23]. Some researchers believe that only the number of "good" ideas is indicative of how well IG succeeds [24]. Our present approach is pluralistic, we analyze all unique ideas in great detail.

The idea assessment process requires reliability and robustness. Thus several independent evaluators are needed, and their agreement should be confirmed using inter-rater agreement assessment (kappa statistic) or consensus. The evaluation process can be iterated a few times particularly if the problem area is new and the ideas are not predicted. However, ultimately the space of potential solutions for the given task becomes mapped extensively, and the assessment of individual ideas is easier as they tend to fall into predefined categories.

3   Essential Findings from Idea Analysis Research

3.1   The Dark Side of Brainstorming

After a long era of investigations into the nature of Brainstorming, its positive and negative aspects have become well-known. They point out that while properly organized Brainstorming does tease out a huge amount of creative potential, there are many kinds of hindrances as well. Group situations generally induce phenomena known as *production blocking, free riding, evaluation apprehension* [3], and different forms of *group think*. Production blocking refers to the fact that once someone else is taking a turn to express an idea, others are expected to listen. This blocks both their output and, if truly paying attention to the presenter, their idea production, as well. The bigger the group, the bigger the loss due to production blocking. It has been repeatedly shown that real groups of more than two people always underperform in comparison to nominal groups [3]. This has motivated the development of different electronic brainstorming platforms, which overcome the production blocking issue and allow multiple people to contribute simultaneously [25].

Social loafing or free riding is a common feature of unequal distribution of effort induced in a group. Participants may decrease their efforts if they think they do not have anything to add, or if they believe others will take care of it. The fear of negative feedback is labeled evaluation apprehension, and mainly effects the communication of ideas. Putting out an idea requires good self esteem or high confidence in the group treating it with an open mind. These three factors directly influence the quantity of work a group can produce. Finally, there are other forms of group think, which affect the qualitative nature of the produced ideas. The best example is the influence of a "group leader" (whether by appointment, charisma, appearance, or social ranking), who can dominate the group opinion. [26]

3.2 Individual Level Effects

Changes in the IG process are commonly described with two distinct types of effects: stimulation and fixation. *Stimulation* is the positive effect that having outside influences prior to or during IG has. Many design practitioners share the view that all sorts of external material can inspire the production of new designs. With some precautions, this seems to be the case [27], although the evidence from design is still scarce [13]. The stimulation is believed to help designers to get started with the ideation process. It helps mainly to create ideas that are similar to the stimuli. For this reason, heterogeneous stimuli are preferred, and have been found to facilitate the production of more diverse ideas.

Fixation, on the other hand, is the down side of stimulation. It refers to the negative impact of external influence. A number of studies under the label of "design fixation" have tried to characterize how fixation shifts the design IG process to a certain, possibly unwanted, direction [28, 29]. They define design fixation as a tendency to unknowingly reproduce parts of the given examples to a new design. This occurs even if the examples contradict the design requirements. In the psychological literature [9, 30] this is called unconscious plagiarism.

IG is an activity that naturally evolves over *time*. A recent review identified three kinds of effects that time can have on idea generation processes [31]. Time pressure, total duration, and time decomposition were each considered important for the effectiveness of an ideation session. Time pressure has a non-linear relationship to productivity and creativity, and the reviewed studies suggest that a hectic warm-up session can affect the productivity of a later, more relaxed real session positively, in terms of productivity and creativity. Interestingly, productivity over time shows an initial burst of productivity when the most common ideas are produced. Time decomposition refers to splitting the ideation session to cope with the time-related issues. For instance, by having a break or by dividing the session between individual and interactive work phases can help.

4 An Investigation of Ideation in the Wild

4.1 Background

In the spirit of user-centered development, a straight line between the world of users and developers is desired. An open question with IG practices in the professional setting is the evident distance between the site of use and design. As a solution, some methods of facilitating the designers' stepping into the users' world have been described. For conceptual design, IDEO has introduced the Deep Dive method, and Oulasvirta et al. have described a Bodystorming ideation method for mobile context [32].

We wanted to investigate the effects of creating new software service concepts in real environments. Using two locations and two tasks, we explored how the change in physical context influenced the resulting design. In the experiment, we had one central research question in mind: Are there systematic differences between the contexts regardless of the ideation task? The cognitive theories of IG would generally make two contradicting assumptions: the stimulation potential of the real environment should qualitatively change the ideas, but it influences designers' thinking to the existing solutions and the specifics of the environment [10, 18].

4.2 Method

Participants and experimental design. We recruited 25 university students participating in a cross-disciplinary product development intensive course (14 male, 11 female). Their average age was 25.9 (st. dev. 4.8 years) and they all represented different disciplinary backgrounds, coming from 14 different nationalities. All subjects evaluated their English skill as fluent or average. After their written consent, they were split into groups of three people. The division was arranged so that the heterogeneity of the groups was intentionally maximized by controlling the mix of backgrounds (design experience), sexes and nationalities.

We examined the effect of context on IG using a factor called *Physical location* trialed across two tasks. For the tasks, we had a Hotel and a Gym task. The former dealt with new services for lobby touch screen points and the latter with an "intelligent gym" in general. Both of the tasks were either carried out at the typical *office* location or at the associated *real-world* site. Participants were not informed about the nature of this experiment, other than that it was a competitive IG practice. Every group worked once in the real setting and once in the office, with a different task at each location. The order of the tasks was counter-balanced between the groups and distributed pseudorandomly.

Tasks and Materials. The participants were asked to create ideas at two consecutive locations. On average, they had a 60 min break between the tasks allocated for transportation (all locations were located within a 500 meter radius). The IG was divided between an individual IG (5 min) and a group IG phase (15 min). Before the IG phases the participants had 5 minutes to explore the context in the real-world setting (see Figure 1) or inspect photos of the real context in the office setting.

Fig. 1. Inspirational photos of the intended use context. The same locations were explored by the participants in the real context situation.

Fig. 2. Participants working in the gym setting (left) or in the office setting (right) in the group brainstorming phase

4.3 Results and Analysis

Here, we only present the results from the individual working phase, in order to avoid potential effects that the type of context had on group social interaction which might obscure the context effect on the produced ideas themselves. We analyzed all qualified ideas (N=139) produced by the individuals according to the dimensions of novelty (NOV), utility (UTI), feasibility (FEA), level of detail (LOD), mode of presentation (MOP) and categories. The assessment scales were defined bottom-up by examining the data on two iterations, so that five evaluators first evaluated a subset of all ideas for both tasks. After this they discussed the suitability of the proposed measures for the data and then converged on the scales. Some unidentifiable ideas were discarded. For all dimensions except MOP, a four step ordered scale used (values 0-3). On the following round, two independent evaluators used the revised scales to score each dimension of the data. The hand-coded data was imported into SPSS statistics software and analyzed using non-parametric statistical tests. Samples of the produced ideas are presented in Figure 3.

Fig. 3. Three examples of documented ideas

We first examined the Gym and the Hotel lobby tasks for between groups effects. Interestingly, no statistically significant differences were discovered in the quantity of the ideas or in the evaluated dimensions. On the individual level, we tested whether subjects differed between the contexts across the two tasks. We found few changes across conditions, the only statistically significant difference occurred in the feasibility dimension, which was generally higher in concepts produced in the real-world setting (see Table 1 below). No changes in the categorical distribution for the mode of presentation were found across the conditions.

Table 1. Results from a Kruskal-Wallis test assessing the differences across all produced ideas in both experiments

	Novelty	Utility	Feasibility	LODetail
Chi-Square	.255	.088	5.332	.918
P-value	.613	.767	**.021**	.338

5 Discussion and Conclusion

In this paper we have presented the paradigm of idea analysis in design IG for the interaction research community. Having first discussed the earlier results found in literature, we described our own experiment on idea generation in the wild. In our study, we witnessed surprisingly small differences in the quantity and quality of ideas produced in different physical contexts. However, we noticed that the feasibility of ideas tended to increase if they were produced in the real use context. This is best understood through the constraining effects of environment. Even though we expected to see positive effects from stimulation, the only observed consequence of the location was getting increased grounding to the present environment. The outcome supports detachment of the target environment as an evidence-based practice in design idea generation during concept development. This is compatible with a theory of idea generation in which all additional pieces of information around the design brief accumulate mentally to design requirements. These design constraints can have a negative effect on a process such as conceptual design which should aim at maximum creativity of the output.

In the future, we plan to make a thorough qualitative analysis of the additional group work data and video material, which were neglected in this study in order to focus on the quantitative results. We see plenty of room and need for similar, theoretically motivated investigations about HCI design methodology in the future. The present investigation had limited statistical power and leaves room for future replications around the topic. The methodology documented in the paper provides for easy follow ups and instructs the reader in quantitative assessment of idea generation methods. We hope that the present study will spark the creation of new approaches on how to leverage exposure to real context environments, to the advantage of designers, maybe through means other than brainstorming *in situ*. However, our study clearly shows that qualitative and quantitative gains cannot be simply assumed to follow.

Acknowledgments. We thank the participants, as well as Juha Forsblom, Viljami Lyytikäinen and Jussi Hannula at the Aalto University Design Factory for helping with the arrangements. The preparation and presentation of this paper was supported by *LUTUS – Creative Practices in Product Design* - research project. The effort of co-authors in the preparation of this paper was equally shared.

References

1. Osborn, A.F.: Applied imagination: principles and procedures of creative problem-solving. Revised edition. Scribner, New York (1957)
2. Shneiderman, B.: Creativity Support Tools. Communications of the ACM 50, 20–32 (2007)

3. Diehl, M., Stroebe, W.: Productivity loss in brainstorming groups: Toward the solution of a riddle. J. Pers. Soc. Psychol. 53, 497–509 (1987)
4. Diehl, M., Stroebe, W.: Productivity loss in idea-generating groups: Tracking down the blocking effect. J. Pers. Soc. Psychol. 61, 392–403 (1991)
5. Gallupe, R.B., Bastianutti, L.M., Cooper, W.H.: Unblocking brainstorms. J. Appl. Psychol. 76, 137–142 (1991)
6. Wang, H.-C., Fussell, S.R., Setlock, L.R.: Cultural Difference and Adaptation of Communication Styles in Computer-Mediated Group Brainstorming. In: SIGCHI Conference on Human Factors in Computing Systems (CHI 2009), Boston, MA (2009)
7. Carroll, E.A., Latulipe, C., Fung, R., Terry, M.: Creativity factor evaluation: towards a standardized survey metric for creativity support. In: ACM Creativity and Cognition (C&C 2009), pp. 127–136. ACM Press, Berkeley (2009)
8. Torrance, P.E.: Torrance Tests of Creative Thinking. Norms-Technical Manual. Ginn and Company, Lexington (1974)
9. Chrysikou, E.G., Weisberg, R.W.: Following the wrong footsteps: Fixation effects of pictorial examples in a design problem-solving task. J. Exp. Psychol.-Learn. Mem. Cogn. 31, 1134–1148 (2005)
10. Nijstad, B.A., Stroebe, W.: How the group affects the mind: A cognitive model of idea generation in groups. Pers. Soc. Psychol. Rev. 10, 186–213 (2006)
11. Volkema, R.J.: Problem Formulation In Planning And Design. Manage. Sci. 29, 639–652 (1983)
12. Shah, J.J., Vargas Hernandez, N., Smith, S.M.: Metrics for measuring ideation effectiveness. Design Stud. 24, 111–134 (2003)
13. Perttula, M., Sipilä, P.: The idea exposure paradigm in design idea generation. J. Eng. Design 18, 93–102 (2007)
14. Kudrowitz, B.M., Wallace, D.R.: Assessing the Quality of Ideas from Prolific, Early-Stage Product Ideation. In: ASME DETC 2010, Montreal (2010)
15. Visser, W.: Design: one, but in different forms. Design Stud. 30, 187–223 (2009)
16. Vargas Hernandez, N., Shah, J.J., Smith, S.M.: Cognitive Models of Design Ideation. In: Design Engineering Technical Conferences (ASME DETC 2007), Las Vegas, NV (2007)
17. Liikkanen, L.A., Perttula, M.: Inspiring design idea generation: Insights from a memory-search perspective. J. Eng. Design 21, 545–560 (2010)
18. Shah, J.J., Kulkarni, S.V., Vargas Hernandez, N.: Evaluation of Idea Generation Methods for Conceptual Design: Effectiveness Metrics and Design of Experiments. J. Mech. Design 122, 377–384 (2000)
19. Cross, N.: Expertise in design: an overview. Design Stud. 25, 427–441 (2004)
20. Dow, S.P., Fortuna, J., Schwartz, D., Altringer, B., Schwartz, D.L., Klemmer, S.R.: Prototyping Dynamics: Sharing Multiple Designs Improves Exploration, Group Rapport, and Results. In: CHI 2011, Vancouver, BC (2011) (in press)
21. Mayer, R.E.: Fifty Years of Creativity Research. In: Sternberg, R.J. (ed.) Handbook of Creativity, pp. 449–460. Cambridge University Press, Cambridge (1999)
22. Srivathsavai, R., Genco, N., Hölttä-Otto, K.: Study OF Existing Metrics Used in Measurement of Ideation Effectiveness. In: ASME DETC 2010, Montreal (2010)
23. Yang, M.C.: Observations on concept generation and sketching in engineering design. Research in Engineering Design 20, 1–11 (2009)
24. Reinig, B.A., Briggs, R.O., Nunamaker, J.F.: On the measurement of ideation quality. J. Manage. Inf. Syst. 23, 143–161 (2007)

25. Shepherd, M.M., Briggs, R.O., Reinig, B.A., Yen, J., Nunamaker, J.F.: Invoking social comparison to improve electronic brainstorming: beyond anonymity. J. Manage. Inf. Syst. 12, 155–170 (1995)
26. Stroebe, W., Diehl, M., Abakoumkin, G.: The Illusion of Group Effectivity. Pers. Soc. Psychol. Bull. 18, 643–650 (1992)
27. Dugosh, K.L., Paulus, P.B., Roland, E.J., Yang, H.C.: Cognitive stimulation in brainstorming. J. Pers. Soc. Psychol. 79, 722–735 (2000)
28. Jansson, D.G., Smith, S.M.: Design fixation. Design Stud. 12, 3–11 (1991)
29. Purcell, A.T., Gero, J.S.: Design and other types of fixation. Design Stud. 17, 363–383 (1996)
30. Smith, S.M., Ward, T.B., Schumacher, J.S.: Constraining Effects Of Examples In A Creative Generation Task. Mem. Cogn. 21, 837–845 (1993)
31. Liikkanen, L.A., Björklund, T., Hämäläinen, M.M., Koskinen, M.: Time Constraints in Design Idea Generation. In: International Conference on Engineering Design, ICED 2009, Palo Alto, CA (2009)
32. Oulasvirta, A., Kurvinen, E., Kankainen, T.: Understanding contexts by being there: case studies in bodystorming. Pers. Ubiquit. Comput. 7, 125–134 (2003)

Usability Standards across the Development Lifecycle

Mary Frances Theofanos and Brian C. Stanton

National Institute of Standards and Technology
100 Bureau Dr
Gaithersburg MD 20878, USA
{maryt,brian.stanton}@nist.gov

Abstract. In 2005 the International Organization for Standardization published ISO/IEC 25062 "Common Industry Format (CIF) for Usability Test Reports." This standard focuses on documenting the results of usability testing in a consistent format in terms of user effectiveness, efficiency and satisfaction that allows comparison among products by purchasers of such systems. However, soon after its publication the user community advocated for additional standards to document the output of usability-related work within the development lifecycle. A second usability CIF, "A General Framework for Usability-related Information" (ISO/IEC Technical Report 25060) is now available that identifies seven outputs of the usability-engineering process. The framework focuses on documenting those elements needed for design and development of usable systems. To successfully use the framework it is critical to understand the relationship of these elements to the human-centered design process and the activities of the system life-cycle processes. These new Common Industry Format standards for usability-related information are a further step in standardizing usability engineering in industry.

Keywords: Usability, User Centered Design, Common Industry Format, standards, lifecycle, software development.

1 Introduction

In October of 1997, the U.S. National Institute of Standards and Technology (NIST) initiated an effort to increase the visibility of software usability, the Industry USability Reporting (IUSR) Project. Cooperating in the IUSR project were prominent suppliers of software and representatives from large consumer organizations. The goals of the initiative were to:

- Reduce uncontrolled overhead costs of software usability problems while improving user productivity and morale.
- Encourage software suppliers and consumer organizations to work together to understand user needs and tasks.
- Define and validate an industry-wide process for characterizing software usability to support product decision-making.

The result of this initial effort was the American National Standards Institute/ InterNational Committee for Information Technology Standards (ANSI/INCITS)-354

M. Kurosu (Ed.): Human Centered Design, HCII 2011, LNCS 6776, pp. 130–137, 2011.

Common Industry Format (CIF) for Usability Test Reports [1]. The CIF, published in 2001, provides a standard method for reporting formal usability tests in which quantitative measurements are collected. The CIF is particularly appropriate for summative/comparative testing [2]. Most importantly, the CIF does not tell how to do usability testing; it tells you how to report on what you did.

Soon after its publication, the user community advocated for additional standards to document the output of usability-related work beyond the scope of the CIF, i.e., within the analysis, design and testing phases of the development life-cycle of interactive systems. For example, an intended purpose of the CIF was to enable consumer organizations to take into account usability when making purchasing decisions. But, organizations must determine whether the usability of the product meets their requirements. Ideally these requirements should be identified in advance. It makes sense to share these requirements with the supplier organization early in the development process so that they can be incorporated into the design, rather than waiting for the results of a usability test. Thus, the second phase of the IUSR project focused on identifying and specifying usability requirements. The results of this effort were published as NIST Internal Report NISTIR:7342 Common Industry Specification for Usability-Requirements (CISU-R) [3] in 2007.

NIST also found that organizations and usability practitioners were modifying the CIF for use in reporting other types of usability testing, particularly testing that informed the improvement of a product, so-called formative testing. The IUSR project participants agreed that this area required further investigation and held a series of workshops starting in 2004. They developed a set of guidelines to assist usability practitioners in the creation of reports that communicate effectively to guide the improvement of products [4]. The project resulted in guidance for reporting, rather than a formal template [5].

In 2005, the International Organization for Standardization (ISO) adopted ANSI CIF as the ISO/Electrotechnical Commission (ISO/IEC) 25062: Software engineering – Software product Quality Requirements and Evaluation (SQuaRE) – Common Industry Format (CIF) for usability test reports [6]. The ISO working group was aware of the IUSR group's additional products. The working group also recognized the user communities' desires for additional standards to document the output of usability-related work. As a result a Joint Working Group (JWG) comprised of members of the ISO Ergonomics of Human – System Interaction Subcommittee and the ISO Software and Systems Engineering, Evaluation and Metrics Subcommittee was established to document usability-related work products within the analysis, design and evaluation phases of the development life-cycle of interactive systems.

2 New Standards Development

The JWG began its charge by defining a framework for usability-related information, ISO/IEC TR 25060: Software engineering – Software product Quality Requirements and Evaluation (SQuaRE) – General framework for usability-related information [7]. This technical report defines a framework for an upcoming series of standards to document usability-related information throughout the development life-cycle.

The framework identifies seven outputs of the usability-engineering process "that are essential to provide the data required to allow systematic human-centred design of an interactive system under development." [7]

1. Context of use description
2. User needs report
3. User requirements specification
4. User interaction specification
5. User interface specification
6. Evaluation report
7. Field data report

Development of individual standards for each of these outputs is planned. Initial working drafts for items 1, 2, 3 and 6 are already in progress and should be available as ISO standards within the next 5 years. For each of these items, Technical Report (TR) 25060 provides a high-level description of the content of the information to be documented, as described in Table 1.

Table 1. Proposed Content for Usability Standards

Proposed Standard	Content
Context of use description	• Overall goals of the system • Stakeholder groups who either use the interactive system or are affected by its output throughout the life- cycle of the interactive system • Characteristics of the users • Task goals and task characteristics • Information processed during tasks • Technical environment (hardware, software and materials) • Physical and social environments
User needs report	• Identified, stated, derived and implied user needs (cognitive, physiological, social) across all identified user groups • Results of the user needs analysis relating the described context of use and its development constraints to the tasks of each user group that is affected including any resulting human-system issues or risks
User requirements specification	• Reference to the context of use description intended for the design • Requirements derived from the user needs and the context of use • Requirements arising from relevant ergonomics and user interface knowledge, standards and guidelines • Usability requirements and objectives including measurable effectiveness, efficiency and satisfaction criteria in specific contexts of use • Requirements derived from organizational requirements that directly affect the user

User interaction specification	• Workflow design: the overall interrelationship (including sequences) between tasks and system components on an organizational level, including responsibilities and roles • Task design: all tasks broken down into sets of subtasks and allocation of subtasks to the user and the system and associated requirements • Task-specific detailed usability objectives • Dialogue model: for each task, the appropriate information exchange between user and system including sequence and timing as well as associated interaction objects and high-level selection of dialogue techniques • Information architecture: from the user's perspective
User interface specification	• Task objects and system objects needed to accomplish one or more tasks and the user interface elements that they are composed of • Properties, behaviors and relationships of task and system objects • Dialogue techniques employed for specific tasks (e.g., menus, form-based dialogues, command dialogues, combinations of those) • Graphical look of task objects and system objects for specific tasks, users and user groups
Evaluation report	• Reporting usability problems, derived user requirements and recommendations for improving the usability of the object of evaluation • Reporting a baseline for usability for the whole product • Reporting differences in usability across a set of products (two or more products) • Reporting conformity with user requirements (conformance test report)
Field data report	• Data on actual usage of the product (versus intended usage of the product) as input for upcoming product releases and identifies emergent user requirements. Sources of field data can include observation of use, user satisfaction surveys, usage statistics and help desk data. • Contains the field data and its sources including the actual context of use, the means of collecting the data, the reasons for its collection and any identified user needs and derived user requirements.

3   Relationship to Other Standards

The General Framework for Usability-related Information (ISO TR25060) is intended to complement and advance existing ISO usability standards. As such the framework is based on the human-centered design approach of ISO 9241-210 [8] (previously ISO 13407). ISO 9241-210 focuses on the process. The first step is acknowledging the need for user-centered design and the need to plan for it. Then the four major activities may be enumerated as:

1. Understand and specify the context of use
2. Specify user requirements.
3. Produce design solutions
4. Evaluate designs against requirements.

ISO 9241-210 focuses on the human-centered design process. ISO 25060 complements 9241-210 describing the outputs of the human-centered design process. However, the human-centered design process is generally performed within a larger system- development process. ISO 15288 [9] provides a common process framework covering the system life-cycle. The products from ISO 25060 must not overwhelm or burden developers and must complement the current development processes and environments in order for organizations to encourage adoption. Although the framework focuses on documenting those elements needed for design and development of usable systems, it is critical to understand the relationship of these elements and how they complement the activities of the system life-cycle and processes.

Fig. 1 illustrates the relationship between the CIF family of standards (ISO/IEC TR 25060), the human-centered design activities (ISO 9241:210) and the system life cycle technical processes (ISO IEC 15288). The figure shows where each standard from the CIF family of standards occurs during the human-centered design activities from ISO 9241 part 210 (as indicated by the four circles), as well as where it corresponds to the ISO IEC 15288 system life cycle technical processes from clause 6.4 for the systems engineering technical process. The figure provides a conceptual model of how the three standards coordinate.

The figure depicts the four major activities of the human-centered design process as a set of intersecting circles, each circle representing an activity. The circles overlap to represent that the activities are not separate, but rather, overlapping in time and scope; the outcome of each activity can inform the input of another. The Users are shown in the center referencing the need for user-centered design (UCD)[1] and the need to plan for UCD, as the first step. As each activity can inform any other, there is no start, endpoint, or linear process intended.

The UCD process centers on users. The users, tasks and the organizational and physical environment are identified during the context of use activity and described in the Context of Use Description [10]. User needs are also identified during the context of use activity and can be documented in the User Needs Report [11] (in progress ISO/IEC 25064). ISO/IEC TR 25060 states that "User needs are an intermediate deliverable that link the context of use data to the user requirements." Both documents are developed during the stakeholder requirements definition process as described in ISO/IEC 15288.

The stakeholder requirements definition process also occurs during the user and organization requirements activity. It is during this activity that the requirements analysis process, as described in ISO/IEC 15288, also occurs. As a result of the requirements process, the User Requirements Specification (work in progress ISO/IEC 25065) would be developed. ISO/IEC TR 25060, states that "The user requirements specification provides the basis for design and evaluation of interactive systems to meet the user needs."

---

[1] For the purposes of this document "User-Centered Design" is the same as "Human-Centered Design".

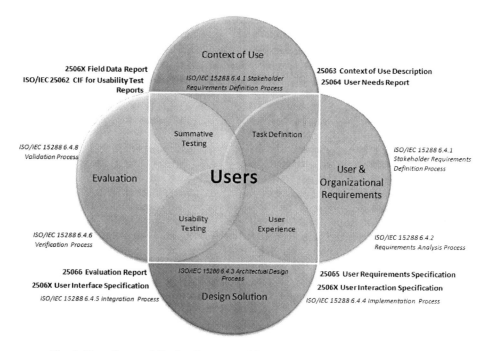

Fig. 1. User-Centered Design Process and Proposed Usability Information Items

The design of the interactive system happens during the design solution activity. This activity encompasses the architectural design, implementation, and integration processes of ISO/IEC 15288. These processes enable the User Interaction Specification and the User Interface Specification to be written. The User Interaction Specification "specifies how users will accomplish tasks with the system at a high level rather than describing what the system should look like. The user interaction specification provides the basis for the design of the user interface, not the design of the user interface itself." (ISO/IEC TR 25060). The actual user interface is documented in the User Interface Specification. This specification "provides the basis for the construction of the user interface as it contains the task and system objects needed to accomplish one or more tasks and the user interface elements that they are composed of." [7]

The evaluate activity can occur at any point in the development process and in an iterative manner (as can the other activities). ISO 9241:210 states that "Even at the earliest stages in the project, design concepts should be evaluated to obtain a better understanding of the user needs." The user needs, the resulting requirements, and the design of the system can be evaluated multiple times as the interactive system is being developed. The verification process, as documented in ISO/IEC 15288, is an evaluation done to ensure that the design requirements are fulfilled by the system. The validation process confirms that the system complies with the stakeholders' requirements, developed during the context of use and the user and organizational requirements activities. From the verification and validation processes the Evaluation Report [13], the CIF for Usability Test Reports [6], and the Field Data Report can be written. The Evaluation Report is a general purpose report format that can be used to document any user-centered evaluation that is performed during the development process. The CIF for Usability Test Reports [6] is a format to be followed when a formal evaluation is performed for comparison or baseline purposes. The Field Data report "provides data on actual usage of the product (versus intended usage of the product) as input for upcoming product releases and identifies emerging requirements."

4 Conclusions

This new Common Industry Format for usability-related information is a further step in standardizing usability engineering in industry. While ISO 9241 provides process-related guidance in the human-centered design process, as well as product-related guidance on dialogue principles, menus, forms, and presentation of information among others, it provides no guidance for deliverables or work products of the usability engineering process. ISO TR25060 and the upcoming standards specified in the technical report address this gap. These standards will enable usability professionals to share a common understanding of deliverables of the human-centered design process. Moreover the standards define the minimum content and to some degree the quality of those deliverables providing for consistency across the practice.

Working drafts for the Context of Use Description, User Needs Report, User Requirements Specification and the Evaluation Report are already in committee. The JWG hopes to have ISO standards available within the next five years for these documents.

References

1. ANSI/NCITS-354 Common Industry Format (CIF) for Usability Test Reports (2001)
2. Theofanos, M.F., Stanton, B., Bevan, N.: A practical guide to the CIF: usability measurements, interactions, pp. 34–37 (November-December 2006)
3. Theofanos, M.F.: NISTIR 7432: Common Industry Specification for Usability – Requirements (2007), http://zing.ncsl.nist.gov/iusr/
4. Theofanos, M.F., Quesenbery, W.: Towards the Design of Effective Formative Test Reports. Journal of Usability Studies (1,1), 27–45 (2005)
5. CIF-Formative Common Reporting Elements Website, http://zing.ncsl.nist.gov/iusr/formative/
6. ISO/IEC 25062, Software engineering - Software product Quality Requirements and Evaluation (SQuaRE) - Common Industry Format (CIF) for usability test reports, ISO (2005)
7. ISO/IEC TR 25060: Software engineering - Software product Quality Requirements and Evaluation (SQuaRE) – Common Industry Format (CIF) for Usability – General Framework for Usability-related Information, ISO (2010)
8. ISO/IEC 9241-210:2009 Ergonomics of human-system interaction - Part 210: Human-centred design process for interactive systems. ISO (replaces ISO 13407:1999) (2009)
9. ISO/IEC 15288. Systems and software engineering - System life cycle processes (2002)
10. ISO/IEC 25063: Software engineering - Software product Quality Requirements and Evaluation (SQuaRE) – Common Industry Format (CIF) for Usability – Context of Use Description (in progress)

11. ISO/IEC 25064: Software engineering - Software product Quality Requirements and Evaluation (SQuaRE) – Common Industry Format (CIF) for Usability – User Needs Report (in progress)
12. ISO/IEC 25065: Software engineering - Software product Quality Requirements and Evaluation (SQuaRE) – Common Industry Format (CIF) for Usability – User Requirements Specification (in progress)
13. ISO/IEC 25066: Software engineering - Software product Quality Requirements and Evaluation (SQuaRE) – Common Industry Format (CIF) for Usability – Evaluation Report (in progress)

Structure of FUN Factors in the Interaction with Products

Sayoko Tominaga[1], Toshihisa Doi[1], Toshiki Yamaoka[2],
Yuka Misyashita[3], and Masayoshi Toriumi[3]

[1] Graduate School of Systems Engineering, Wakayama University
[2] Faculty of Systems Engineering, Wakayama University
[3] Vehicle Test Technology Development Division, NISSAN MOTOR CO., LTD,
930, Sakaedani, Wakayama City, Wakayama, 640-8510, Japan
s105032@sys.wakayama-u.ac.jp

Abstract. In recent years interaction design has looked to questions of most typically positive emotions such as satisfaction, pleasure and delight. This study investigated the factors of FUN as joy and pleasure, which are created in interaction between human and products, and found their relationship. We aim for gaining useful information when we design products. First, the questionnaire regarding SCT (sentence completion test) were conducted and were analyzed by DEMATEL method (Decision MAking Trial & Evaluation Laboratory) As a result, it was found that the factors of FUN were clarified and examined the relationship.

1 Introduction

Lately, not only Usability but also the promotion of users' psychological requirement (pleasure and joy) is needed in product development. Ministry of Economy, Trade and Industry has designated the three years from 2008 to FY 2010 as the "Kansei Value Creation Years and intensively implement a variety of measures to carry out the "Kansei Value Creation" [1]. Therefore, in Japan, "Kansei" would become an increasingly thing in product development.

In this study, the factors of FUN, as joy and pleasure, which are created in interaction between human and products, were investigated and found their relationship. These results would be useful information when we design several products.

Before introducing our full study, the area of our study is explained with the idea of "Hedonomics" [2] In conventional Ergonomics, human-centered design had been practiced to satisfy some requirements (e.g., safety, Functionality and usability). However, these were not enough requirements for users to satisfy their needs. Then, the idea of Hidonomics has come (fig.1). Hidonomics aims for the promotion of pleasure and joy in user interaction with safety, functionality and usability. In addition, it is said that Hidonomics can completely satisfy user requirements by individualized customizing. Figure 1 derived from Maslow's conception. The fact that these design imperatives match the social edict of "life, liberty and the pursuit of happiness" has not escaped our attention (see Hancock, 1999). In this study, "FUN" is defined the interaction factors which roused pleasure and joy to create "Pleasurable Experience" in the idea of Hedonomics.

M. Kurosu (Ed.): Human Centered Design, HCII 2011, LNCS 6776, pp. 138–143, 2011.

"Kansei" which was used this paper means a sequence of information flow (sense→awareness/cognitive→emotion/affectivity→the expressions of language) which occur after exterior stimulus is communicated to sensory receptor.

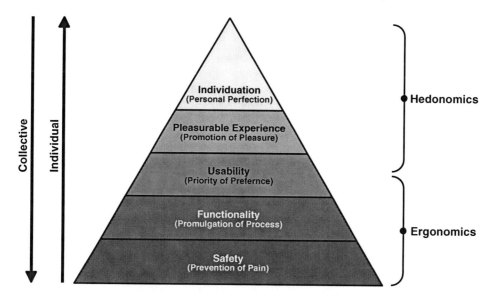

Fig. 1. A hierarchy of the relationship between Ergonomics and Hedonomics [2]

2   Conducting a Questionnaire

2.1   Summary of a Questionnaire

This study investigated the factors of FUN and found their relationship. We aim for gaining useful information when we design products. First, the questionnaire regarding SCT (sentence completion test) were conducted and the results of the questionnaire is analyzed by DEMATEL method (Decision MAking Trial & Evaluation Laboratory) To investigated the factors of FUN and their relationship, SCT by a questionnaire was conducted to some students and workers. We will grasp the factors of FUN which human generally feels from products and analyze their relationship by DEMATEL method.

2.2   Participants

Study participants included 33 adults from twenty-two to forty-nine. (Average: 31, SD:9.38, man:24, woman:9) It is very difficult for ordinary people to answer our questionnaire. Therefore, in this study, we confined the participants to only users who know HMI (Human Machine Interface), product design, or user interaction design. In doing so, we concerned that better answer can be gotten.

2.3 Contents of the Questionnaire

SCT is one of the Projective Technique method. SCT is a method that participants are asked to fill in the blanks on the sentence to complete with the words which they thought. [3]. The Projective Technique method is a method that participants are showed ambiguous stimulus and asked to interpret it and to explain it. The more ambiguous the stimulus is, the more participants tend to express their hidden viewpoint and emotion in one way or another. Therefore, this method aims to have participants express their genuine thought.

In this study, we conducted the following questionnaire to find the factors of FUN. The question is "Please show me the products which make user feel pleasure and joy through the interaction between products and users?" Participants are asked to fill in the blanks on the sentence which shows "Because (   ), (   ). Then, (   )." with their thought. Figure 2 shows a questionnaire actually used in this study.

■ Question

Please show me the products which make user feel pleasure and joy through the interaction between products and users?

Q9   :   Because (*this feedback is enjoyable*), (*I enjoy using this product*).
          Then, (*I like this product*).

Q10 :   Because (*the weight of this product is proper*), (*I don' t feel strain*).
          Then, (*I can use natulally*).

Q11 :   Because (*design is nice*), (*I want to take it anywher and anytime*).
          Then, (*I want to show it to everyone*).

Q12 :   Because (texture is good), (I feel nice).
          Then, (I want to touch it more).

Fig. 2. Questionnaire of SCT

3 Results of SCT

To visualize the FUN factor's relationship, data obtained from SCT were analyzed with DEMATEL method. DEMATEL method was developed to grasp the difficult and complicated problem structure. As its feature, the method can represent not only the existence of factors' relationship but also the strength of their relationship among factors.

Before starting to analyze, the matrix needed the DEMATEL method was made from gained sentence as explained next. First, words and phrases gained from a questionnaire were coded and made a classification of the coded words on the basis of similar significance. Second, the keywords from one sentence were divided into cause and effect and two causations were made (fig3). (Ex: "Because (A), (B). Then, (C)." was changed into "A-B" and "B-C"). Finally, an n times n matrix is made from the degree of the each causation (Fig4).

① The sentence gained from a questionnaire

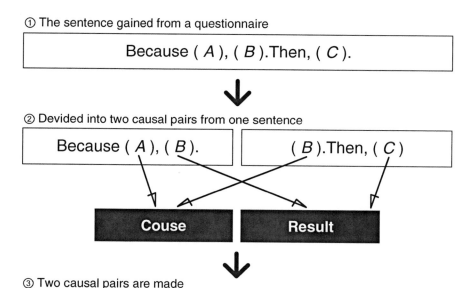

Fig. 3. How to make two causations from one sentence

		Keyword						[D]
		A	B	C	D	E	F	
Keyword	A	0	1	1	3	0	0	5
	B	2	0	0	0	1	0	3
	C	0	0	0	0	1	1	2
	D	1	1	3	2	0	0	7
	E	0	0	0	1	1	0	2
	F	0	2	1	1	1	0	5
[R]		3	4	5	7	4	1	24

Fig. 4. Matrix to analyze with the DEMATEL method (sample)

The row sum D of the matrix represents the total sum of influence level (influence rate) from a certain keyword to another keyword, while the rank sum R of the matrix represents the total sum of influenced level (influenced rate) from a certain keyword to another keyword.

D plus R means "Degree of association". "Degree of association" represents how much the keyword plays a key role in this structure of the problem because D plus R is the sum of the influence rate and influenced rate. On the other hand, D minus R represents the difference of the two. D minus R means "Influence rate". "Influence rate" represents how much the keyword influence others keywords as a cause. If this value is plus, it shows that the keyword influence another keywords. If this value is minus, it shows that the keyword is influenced from other keywords.

Additionally, we conducted the hierarchical clustering to consider the relationship among gained keywords from their Degree of association and Influence rate. Gained keywords were divided into five clusters. As a result, it found that Cluster A has five keywords, Cluster B has eleven keywords, Cluster C has twenty-one keywords, Cluster D has thirteen keywords and Cluster E has eight keywords. The following figure which is shown the Degree of association as x-axis and the Influence rate as y-axis represents every gained keywords plotted.

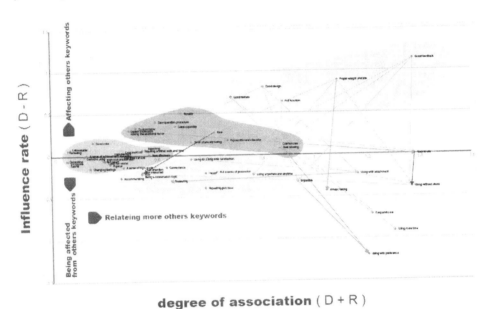

Fig. 5. Visualization of the result of the DEMATEL method

4 Discussion

We will explain each group classified into five clusters. Then, we will discuss structure as a whole.

Cluster A is a cluster which has high Degree of association and high Influence rate. In other words, each words of Cluster A relate more others keywords and affect them. The words regarding "Product's attribution" were seen from Cluster A. They are "Good feedback", "Good texture", "Proper weight and size", "Good design" and "Full function".

Cluster B is a cluster which has some Degree of association and Influence rate but both of numbers of Cluster B are not higher than those of Cluster A. In other words, each words of Cluster B relate some keywords and affect them. The words regarding "Concrete usability and convenient" were seen from Cluster B. They are "Novelty", "Customizable", "Explicit expression", and "Comfortable".

Cluster C is a cluster which Degree of association is the lowest of the five clusters. Influence rate of Cluster C is neither low nor high. In other words, each words of Cluster C are individual solution so that they don't particularly relate other keywords. Additionally, it seems that Cluster C is a neutral position cluster because their words have the possibility to affect other words and to be affected from others words. The words regarding "Emotion" were seen from Cluster C. They are "Changing feelings", "Hankering", "Reliability", and "Impressive". It is considered that Influence rate of Cluster C is the lowest so that emotion is the personal factor.

Cluster D is a cluster which Degree of association is as much numbers as Cluster B and Influence rate is second lowest cluster of the five. In other words, Cluster D is a cluster which is affected from other clusters. The words regarding "Relation to other people" were seen from Cluster D. They are "Get attention", "Being a conversation topic", "Recommending", "Satisfaction" and "Using for a long time". Cluster D is affected from Cluster A or Cluster B. For this reason, it seems that Cluster D also means a people's next action which will occur in filling situation of Cluster A or Cluster B.

Cluster E is a cluster which Degree of association is as high numbers as Cluster A and Influence rate is the lowest cluster of the five. In other words, each words of Cluster E relate more other keywords and are affected from other keywords. The words regarding "Frequency of use and Status of use" were seen from Cluster E. They are "Using with preference", "Using more time", "Using without stress", "Using with attachment" and "Always taking".

Next, we will explain the whole causal relation. According to each discussion, it found that the concept starts Cluster A (Product's attribution) as high conception and connects to Cluster B (Concrete usability and convenient) and next connect to Cluster C (Emotion) and Cluster D (Relation to other people), then, last connects to Cluster E (Frequency of use and Status of use). For this reason, it would appear the following discussion. The enchantments of product's attribution give good Usability to user. Next, positive emotion occurs to users and users try to relation to other people. Finally, it happened that frequency of use becomes more and the tendency of use becomes more positive.

As another consideration, it found that the top 3 keywords which have high Degree of association are the keywords regarding usability. For instance, "Easy to use", "Good feedback" and "Using without stress". For this reason, it is again found the view of usability is very important for people to feel pleasure and enjoyment on interaction between human and products.

References

1. Ministry of Economy, Trade and Industry: Kansei Value Creation Initiative (2007), http://www.meti.go.jp/press/20070522001/20070522001.html
2. Hancock, P., Pepe, A., Murphy, L.: Hedonomics: The Power of Positive and Pleasurable Ergonomics, p. 11. Human Factors and Ergonomics Society (2005)
3. Usability Handbook editorial committee: Usability Handbook. Kyouritsu-Pub. Co., p. 686, p. 564 (2007)

Extraction of User Interaction Patterns
for Low-Usability Web Pages

Toshiya Yamada[1], Noboru Nakamichi[2], and Tomoko Matsui[3]

[1] Department of Statistical Science, The Graduate University for Advanced Studies,
10-3 Midiri-cho Tachikawa Tokyo Japan
[2] Faculty of Information Sciences and Engineering, Nanzan University,
27 Seirei-cho, Seto, Aichi Japan
[3] The Institute of Statistical Mathematics
10-3 Midiri-cho Tachikawa Tokyo Japan
tyamada@ism.ac.jp

Abstract. Our goal is to point out usability problems in web pages in order to improve the web usability. We investigate the relation between user interaction behaviors in web-viewing and evaluation results of web usability by subjects. And we extract discriminative patterns for user interaction behaviors in visited web pages with low usability by using the PrefixSpan based subsequence boosting (Pboost).

Keywords: Web Usability, PrefixSpan Boosting (Pboost), User Interaction, Machine learning.

1 Introduction

The usability of web sites is important because it reach sales and user interest. To improve usability, methods for evaluating web sites pages are required. A well known one is usability testing. In this method, multiple subjects actually visit the target web site and evaluators then analyze the interactions. The evaluators often discover unpredictable problems with the web site. To date, usability testing evaluators have used recording devices such as a video tape recorder and voice recorder to record the subjects' interactions. They have interviewed subjects asking them several questions about the web site usability in order to collect subjects' evaluations. Usability testing has a problem that evaluators must spend a huge amount of time on the interview and analysis replaying the recorded data.

Our goal is to identify problems on web pages that have low usability in order to reduce such prodigious labor for evaluators. We investigated the relation between user interaction behaviors in web-viewing and the subjects' evaluations. And we extracted discriminative patterns for user interaction behaviors in viewing low-usability web pages by using PrefixSpan-based subsequence boosting (Pboost) [S. Nowozin: 2007].

For our future tasks, we should do a lot of tests because we can evaluate the usability of web sites more easily. It is expected that the interaction that causes the new problem of usability can be discovered.

M. Kurosu (Ed.): Human Centered Design, HCII 2011, LNCS 6776, pp. 144–152, 2011.
© Springer-Verlag Berlin Heidelberg 2011

2 Related Works

In this chapter, we describe methods of supporting evaluation objectively using quantitative data such as browsing time, mouse movement and eye movement.

Laila [Laila and Fabio 2002] analyzed the execution situation of a task from a user's operation event recorded by Java script. They supported the analysis based on quantitative data, such as page reference time and task execution time. They analyzed the usability of a Web page based on task execution time totaled for every Web page. Okada [Okada and Asahi 1999] developed the GUI-TESTER which extracts a common operation pattern from two or more users' operation history. If the tool is used, the operation pattern for mistaken operation can be extracted. And when the moving distance of a mouse cursor is long and an operation time interval is long, they suggest a possibility that a screen layout is bad.

WebTracer [Nakamichi et al. 2007] can collect the operation log of users on the Web pages. Collectable data include the information on users' sight line (the coordinates of the gazing point on the computer screen), operation log of a mouse, and the displayed screen images, together with their time information. The data collected by WebTracer characterize Web pages and have the possibility of being used for supporting the usability evaluation. However, the relation between such data and the problems in the Web usability was merely an example of the characteristics of the Web pages. Quantitative evaluation of the relation to the usability of Web pages was not done.

Heatmaps from user eyetracking studies based on fixations are used for observing in detail. Nielsen found that users' main reading behavior was fairly consistent across many different sites and tasks. [Nielsen 2006]. Eyetracking visualizations show that users often read Web pages in an F-shaped pattern: two horizontal stripes followed by a vertical stripe. Specialists may evaluate using eyetracking heatmaps. However it doesn't lead to the cost reduction of the web usability evaluation.

Eye-tracking methodologies are applying in the domain of Web search because gaze can be used as a proxy for a user's attention. Eye-tracking measures include pupil dilation, fixation information, and sequence information such as scan paths [Guan and Cutrell 2007]. They relied on measures related to gaze fixations with a minimum threshold of 100 ms in areas of interest. They found that as they increased the length of the query-dependent contextual snippet in search results, performance improved for informational queries but degraded for navigational queries. Analysis of eye movements showed that the decrease in search performance was partially due to the fact that users rarely looked at lower ranking results. Matsuda measured users' eye movements during web search tasks to analyze how long users spend on each result of the results pages [Matsuda et al. 2009]. They found the results displayed on the bottom of the page were viewed for a shorter time than the results displayed on the top of the next page.

In these conventional researches, the specialist had discriminated usability only using certain quantitative data. However, the effectiveness of combinational data was not verified statistically.

3   Experiment of Usability Testing

3.1   Quantitative Data of Users' Behavior

Browsing time, mouse movement, and eye movement are the quantitative data about users' behavior mainly used for web usability evaluation. This experiment recorded the interaction data for every Web page. Here, the interaction data is a record of the user's behavior during browsing such as "Where on the page is the user looking?" and "Where is the user's mouse cursor?". In this study, it adopts PageView (PV) as the count way of the Web page. We count the Web page which required of once with the browser and read from the web server as 1PV.

The interaction data is the vector which composed of the gazing point coordinates and the mouse cursor coordinates on the whole time of a subject browsing web 1PV. Gazing point is the point at the intersection of the users' look with the target screen.

3.2   Experimental Environment

The experiment environment used by this research is as follows.

- Display: 21 inches (Viewable screen size: H30 x W40cm)
- Device for measurement of sight line: NAC, EMR-NC (View angle: 0.28, resolution on the screen: approx. 2.4mm)
- Recording and playing of sight-line data: WebTracer (Sampling rate: 10 times per second)

WebTracer [Nakamichi et al. 2007] is an environment for recording and analyzing the users' operations in Web pages.

3.3   Experimental Procedure

We experimented with usability evaluation in the following procedures to five tasks. Subjects are 15 frequent users of the Internet. They have never visited the sites used in the experiment. We requested the subject to perform five tasks of looking for the starting salary of a master from the site of five companies, as a main experiment.

Procedure 1: The Web page for an experiment linked to the top page of each company is displayed by a subject. And the experiment is started from the time of a subject clicking the link.

Procedure 2: While subjects are doing the tasks, several types of quantitative data are recorded using WebTracer.

Procedure 3: The Web pages that subjects visited are displayed. We requested the subject to choose the ease of use for every visited Web page from the following five levels. We defines a low usability page as a page that a subject choose "hard to use" from four levels of the questionnaire.

1. Hard to use
2. Relatively hard to use
3. Relatively easy to use
4. Easy to use

Procedure 4: We reproduce the operation history recorded by WebTracer, and a subject checks all the visited Web pages. At that time, we interviewed the subjects about the situation of their search.

We recorded the quantitative data for 275 pages which the subjects visited. We were not able to record correctly about 12 pages of them. The cause is a frequent blink and head movement. Moreover, there were 8 pages which the subjects answered "don't know" about the usability of the Web page. We measured the quantitative data in 263 pages except these pages: 20 pages of them were class 1 (hard-to-use group) and 243 pages were class 0 (other evaluation group).

4   Pattern Extraction Using Pboost

4.1   Pboost

We used Pboost to extract interaction patterns that were able to discriminate a web page evaluated as having low usability from other pages. Pboost was originally developed for sequential data classification which was proposed by Nowozin [9].

We assume that "S" is whole interaction pattern, and "s" is subspace of interaction pattern "S": $s \in S$. The input data x_n is a vector of interaction data of a web page " n " ($n = 1 \cdots l$). And class label $y_n \in \{-1,1\}$ is input data too. Discriminant function $f(x)$ is follows:

$$f(x) = \sum \alpha_{s,\omega} h(x; s, \omega)$$

$$h(x; s, \omega) = \begin{cases} \omega & s \subseteq x \\ -\omega & otherwise \end{cases} \quad \omega \in \{-1,1\}$$

Where, $\alpha_{s,\omega}$ is called weight for pattern s . $h(x; s, \omega)$ are hypothesis function, an extra variable ω can decide for either class decision. To obtain an evaluable discriminant function, we need solve the optimization problem that formulated as linear problem. Using PrefixSpan algorithm as clever search strategy, we can solve this problem optimally.

4.2   Input Data

The interaction data with a class label is used as input data for Pboost. For the class label, we asked the subject to choose the ease of use for every visited web page from four levels: (1) hard to use, (2) relatively hard to use, (3) relatively easy to use, and (4) easy to use. We defined class 1 for low-usability pages for which a subject chose (1) and class 0 for the other pages as the other evaluation group.

Table 1. 10 criteria of dividing equally a screen

Criterion No.	Number of dividing equally a screen	Size of the division cell
1	2	512 pixel × 768 pixel
2	2	1024 pixel × 384 pixel
3	4	512 pixel × 384 pixel
4	12	256 pixel × 256 pixel
5	25	205 pixel × 154 pixel
6	48	128 pixel × 128 pixel
7	100	102 pixel × 77 pixel
8	192	64 pixel × 64 pixel
9	400	51 pixel × 38 pixel
10	768	32 pixel × 32 pixel

Meanwhile, it is possible to use only the data of the discrete-value as the input-data of Pboost but our recorded data is a continuation value. Due to this, we need to transform the interaction data from continuation value into discrete-value. First, we divide a screen into some the same size. Next, we give an integer value to each division cell. Lastly, it makes the coordinates of our recorded data correspond to the integer number of the contained cell. However, in this way, "the similarity between data" which could be expressed in the original coordinate-value cannot be expressed. Therefore, we used 10 criteria when dividing equally a screen. Then, we gave 10 values which were put by those 10 criteria to one interaction data. Table 1 shows these 10 criteria.

We show the example which was transformed into the integer value using these. In case of the interaction data such as $(x_{eye}, y_{eye}, x_{mouse}, y_{mouse}) = (250,1,250,1)$, the changed integer value vector becomes [1, 1003, 2001, 3001, 4002, 5002, 6003, 7004, 8005, 9008, 10001, 11003, 12001, 13001, 14002, 15002, 16003, 17004, 18005, 19008]. This vector consists of 20 elements about viewpoint and mouse cursor. When changing in this way, we can express " the similarity between data" as the number of the same elements of vectors.

We use the data which did such a change as the input data for Pboost. Our method using Pboost has two advantages: (i) the evaluator does not need to replay recorded data in order to analyze the subjects' interactions and (ii) the subjects' only need to answer simple questions and the burden on them is small.

5 Experiments

We extracted user interaction patterns for the Web page with low usability. All the data was used as input-data to train Pboost. We could completely classify the interaction data by using Pboost and extracted the interaction patterns. Total numbers of patterns extracted were 76, of which 40 were positively related to the identification of class 1 and the remaining 36 were related to class 0 identification. Figure 1 and the 2 show 10 visualized extraction patterns for each class.

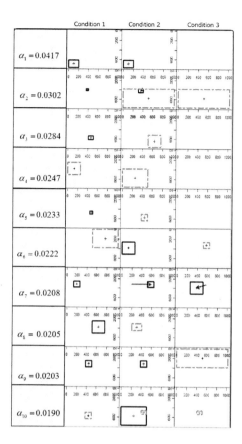

Fig. 1. Top 10 of the interaction pattern which has an influence on the direction which is identified by class 0

In the Figure 1 and 2, 10 patterns are showed at the longitudinal-direction. Also, "the condition" transition of each pattern is pictured in the crosswise direction. The outer frame with the scale shows the whole screen, the black square frame shows the area which a viewpoint is contained in and the gray chain line frame shows the area which the mouse cursor is contained in. Now, "the condition" is that a view point or the coordinates of the mouse are stored in the area in the screen. Moreover, it counts even if it doesn't transfer continuously between each condition and the condition. α_i shows the strength of the influence of each pattern. The bigger $|\alpha_i|$ is, the stronger the influence is.

These relations were found from the weight values for the interaction patterns in the discriminant function of Pboost. Many of the interaction patterns included mouse information.

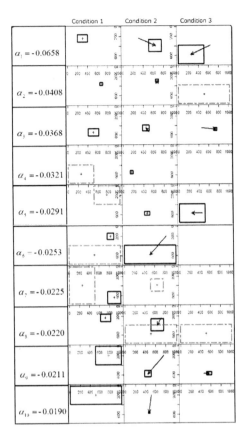

Fig. 2. Top 10 of the interaction pattern which has an influence on the direction which is identified by class 1

6   Conclusion Remarks

By using the proposal method, we were able to extract a total of 76 interaction patterns for the hard-to-use group and other evaluation group. Our method does not need the evaluator to reproduce of subjects' interactions and is effective at reducing the evaluator's load. Moreover, subjects only need to answer simple questions. Therefore, the method is effective at reducing the subject's load.

Figure 3 show two discriminant-function values $f(x)$ of Pboost when a user browses a "hard to use" page (class 1) and an "easy to use" page (class 0), where the horizontal axis shows browsing time and the vertical axis shows a discriminant-function value. If the discriminant-function value of a page i is $f(x_i) \leq 0$, it is classified into class1. We see from Figure 3 that this page was judged "hard to use" page 6 seconds later after beginning to browse it. Thus, we can understand when the web page was judge into "hard to use".

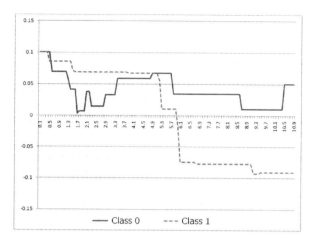

Fig. 3. Line graph of discriminant-function value

In addition to that it is possible to do extraction interaction patterns by using the proposed method, it is possible to understand when or why the user feels that the web page is "hard to use".

Acknowledgements. This work was supported in part by the Center for the Promotion of Integrated Sciences (CPIS) of Sokendai.

References

1. Guan, Z., Cutrell, E.: An eye tracking study of the effect of target rank on web search. In: Proceedings of the SIGCHI Conference on Human Factors in Computing Systems, pp. 417–420. ACM, New York (2007)
2. Joseph, S.D., Redish, A.P.: A Practical Guide to Usability Testing. Ablex Publishing, New Jersey (1993)
3. Kelly, G., Emily, G.: Web ReDesign. Pearson Education, London (2002)
4. Laila, P., Fabio, P.: Intelligent analysis of user interactions with web applications. In: Proceedings of the 7th International Conference on Intelligent User Interfaces, pp. 111–118. ACM, New York (2002)
5. Matsuda, Y., Uwano, H., Ohira, M., Matsumoto, K.: An Analysis of Eye Movements during Browsing Multiple Search Results Pages. In: Proceedings of the 13th International Conference on Human Computer Interaction, pp. 121–130. Springer, Heidelberg (2009)
6. Nakamichi, N., Sakai, M., Hu, J., Shima, K., Matsumoto, K.: Webtracer: A New Web Usability Evaluation Environment Using Gazing Point Information. In: Electronic Commerce Research and Applications, vol. 6(1), pp. 63–73. Elsevier Science, Amsterdam (2007)
7. Nakamichi, N., Sakai, M., Shima, K., Hu, J., Matsumoto, K.: Webtracer: A New Web Usability Evaluation Environment Using Gazing Point Information. Electronic Commerce Research and Applications 6(1), 63–73 (2001)
8. Nielsen, J.: Designing Web usability. New Riders, Indianapolis (2000)

9. Nielsen, J.: F-shaped pattern for reading web content. Jakob Nielsen's Alertbox (April 17, 2006), http://www.useit.com/alertbox/reading_pattern.html

10. Nowozin, S., Bakir, G., Tsuda, K.: Discriminative subsequence mining for action classification. In: Proceedings of the 11th IEEE International Conference on Computer Vision. IEEE Computer Society, Los Alamitos (2007)

11. Okada, H., Asahi, T.: GUITESTER: a log based usability testing tool for graphical user interfaces. IEICE-Trans. on Information and systems E82-D(6), IEICE, 1030–1041 (1999)

12. Oliveira, T.P.F., Aula, A., Russell, M.D.: Discriminating the relevance of web search results with measures of pupil size. In: Proceedings of the 27th International Conference on Human Factors in Computing Systems, pp. 2209–2212. ACM, New York (2009)

A Proposal of Service Design Evaluation Method

Toshiki Yamaoka

Wakayama University, Sakaedani 930, Wakayama, Japan
yamaoka@ja2.so-net.ne.jp

Abstract. This paper describes service evaluation method After a structure of service are explained, the service design are shown concretely. A structure of service consists of 4 factors of environment, machine, customer and employee. The relationship of customer versus environment, machine, and service employee are defined. The structure of service design is constructed based on 5 factors: service organization system, service design concept, interaction between customers and service employees / machine, produced good service quality and increased service productivity. Finally the two service design evaluation methods based on above-mentioned service items are proposed. The one is evaluated from viewpoint of customer's expectation and evaluation. The another one is the checklist consisted of seven questionnaires.

Keywords: service design evaluation, structure of service, structure of service design.

1 Introduction

The service business is getting very important in 21 century. Generally speaking, quality of service depends on person, and service design and the evaluation are not constructed systematically and efficiently. Especially the service design evaluation method in this paper is described in detail. The service design evaluation items are extracted from four factors of a structure of service and five factors of a structure of service design. Two proposed service evaluation method was constructed based on above-mentioned service items of factors.

2 The Structure of Service [1]

Service activities are analyzed from viewpoint of business relationships between B (Business) to B and B to C (Consumer). Final target user in B to B and B to C is consumer. This paper focuses on the B to C. The structure of service in the B to C consists of four factors: environment, machine, customer and service employee. The six relationships among them are necessary for constructing service design. The six relationships are as follows.

1. The relationship between customer and machine

(a) **Effective acquirement of information**
(b) **Ease of understanding and judgment**
(c) **comfortable operation**

M. Kurosu (Ed.): Human Centered Design, HCII 2011, LNCS 6776, pp. 153–159, 2011.

They are important items for designing and evaluating the relationship between customer and machine in the human information processing sequence.

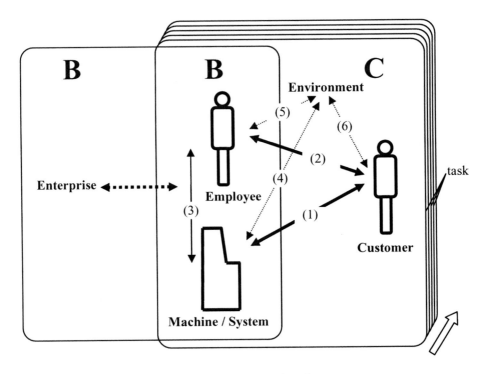

Fig. 1. The structure of service

2. The relationship between customer and service employee

(a) **Careful consideration**
    It is related to grasp situation and feel sympathy.

(b) **Correspondence to the situation**
    It means to correspond quickly, flexibly, precisely, equally and at ease, and confirmation such as an order in a restaurant and so on is also important in interaction between user and employee.

(c) **Caring attitude**
    It means making a good impression, getting sympathy, getting trust and being tolerant of mistakes.
    These three items correspond to the three items between customer and machine.
The services between customer and service employee consist of main service and support service. The main service means the value and utility of system.

3. The relationship between machine and employee
    Maintenance is very important job for employees.

4. The relationship between machine and environment
    Environmental conditions for machine are temperature and so on.

5. The relationship between user and environment

Environmental conditions for user are temperature/humidity, illumination and noise and so on.

6. The relationship between employee and environment

Environmental conditions for employee are temperature/humidity, illumination and noise and so on.

These items are needed for (1) motivation for service, (2) satisfaction with service and (3) incline to receive service again in service process.

3   The Structure of Service Design [1]

The structure of service design consists of five aspects: (1) service organization system, (2) service design concept, (3) interaction between customers and service employees / machine, (4) produced good service quality and (5) increased service productivity.

The five aspects are as follows.

1. Service organization system

The following three items are important to construct the service organization system.

(a) **Organizational policy**

(b) **Sharing information among service employees**

(c) **Motivation of employees**

As the policy of service organization system influences the service design concept, the policy is very important especially.

2. Service design concept

The service concept is structured. The concept items on the second layer of service design concept are weighted for making the service design concept concrete and clear. This idea is constructed based on Human Design Technology (HDT) [2][3].

3. Interaction between customers and service employees / machine.

The relationships between them are already above-mentioned.

The relationship between customer and machine

(a) **Effective acquirement of information**

(b) **Ease of understanding and judgment**

(c) **Comfortable operation**

The relationship between customer and service employee

(a) **Careful consideration**

(b) **Correspondence to the situation**

(c) **Caring attitude**

4. Produced good service quality

As good service is imitated sometimes and customers are tired of the service, a new service should be developed.

(a) **New service should be developed.**

(b) **Uneven quality of service should be reduced.**

(c) **Improvement of customer satisfaction**

5. Increased service productivity

The service productivity is important for service business. The following items are useful for increased service productivity.

(a) **Making service efficient**
(b) **Adjusting supply and demand**
(c) **Using IT (Information Technology)**

6. Service design

The service design in service encounter is done based on the service design concept mainly. If customers' evaluation for service is better than their expectation in advance, they feel satisfaction with the service.

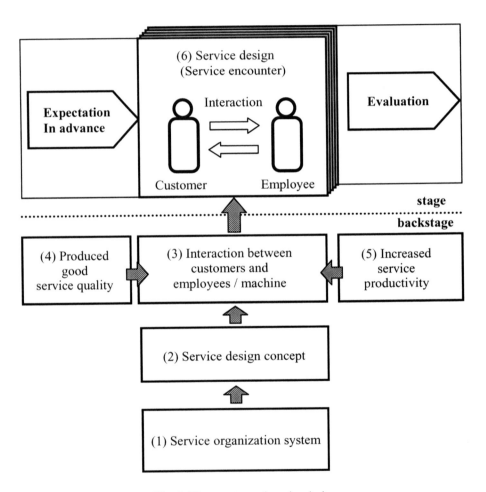

Fig. 2. The structure of service design

4 Two Service Evaluation Method

As the service evaluation method should be practical and easy to use in business, the two service evaluation methods proposed were constructed to use intuitively and easily. The two evaluation methods are as follows.

1. The evaluation based on the difference between user's expectation in advance and evaluation

Generally speaking, ordinary users feel satisfied when they evaluated a service evaluation was good than expectation in advance before using the service. Contrarily they feel dissatisfied when they evaluated a service evaluation was not good than expectation in advance.

The method evaluates a service from viewpoint of expectation in advance and evaluation. Testers estimate and weight a service and a price for a service system such as hotels, restaurants and so on. For an example of a coffee shop, the service weight is 0.8 and the price is 0.2. The weight of the service is divided into the main service and the support service. In case of the coffee shop, the main service and the price were estimated to 0.7 and 0.2. The service and price of the expectation in advance of the coffee shop is evaluated from 5-point Likert scale (table 1). The values of expectation in advance shows 3.2 in table 1 which is summed up according to each weighted values.

Table 1. The expectation in advance

Expectation in advance (S):Service, (P):Price	Service (weigh:0.8)		price [weight:0.2]	grades
	Main service [weight : 0.7]	Support service [weight : 0.1]		
(S):very expected(5 grades) (P):very cheap(5grades)				
(S): expected (4 grades) (P): cheap (4gardes)			0.8	0.8
(S):neither expected or unexpected (3grades) (P):neither cheap or expensive(3 grades)	2.1	0.3		2.4
(S): unexpected (2 grades) (P): expensive (2gardes)				
(S):very unexpected(1 grades) (P): very expensive (1gardes)				
total				3.2

The evaluation after receiving services was done in the same way as estimating the expectation in advance (table 2).

Table 2. Evaluation

Evaluation (S):Service, (P):Price	Service (weigh:0.8)		price [weight:0.2]	grades
	Main service [weight : 0.7]	Support service [weight : 0.1]		
(S):very satisfied(5 grades) (P):very cheap(5grades)		0.5		0.5
(S): satisfied (4 grades) (P): cheap (4gardes)	2.8		0.8	3.6

(S):neither satisfied or
  dissatisfied (3grades)
(P):neither cheap or
  expensive(3 grades)				

(S): dissatisfied (2 grades) (P): expensive (2gardes)				
(S):very dissatisfied(1 grades) (P): very expensive (1gardes)				
total				4.1

2. Service checklist. Checklists are very easy method to extract problems. They are also useful in case of service business. Very simple 7 items as a new service checklist (table 3) are proposed based on the above-mentioned service items.

Table 3. Service checklist

Evaluation items	Check or 5 point scale
1. Consider customers not to feel uneasy when receiving services.	good V bad
2. Service providers have skills or knowledge.	
3. Service providers have a good attitude.	
4. Main services are good.	V
5. Services are efficient.	V
6. Environment is good.	
7. Equipment (Machine) is good.	V

5-point Likert sacle instead of check is also available. 5-point: (1) Strongly Agree, (2) Agree, (3) Undecided, (4) Disagree, (5) Strongly Disagree. The snake plot in the table 3 shows characteristics of a service provided system.

5 Summary

The service evaluation items were extracted from the structure of service and service design. The two evaluation methods were proposed based on these items and the service structure. The one evaluation method is planned to measure costumer's satisfaction using the difference between the expectation in advance and evaluation.

The another method is planned to check or measure seven important items in order to evaluate service provided systems easily and quickly.

References

1. Yamaoka, T.: A study on service design method based on Human Design Technology. In: Proceedings of the 2nd International service Inovation design Conference, pp. 107–112 (2010)
2. Yamaoka, T.: An introduction to Human Design Technology (in Japanese). Morikita publishing company, Tokyo (2003)
3. Yamaoka, T., Matsunobe, T.: Making products user-friendly and charming using Human Design Technology. In: Proceeding of HCI International Universal Access in HCI, vol. 3, pp. 55–59 (2001)

Idea Creation Method for HCD

Kazuhiro Yamazaki

Chiba Institute of Technology,
2-17-1 Tsudanuma, Narashino, Chiba, 275-0016, Japan
designkaz@gmail.com

Abstract. The purpose of this study is to discover a design methodology for user experience design. This paper focuses on design creation method on UCD (User Centered Design) process. After proposing an approach to utilize design creation method, author utilized this method on UCD education for 3rd grade student on university to evaluate propose method. After the result of this education, author got several findings.

Keywords: Design Creation, User Centered Design, User Experience Design, Photo Essay.

1 Introduction

User experience design is becoming more and more important for designing interactive product, IT system and services, but its practical design method has not yet been established.

The purpose of this study is to discover a design methodology for user experience design. User Centered Design (UCD) is one of the best method for User experience design. This paper focuses on design creation method on UCD process. Author categorized 4 approaches for idea creation and described each approaches. Based on this approach, author proposed 3 idea creation method.

2 Design Method for Design Concept

Design concept is a basic design idea across all design process. For example, design concept is including target user, target market, value for target user, user experience, product image, draft product specification, etc.

Following is the design process for design concept;

1. Gathering Information and Clear Goal
2. Drafting Design Concept and Idea Creation
3. Visualizing Design Concept
4. Evaluation for Design Concept
5. Specified Design Concept

I summarized the process and design method for design concept as Table 1.

M. Kurosu (Ed.): Human Centered Design, HCII 2011, LNCS 6776, pp. 160–165, 2011.
© Springer-Verlag Berlin Heidelberg 2011

Table 1. Process and Method for Design Concept

Design Process	Design Method
1. Gathering Information and Clear Goal	Observation Method Interview Method Persona Method Scenario Method
2. Drafting Design Concept and Idea Creation	Problem Solving Approach Vision Proposed Approach Convergent Thinking Support Divergent Thinking Support
3. Visualizing Design Concept	Concept Sketch Paper Prototyping Acting Out Diagram
4. Evaluation for Design Concept	Check List Usability Testing Heuristic Evaluations Impression Evaluation
5. Specified Design Concept	Concept Sheet Concept Catalog Design Concept Specification

3   Approaches for Idea Creation

As showed Fig 1., I identified four approaches for idea creation from human and product viewpoint such as subjective point of view in human, objective point of view in human, subjective point of view in product, objective point of view in product. This paper does not include idea creation by change. The approach for idea creation identified thorough my experience are as follows:

1. The approach for idea creation by human viewpoint subjectively. During user research phase, we will find out many things. All finding is the source of idea.
2. The approach for idea creation by human viewpoint objectively. It is popular to define the persona after user research. The persona is the source of idea and we will be able to create idea by using persona and scenario.
3. The approach for idea creation by product viewpoint subjectively. During watching the product or technology, we will find out many things. All findings is the source of idea.
4. The approach for idea creation by product viewpoint objectively. It is popular to make product road map or technology road map after researching product or technology. The road map helps to create new idea.

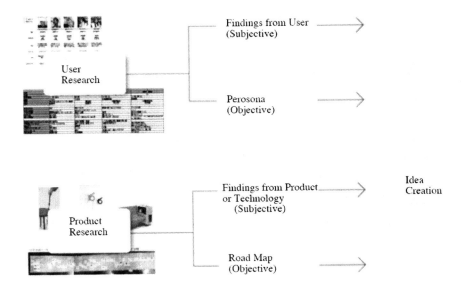

Fig. 1. Approach for idea creation

4   Idea Creation Method

Next, we defined the idea creation method based on the approach for idea creation as follows.

4.1   Utilize User Experience Viewpoint

During idea creation phases, To extend idea, it is effective to make idea from user experience viewpoint. Because, it helps to avoid the idea by stereotypes. For example, if we start to think about new flower base, we already have stereotype of flower base, such as regular shape of flower base. But if we start to think about person's life with flower, we will be able to make many ideas.

To extend idea, one of the approach is to think about idea under specified the context such as time, people, place, reason etc. For example, not to create idea for new flower base under general situation, but also to create idea under the situation that my mother enjoy color of flower at bathroom.

4.2   Utilize User Experience Matrix

User Experience Matrix is the method to create ideas from user experience view point with the matrix. At first, two elements should be selected from basic user experience element such as person, time, place, purpose etc. After selecting tow elements, the matrix should be made by selected two elements. And the cell of this matrix is the place to put the idea under each situation of the column and the row.

As shown in Table 2, this column of matrix was assigned each person and this row of matrix was assigned the location such as living room, kitchen and bathroom. And each cell is the place to put the idea under the situation of the column and the row.

Table 2. Example of User Experience Matrix

	Living room	Kitchen	Bathroom
Grand Mother			
Father			
Myself			
Younger brother			

4.3 Utilize User Scenario

User scenario is the story about user behavior on selected environment and event. The scenario-based design method is easy for designer and developer to understand the relation with user behavior and system. It is utilized to think about new idea and also evaluate idea.

5 Experiments

5.1 Experimental Design

After proposing an approach to utilize design creation method, author utilized this method on UCD education for 3rd grade student on university. To evaluate the proposed idea creation method, 35 students took part in the evaluation experiments. All students are learning design but limited knowledge of eco design. The students performed the idea creation task using three idea creation methods for eco design workshop.

Following is the process to create several ideas for eco design;

- Step1: Creating ideas with sketches from user experience view point
- Step2: Creating ideas with text from scenario design view point
- Step3: Creating ideas with sketches from user experience matrix
  After the experiment, the participants wrote the questionnaire about this workshop.

5.2 Utilize User Experience Viewpoint

To learn idea creation method from user experience viewpoint, students made product sketch and scene sketch. At first, student made product sketch for the flower base which he or she likes. And then, after writing context of the situation, student made scene sketch for the life which someone enjoy flower in his/her home.

The result of this workshop, most students made unique sketch by user experience viewpoint as shown Fig 2.

Fig. 2. Example of Scene Sketch

5.3 Utilize User Scenarios

To learn idea creation method from user scenario, students learned user scenario method and process. At first, student made product sketch for the flower base, which he or she likes. And then, after writing context of the situation, student made scene sketch for the life which someone enjoy flower in his/her home.

Following is the process to create several ideas for eco design;

- Step1: Selecting future scenario from eco design view point
- Step2: Creating ideas with text from scenario design viewpoint based on Step1
- Step3: Creating ideas with sketches for the product which save the world

The result of this workshop, most students made unique idea by user scenario viewpoint.

5.4 Utilize User Experience Matrix

To learn idea creation method from user experience matrix, students selected two element from user experience viewpoint, and then created idea sketch after making user experience matrix. For this workshop, theme was assigned to student as creating 9 sketches for eco design after making user experience matrix.

Following is the process to create several ideas for eco design;

- Step1: Selecting future scenario from eco design view point
- Step2: Creating ideas with text from scenario design viewpoint based on Step1
- Step3: Creating ideas with sketches for the product which save the world

The result of this workshop, most student's created minimum 6 idea sketches by user experience matrix as shown Fig 3.

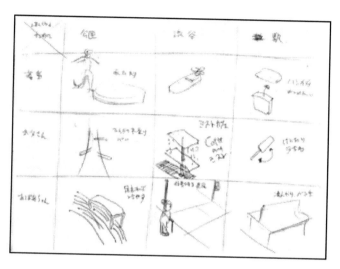

Fig. 3. Example of User Experience Matrix

6   Conclusion

After proposing an approach to utilize design creation method, author utilized this method on UCD education for 3rd grade student on university to evaluate propose method. After the result of this education, author got several findings. In especially, user experience matrix helps student to extend idea and make several ideas.

References

1. HCD-Net Website, http://www.hcdnet.org/en/index.html
2. Nielsen, J.: Usability Engineering. Academic Press, US (1993)
3. Carroll, J.M.: Scenario-based Design–envisioning work and technology in system development. Wiley, US (1996)
4. Carroll, J.M.: Making Use of Scenario-based design of human-computer interactions. The MIT Press, US (2000)
5. Yamazaki, K., Frusta, K.: Proposal for design method considering user experience. In: 11th International Conference on Human-Computer Interaction, Las Vegas (2005)

Vision-Proposal Design Method

Koji Yanagida[1], Yoshihiro Ueda[2], Kentaro Go[3], Katsumi Takahashi[4],
Seiji Hayakawa[5], and Kazuhiko Yamazaki[6]

[1] Kurashiki University of Science and the Arts, Kurashiki, 712-8505, Japan
yanagida@arts.kusa.ac.jp
[2] Fujitsu Design, Ltd., Kawasaki, 211-8588, Japan
y.ueda@jp.fujitsu.com
[3] University of Yamanashi, Kofu, 400-8511, Japan
go@yamanashi.ac.jp
[4] Holon Create Inc., Yokohama, 222-0033, Japan
takahasi@hol-on.co.jp
[5] Ricoh Company, Ltd., Yokohama, 222-8530, Japan
hayakawa@rdc.ricoh.co.jp
[6] Chiba Institute of Technology, Narashino, 275-0016, Japan
designkaz@gmail.com

Abstract. The "Vision-proposal Design Method" discussed in this paper is a practical method for designing in an age of ubiquitous computing. This comprehensive method makes possible new and innovative services, systems and products that are currently unavailable, as well as proposing advances for those that currently exist. It encompasses the entire HCD (Human-Centered Design) process, and presents a new vision with experiential value for both user and business from an HCD viewpoint. It creates specific ideas for services, systems and products while also delivering their specifications. This paper reviews evaluation results of its utility and effectiveness through a brief summary of the method with examples of its application.

Keywords: Structured scenario-based design method, vision, scenario, persona.

1 Introduction

In the present day, product development for matured markets requires a research method of user needs that even users do not yet anticipate. In the case of products using ICT (Information Communication Technology), considered to be ubiquitous computing, a designing method is needed which meets users' intrinsic needs while avoiding functions and performance which do not engage those needs, such as those referred to as "Osekkai" [1]. Furthermore, in order to create attractive experiential value, it is necessary to develop services, systems and products not from the viewpoint of technical elements, but from the viewpoint of value to be provided.

Under such circumstances, it often happens that the problem-solving design approach for existing services, systems and products no longer works sufficiently, and therefore a new design approach is expected as a complement. This is a vision-proposal design approach that can create new services, systems and products which

M. Kurosu (Ed.): Human Centered Design, HCII 2011, LNCS 6776, pp. 166–174, 2011.

are sure to be introduced and attractive to people and society in general. It proposes new visions of provisional values that meet users' intrinsic needs from the viewpoint of HCD (Human-Centered Design).

Since 2007, the authors organized a working group within the Ergonomic Design Research Group of the Japan Ergonomics Society. This research has established the vision-proposal design approach, making it a practical and serviceable methodology useful for development of future generation services, systems, and products from the viewpoint of HCD. In 2009, we completed development of the "Structured Scenario-based Design Method" [2], in which a scenario was utilized consistently for the development process. This method helps to create ideas for services, systems and products from the provision values, and provides specifications utilizing personas and structured scenarios in three stages.

In addition to the Structured Scenario-based Design Method, the various methods effective for the vision-proposal design approach are widely used in the field of design, especially in information design. They include interview and observation methods for getting user data, the Superior-Subordinate Relationship Analysis Method [3] and the KJ Method[4] used for abstracting and/or structuring the users' intrinsic needs, and the Persona Method or Scenario Method [5] which create new provision values and user experiences. However, since they are used for only a portion of the design processes, it is difficult to consistently utilize HCD activities. Also, how and where in the design process each method can be used appropriately is another issue.

Therefore, we continued research with the aim of developing a comprehensive design method which consistently allowed for the introduction of HCD into the design process by advancing our Structured Scenario-based Design Method. The result is the "Vision-proposal Design Method." We discuss the outline of this method and evaluation results for successful findings.

2   Vision-Proposal Design Method

2.1   Characteristics of the Vision-Proposal Design Method

The Vision-proposal Design Method is a comprehensive design method which covers an entire HCD process. It is used for proposing new services, systems and products, as well as for making new proposals for those that currently exist. This method facilitates the creation of new visions of experiential values that can be embraced by people and society by offering practical ideas for services, systems, products and specifications based on HCD consistently. It emphasizes values that both parties, provider and receiver, can share. In particular, this method intends to satisfy both viewpoints of business policies (profits for business) and users' intrinsic needs (user experience). Further, a repeating process of emanation (creation) and convergence (evaluation) is stressed to arrive at new ideas. In this way, personas and scenarios are utilized proactively as a means of expression for an exact sharing of ideas. In addition, there are a sufficient number of templates prepared as hands-on tools.

The following are summarized characteristics of the method:

1. Gain ideas from value level.
2. Clarify the users' intrinsic needs.
3. Drive the specifications of services, systems and products from users' intrinsic needs.
4. Repeat visualization and evaluation by stages.
5. Consider collaboration with experts having specialties in other fields (user viewpoint / business viewpoint).

2.2   Basic Model for the Vision-Proposal Design Method

At its core, the basic model for the Vision-proposal Design Method uses the Structured Scenario-based Design Method. The user image is gradually elaborated to the personas along with users' intrinsic needs and business offering policies, then input to the core. Utilizing personas and scenarios through the Structured Scenario-based Design Method, particular ideas for services, systems, and products as well as their specifications are output. Figure-1 illustrates the basic model for the Vision-proposal Design Method.

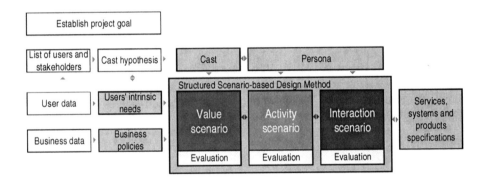

Fig. 1. Basic model for the Vision-proposal Design Method

2.3   Process and Elements of the Vision-Proposal Design Method

The Vision-proposal Design Method can be accomplished in five major steps. Various existing methods used for HCD are utilized in each process. An outline of each process and its relative activities are shown together with a sampling of the templates developed as hands-on tools. (Fig. 2-7)

1) Establishing project goal
At the beginning of a project, user experiences and business themes must be confirmed to ensure accuracy of the project goal.

2) Establishing users' intrinsic needs and business policies
The users' intrinsic needs are identified by user-related research data. Various existing research methods can be applied. For example, questionnaires as a

quantitative approach, or photo diaries, photo essays [6], interviews, or observation as a qualitative approach. Analyzing user data using appropriate methods such as the Superior-Subordinate Relationship Analysis Method to give a hierarchy of needs, users' intrinsic needs are clarified at its uppermost realm. The business offering policies are identified as project policies based on business data including business domains, possessive technologies, business strategies and business environments.

3) Establishing target users and creating personas

In the Vision-proposal Design Method, ideas are developed in the sequence of "project target setting," followed by "values to be provided," then "user activities" and finally "interaction." Target users are elaborated in the sequence of user list, hypothetical cast, cast and Persona. By linking the target users and each scenario in respective stages, the process of emanation and convergence of idea generation can be achieved with creativity. Persona is especially useful when embodying the user's experience into activity scenarios and interaction scenarios. Personas help to describe concrete users' activities, behavior patterns and goals in the scenarios.

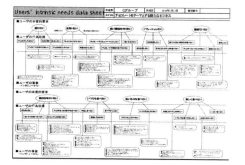

Fig. 2. Template: Users' intrinsic needs sheet

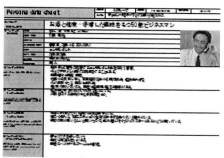

Fig. 3. Template: Persona sheet

4) Creating and evaluating ideas using Structured Scenario-based Design Method

The Structured Scenario-based Design Method is used to create new ideas for services, systems and products from the view points of HCD and to deliver their specifications by illustrating scenarios meeting the three structured layers step-by-step. After confirming requirements, creation of ideas, scenario development and evaluation are conducted in each stage.

i) Creation of ideas

Ideas are created hierarchically in three steps through the scenario development. Vision-proposed ideas that go beyond problem solving can be created from the viewpoint of superior provisional values. These ideas confirm consistency with the current scenario as well as those before and after it through repeating processes.

ii) Scenario development

Ideas are visualized in the form of a scenario. Scenarios are described by three steps; the "value scenario" describes the values provided to users by business, the "activity scenario" focuses on users' experiences, and the "interaction scenario" shows the interaction between users and systems and products. The activity scenarios and interaction scenarios illustrate user experiences in detail through personas' activities and interactions. Furthermore, user experience is visualized effectively using a wide range of expression, including sketching, paper prototyping and acting out.

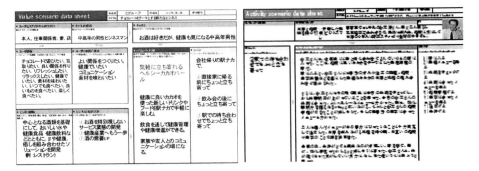

Fig. 4. Template: Value scenario data sheet   **Fig. 5.** Template: Activity scenario data sheet

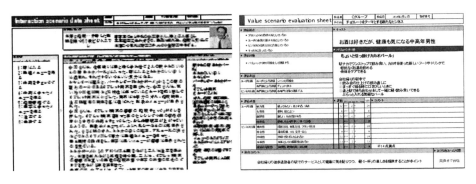

Fig. 6. Template: Interaction scenario data sheet   **Fig. 7.** Template: Value scenario evaluation sheet

iii) Evaluation

The evaluation of ideas is a key process which determines the achievement of the project. The objects of evaluation are structured scenarios characterized and described in three layers. Each scenario is evaluated at each stage, then revised after reflecting on the results of evaluations. The evaluation criteria consist of two aspects: the user's viewpoint, including attractiveness, effectiveness and efficiency, and the business viewpoint, including strategic characteristics, sociality, marketability and business feasibility. The weight of evaluation criteria of each aspect can be changed to meet the project goal. User evaluation and expert evaluation, such as check lists, are available as evaluation methods. The ideas will be fitted to the project goal by repeating the creation and evaluation of ideas using the structured scenarios. There are five points of evaluation, as follows:

1. Evaluate from the upper stage
2. Evaluate structured scenarios by each stage
3. Visualize ideas for evaluation
4. Clarify the evaluation criteria
5. Evaluate from both user viewpoint and business viewpoint

Since scenarios are illustrated hierarchically by each persona, a high volume of scenario descriptions are required if we attempt to systematically illustrate all possibilities. However, the purpose of this method is not to cover all potential solutions, but rather to create attractive ideas that satisfy the project goal from the viewpoint of HCD. At this point, we narrow down on the attractive ideas generated by the results of evaluations.

5) Realizing specifications

Based on the requirements from interaction scenarios, the services, systems and products specifications are discussed while considering consistency with the three structured scenarios. For example, use case scenarios can be used. The major objects of description are technical elements, which are evaluated from a technical perspective.

2.4 Expected Effects of Vision-Proposal Design Method

Benefits of the Vision-proposal Design Method include;

1. Easy-to-use products and services can be developed
2. Customer value can be discovered as a resource for effective competition in future generations.
3. Acceleration of development and cost savings can be designed.
4. Development of high sales merchandise and attractive services become possible.
5. Vectors of business become clearly defined, promoting effective business management and forecasting.

3 Application of Vision-Proposal Design Method

We conducted an evaluation workshop to apply and to verify the utility and effectiveness of the method.

3.1 Overview of Evaluation Workshop

An outline of the evaluation workshop is shown below:

1. Theme: Proposal of design of new services, systems and products using "Chocolate"
2. Participants: 43 individuals (including product designers, user interface designers, information designers, usability and computer experts, university faculty and students) divided into six groups.
3. Process: After a lecture about the method, people split into groups and worked. Using fourteen types of templates prepared for different themes, practical ideas were created and presented in accordance with the processes of

this method. Due to time constraints, the policies of provision of Persona and business were prepared in advance.

4. Term: Two days (Afternoon on the first day and morning of the second day)
5. Method of evaluation: Subjective evaluation using questionnaires for all participants of the workshop after practicing the method, and participant observation by one member of each group, those being the individuals who developed the method.

Following is one of the design proposals presented from the six groups which gained the highest score as a result of mutual evaluation among participants.

Value scenario: "Healthy Cacao Bar – Right around the corner." It is possible to enjoy healthy food and drinks containing cacao in a convenient location, such as "Ekinaka," the shops found in train stations, and promote good health through food and drinks. It can also be a good place to meet and communicate with family and friends.

Activity scenario: (abbreviated) When Mr. and Mrs. Yamamoto entered the bar, the bar staff recommended a combination of dishes with an appropriate Cacao liquor to enjoy with their meal, based on their preferences, record of past orders, and health conditions. Furthermore, Mr. Yamamoto came to know that his body fat percentage had slightly increased after performing a simple physical check at the bar counter. He understood that the menu recommendation for him was healthier based on his current state of health. (Abbreviated) Satisfied with the personalized menu, they placed their orders accordingly. (Abbreviated)

Interaction scenario: (Abbreviated) The "Recommended Menu," which was selected based on information such as preferences, health information and order history, was displayed on the tablet terminal. Mr. Yamamoto pressed a button labeled "Physical Check-up" on the screen. By gripping the handles of the tablet and pointing it towards his face, the instrument analyzed his palms along with his face image and presented a reading of his current physical body condition. His body fat seemed to have increased, and the "Recommended Menu" was updated to display healthier items. He customized the listing and created a menu optimized to his preference, type of cacao flavor, and strength of alcohol. (Abbreviated)

The Fig. 8 illustrates this idea based on the delivered specifications. It realizes a new vision and an experience at a bar concerned with health and entertainment. Food and drinks may be ordered based on both users' intrinsic needs and business policies.

Fig. 8. Image of the case example

3.2 Evaluation Results

Among all workshop participants, 32 people answered the questionnaire. The effectiveness and satisfaction of respective processes were evaluated using five rankings. The percentages of grade-four or higher are shown below:

- Mutual interview and laddering :Effectiveness: 84.3%,   Satisfaction: 74.9%
- Business offering policies   :Effectiveness: 65.5%,   Satisfaction: 53.0%
- Value scenario   :Effectiveness: 71.8%,   Satisfaction: 71.8%
- Activity scenario   :Effectiveness: 65.5%,   Satisfaction: 53.0%
- Interaction scenario   :Effectiveness: 71.8%,   Satisfaction: 59.3%

These results show that the participants recognized high-level estimation for effectiveness and satisfaction. Also, in answer to the question, "Do you wish to implement the Vision-proposal Design Method at your workplace or school?," 78.1% answered "Yes," 0% answered "No," and 9.3% answered "No opinion." It was understood from this response that people highly evaluated the general effectiveness and efficiency of the method.

According to the free comment column, the following points were especially valued: "It is a method based on HCD, one searching for users' intrinsic needs and utilization of the Persona scenario, where both user and business aspects have been considered. Further, it is a comprehensive method where specifications are output when user and business data are input." However, some opinions were as follows: "It takes time to implement the entire method. It becomes confusing when dealing with the evaluation criteria of the scenario, and the relationship between templates is difficult to understand. The method itself is fundamentally sound, but there are issues with the templates which must be considered for future improvement."

In findings from participant observation, there was no deviation in their subjective evaluation. There were some difficulties, such as describing three scenarios differently and then performing scenario evaluation. These can be improved by more effectively guiding the procedure among the functions of templates. As for the size of the entire method being too large, development of simplified versions may be required, along with a guide function enabling simple customization of processes and methods within various projects.

The results of the evaluation workshop show that features of this method are realized, and together with the above-mentioned results, the utility and effectiveness of the method has been confirmed.

4 Summary and Future Work

This paper has introduced the Vision-proposal Design Method, a comprehensive design method which illustrates a vision having new experiential value with regard to both user and business aspects from the viewpoint of HCD. It creates specific ideas for services, systems and products, while also delivering their specifications. In addition, the results of application using templates as practical tools, along with its utility and effectiveness were confirmed.

We continue our activities to verify the method and to further its usefulness by setting up a SIG as well as holding practical workshops and exchanging information.

The application of this method is already being used in various locations. In education, there are currently five universities utilizing this method in their design classes with more considering its adoption. In business, its utilization is spreading as well. We intend to advance this method through a wide range of practices.

References

1. Ueda, Y., Takahashi, K., Go, K., Hayakawa, S., Yanagida, K., Yamazaki, K.: Perspective of the Structured Scenario-based Approach for User Experiences. The Japanese Journal of Ergonomics 46 Supplement, 108–109 (2010)
2. Yanagida, K., Ueda, Y., Go, K., Takahashi, K., Hayakawa, S., Yamazaki, K.: Structured scenario-based design method. In: Kurosu, M. (ed.) HCD 2009. LNCS, vol. 5619, pp. 374–380. Springer, Heidelberg (2009)
3. Marketing Concept House: Group Dynamic Interview (in Japanese). Dobunkan Shuppan, Tokyo (2005)
4. Kawakita, J.: The way of conception (in Japanese). Chuokoron-shinsha, Tokyo, pp. 151-187 (1967)
5. Carroll, J.: Making Use: Scenario-Based Design of Human-Computer Interactions. The MIT Press, Cambridge (2000)
6. Go, K.: Scenario-Based Design for Services and Content in the Ubiquitous Era. Journal of the Society of Instrument and Control Engineers 47, 82–87 (2008)

Part II

Mobile and Ubiquitous Interaction

Possibilities for Cultural Customization of Mobile Communication Devices: The Case of Iranian Mobile Users

Bijan Aryana[1], Casper Boks[1], and Azadeh Navabi[2]

[1] Department of Product Design, Norwegian University of Science and Technology,
Kolbjørn Hejes Vei 2B, 7491 Trondheim, Norway
[2] Department of Computer Science and Engineering, Chalmers University of Technology,
SE-412 96 Göteborg, Sweden
{Bijan.Aryana,Casper.Boks}@ntnu.no, azadehn@student.chalmers.se

Abstract. Global producers of mobile communication devices recognize the importance of cultural differences in the emerging markets; however it seems that the main concentration in both academic and business areas is on the large number of users with low incomes, while users from other classes of these societies are not studied well. In this study after set of integrated reviews on areas of Mass Customization, New Product Development and Mobile HCI an experiment was planned based on the unexplored aspects of users' culture and mobile communication devices relationships. A number of young educated users from middle class tested a new smart phone during its marketing process in Iran. They were sampled based on a global producer's marketing program. After a phase of self documentation, users selected two applications of the device for the usability tests and found culture related usability problems during the tests. Finally they proposed solutions in a participatory design process.

Keywords: Cultural Customization, Mobile HCI, New Product Development, Mass Customization, Smart Phones, Emerging Markets.

1 Introduction

Nowadays mobile communication devices are not only tools for mobile communication, but also means of mobile computation. In recent years global producers have concentrated on exploring markets in developing countries and their specifications. A reason for this attention is the saturation of mature markets in developed countries. Also the phenomenon of the 'next billion mobile users' in emerging markets (EMs) is an opportunity for producers of mobile communication devices [1]. Because of the cultural and economic heterogeneity of EMs and developing countries, cultural differences became an important issue in the user research of mobile devices. At the same time there is a similar interest among academics, especially in the Base of the Pyramid (BOP) concept which represents the low-income and high-population potential users in developing countries [2]. However in some developing countries, *middle class* users who usually live in urban areas are a major part of the mobile users. These users do not fit into the definition of BOP users as they have their own cultural specifications which

M. Kurosu (Ed.): Human Centered Design, HCII 2011, LNCS 6776, pp. 177–186, 2011.
© Springer-Verlag Berlin Heidelberg 2011

influence their use of mobile devices. It seems that this group of users has received limited focus in culture related studies. Furthermore, according to an integrated review about culture and mobile HCI, there are interesting references about the influence of users' cultural background in both HCI and mobile HCI domains, but only few of those research studies suggest processes or models for culture oriented design [3]. This paper aims to contribute to the lack of academic literature discussing cultural aspects in product design, especially design of mobile communication devices, by presenting a case study focusing on Iranian users. Iran is a country with 68 percent urban population, ranked 19th in the world in terms of purchasing power parity [4]. These characteristics along with specific cultural, social and politic characteristics make Iran a suitable alternative for studying middle class mobile users in the context of cultural differences. The case study has been done in cooperation with the marketing department of a global producer of mobile communication devices, providing access to lead users of this product category.

The main research questions for this study were formulated as follows:

1. What may be, for global producers of mobile devices, the most important aspects to focus on in terms of market segments, products, product characteristics or components, when they want to consider cultural differences in Iran in the design of their products?
2. What are the opportunities for the cultural customization from the perspective of a global producer? Are there any specific needs, characteristics or problems that should be addressed in pursuing these opportunities?

Answers to the above questions are used as input for a more general discussion on cultural customization of mobile communication devices for middleclass users in Non-Western markets, considering the current marketing and manufacturing contexts.

2 Case Study Approach

In preparation of the case study, a literature review was done to explore current the understanding of how the cultural backgrounds of users, in particular those in emerging economies, may play a role in their use and appreciation of mobile communication devices such as mobile phone, digital assistants, etc. The review, presented in section 3, suggested unexplored areas of research, and provided as such input for the case study reported on in section 4. In the case study, a combination of semi-structured interviews, self-documentation and usability tests has been used, and where selected based on available resources and time. The case study was done in the context of the launch of a new smart phone in Iran by a global manufacturer, allowing for access to lead users and participation in user studies.

3 Literature Review

How culture may affect the design of mobile communication devices, and vice versa, can be viewed from different perspectives, such as Mass Customization, New Product Development (NPD), mobile Human Computer Interaction (HCI), and this section explains how elements from each perspective have been used for design the case study presented in this paper.

3.1 Mass Customization and Cultural Customization

The shift from Mass Production to Mass Customization shows that users' needs and desires are key parameters that not only shape the design of products, but also manufacturing systems [5]. Mass Customization has two main requirements for it to become relevant: modularity of the products' components and existence of a configuration system which can manage the production of customized products based on mass produced components [5]. Because of its specifications, the mobile communication device industry can be a good candidate for mass customization, as both of above characteristics of mass customization potentially exist in this industry [6]. In the context of finding solutions for considering users' cultural specifications in mass customization, Marcus has suggested a number of recommendations for individual components of user interface design of websites, according to the cultural context [7], using Hofstede's cultural model. This model represents cultures by five dimensions appointing a set of scores for these dimensions for different countries, allowing comparison of different countries by their scores [8]. Marcus' idea was later used in a more comprehensive model by Röse [9] for human machine systems. In both models, there is a concentration on individual components of the design. However in an experiment about cultural customization of mobile communication devices with a similar approach, results showed that it is not possible to predict users' final decisions about a product, only by measuring their tendencies towards individual product components [6]. Users usually look at the product as an independent entity, which is different from a simple combination of its components. That is why the case study presented in this paper focuses on qualitative research with an in-depth study on users.

3.2 New Product Development and Culture

A literature review of scientific articles was done by the authors, of which publication is forthcoming. The review suggests that consumer culture is often considered in the last phases of NPD, when a product is developed and is going to be launched in the market. Especially the diffusion process is a core part of these research studies. In the diffusion process a new product or idea is gradually accepted by the customers in a market [10]. Customers can be categorized as innovators, early adopters, early majority, late majority, and laggards according to the time that they adopt and accept a new product. Other studies have explored other phases of NPD such as the design and conceptualization of products. These studies suggest methods like concept testing [11] for understanding users with specific cultural backgrounds, instead of attribute based methods such as Conjoint Analysis or Quality Function Deployment (QFD). Base of the Pyramid (BOP) consumers are of great importance to studies which target EMs. These consumers usually cover a large population and have low income [2]. However, it seems that there is less attention to more elite consumer segments in these countries which usually have higher incomes but represent a smaller segment of the population [2]. This is one of the unexplored areas which is considered in the experimental phase of this research.

3.3 Mobile HCI and Culture

A similar review was done of articles in the mobile HCI domain [3], targeting HCI and mobile HCI articles that discuss users' cultural differences. The review suggested that a large number of research studies relied on the Hofstede's cultural dimensions for defining culture, while there was just one example of defining specific dimensions and attributes for defining the culture in the context of consumer electronics [12].

In addition, a majority of the data which is used in case studies in these articles is collected using conventional research methods like questionnaires. Methods like user research through observation, verbal protocols, heuristics, cognitive walkthrough, and post-event protocols were only used in few studies. In short, the majority of articles reviewed did not propose a solution or model for culture oriented design. In addition there are few examples of user research in this area of research, although it is recognized that more samples may exist in the industry which are not published in academic literature. To go a step further, as it will be explained in the next section, the user research phase in this study includes observations and usability tests. The user studies were done in the context of an actual new product launch, providing opportunities to obtain a good understanding of the business and industry context.

3.4 Guidelines Resulted from the Theoretical Review

The literature research led to the following aspects that were considered in the design of the case study presented in this paper:

1. Mobile users in EMs who do not belong to the BOP were not a core part of the studies. This may however be an area for further research.
2. Understanding the architecture of products and their components is necessary for developing a customization process, however in order to consider users' cultural specifications relying on users' tendencies towards attribute and components is not enough. It is important to study the way that users interact and think about the product as a "whole" (not just a combination of its attributes).
3. Conducting user research in collaboration with industry is a relatively unexplored area in the context of addressing cultural aspects in product design, which can bring interesting opportunities for investigating users' culture and mobile communication devices relationships.

4 Case Study Design

In December 2010 and January 2011, a case study was done in cooperation with a large global manufacturer of mobile communication devices. This Original Equipment Manufacturer (OEM) has a representative office in Iran, which has as main responsibility to market its new products and to provide after sales services and support. The cooperation allowed for interviews with the company's marketing team, and access to users that were selected from a sample of users known to be early adopters of new technologies. They typically belong to middle and upper middle economic ranks, and not the BOP. They usually have an influence on other market segments markets; therefore they are good choices for marketing research when targeting all segments of the market is not feasible.

4.1 Semi-structured Interviews with the Company's Mobile Marketing Team in Iran

In order to get a deeper understanding of the company's industry and business context, four marketing managers for mobile phones in Iran were interviewed. The main insights from these interviews included:

- **Relationships between Cultural Specifications and Market Behavior**

No examples exist of intentional customization of mobile communication devices for the Iranian market. There are however a few cases where a product was adopted very fast in Iran and achieved high sales rates in comparison to other countries in the world. This success (which was not predicted) made that the Iranian market was selected as the lead market for the next generation of that product, and therefore the main consumer research activities were conducted in Iran. However, even in these cases, the products that were developed based on this consumer research were presented globally and not only in the regional Iranian market.

- **Main Categories of Users in EMs, and Their Situation in Iran**

The company's main marketing strategy for mobile phones is the early concentrating on the "Innovator" and "Early adopter" segments of consumers. Tehran is the lead regional market in which the products are launched first and usually products which are successful in Tehran will be successful on the national level as well. Therefore users who belong to the BOP are not a core part of the marketing activities. Unlike countries like India or China these users do not cover an important segment of Iran's market in terms of population, and the middle class has a priority.

- **Smart Phones and Ordinary mobile phones in Iran**

For the company, a relatively new brand in Iran, it is relatively hard to find its way in this segment of the market for ordinary mobile phones, given the dominant presence of two other globl OEMs. Therefore marketing activities are concentrated on smart phones as this market has more space for new players. At the time of the interviews, a Word Of Mouth (WOM) program was developed for one of the company's smart phones. This product is intended for users who want to shift from ordinary mobile phones to smart phones. In the WOM program, 200 users aged between 18 and 35 were provided with a free smart phone. These users were selected based on their social networking capabilities, their interest in using consumer electronics and the average time that they usually spend for the electronic communication during a day. In addition, all these users were first time users of smart phones.

- **Recommendations for the Main Areas of Customization**

According to the initial marketing research on smart phones, the main competition is in the area of secondary features of smart phones. Secondary features usually deal with non-communication functions of mobile phones, such as multimedia, web and entertainment. Consequently, primary features include functions such as call and SMS which are basic communication requirements in a mobile phone. For most producers, after several years of development these primary functions are quite mature today and the computation side of mobile communication devices is more in demand. Moreover, because of the nature of smart phones, software components were recommended for the customization process, since in their opinion the scale of market in Iran is not large enough for customized hardware components.

These insights from the interviews provided answers to the first research question, and as such the background for the user tests which were done to obtain insights in the opportunities for cultural customization of smart phones for the Iranian market.

4.2   Usability Testing

A team of 15 volunteers selected from the 200 users in WOM program participated in the experiment, before WOM program was started. While these volunteers had the general specifications of the participants of the WOM program, they have also an additional characteristic, which was their educational background. In order to play the user/designer role, they had backgrounds in computer engineering, industrial design and graphic design. They participated in the usability tests which were of a participatory design nature [13]. The participants were invited to share their design ideas before and after the tests, revealing a number of solutions for the existing deficiencies of the smart phone that were identified through the tests.

The next subsections will present the steps of participatory design process.

Self-documentation and Selection of Features for Cultural Customization. Before starting the main tests, the participants were asked to explore similar products without any limitations. It should be mentioned that the test product was not available in Iran's market at that time, which was two months before the start of WOM program.

The participants used different ways for the exploration like testing similar products in showrooms, informal interviews with friends and relatives who were using these products, and reading online reviews. At the end of their research which took two weeks, they self-documented their results and gave suggestions about which components or features might be suitable for customization on the Iranian market. At the beginning of the user tests of the main smart phone model under consideration, a summary of self-documentation reports was presented to the participants, as well background information from the company itself. These two sources include a set of components and features which can be potentially customized for Iranian users. The main suggestions were SMS, music player, maps, GPS, and security features. After a discussion and voting, SMS and music applications were selected for the cultural customization process by the participatory design team. Although the marketing team preferred the secondary features for the study, the self-documentation studies showed that Iranian mobile users use SMS as a social networking tool, and find current SMS features inefficient. Another issue indicated by the self-documentation studies were problems with music files in Iran, which usually do not have correct and complete music tag information (such as artist's and album's names, year, and genre). Therefore Iranian users usually use folder management to browse and play music files, while many recent music player applications of mobile communication devices use music tag information for browsing and playing tasks.

Aim of the Usability Tests. The experiences from the self-documentation studies were an important part of the usability tests. As the main findings from using a range of different smart phones were that 1) Iranian users encounter problems with the current

SMS application particularly in forwarding messages, and managing the previous sent or received messages, and 2) Iranian users do not have access to the music files by folders in current music applications on the device and this will cause some usability problems, two tasks were designed for the usability tests accordingly. In addition, short structured interviews were designed focusing on their current use of SMS and music player applications on their phones and portable music players. The tasks and interviews were modified and finalized after a pretest.

Usability Tests and Structured Interviews. Participants were tested under the same situations. For each test, a representative from the research team was responsible for guiding the participants' behavior while he or she was performing the tasks, whilst another facilitator recorded a video during the tests. It was tried to capture the users' fingers and the smart phone interface when they perform the tasks. The camera was fixed on a tripod focusing on the smart phone interface and the users' fingers when they performed the tasks. In addition, short structured interviews were done before each test.

Analysis of the Results. A review of the results of the interviews with the participants showed that the majority of participants use SMS for social networking, and similarly they prefer using folder management for browsing music on their current portable music players instead of using tag information. After the analysis of the tests' recorded videos, a number of common errors were observed. There were 12 common errors in the SMS tests, and 9 common errors in the music task. These errors not only confirmed the usability problems found through the self-documentation studies, but identified several other problems.

For instance, although forwarding the messages is a common habit in SMS social networking, the forward function is just accessible by keeping the finger on a message for few seconds, and it is not available by a single touch or using the options button. Therefore none of the participants were able to forward a message during their first experience with the smart phone, and other SMS related functions were more accessible. Apprently, the forward function was considered as a function which is not going to be used frequently by users. However, according to the self documentation by the participants, and also according to the interviews, Irians users often prefer a kind of group SMS communication instead of pair communication. This can be related to some cultural specifications such as collectivism.

In the music related task, all participants had problems in finding and playing all tracks of an artist, as the music files did not have standard tag information.

Gathering of Requirements and Survey Design. The same participants attended in two problem solving sessions. In these sessions they were asked to 1) sketch new structures for SMS and music applications in which current features and functions are better arranged; and 2) create new ideas for features related to these tasks. For example, participants classified all of the functions in the SMS application in three access levels in a way that social networking functions (such as forward) were more accessible. They also propose a new feature for sending fiestas' or national ceremonies' greetings, in which the device automatically adds the recipients' names to the beginning of a message which is sent to multiple contacts. An option for browsing music files through folders was also suggested for the music player application. Finally five modifications and five new features were proposed. The results of requirement gathering step were then translated to a survey which will be answered by the first 200 users of the device, who are selected for the WOM program. Five modifications are presented by modified user interface designs and five new ideas are described by text in the survey. The survey is designed in the form of a scalar questionnaire so users will show their tendencies towards each new or modified item. This article was written when the results of the survey were not available yet.

5   Discussion

In the real conditions, global producers possibly target the more innovative groups of users who can influence other parts of the society. Therefore new products which are accepted by these users, have more chances to be used by other groups. This is a point that user research, even in small scales can have a considerable influence. Another advantage of concentrating on more innovative users is more variety in possible research methods. Because of their relatively high level of education and knowledge, it is feasible to find participatory design teams in which participators can play designer-user roles. Recommendations by the marketing and participatory design team showed the domination of software side in smart phones. Flexibility of mobile applications and mobile operating systems for customization can bring more opportunities for cultural customization as well, even after purchase. Looking at the wide range of features in a smart phone, usability tests with only two features showed that there are a large number of usability problems which can be connected to the cultural or regional differences. These problems sometimes are tied together in a way that it is not possible to draw a border line between regional and cultural ones. For example, some participants who mentioned SMS social networking in their self documentation reports, referred to Iran's government restrictions for social networking websites as a possible reason for this behavior. However, also the tendency of Iranian culture towards collectivism may be a contributing factor [8].This combination of intercultural variables and cultural specifications has been noted before. In her solution for culture oriented design, Kerstin Röse recommended two phases for development of a culture-oriented human machine system. The first phase can cover intercultural variables (like the restrictions by Iran's government), and the second one includes looking at cultural specifications (such as collectivism) [9].

6   Conclusion

Although at the time of writing this paper, the final survey results were not available yet, the study suggests that through human centred design can reveal clues for cultural customization of mobile communication devices. Using a combination of different human centred design research methods, such as self-documentation, video recording, and interviews, a number of usability problems were identified for Iranian users that may be attributed to cultural factors. Using the same user sample as was been used for marketing research purposes facilitated the access to users whose characteristics were very suitable for the user studies As it was shown in this study, the global OEM, focused on more innovative users for a WOM program mainly because of their influence on other segments of the market. This WOM program was used to arrange a participatory design team which was able to participate in the customization process. This is a process that can be experienced in similar situations. While the marketing team samples a group of innovative consumers, the same sampling can be used for a participatory design team. The new product can be tested and recommendations for the customization will be available before the launch of the product in the market. Considering the flexible software features on smart phones, applying these recommendations in the final version of the product can be feasible.

To conclude, the research questions can be answered as follows:

1. This example shows how one of the global producers in Iran used marketing strategies which were more focused on innovative users and advanced products. Although other producers can have different strategies, it is possible to conclude that in the case of mobile communication devices, the main concentration of global producers in EMs is not only on ordinary mobile phones and BOP users. Advanced products and innovative users are also important in these markets.
2. Because of the computation abilities of smart phones and flexibility of applications and operating systems, these devices are more flexible for cultural customization. The tests showed that there are usability issues in both entertainment and primary communication features in the selected device that can be related to the cultural and regional specifications. This can be true for other similar products as well.

Further research studies can be done in multiple countries to make comparisons possible.

Acknowledgments. The authors would like to appreciate Mr. Farzin Shariati Corporate Marketing Director and Mr. Babak Amjadi MC Marketing Manager of LG Electronics, LGEIR, Tehran Office because of their great help and support during the study.

References

[1] White, G.: Designing for the Last Billion. Interactions XV.1, 56–58 (2008)
[2] Alden, D.L., Steenkamp, J.-B.E.M., Batra, R.: Consumer attitudes toward marketplace globalization: Structure, antecedents, and consequences. International Journal of Research in Marketing 23(3), 227–239 (2006)
[3] Aryana, B., Øritsland, T.A.: Culture and Mobile HCI: A Review. In: Norddesign 2010 Conference, vol. 2, pp. 217–226. Gothenburg (2010)
[4] Hvam, L., Mortensen, N.H., Riis, J.: Product Customization. Springer, New York (2008)
[5] CIA world fact book,
   http://www.odci.gov/cia/publications/factbook/country.htm
[6] Aryana, B., Boks, C.: Cultural customization of mobile communication devices' components. In: International Design Conference - Design 2010, vol. 1, pp. 137–146. Dubrovnik (2010)
[7] Marcus, A.: User-interface design, culture, and the future. In: Working Conference on Advanced Visual Interfaces, AVI, pp. 15–27 (2002)
[8] Hofstede, G.: Culture's Consequences: Comparing Values, Behaviors, Institutions and Organizations across Nations. Sage Publications, Thousand Oaks (2002)

[9] Rose, K.: The Development of Culture-Oriented Human Machine Systems: Specification, Analysis and Integration of relevant Intercultural Variables. In: Advances in Human Performance and Cognitive Engineering Research, vol. 4, pp. 61–103. Elsevier, Amesterdam (2004)

[10] Rogers, E.: Diffusion of Innovations, 5th edn. Free Press, New York (2003)

[11] Vallaster, C., Hasenöhrl, S.: Assessing new product potential in an international context: lessons learned in Thailand. J. Consumer Marketing 23(2), 67–76 (2006)

[12] Choi, B., Lee, I., Kim, J.: Culturability in Mobile Data Services: A Qualitative Study of the Relationship between Cultural Characteristics and User-Experience Attributes. International Journal of Human-Computer Interaction 20(3), 171–203 (2006)

[13] Rosson, M.B., Carroll, J.M.: Usability engineering: scenario-based development of human-computer interaction. Morgan Kaufmann Publishers, San Mateo (2002)

Smart Sol – Bringing User Experience to Facility Management: Designing the User Interaction of a Solar Control Unit

Patricia Böhm, Tim Schneidermeier, and Christian Wolff

University of Regensburg, Department of Media Informatics, Universitätsstrasse 31, 93053 Regensburg, Germany
Patricia.Boehm@stud.uni-regensburg.de,
{Tim.Schneidermeier,Christian.wolff}@sprachlit.uni-regensburg.de

Abstract. While a lot of attention is paid to the design of consumer electronics like mobile phones, various other domains have been neglected so far when it comes to user experience. In this paper a user-centered design approach for designing the user interface of a controller for solar thermal plants and heat exchanger stations – called *smart sol* – is described. The design process is characterized by the cooperation of user experience designers on the one hand and engineers and programmers on the other hand.

Keywords: user experience, user-centered design, user interface design, human-machine interaction, nontraditional user interfaces, facility management.

1 Introduction

Electronic appliances can be found throughout our homes and are increasingly influencing our everyday lives. We are not just affected by consumer electronics like mobile phones, personal computers or flat screen televisions, but by a still growing number of information appliances like digital watches, heat regulators or control units for microwaves or ovens that can be characterized by small displays and hardkey interactions. Almost everyone interacts with this kind of devices on a daily basis [1], [2]. The way we deal with them is to a large extent affected by the usability and user experience (UX) they provide: Creating pleasurable designs is different from eliminating frustration: "This is where designing an experience – not just a product – begins." [3] User experience can be defined by the attributes *useful, usable, valuable, findable, credible, desirable* and *accessible* [4]. Research in the last few years has focused on user experience design for consumer electronics, e.g. [5], [6], [7], [8]. At the same time other everyday information appliances have received little attention in this respect [9], [10].

User interface guidelines and sets of predefined input controls limit the design space for desktop and mobile interfaces, e.g. [11]. While each application provides different functionality, users can expect consistent controls for standard operations such as opening, closing or setting preferences. This predictability is completely absent in nontraditional interfaces [12].

M. Kurosu (Ed.): Human Centered Design, HCII 2011, LNCS 6776, pp. 187–196, 2011.
© Springer-Verlag Berlin Heidelberg 2011

In this paper we describe a user-centered design (UCD) approach for the solar control unit *smart sol* within a research, which is funded by the Bavarian Ministry of Economic Development[1] and involving user experience designers of the *University of Regensburg* on the one hand and engineers and programmers of the domestic and environmental engineering company *emz Hanauer*.[2] In the first section we start with a characterization of nontraditional interfaces taking a closer look at small-screen devices and the diversity of input methods for such nontraditional interfaces. In a second step, state of the art user interfaces of control units for solar heating systems for home environments are described. This is followed by a case study, which describes the user-centered design process of the *smart sol* user interface.

2 Nontraditional User Interfaces

[12] characterizes nontraditional interfaces as "beyond the GUI". This definition includes all types of interfaces, which differ from the windows, icons, menus and pointer metaphor (WIMP) of standard personal computer software. Because of the huge variety of context of use, input and output devices, nontraditional interfaces have a lack of standards and common guidelines [1].

2.1 Small-Screen Interfaces

Examples of small-display devices are mobile phones, digital audio players, car entertainment and navigation systems or control interfaces for home appliances [2]. Designing for small-screen user interfaces is affected by particular constraints and characteristics including non-standardized input methods, limited screen-size and amount of information that can be displayed, the specific techniques for data input and in general the lack of standardized user interfaces [1], [2], [13]. Designers have to pay special attention to the way information is presented on the screen (visual angle, viewing distance, luminance, character and font attributes as well as display placement) as well as on the organization of the information (menu structure and depth, icons) due to the screen-size limitations and the variety of the context of use (private vs. public space, bright vs. dark environment etc.) [14], [2]. Due to the market penetration of modern mobile phones with their specific interaction design (touchscreens, gesture-based interactions), users' expectations for other application fields have been raised. One of our observations is that user experience issues will become a decisive factor for a broad range of appliances.

2.2 Input Methods for Nontraditional User Interfaces

Nontraditional interfaces lack standardized input devices like mouse or keyboard known from human-computer interaction. Thus not just designing the elements of the user interface, but all aspects of the way the user interacts with the appliance have to be considered [15]. Therefore interaction design is challenged by a reasonable and

---

[1] Grant IUK 0910-0003.
[2] http://www.emz-hanauer.de, a leading producer of heating, water and solar installation, based in Nabburg, Bavaria [accessed February 2011].

usable combination of hard- and software parts. Additionally, designing for a good user experience requires that the user interface is not just evaluated in terms of simplicity, ease of use, efficiency and effectiveness to complete a specific task, but also with respect to aesthetic and emotional aspects of design that influences users' perception of appliances as valuable or desirable [7], [16], [17].

An informative overview of the variety of input methods for nontraditional interfaces can be found in [18]. Common input controls used in conjunction with small displays are buttons, joysticks, arrow keys and touch interfaces [18]. Typical combinations involve sparsely labeled small buttons to be used for controlling the device along with small alphanumeric displays (e.g. seven-segment LCD displays).

3   Need for Action: HCI for Control Units in Facility Management

Facility management ensures the functionality of the built environment by involving multiple disciplines and integrating people, place, process and technology [19]. Facility management has to address private homes, housing estates and industrial environments individually, since their needs of management are fairly different. In this paper we will focus on facility management technology of home environments.

Fig. 1. Examples of a state of the art solar control units [20]

Facility management control devices are characterized by growing complexity: Continuously increasing feature sets are accompanied by large menu structures and heterogeneous user groups. Typically, user interface development lacks an (professional) outside perspective: The user interface of state of the art control units is designed and implemented by the very same engineers that are responsible for the features of the system, typically engineers or computer scientists without formal training in human-computer interaction. End user requirements (user-centered design) are not integrated into the development process. In many cases, usability and user experience are considered as nice-to-haves, but mostly ignored due to financial and time constraints or lack of awareness [9]. As a result, most user interfaces of solar control units offer a small monochrome display and the interaction is based on buttons or a combination of buttons and knobs. Complex and counterintuitive interaction steps which include pressing two or more buttons simultaneously for some seconds in order to de-/ activate a certain feature are widespread (see figure 1).

Because of their poorly designed interfaces, such control units are hardly ever used. The interfaces are not intuitive and hard to learn. As these devices are not used on a day-to-day basis, users will not remember complicated interaction steps and never touch them again.[3] Most users are still unaware that there are usability problems, but blame themselves when they cannot achieve their intended goal [9]. There are many reasons why such devices should actually be used by consumers after installation like changing user needs, necessary adaptions or saving energy. It is important for everybody to make a contribution to saving the environment. By optimizing the energy management of a facility, e.g. a family home, a small step can be done [21].

4   Case Study: *Smart Sol* - Designing for User Experience

In this chapter a user-centered approach for designing the user interface of the solar control unit *smart sol* is described. The design process implemented for this product involves the cooperation of user experience designers on the one hand and engineers and programmers on the other hand.

The philosophy of a user-centered design process easily spoken is that *the user knows best*. The process focuses on users` needs and goals they want to accomplish. Ideally users are involved in every stage of the iterative design process [22]. Figure 2 shows the applied user-centered design process (cf. [23]).

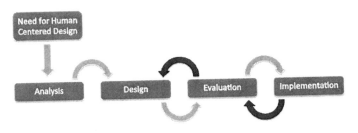

Fig. 2. User-centered design process

Implementing those needs in designing for an overall good user experience involves a wider range of parameters than traditional usability engineering, like performance and an easy to use interface. It also includes factors relating to affect, interpretation and meaning (e.g. aesthetic and emotional aspects) [24].

4.1   Need for Human Centered Design

To encourage people to use a solar control we detected three main goals.[4] Besides a high functional quality, it is essential to design an easy to use and intuitive interaction paradigm [7]. Our goal was to develop a product that can be used without reading the

---

[3] Results of market research and user and expert interviews conducted by the authors in 2009/ 2010.
[4] Results of user and expert interviews conducted by the authors in 2009/ 2010.

manual. To design for an overall good user experience, hedonic qualities also had to be considered [25], [26], [27]: "The aesthetic-usability effect concludes that more aesthetic designs appear easier to use – regardless of whether they are or not." [7]

Reaching these goals and providing a pleasant and welcoming feeling for the user, an iterative and user-centered design approach was used. We could identify two main user groups: the installer and the end user. Both groups have different needs. The installer is responsible for setting up the device and to handle errors when contacted by the end user. The end user needs a device he can rely on and customize to his own needs.

4.2 Analysis

First, usability professionals, programmers and engineers analyzed the functional specifications in terms of the required human-machine interaction. Based on this information a competitor analysis was conducted, which focused on the usability of state of the art user interfaces. Additionally, users were interviewed after asked to have a go on the controller about their impressions. The results indicate that most of the interfaces tested have very poor usability as the users were confused right after starting to use the interface. Among the reasons named are ill-designed or non-existing menu navigation, misleading onscreen information presentation and the hard-to-use interface due to non self-explanatory controls.

4.3 Design and Evaluation

Taking these results as a basis, the user interface was designed in an iterative process. In the early stage we used paper prototypes and wireframes for the design alternatives, asking users about their thoughts and opinions. For evaluating different color designs, fonts and font sizes mockups were chosen. The users` best-rated designs were implemented in click-through html-based vertical prototypes for usability testing. User interviews and post-test questionnaires were additionally used for evaluation of the interface. End users and installers participated in a closed card sort for defining the menu structure. For more detailed information on the conducted usability tests see [28].

4.4 Implementation

Results of the questionnaires, card sorting and user tests were translated into the final design.

Hardware and Navigation Design. Due to time constraints imposed by the production period, the device hardware had to be designed first. In cooperation with a product design company, user experience requirements were accounted for. A full color display improves the strictly technical image right from the start and communicates a pleasant impression [29]. 90% of the users prefer a color display and feel more comfortable using this type of interface.[5]

---

[5] Based on the results of the usability post-test questionnaire.

Fig. 3. *Smart sol* solar control unit

The interaction is based on a click wheel for navigation as well as confirmation and an escape button for going back (fig. 3). Click wheel navigation provides fast navigation and data input (e.g. target temperature settings) and can be handled efficiently [18]. Click wheels are increasingly used in consumer electronics like digital audio-players and car entertainment systems (e.g. *Apple iPod, BMW*). The escape button gives the user the opportunity to leave the menu at any time and to return to a safe starting point. When pushed, it will go up one hierarchy level at a time until the main menu is reached. When changing parameters, e.g. the target temperature of the solar storage tank, the user has the possibility to undo his action or set the parameters back to factory settings with one click at any time. This is to guarantee the easy reversal of actions. Thus the users are encouraged to explore the functionality of the controller.

Screen Design. Based on guidelines for screen and small screen design (e.g. [30], [31], [2]) as well as on the results of our iterative design process and user interviews the final screen design is shown in figure 4.

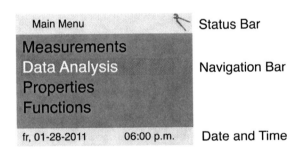

Fig. 4. *Smart sol* screen design

The basic design approach is derived from standard graphical user interfaces: Using familiar elements and well-known overall screen design, the user interface is easier to use because of previous experience [2]. The screen design of the smart sol unit follows a common trisection of everyday small-screen interfaces like smartphones. The overall screen and dialogue design is kept as simple as possible following the constraints of the small display. We use a full color display with a good contrast, a blue background and green highlighting. A small set of icons additionally indicates the type of assistant the users currently interacts with. This combination provides consistently positive and comfortable feelings [17]. Being asked in the post-test questionnaire, 90% of the users approved the final design as intuitive, pleasing and well-arranged.

Menu Design. The menu design represents the functionality of the device and should be carefully designed, so that basic and frequently used functions can be reached within a few steps [2]. Menu selection errors are most often due to miscategorization, overlapping and/or vague category titles [32]. We used card sorting in order to correctly represent the users' mental model in the menu structure [28], [33]. The final menu was designed with the help of common guidelines for menu design [2], [34], [35] and taking the particular challenges of small-screen devices into account.

Interaction Assistant Approach. To optimize the user experience of both user groups, the installer as well as the end user, we designed three intuitive interaction assistants: an assistant for a guided installation, a service assistant for error handling and an information assistant. These assistants are based on a survey on the most frequent problems end users have when using a solar control, which was also confirmed by interviews with installers.

The guided installation assistant simplifies the setting up of the solar thermal plant using a plug and play technique and offering a click-through guided installation. With this approach we mainly address the needs of the installer to quickly set up the solar thermal plant. The controller automatically recognizes the number of connected outlets and after selection of the kind of connected pumps a graphically representation shows a scheme of the possible solar plant settings (fig. 5).

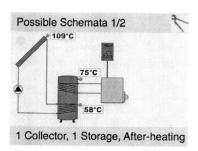

Fig. 5. *Smart* sol guided installation

When errors occur, e.g. a dysfunction, the service assistant guides the user through all possible sources of error and points out the solutions onscreen (fig. 6). No looking up in the manual is needed and hence a fundamentally faster problem solving process should be given.[6] The user is supported and given confidence in dealing with the situation. Special attention is paid on applying non-technical and easy to understand terminology.

---

[6] Based on the results of the usability post-test questionnaire.

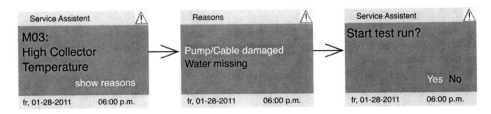

Fig. 6. *Smart sol* service assistant

Additionally a context-sensitive information assistant explains the user certain features of the control unit and describes the impact of his actions, e.g. activating the holiday function to support the user at any time to reach his goal.

5 Conclusion

In this paper a user-centered approach for designing the user interface of solar control unit is described. UX factors could be adopted successfully for use in the facility management domain. Presented at the world`s largest trade show for solar products, *Intersolar Europe 2010*[7], the overall design and in particular the interaction assistant approach were rated a success. The need for domain-specific usability engineering can also be seen in [36] and in the holding of the *1st European Workshop on HCI Design and Evaluation*[8] focusing on the influence of domains. While these first user studies show that the integration of user-centered design has been successful so far, the long term usage is hard to measure empirically. Continuing with the results of this project, the authors' further research will focus on an efficient approach to manage variability in interaction design for nontraditional interfaces, transferring research on software product families and product lines from software engineering.

Acknowledgments. This work has been conducted in the context of the research project Modino that is funded by the Bavarian Ministry of Economic Development (Grant IUK 0910-0003).

References

1. Baumann, K.: Introduction. In: Baumann, K. (ed.) User Interface Design for Electronic Appliances, pp. 6–29. CRC Press, Boca Raton (2001)
2. Mauney, D.W., Masterton, C.: Small-Screen Interfaces. In: Kortum, P. (ed.) HCI Beyond the GUI: Design for Haptic, Speech, Olfactory, and Other Nontraditional Interfaces, pp. 307–358. Morgan Kaufmann, San Francisco (2008)
3. Chisnell, D.: Pleasant Things Work Better (accessed, February 2011),
   http://uxmagazine.com/design/pleasant-things-work-better

---

[7] http://www.intersolar.de [accessed February 2011].
[8] http://sites.google.com/site/ehcide/, taking place in Limassol, Cyprus in April 2011 [accessed February 2011].

4. Morville, P.: User Experience Design (accessed, February 2011), http://semanticstudios.com/publications/semantics/000029.php
5. Hallnäs, L., Redström, J.: From Use to Presence: On Expressing and Aesthetics of Everyday Computational Things. ACM Trans. Comput.-Hum. Interact. 9, 106–124 (2002)
6. Han, S.H., Yun, M.H., Kwahk, J., Hong, S.W.: Usability of Consumer Electronic Products. International Journal of Industrial Ergonomics 28, 143–151 (2001)
7. Klauser, K., Walker, V.: It's About Time: An Affective and Desirable Alarm Clock. In: Proceedings of the 2007 Conference on Designing Pleasurable Products and Interfaces, pp. 407–420. ACM, Helsinki (2007)
8. Petersen, M.G., Madsen, K.H., Kjær, A.: The Usability of Everyday Technology: Emerging and Fading Opportunities. ACM Trans. Comput.-Hum. Interact. 9, 74–105 (2002)
9. Thimbleby, H.: The Computer Science of Everyday Things. In: Proceedings of the 2nd Australasian Conference on User Interface, pp. 3–12. IEEE Computer Society, Queensland (2001)
10. Norman, D.: The Design of Everyday Things. Perseus Books (2002)
11. Apple Inc.: iOS Human Interface Guidelines (2010) (accessed, February 2011), http://developer.apple.com/library/ios/#documentation/userexperience/conceptual/mobilehig/Introduction/Introduction.html
12. Kortum, P.: Introduction to the Human Factors of Nontraditional Interfaces. In: Kortum, P. (ed.) HCI Beyond the GUI: Design for Haptic, Speech, Olfactory, and Other Nontraditional Interfaces, pp. 1–24. Morgan Kaufmann, San Francisco (2008)
13. Mavrommati, I.: Design of On-Screen User Interfaces. In: Baumann, K. (ed.) User Interface Design for Electronic Appliances, pp. 108–128. CRC Press, Boca Raton (2001)
14. Kärkkäinen, L., Laarni, J.: Designing for small display screens. In: Proceedings of the Second Nordic Conference on Human-Computer Interaction, pp. 227–230. ACM, New York (2002)
15. Garrett, J.J.: The Elements of User Experience. New Riders, Indianapolis (2002)
16. Norman, D.A.: Emotional Design: Why We Love (or Hate) Everyday Things. Basic Books (2003)
17. Stone, T., Adams, S., Morioka, N.: Color Design Workbook: A Real-World Guide to Using Color in Graphic Design. Rockport Publishers Inc. (2006)
18. Baumann, K.: Controls. In: Baumann, K. (ed.) User Interface Design for Electronic Appliances, pp. 131–161. CRC Press, Boca Raton (2001)
19. International Facility Management Association: IFMA/ What is FM (accessed, February 2011), http://www.ifma.org/what_is_fm/index.cfm
20. Mare Solar: Solar Regler (accessed, February 2011), http://www.mare-solar.com/shop/solarthermie-solar-regler-c-67_402.html
21. Pierce, J., Roedl, D.: Cover Story: Changing energy use through design (2008)
22. Saffer, D.: Designing for interaction. New Riders, Berkeley (2007)
23. Usability Professionals' Association: What is User-Centered Design: About Usability: UPA Resources (accessed, February 2011), http://www.upassoc.org/usability_resources/about_usability/what_is_ucd.html
24. Roto, V., Law, E., Vermeeren, A., Hoonhout, J. (eds.): User Experience White Paper. Dagstuhl seminar on Demarcating User Experience (2010), (accessed, February 2011), http://www.allaboutux.org/files/UX-WhitePaper.pdf
25. Hassenzahl, M., Beu, A., Burmester, M.: Engineering Joy. IEEE Softw. 18, 70–76 (2001)

26. Hassenzahl, M.: The Interplay of Beauty, Goodness, and Usability in Interactive Products. Hum.-Comput. Interact. 19, 319–349 (2008)
27. Zhou, H., Fu, X.: Understanding, Measuring, and Designing User Experience: The Causal Relationship Between the Aesthetic Quality of Products and User Affect. In: Jacko, J.A. (ed.) HCI 2007. LNCS, vol. 4550, pp. 340–349. Springer, Heidelberg (2007)
28. Böhm, P., Schneidermeier, T., Wolff, C.: Customized Usability Engineering for a Solar Control: Adapting Traditional Methods to Domain and Project Constraints. To Appear in: Proceedings of HCI International (2011)
29. Schmidt, A., Terrenghi, L.: Methods and Guidelines for the Design and Development of Domestic Ubiquitous Computing Applications. In: Proceedings of the 2006 ACM Symposium on Applied Computing, pp. 1928–1929. ACM Press, New York (2006)
30. Baumann, K.: Summary of Guidelines. In: Baumann, K. (ed.) User Interface Design for Electronic Appliances, pp. 345–354. CRC Press, Boca Raton (2001)
31. Lieberman, H., Espinosa, J.: A Goal-Oriented Interface to Consumer Electronics Using Planning and Commonsense Reasoning. Know.-Based Syst. 20, 592–606 (2007)
32. Lee, E., Whalen, T., Mcewen, S., Latrémouille, S.: Optimizing the design of menu pages for information retrieval*. Ergonomics 27, 1051–1069 (1984)
33. Tullis, T.: Measuring the User Experience. Elsevier, Amsterdam (2008)
34. Shneiderman, B., Plaisant, C.: Designing the user interface: Strategies for effective human-computer interaction. Addison-Wesley/Pearson, Upper Saddle River (2010)
35. Tidwell, J.: Designing interfaces. O'Reilly, Beijing (2006)
36. Chilana, P.K., Wobbrock, J.O., Ko, A.J.: Understanding usability practices in complex domains. In: Proceedings of the 28th International Conference on Human Factors in Computing Systems, pp. 2337–2346. ACM, New York (2010)

Co-simulation and Multi-models for Pervasive Computing as a Complex System

Laurent P. Ciarletta

Madynes Research Team, INRIA Nancy Grand-Est
Université de Lorraine, INPL, ENSMN, France
Laurent.Ciarletta@loria.fr

Abstract. Pervasive Computing is about interconnected and situated computing resources providing us(ers) with contextual services. These systems, embedded in the fabric of our daily lives, are complex: numerous interconnected and heterogeneous entities are exhibiting a global behavior impossible to forecast by merely observing individual properties. Firstly, users physical interactions and behaviors have to be considered. They are influenced and influence the environment. Secondly, the potential multiplicity and heterogeneity of devices, services, communication protocols, and the constant mobility and reorganization also need to be addressed. This article summarizes our research on this field towards both closing the loop between humans and systems and taming the complexity, using multi-modeling (to combine the best of each domain specific model) and co-simulation (to design, develop and evaluate) as part of a global conceptual and practical toolbox. We share our vision for a strong research (and development) leading to the realization of Pervasive Computing.

Keywords: Pervasive Computing, Ubiquitous Computing, Ambient Intelligence, Human-in-the-loop, Distributed Simulation, Co-simulation, Multi-model, Emulation, Benchmarks, Multi-Agent System.

1 Introduction

Pervasive Computing [16], Ubiquitous Computing, or Ambient Intelligence is about interconnected and situated (or contextual) computing resources. They are embedded in the fabric of our daily lives and provide adapted applications and services to users in a changing context and environment.

It is at the convergence zone of four mature domains of traditional computing: personal computing, embedded systems, human-computer interaction (and artificial intelligence in some sort of way) and computer networking. In addition to this cross-domain nature, Pervasive Computing has other characteristics and constraints that are often difficult to combine and satisfy at the same time:

- Ubiquity (low-cost, embedded, distributed and non-intrusive)
- Interconnectedness (internetworking by wired and wireless technologies)
- And dynamism (a result of its mobile and adaptive applications that are able to automatically discover and use remote services).

M. Kurosu (Ed.): Human Centered Design, HCII 2011, LNCS 6776, pp. 197–206, 2011.

The notion of information appliances (small, specially designed computing devices, such as Internet Service Providers boxes, game consoles, connected TVs, and other Digital Video Recorder) is important to the success of pervasive computing. And the exponential rise of smart-phones, as the result of the convergence of Personal Digital Appliances (PDA) and mobile phones, and other tablets, is but one indication of this new computing paradigm coming of age.

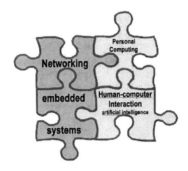

Fig. 1. Pervasive Computing: convergence of *Networking, Personal Computing, Embedded Systems* and *HCI*

Also the vocabulary depends on the community and its concerns, the overall statement is that our (as in "we, humans") environment is going to be filled with technology that will orchestrate their services to offer an improved / augmented / useful world and Internet of things. The 2 concepts, "ambient" and "intelligence", as shown in figure 2, are both relative to the human.

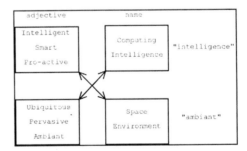

Fig. 2. Pervasive Computing, Ubiquitous Computing, Ambient Intelligence, Smart Space, Pro-active environment: a combination of some sort of intelligence provided by technology in (our) human environment

Although it may be easier to minimize or simplify his interactions and therefore his impact on the system, the "human" needs to be considered both at design and evaluation levels. He can't just be abstracted away. Another key element is related to the social side of many human activities. Not only should his physical, cognitive and biological properties be considered, but also his social interaction, cultural and scientific knowledge and abilities etc.

The remaining of this paper is organized as follows. In Section 2, we present our past proposition on how to put the human in-the-loop at every step and level, and how to model these complex systems. Section 3 details our proposition regarding the human part, starting from cognitive agents going to the Multi-Agent paradigm. Section 4 will give a short presentation of our solution applied to 2 use-cases. Section 5 opens up to new ideas and ongoing and future work, and try to synthesize our vision.

2 Pervasive Computing: Modeling a Human-Centric and Complex System

Pervasive Computing is in essence augmenting the human environment with smart technology to provide advanced services. It can be considered human-centric, since the most important evaluation metric will be the human acceptance and the impact on his activities and behavior. But it is at the same time a human in-the-loop system, where several interacting entities may exhibit non-predictable and non-controllable behaviors, and where the functional interaction should be well specified.

2.1 A Complex System

In future fully and commercially viable embodiments of Pervasive Computing, a large number of heterogeneous and autonomous objects, users, and systems will interact in a continuously changing environment. Mostly rooted in Social Science, Natural Science and Mathematics, a complex system is a set of interconnected elements that has global properties that can't be inferred from the properties of the individual elements, whether it comes form the large number of entities, their dynamicity, their heterogeneity, their autonomy etc.

2.2 Layered Pervasive Computing Model

The LPC [3] (Layered Pervasive Computing) model is inspired by the ISO-OSI [1] and TCP/IP networking models and their abstraction layers. It was developed in the NIST Aroma project [2]. It gives a first conceptual framework that aims at classifying issues raised during design, evaluation or analysis of a Pervasive Computing system.

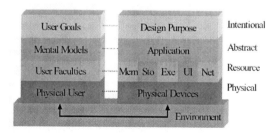

Fig. 3. The LPC (Layered Pervasive Computing) model, both the *user*(s) and the *environment* are considered in addition to the technological system

The first two layers are about the physical and physiological properties of the actors and their environment. The environment must be factored into the model: issues beneath the physical layer of the computer and users can be crucial. The growing interest for Sensor (and Actuator) networks and all the environmental information needed for pertinent context assessment justifies this layer (temperature in a smart home for a simple example). And the environment is also the medium through which all the other layers will interact. The physical layer deals with compatibilities. For example a device should match the physical capabilities of its users, or adapt to his disability. Simple mistakes such as inadequacy of a touch screen with big fingers could be avoided by carefully looking into the issues raised in these layers.

The resource layer essentially tries to match the computational resources (memory, processing power, storage etc.) with the potential user skills and abilities (his education or skills, is language or is temperament). Too often, the developer makes wrong assumptions about the future users of his system which can be time and money consuming, since the product or system will need reengineering at a sooner stage. Users have tasks or goals that should not be frustrated by a poor design. We can put in this layer services such as auto-configuration and self-management and concerns such as ergonomic. The abstract layer is equivalent to the application layer in the OSI model, where application is matched with the users mental models, which in turn are greatly dependent on their faculties. There are limits to the complexity the various users can cope with in Pervasive Computing environment. They won't waste time learning how to use a program designed by a computer scientist with his own vision of the product and its usage, but reuse the mental models they have acquired from former experience with similar designs. There is a need for consistency here. Finally, the intentional layer represents the purpose of an application that should be in harmony with the users goals. The user's mental representation of the underlying layer is directly linked with his goals. The requirements phase of any project should be where all starts and ends.

In the Aroma project at NIST, we've used this model to study a smart space populated with discoverable projectors and smart objects, which reveals quickly issues that should have been addressed to realize our prototypes as commercial products.

The model is still conceptual, and in order to be valid, the inter layers and same level layers interaction should be more formally described. Problems such as dimensional consistency (time, space, etc.) and formalizing interactions have to be resolved.

3   From Using Cognitive-Agents to Multi-agent Systems to Assess and Develop Pervasive Computing Systems

Depending on the application and the situation or context, human physical or cognitive abilities, as well as his behavior (considering the situation or context, acting upon his environment) should be considered. While this is the case for user interfaces (UI) or interfaces in general, it is quite not the case in other domains when considering computing elements such as communication technologies (network, services).

In the Aroma project, we have also developed EXiST, the Experimental Simulation Tool. It has being created to explore Pervasive Computing requirements and use cases. We were exploring the use of Intelligent Agents firstly for representing the users in our simulation [4]. Intelligent or cognitive agents can learn or even teach how to use resources by interacting with human users, providing a higher level of intelligence to the environment, but they also can be used as human models in simulations or in expert systems. We used a cognitive agent modeling toolkit with EXiST to create unit agents to help define usability metrics in Pervasive Computing.

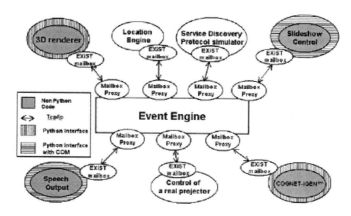

Fig. 4. EXiST co-simulating an Aroma projection room, with simulated users and environment and real projector and applications

Cognitive agents are full blown agents that can be used to model reactive to strategic human actions, but in some scenarios where we deal with hundreds, thousands or more users, for resources reasons (both computing and scientist), simpler agents could be used.

For example, in the domain of dynamic networks (adhoc, P2P networks), which is one of our case study (see next section), the number of nodes may be quite high and two users' properties need to be integrated and their impact evaluated: movement or mobility and application usage. In that type of situation, where individual behavior needs to be modeled, but where global behavior can't be forecasted, the multi-agent paradigm is well fitted. It can model sets of situated autonomous proactive entities interacting together, and with and within an environment. It is widely used in human and social sciences, ecology or in robotics. The system is described with (at least) 3 components: agents, environment, and interactions. Agents only have a partial (local) view of the environment and decide which action to take according to their perceptions and reasoning.

MAS (Multi-Agent Simulation) are adapted when we want to model users' behavior, goals and actions. Instead of using a global equation to model users' trajectories, we can, via the agent based model, re-create the way users move. It means that we can directly model behaviors such as "if there are obstacles, avoid them" or "follow a target during five minutes and then stop". Using this tool in our global strategy, we can also model complex behaviors such as willingness to use an application and share a resource depending on the available bandwidth or the generosity of a user. We can also deal with reactions to unlikely events.

4 Multi-modeling and Co-simulation Use Cases: Smart Environment and Mobility Models

We've defended the case for co-simulation in [5], stating that like in every field of science, simulation is a key tool for researchers, developers and designers when used in a co-simulation setting. By co-simulation we mean mixing real or prototype devices or application or users within complex simulated environment, and devices and users, in a refining development cycle.

As seen in figure 3, we've coupled several simulators and real devices using EXiST, focusing on the co-simulation aspect. With this setting we were able to test prototype applications and devices such as a smart wireless projector and a wireless sharing service. Combine with the LPC model, we have been able to pinpoint scientific and technological challenging issues. But we haven't considered the modeling and coherence issues in that first set of experiments.

In the frame of the SARAH (Advanced Services for Adhoc Networks) French research ANR project, we've tried to study such problems, with a strong focus on wireless technologies and services, context awareness and users' behavior.

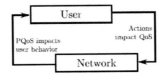

Fig. 5. Multi-modeling: user and network in a closed-loop

In wireless technologies, ad hoc or mesh routing protocols, or ubiquitous services are often designed and evaluated using network simulators. More specifically, when studying MANETs (Mobile Ad hoc NETworks), real world experimentations with a representing set of devices is excessively time and money consuming, or even is scientifically of little relevance since reproducing a scenario / an experiment is not possible, network simulators are almost mandatory. In simulation, nodes move according to a mobility model [9] but for real dynamicity, we need to close the loop between the user's actions (mobility being one of them) and network performance (the perceived quality of service seen in figure 4). We're using the work on pedestrian [7] and distributed flock and herd behaviors [6] as basis for our mobility models.

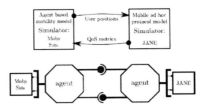

Fig. 6. The AA4MM framework used in multi-modeling and multi-simulation of mobility models in adhoc networks. Coupling *Jane* and Masdynes.

The design as in EXiST is inspired by the DoD HLA (High Level Architecture) which main goal is re- usability and interoperability of different simulators. The framework fosters the use of existing simulators (or other entities e.g. visualization systems) and combines their specialization. It is therefore a multi-model simulation and evaluation environment. Each domain retains its specialist and tools and replacing a simulator does not require major changes on the existing architecture.

The AA4MM (Agent and Artefact for Multiple Models) [11] framework is used to combine the JANE[10] (Java Ad Hoc simulator) Network Simulator and our own Multi-Agent Simulator for the users' behavior. AA4MM itself is grounded in the Multi-Agent paradigms, since the multi-models and their corresponding simulators are seen as a society of entities. The dimensional discrepancies are dealt within the framework that acts as a synchronization and conversion middleware. The actual implementation is centralized using JMS, but in principles it is distributed and has no explicit synchronization entity (global clock). In our case, nodes are modeled as agents with goals and tactics, and the network layer provides feedback.

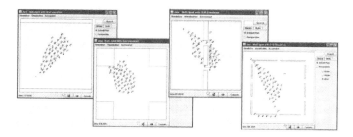

Fig. 7. Mobility models: several environments with same behavior

From simple new mobility models ... to new functionalities and benchmarks.
Typical scenarios and simulation, as shown in figure 6, where the advanced mobility model is used in different environments (Corridor, Crossroads, Walls and Museum), shows how simple is it to tweak or integrate new mobility models, i.e. new behavior in our networking simulations. Furthermore, our approach allows the creation of mobility models that interact with the network. For example, depending on the signal strength, a user may decide to stay put or move. In a middle term, it may be interesting to present a set of mobility models, a set of environments and their combination that would be both virtualized (modeled and simulated) and have a real setup (a typical existing room or building or city modeled in 3D for example). They would serve as references that could be used to evaluate the performances and applicability of a solution, and validate it in certain contexts, thus providing the pervasive computing community with a benchmark evaluation toolkit.

The AA4MM conceptual framework and prototype implementations applied for mobility modeling exhibit the following results

- The Multi-Model approach is a separation of concern (and problems) approach
- The tools allow even non-specialists to scientifically implement and validate their solutions, and the low-levels designers to give "real-life" example for their technology or protocols

- While it still provides the usual mobility models, it is very simple to design, fine-tune, redesign those models or even design new ones.
- The new mobility models can take into account networks or more generally environment inputs, basically having a closed-loop system where something closer to the "human behavior and real-life" is considered.

Therefore, our approach offers a basis for valid comparison of wireless technologies and services but it can be extended to any dynamic environment, such as P2P or more complex pervasive computing networks for example and basically for every situation where there are interactions between the users and the (networking) environment.

5   Ongoing and Future Work

First of all, technically, our platform will be extended:

- The core distributed engine will be rework to handle both horizontal (scalability) and vertical (across abstractions and domains) co-simulations
- More standard and novel mobility models (node/users behaviors), and reference environments will be developed
- Testing and visualization tools should be integrated in the near future, and should be extended to the analysis of security

And our work needs to be confronted with real environments, users and usage. This is planned in a joint-work with Supélec Metz and its Smart Room, where advanced services and complex user behaviors will be tested.

Another interesting playing field is the gaming industry [8,15]. Massively Multiplayers games and Pervasive gaming would provide real-life, large-scale, complex experiment and results.

Finally, we will explore the fundamental needs behind the fabric and the building of a sound by design (provable) Pervasive Computing system:

- On the domain of formal methods: they are starting to look into multi-modeling and co-simulation [14]
- On the modeling and mathematical theory of integration between humans and system [12,13].

5.1   The Best is Yet to Come ...

After surveying the situation in Pervasive Computing, or example when surveying major scientific or technological conferences in the field, it seems that:

- The technologies has reached a certain level of maturity, the computing power embedded in our smartphones or the new interaction devices such as the Microsoft Kinect™
- But the community still focuses on technical challenges, like location and awareness or novel interactions

On a large system view, the engineering processes and tools are yet to be tangible. It is still largely more the work of a scene, of some crazy scientist and hackers.

6 Final Thoughts

This article explores more than 10 years of Pervasive Computing research driven by networking and services management concerns. We've proposed a conceptual model (LPC) and seen that formal methods and mathematical theories may help us engineer better applications. We believe that using well-defined models and tools from specialist in every domain is the good path to follow in the near future, and we are building a conceptual framework (AA4MM) based on multi-agent systems for multi-modeling as well as tools (EXiST–AA4MM) for the evaluation and development using co-simulation.

Our goal is to provide a framework where technologies could be evaluated and ideally certified "for a standard type of environment". This would give the ability for a technology designer to say: "my service, my device, my protocol will or wont' work in those environments with those type of users". About a specific solution, its designer could still satisfy his scientific colleagues with:" it works well with high density, low mobility scenarios and conforms to the formal specification", but he could also add: "it works well in downtown scenarios at rush hour, but not in a shopping mall on Saturday afternoon" and we believe that is what will make all the difference.

During this period, the technologies have progressed tremendously and a future a la Harry Potter can be envision, where our gesture and thoughts will be seen and interpreted by our augmented environment (full or useful services, robots and clouds). But when looking into the conceptual matter, cross-layer issues and the engineering side of the domain, we can only see that we are still a long way from the realization of commercially sound-by-design, large-scale, fully-integrated, cross-layer, failed-proof secure Pervasive Computing environments.

References

1. Day, J.D., Zimmerman, H.: The OSI Reference Model. Proceedings of the IEEE 71, 1334–1340 (2003)
2. Mills, K.L.: AirJava: Networking for Smart Spaces. In: Usenix Embedded Systems Workshop (March 1999)
3. Ciarletta, L., Dima, A.: A Conceptual Model for Pervasive Computing. In: International Workshop on Parallel Processing (2000)
4. Ciarletta, L., Dima, A., Iordanov, V.: Using Intelligent Agents to assess Pervasive Computing Technologies. In: IAWTIC 2001 (2001)
5. Ciarletta, L.: Emulating the Future with/of Pervasive Computing Research and Development in What make for good application-led research. Workshop Pervasive (2005)
6. Reynolds, C.W.: Flocks, herds, and schools: A distributed behavioral model. Computer Graphics 21, 25–34 (1987)
7. Teknomo, K., Takeyama, Y., Inamura, H.: Review on microscopic pedestrian simulation model. In: Proceedings Japan Society of Civil Engineering Conference (2000)
8. http://www.massivesoftware.com/
9. Bai, F., Helmy, A.: A Survey of Mobility Models. In: Wireless Adhoc Networks. University of Southern California, U.S.A (2004)

10. Gorgen, D., Frey, H., Hiedels, C.: Jane - the java ad hoc network development environment. In: ANSS 2007: Proceedings of the 40th Annual Simulation Symposium, pp. 163–176. IEEE Computer Society, Washington, DC (2007)
11. Leclerc, T., Siebert, J., Chevrier, V., Ciarletta, L., Festor, O.: Multi-Modeling and Co-Simulation-based Mobile Ubiquitous Protocols and Services Development and Assessment. In: 7th International ICST Conference on Mobile and Ubiquitous Systems (2010)
12. Fass, D.: Rationale for a model of human systems integration: The need of a theoretical framework. Journal of Integrative Neuroscience 5(3), 333–354 (2006)
13. Chauvet, G.A.: Hierarchical functional organization of formal biological systems: a dynamical approach. I. An increase of complexity by self-association increases the domain of stability of a biological system. Phil. Trans. Roy Soc. London B 339, 425–444 (1993)
14. Fitzgerald, J., Larsen, P.G., Pierce, K., Verhoef, M., Wolff, S.: Collaborative Modelling and Co-simulation in the Development of Dependable Embedded Systems. In: Méry, D., Merz, S. (eds.) IFM 2010. LNCS, vol. 6396, pp. 12–26. Springer, Heidelberg (2010)
15. McGonigal, J.E.: This Might Be a Game: Ubiquitous Play and Performance at the Turn of the Twenty-First Century, doctoral dissertation. University of California, Berkeley (2006)
16. Weiser, M.: The Computer for the 21st Century, Scientific American Special Issue on Communications, Computers, and Networks (September 1991)

Design and Development of Eyes- and Hands-Free Voice Interface for Mobile Phone

Kengo Fujita and Tsuneo Kato

KDDI R&D Laboratories Inc., 2-1-15 Ohara, Fujimino, Saitama, 356-8502, Japan
{fujita,tkato}@kddilabs.jp

Abstract. This paper describes the design and development process of our new eyes- and hands-free interface which provides the fundamental functions of a mobile phone by voice interaction through a Bluetooth headset. We first identify four conditions which must be met in order to make the interface acceptable to Japanese users. Next, we define design guides which address each of these conditions. In accordance with the design guides, we propose and implement the interface system. To assess the effectiveness of the proposed interface, we had participants operate a mobile phone while walking while simultaneously confirming a switching signal which either permitted or forbade them to walk. The experimental results showed that the proposed interface was more effective than the conventional interfaces for operating a mobile phone while simultaneously performing other tasks. The participants pointed out some problems during the interviews, and we address these problems.

Keywords: Mobile Phone, Voice Interface, Eyes-free Operation, Hands-free Operation, Bluetooth Headset, Design Process, Japanese users.

1 Introduction

We have been aiming to make mobile phones easy to use in any situation. In Japan, distinctive culture of mobile phone use has evolved due to the early spread of mobile Internet connection services and third generation (3G) mobile phones [1]. More than 75% of Japanese users use a mobile Internet connection service compared to around 40% in the U.S. and Europe [2]. According to a survey on the daily use of mobile phone functions and services in Japan, e-mail and Internet search are the most common [3]. Japanese users, especially teenagers and those in their 20s, prefer e-mail to phone call as a means of communication. The reasons can include that Japanese uses do not like to be heard a mobile phone conversation by others, that conversely, they are annoyed to hear a mobile phone conversation conducted in public, that they do not want to disturb the receiver by making a call and that the use of monthly fixed-rate data communication service saves their mobile phone bills. With an increase in use of the e-mail function, a number of additional functions for e-mail such as moving pictograms or decorated messages have been developed. There is an unwritten rule that receivers have to reply to received e-mails as quickly as possible. When a quick reply is impossible, both the sender and receiver feel stress. The users want to respond

M. Kurosu (Ed.): Human Centered Design, HCII 2011, LNCS 6776, pp. 207–216, 2011.
© Springer-Verlag Berlin Heidelberg 2011

to received e-mails even if their messages are simple. They always have their mobile phones at hand and are ready to reply while doing other tasks, such as watching TV, playing a video game, cooking, chatting and doing their nails. However, generally speaking, they have to look at the display panel and input through their keypads or touchpads. It is more difficult to write e-mails while performing other tasks as the users cannot concentrate on just writing. As a result, more input errors and delays occur.

Because most Japanese commute by train, they also often take their mobile phones out of their pockets or bags in order to search the Internet or to exchange e-mails while walking to/from a station, waiting on the platform and while on the train. In such situations, many users search destinations or train connections. Therefore, we have developed a novel voice interface applicable to a navigation service which allows users to search destinations or train connections specifying the date and hour using their mobile phones. Since 2006 in Japan, the interface employing a distributed speech recognition (DSR) system has been commercialized as "Voice de Input" [4]. This was the world's first DSR based speech recognition function on 3G mobile phones. After the release of smartphones onto the market, voice search applications such as "Google Voice Search" [5] have become increasingly popular.

If a browsing talk type voice interface (e.g., "Voice de Input" or "Google Voice Search"), operated by holding a mobile phone and looking at the display panel, were also to be employed for the e-mail function, it would make it easier for users to use the function. However, even if such interfaces were to be employed, the users would still have difficulty when they did not want to take their eyes and hands off other tasks. In contrast, if auditory confirmation of the inputted addressee and message were possible, the users would be able to operate the e-mail function in an eyes- and hands-free manner. The "iPhone Voice Control" [6] or some Bluetooth headsets employing a speech recognition system [7] provide voice control of mobile phones with a headphone. These conventional interfaces do not provide an e-mail function though they provide a phone call function. On the other hand, although Bluetooth headsets are seldom used for phone calls in Japan, listening to music with headphones is popular. A Bluetooth headset which could provide both e-mail and music player functions would fit into Japanese culture and become widely accepted in Japan. Therefore, we designed an eyes- and hands-free voice interface for mobile phones with a Bluetooth headset. This paper focuses on the voice control of the e-mail function.

2   Related Works

A number of interfaces for mobile terminals have been proposed for eyes-free operation. Key operation methods enabling the input of characters or alternative selection without visual confirmation have been proposed [8, 9]. These methods require users to hold the mobile phone in their hands. Moreover, adequate training is essential to master the operation. Other methods have been proposed that employ gesture input based on sensor-acquired data such as acceleration. While using the methods which detect the motion of a mobile terminal [10, 11], users have to hold the terminal in their hand. As waving the terminal itself in the air is required, these methods are not suited to the complicated operation of writing e-mails. For hands-free operation, some interfaces employing foot gesture [12, 13] or head tilting [14] have been proposed. However, such interfaces are also unsuited to the creation of e-mail messages. Users have difficulty employing such gesture-based input methods as they require the users to perform unusual actions which are not used in daily life.

3 Conditions for Design of New Interface

We first identified four conditions for a new eyes- and hands-free voice interface which would enable operation of the e-mail function of a mobile phone employing a Bluetooth headset and which would also be widely accepted by Japanese users. The conditions identified thorough our survey are as follows:

1. The new interface of the mobile phone must make access to the e-mail function easy to be widely accepted by Japanese users.
2. The users do not want to feel unsure of the behavior of the mobile phone due to a new voice interaction operation method which is different from the conventional key-based methods.
3. The operation of the e-mail function requires the user's concentration even if using only voice interaction. The users want to be able to use the mobile phone functions with which they are already familiar when they do not have other tasks to perform simultaneously.
4. The users want to minimize the costs of using the new interface. It is desirable for the interface system to work on the user's own mobile phone.

4 Design Guides

Next, we defined the design guides for the new interface corresponding to the above conditions:

1. The new voice interface allows the users to send e-mails specifying both addressee and message and to read out the received e-mails with the text-to-speech function.
2. The interface responds quickly to the user's voice commands and requires their confirmation before execution in order to prevent incorrect operation due to misrecognition. A single trigger key, which the users can press without having to look at, for voice input is installed on the headset to prevent response to sounds other than the voice commands.
3. The interface receives a single sentence as an e-mail message so that a user says the sentence at the same time as the addressee. Double-clicking of the trigger key on the headset enables the user to control establishment and termination of the connection between the headset and the mobile phone. The conventional key operation is available when disconnected.

4. The interface transmits the voice input and output data between the headset and the mobile phone over SPP (Serial Port Profile) so that the interface works on any Bluetooth-capable model of our commercial mobile phones.

5   Design of New Interface

In accordance with the design guides, we designed a new interface system consisting of a Bluetooth headset and a mobile phone as shown in Fig. 1. We employed our proprietary embedded speech recognition (ESR) engine, which recognizes several thousand words in real time, for the Qualcomm BREW mobile platform [15]. Although a speech recognition dictionary, which has a vocabulary for the target recognition task, generally requires a large amount of throughput and memory for it to be generated and updated, our engine is able to update the dictionary with the low throughput and memory of a mobile phone. Therefore, the dictionary is instantly updated in accordance with user's editing of the address book. The short e-mail messages are maintained in a list and can be freely added to, deleted or edited by the user for personalization. The e-mail message dictionary is also updated in accordance with edition of the list so that the messages can be applied to voice commands immediately.

The interface enables access to the e-mail function by voice interaction. An example of voice interaction for sending an e-mail is shown in Fig. 2. We employed SPP for transmission of input and output data so that data other than voice can be transferred in the future (e.g., transmission of the acoustic feature data extracted for speech recognition from voice input data in the headset to the mobile phone). Because many models of our commercial mobile phones are unable to bi-directionally transmit the data over SPP in parallel, we designed the flow of data transfer between the headset and the mobile phone as depicted in Fig. 3. The interface also provides voice dialing by the names in the address book or any phone number. To prevent parallel multi-profile connection, the interface automatically switches the Bluetooth connection between SPP for the voice interaction and HFP (Hands-Free Profile) for the phone call.

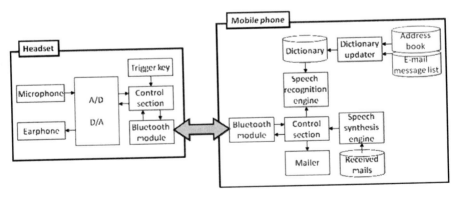

Fig. 1. Configuration diagram of the proposed interface system

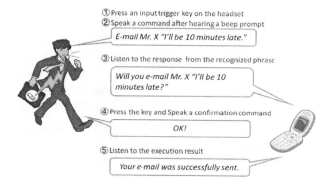

Fig. 2. Example of voice interaction using the proposed interface for sending e-mail

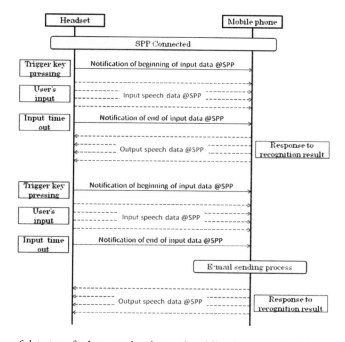

Fig. 3. Flow of data transfer between headset and mobile phone over SPP for sending e-mail

6   Evaluation Experiments

6.1   Experimental Design

To evaluate the usability of eyes- and hands-free operation using the proposed interface, 4 male and 8 female participants in their 20s to 50s took part in the evaluation experiments. The participants performed the e-mail task using three other operation methods while walking. The three methods comprise the conventional key operation, the browsing talk type operation and the eyes- and hands-free operation using the proposed interface. Table 1 shows the operation procedure of each method, respectively. The same model of mobile phone "SH002" as depicted in Fig. 4 was used for all the operation methods. Each participant held the mobile phone in his/her hand while performing the tasks using either the key operation or the browsing talk type operation. On the other hand, the mobile phone was put in his/her pocket while using the proposed interface. For each operation method, e-mail tasks with three pairs of an addressee and a message were performed. The participants walked 35 meters along a straight route while watching a signal switching between green and red at five-second intervals (see Fig. 5). The green and red signals alternately permitted and prohibited the participants from walking. Walking time was measured, and the number of times which the participants erroneously walked on the red signal was counted while each task was being performed. For comparison, the walking time and the number of missed red signals were also examined in the situation where each participant just walked at the beginning and the end of the experiment without performing any task. Before the experiments were conducted, the participants received instructions and sufficient training for the browsing talk type operation and the proposed eyes- and hands-free operation.

After the experiment, the participants were interviewed about the usability of the operation using the proposed interface.

Table 1. Procedure for each operation method used in the experiments

	Key operation	Browsing talk type operation	Proposed eyes- and hands-free operation
1	Press the "E-mail" key on the mobile phone	Press the input trigger key on the mobile phone	Press the input trigger key on the headset
2	Select the addressee from the address book using the arrow keys	State the addressee and message by speaking into the mobile phone microphone after hearing a beep prompt	State the addressee and message by speaking into the headset microphone after hearing a beep prompt
3	Enter the message with the numerical keypad	Confirm the recognition results in the display panel	Listen to the synthesized speech of the response from the headset earphone
4	Press the "Send" key	Press the "Send" key	Press the key and utter a confirmation command

Fig. 4. Mobile phone used in the experiments (au SH002)

Fig. 5. Route walked by participants

6.2 Experimental Results

Figure 6 shows the average walking time for just walking and walking while performing the e-mail tasks using the three operations. Each error bar indicates a 95% confidence interval. The key operation was associated with a large increase in the walking time compared to just walking because the participants had to pay attention to the display panel and the keypad in addition to the signal. Due to the need to perform another task in addition to confirmation of the signal, the walking time for the browsing talk type operation and the proposed operation also became longer compared to just walking. We confirmed significant differences between the walking time for each method and that for just walking by applying t-test (t < 0.05). The walking time for the browsing talk type operation and the proposed operation was significantly less compared with the key operation. This fact indicates that less concentration was required for the e-mail tasks performed using the voice operation than by the key operation. The participants often had to stop in order to concentrate on selecting an addressee from the address book or inputting e-mail messages while performing the key operation. The browsing type operation and the proposed operation had a little effect in terms of forcing the participants to stop to select an addressee and input e-mail messages as neither operation requires visual confirmation. On the other hand, there was no significant difference between the browsing talk type operation and the proposed operation. While performing the browsing talk type operation, the participants only had to press two keys. As the participants had become familiar with pressing the keys without visual confirmation as a result of training, they were able to press the keys easily while walking. Therefore, the walking times for the browsing talk type operation and the proposed operation were almost equal.

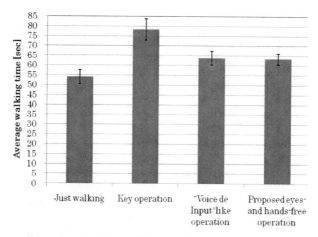

Fig. 6. Average walking time for just walking and while performing e-mail tasks using the three mobile phone operation methods

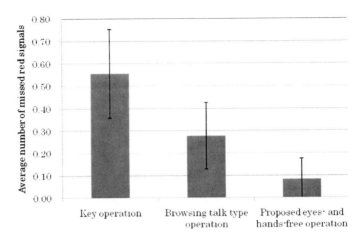

Fig. 7. Average number of missed red signals while performing e-mail tasks using the three mobile phone operation methods

Figure 7 depicts the average numbers of missed red signals for the three operations. No participants missed any red signals while just walking. The number for the browsing talk type operation was significantly less compared to the key operation. This result can be mainly attributed to the fact that the only reason the participants took their eyes off the signal was to confirm the recognition results. The number for the proposed operation was significantly smaller than that for the browsing talk type operation because there was no need for the participants to take their eyes off the signal while performing the proposed operation. This fact indicates that the proposed eyes- and hands-free operation is effective in allowing a mobile phone to be operated while simultaneously performing other tasks.

During the interview, 11 out of 12 participants stated that the proposed operation made it easier to watch the signal compared to the two other operations. Moreover, the other 11 participants stated that the interactive step for obtaining confirmation before executing the commands eliminated their anxiety about incorrect operations. On the other hand, some participants complained about the delay until they heard the response after uttering the confirmation command. There were suggestions that push-to-speak operation, which allows spoken voice commands only while pressing the input trigger key, is more convenient than the proposed push-and-release-to-speak operation.

7   Further Improvements

We implemented changes to improve the usability of the proposed interface based on the interviews. The headset received the user's command at the same time interval (5 seconds) after the input trigger key was pressed. As shown in Fig. 3, it was not until time-out occurred that the mobile phone began to transmit the response data. The participants were unconcerned because the remaining time until time-out was short when inputting operation commands for the e-mail or phone call functions. On the

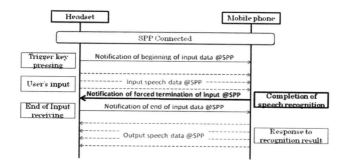

Fig. 8. Flow of modified data transfer between headset and mobile phone over SPP

other hand, as uttering the confirmation commands, such as "Yes" or "OK," requires a small amount of time, the remaining time lengthened. This fact could make the participants feel that they had to wait for a long time before hearing the responses. Therefore, we modified the flow of data transfer between the headset and the mobile phone as shown in Fig. 8 so that the input-receiving through the headset is forcibly terminated as soon as the speech recognition process is finished. This modification quickens the response to the user's confirmation commands.

We implemented the push-to-speak operation which works when the input trigger key is continuously held down for a predetermined time. Users are able to choose from two voice input starting methods by holding down the trigger key for different periods of time.

8 Conclusion

Taking into account Japanese mobile phone culture, we proposed and implemented a new eyes- and hands-free interface which enables operation of the e-mail function of a mobile phone by voice interaction through a Bluetooth headset. We first identified four conditions which would make the interface acceptable to Japanese users. Next, we defined design guides corresponding to the conditions which we identified. In accordance with the design guides, we proposed and implemented the interface system. In an experiment to assess the usability of the proposed interface, we had participants operate the e-mail function of a mobile phone while walking while at the same time confirming a signal switching between green and red, and measured the elapsed time and number of missed red signals. The proposed interface reduced both the walking time and the number of missed red signals compared to the conventional key operation. Furthermore, there were fewer missed red signals for the proposed interface compared with the browsing talk type operation. However, there was no significant difference in walking time between the two interfaces. Therefore, the proposed interface was shown to be effective in allowing operation of a mobile phone while performing other tasks at the same time. During the interviews, most participants stated that watching the signal during eyes- and hands-free operation using the proposed interface was easier than when using the other two conventional interfaces. However, some participants pointed out two problems which needed to be improved. One was the delay until they hear the responses after uttering the confirmation commands. The other was employing another voice input starting method. We modified the proposed interface to address these problems and improve the usability.

References

1. The New York Times: Why Japan's Cellphones Haven't Gone Global (2009),
   http://www.nytimes.com/2009/07/20/technology/20cell.html
2. comScore Press Release: comScore Releases First Comparative Report on Mobile Usage in Japan, United States and Europe (2010),
   http://www.comscore.com/Press_Events/Press_Releases/2010/10/
   comScore_Release_First_Comparative_Report_on_Mobile_Usage_
   in_Japan_United_States_and_Europe
3. Communication and Information Network Association of Japan: CIAJ Releases Report on the Study of Cellular Phone Use (2010),
   http://www.ciaj.or.jp/en/news/news2010/2010/07/28/484/
4. KDDI Corporate Information: KDDI Announces World's First Distributed Speech Recognition Function with "Voice de Input" Carried on au Cellular Phones (2006),
   http://www.kddi.com/english/corporate/news_release/2006/
   0112/index.html
5. Google Mobile App,
   http://www.google.com/mobile/google-mobile-app/
6. iPhone Voice Control, http://www.apple.com/iphone/features/voice-control.html
7. BlueAnt V1x Voice Controlled Headset,
   http://www.myblueant.com/products/headsets/v1x/
8. Ji, H., Kim, T.: CLURD: A New Character-Inputting System Using One 5-Way Key Module. In: Jacko, J.A. (ed.) HCI International 2009. LNCS, vol. 5612, pp. 39–47. Springer, Heidelberg (2009)
9. Li, K.A., Baudisch, P., Hinckley, K.: BlindsSight: Eyes-Free Access to Mobile Phones. In: 26th International Conference on Human Factors in Computing Systems, pp. 1389–1398. ACM, New York (2008)
10. Niezen, G., Hancke, G.P.: Evaluating and Optimizing Accelerometer-based Gesture Recognition Techniques for Mobile Devices. In: The 9th IEEE AFRICON, pp. 1–6. IEEE Press, New York (2009)
11. Joselli, M., Clua, E.: gRmobile: A Framework for Touch and Accelerometer Gesture Recognition for Mobile Games. In: VIII Brazilian Symposium on Games and Digital Entertainment, pp. 141–150. IEEE Press, New York (2009)
12. Scott, J., Dearman, D., Yatani, K., Truong, K.N.: Sensing Foot Gestures from the Pocket. In: 23rd ACM Symposium on User Interface Software and Technology, pp. 199–208. ACM, New York (2010)
13. Yamamoto, T., Tsukamoto, M., Yoshihisa, T.: Foot-Step Input Method for Operating Information Devices While Jogging. In: 2008 International Symposium on Applications and the Internet, pp. 173–176. IEEE Press, New York (2008)
14. Crossan, A., McGill, M., Brewster, S., Murray-Smith, R.: Head Tilting for Interaction in Mobile Contexts. In: 11th International Conference on Human-Computer Interaction with Mobile Devices and Services. ACM, New York (2009)
15. Brew Mobile Platform, http://www.brewmp.com/

Influence of a Multimodal Assistance Supporting Anticipatory Driving on the Driving Behavior and Driver's Acceptance

Hermann Hajek[1], Daria Popiv[1], Mariana Just[2], and Klaus Bengler[1]

[1] Lehrstuhl für Ergonomie, Technische Universität München, Boltzmannstraße 15, 85747 Garching bei München, Germany
[2] BMW Group Forschung und Technik, Hanauerstraße 46, 80992 Munich, Germany
{hajek,popiv,bengler}@lfe.mw.tum.de, Mariana.Just@bmw.de

Abstract. This work presents an investigation of a multimodal human-machine interface (HMI) of an anticipatory driver assistance system. The HMI of the system consists of visual indicators displayed in the digital instrument cluster and discrete impulses of an active gas pedal (AGP). The assistance recognizes the upcoming driving situation, informs the driver about its emergence, and suggests a driving action, which execution assures significant reduction in fuel consumption. The experiment is performed in the fixed-base driving simulator. Results show that during assisted drives an average reduction in fuel consumption amounts to 7.5%, in comparison to the drives without assistance. In 50% and 80% of all the cases, participants release the accelerator correspondingly within 1.2 and 2 seconds after receiving the first information. Two thirds of the test subjects grade the concept as "good" and "very good". The participants appreciate AGP discrete feedback especially in rare, unexpected, and potentially critical situations.

Keywords: Advanced driver assistance system, multimodal human-machine interface, anticipatory driving, active gas pedal.

1 Introduction

Due to the continuing development of traffic information sources, advanced driver assistance systems (ADAS) gain the possibility to inform drivers about an upcoming traffic situation before it becomes visible. These sources are e.g. detailed digital maps, far-field radar systems, car-to-car (C2C), and car-to-infrastructure (C2X) communication systems [1], [2]. This early information via ADAS can help to reduce fuel and increase safety, i.e. by supporting coasting[1] phases and alerting the driver about upcoming safety-critical situations.

The objective of this study is to investigate the effect on driving behavior of a multimodal assistance system, which expands the driver's natural anticipation horizon on the maneuvering level [3] by presenting information about upcoming deceleration situations and suggesting beneficial driving action even if the situation cannot be seen

---

[1] Coasting a vehicle – exploitation of motor torque during deceleration phases via releasing the accelerator and not depressing the brake pedal.

M. Kurosu (Ed.): Human Centered Design, HCII 2011, LNCS 6776, pp. 217–226, 2011.

yet. The benefit is defined in this work as increase in efficiency (reduction in fuel consumption), comfort (reduction of strong decelerations), and safety (reduction of extreme decelerations and collisions).

2 Experimental Design

In the following, information about the investigated assistance concept, test subjects, hardware and software tools used for the implementation of the test course and analysis, as well as descriptions of all investigated deceleration situations are provided.

2.1 Investigated Assistance Concept

The concept consists of a visual assistance based on a bird's-eye view perspective on the situation (in the following called as "visual Bird's-Eye View HMI concept"), and discrete impulses of the accelerator. The visual part of the HMI is presented to the driver in the instrument cluster. It is based on the preceding work of [4]. A similar concept of a discrete AGP is also investigated in the work of [5].

The so called Bird's-Eye View HMI depicts a virtual road scenery including the ego vehicle, and the emerging deceleration situation. This HMI possesses continuous characteristics: the presentation of the ego vehicle on the occupied lane is permanently displayed in the instrument cluster. The deceleration situation is superimposed on the virtual road when such is detected, e.g. in Fig.1 a situation involving a slower preceding vehicle and a traffic light.

Fig. 1. Bird's-Eye HMI

The legitimate traffic sign corresponding to the deceleration situation is shown at the side of the virtual road to enhance the comprehensibility of the emerging situation. This information appears at the point of time when the so-called beginning of the optimal coasting phase should start. It is an efficiency optimized action, which assures sufficient reduction of speed solely via engine braking. Also the color coding of the displayed driven vehicle from white to green suggests the start of coasting.

If pure coasting is not sufficient to reach the required lower speed, the color of the ego vehicle changes to orange to suggest active braking by depressing the brake pedal. It is left to the driver to decide with which strength to brake. Further information about the visual concept can be found in [6].

Simultaneously with the activation of the Bird's Eye HMI the driver receives a discrete impulse of the AGP (see Fig.2). This haptic component of the multimodal assistance indicates an oncoming change of the current traffic situation and it encourages the driver to release the gas pedal. As it is shown in [7], the AGP activation helps drivers to react more rapidly in comparison to a HMI concept, which only consists of the described visual interface without AGP support (reviewed in [8]).

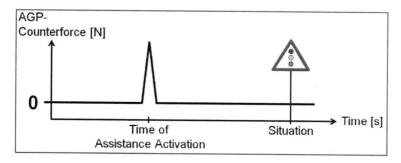

Fig. 2. Operating mode of the Active Gas Pedal

2.2 Subjects

Thirty participants (twenty five male and five female) took part in the experiment. All of them hold valid category B German driving licenses. The average age of the test subjects is thirty seven years (standard deviation, sd = 15.2 years). The driving experience varies: nine participants drive less than 10.000 km per year, fourteen – between 10.000 km and 20.000 km, and seven – more than 20.000 km per year.

2.3 Hardware and Software Tools

The experiment is performed at the fixed-base simulator located at Lehrstuhl für Ergonomie, Technische Universität München. A field of the driver's front view of 180° is used. The landscape and driving environment are simulated using SILAB software [9], which allows flexible and precise creation of the driving situations including the control over simulated traffic. The driving data of the test vehicle as well as relevant situational data, e.g. distance and speed of other traffic participants, are recorded at 60 Hz within the SILAB framework. The descriptive analysis of the driving data is done with MATLAB and Excel, the statistical analysis is performed using SPSS.

2.4 Simulated Test Course

Each one of the test subjects drives the simulated test course two times in permuted order: without the assistance (in baseline condition) and with the multimodal assistance. The goal and functionality of the visual and haptic assistance is explained to the test subject by a handout before the experiment starts. Moreover all questions concerning the assistance concept are answered. The test subjects pass an introductory drive before the experiment starts. One experiment drive lasts between seventeen and twenty minutes, during which the test subjects cover a 24.5 km long drive and are confronted with twelve different deceleration situations: Seven situations on a rural road (RR), three on a highway road (HR) and two urban situations (UR). The situations systematically differ in the length of the suggested coasting phase and their criticality. The order of deceleration situations and surrounding landscape is changed in the two drives to avoid recognition effects. In the following the description of all investigated deceleration situations is provided.

Situations on a Rural Road. "RR Construction site behind a curve on a straight segment". This situation occurs on a two-lane rural road, where the permissible speed is 100 km/h if not explicitly changed by other speed limit signs.

The test subject has to decelerate in front of a construction site located on the driven lane in order to let the oncoming cars pass. The site is located 200m after a right curve. The situation becomes visible at the distance of 200m-250m before the construction site, while the optimal coasting phase assuring efficiency benefit lasts 600m.

"RR Construction site in a right curve". The difference to the previously described situation is that the construction site is located directly in the curve. The driver is able to see the situation at the distance of 450m before the site.

"RR Speed limit". The driver has to decrease the driven speed down to 70 km/h due to an incoming sharp curve in this situation. The sign becomes visible approximately 170-200 m before it is reached, while coasting from 100 km/h down to 70 km/h lasts 500 m.

"RR Speed limit and slower lead vehicle". The driver has to decrease the driven speed down to 70 km/h. The speed limit sign becomes visible approximately 100 m before it is reached, because the sign is located in a curve. At the time when the driver sees the sign he also recognizes a slower leading vehicle driving also at 70 km/h.

"RR Town entrance". The driver has to decrease speed to 50 km/h when entering an urban area according to German traffic regulation rules. Even though this investigated situation is well-visible at larger distances, it is still unclear if the driver without assistance starts coasting early enough to perform the efficient deceleration maneuver. Coasting from 100 km/h to 50 km/h lasts 800 m.

"RR Slower preceding vehicle in the vicinity of prohibited overtaking". On a two lane rural road, the driver is confronted with a slower vehicle driving 80 km/h in the vicinity of prohibited overtaking. The situation is well-visible, optimal coasting lasts 110 m.

"RR Slower preceding vehicle and oncoming traffic". Drivers approach a vehicle driving 60 km/h on a rural road. Optimal coasting lasts 220 m. They are allowed to overtake it after the opposite lane is free from the oncoming traffic.

Situations on a Highway. "HR Speed limit". The allowed speed before this highway segment is 130 km/h. In the situation, the allowed speed is set down to 100 km/h. The Situation becomes visible about 300 m after the beginning of the optimal coasting phase (which is 450 m long) should have taken place.

"HR Stagnant traffic". The driver approaches a traffic congestion moving at 60km/h on the highway. This situation is well-visible and occurs shortly after "Speed limit on the highway". Optimal coasting lasts 220 m.

"HR Highway jam". This is the most critical situation investigated in the experiment. Drivers are driving on the highway with a speed of 130 km/h. They approach a curve behind which idle vehicles are located on all of the lanes. This jam tail becomes visible 250-300 m before the driver has to come to a stop.

Situations in Urban Environment. "UR Parking car and oncoming traffic". The permissible speed limit is 50 km/h in a town. In this situation, the driver has to decelerate because of a parking car occupying the driven lane. After the oncoming traffic on the opposite lane has passed, the driver can overtake the obstacle. The situation becomes visible at the distance of 140 m, so the driver has to brake slightly.

"UR Red traffic light". The traffic lights are in red phase before the driver approaches them. So he has to stop in front of them before he can pass. The suggested coasting phase lasts around 450 m.

2.5 Dependent Variables

The focus of the presented analysis is put on the reduction of the estimated fuel consumption and reaction times in the investigated situations.

The estimated fuel consumption is calculated for the entire drive course, and situation segments. An analyzed situation segment starts where the optimal coasting should begin when driving the permissible speed plus 100 m to take faster driving test subjects into account. It ends when the situation is bypassed.

In a safety critical situation, the analyzed segment begins 1000 m before it, which is the point at which an early comfortable braking not exceeding -3m/s² [10] should take place in order to avoid collisions.

The reaction time is defined as the time gap between the activation of the multimodal system and the point of time at which the driver steps of the accelator.

3 Results

This chapter presents the summary of the efficiency benefit reached during assisted drives and reaction times of the drivers on the assistance suggestions. Furthermore the general user acceptance of the haptic feedback via AGP, as well as detailed description of the results in every of the investigated deceleration situations is presented.

3.1 Situational Analysis of the Driving Behavior

Generally the release of the gas pedal after the corresponding multimodal advice is being issued by the system follows within the next 0.8 s[2] to 4.0 s[3]. In nine out of the twelve situations the 75% percentile of the reaction time is 1.75 s. These nine situations cover all highway and rural situations except "RR Slower preceding vehicle in the vicinity of prohibited overtaking". In this situation the 75% percentile is 3.99 s while the 50% percentile is 1.04 s. The slower preceding vehicle is visible to the

---

[2] Minimum of 50% percentile.
[3] Maximum of 75% percentile.

driver for a long time so the driver is able to react proper in this common situation without assistance. In the two urban situations the reaction takes place within 2.3 s in the 75% percentile.

The fast reaction times and, as a consequence, the long coasting phases lead to a significant reduction in fuel consumption. So throughout the entire drive, the test subjects are able to reduce fuel consumption on average 7.5% with activated assistance (estimated fuel consumption during baseline drives: Ø=100.0%, sd= 16.2%; assisted drives: Ø=92.4%, sd= 12.0%).

The visual distraction due to the proposed visual concept of the assistance system is investigated in a previous study and does not possess any critical or driving endangering character [6].

In most of the analyzed situations, the average reduction of the fuel consumption while driving with assistance ranges from 15% to 35%. This occurs due to the longer coasting phases which are being proposed by the assistance system, and, as a general rule, are reflected by the driver's corresponding actions as it is shown above. Exceptions are the situations which are visible to the driver before the optimal beginning of a coasting phase should take place and its length is below 200 m, and therefore also without the assistance drivers act efficiently: "RR Slower lead vehicle in the vicinity of prohibited overtaking", "HR Stagnant traffic", and "UR Parking car and oncoming traffic".

A noticeable benefit is established in situations, in which drivers get the coasting suggestion before they are able to perceive the situation on their own. 30-35% of fuel is saved in the situations "RR Construction site behind a curve on a straight segment", "RR Speed limit", "RR Speed limit and slower lead vehicle". Approximately 25% of the fuel is reduced due to the system's advice in "UR Red traffic light" situation, 20% - in "RR Construction site behind in a right curve", 15% - in "RR Slower preceding vehicle and oncoming traffic".

The greatest reduction of 60% by following the system's advice is achieved in the situation "RR Town entrance". The town is visible to the driver from larger distances, but during the baseline drives nobody prefers to decelerate to the target speed purely through coasting. The experienced coasting phase with the activated system of 800 m is subjectively perceived by test subjects as extremely prolonged, and AGP feedback has a vast patronizing character (discussed further in this paper). However, the test subjects decide to follow the advice in the experiment drive.

The results regarding "HR Speed limit" are following: baseline drives Ø=100%, sd=50%, assisted drives Ø=47%, sd=118%. Eleven test subjects decide to accelerate during assisted drives again after initially performing the coasting phase, which is reflected in standard deviation.

Detailed explanation of the driving behavior can be found in [7].

3.2 Results Regarding the Drivers Acceptance

In Fig. 3, the number of participants who evaluate the influence of the multimodal assistance system in particular situations as helpful, neither helpful nor irrelevant, and irrelevant are provided. In the very critical situations "HR Highway jam" none of the drivers evaluates the system negative. Other critical situations like the construction sites on the RR, the "HR Stagnant traffic" situation and the noncritical situations with

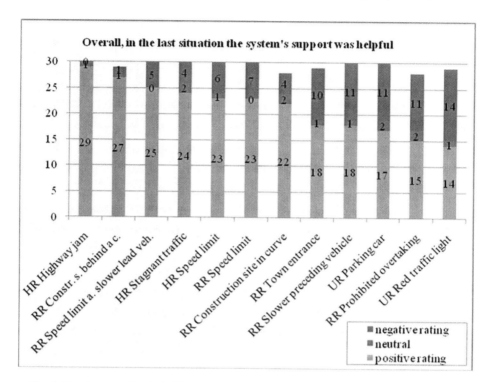

Fig. 3. Results regarding helpfulness of AGP activation in different deceleration situations

speed limits are also assessed very positive by the majority of the drivers. More than half of drivers estimate the system as helpful in "RR Town entrance" and "RR Slower proceeding vehicle and oncoming traffic". The acceptance of the system declines in the two analyzed urban situations and "RR Slower leading vehicle in the vicinity of prohibited overtaking". In the urban situation with a parking car and the prohibited overtaking on a rural road the suggested coasting phases are less than 200 m and the drivers are able to perform these well by their own without assistance. The situation "UR Red traffic light" suggests a very long coasting phase of 450 m and therefore the acceptance drops down.

The results regarding AGP appeal to the test subjects in deceleration situations are following: eight test subjects do not show any interest in the currently proposed concept of AGP assistance (two of whom clearly state their dissatisfaction), 21 participants find the concept as "good" or "very good", and one test subject cannot decide. However, the comments of four test subjects which do not like the current AGP activation strategy show that they could imagine great benefit of the discrete AGP assistance in very rare, critical situations. This also complies with the summary of the rest of the comments, in which the participants value the AGP activation especially in seldom, unexpected, and critical situations. Overall, it can be stated that test subjects perceive haptic assistance in form of discrete impulses of the gas pedal in deceleration situations rather as a warning, than as information indications for a fuel saving strategy. However, inter-subjective opinion on which situations one should be warned and AGP is "not patronizing" differs. Almost in every situation the majority of test subjects admit the helpfulness of the AGP feedback for reduction of fuel consumption, though in urban situations this number considerably decreases.

An important factor influencing the acceptance of AGP is the negative feeling the user gets when feeling patronized by the system (Fig. 4.). In urban situations and situations which can be seen early enough to undertake an efficient deceleration strategy without any assistance the negative patronizing effect of AGP is especially obvious ("UR Parking car and oncoming traffic", "RR Town Entrance", "UR Red traffic light", "RR Slower preceding vehicle in the vicinity of prohibited overtaking", "RR Slower preceding vehicle and oncoming traffic"). The majority of the test subjects in these situations either explicitly stated their dissatisfaction, or never experienced the assistance activation because they themselves were taking the optimal course of driving actions. The presence of urban situations in this category should not lead to a conclusion that no assistance would be accepted in the investigated situations.

A high quota of participants felt patronized by AGP when approaching speed limit signs: 14 test subjects in both situations "RR Speed limit and slower lead vehicle" and "HR Speed limit", 13 – in the situation "RR Speed limit".

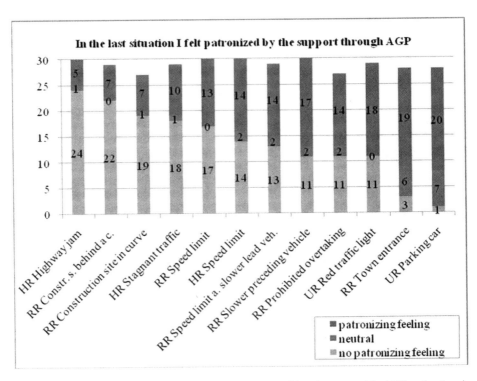

Fig. 4. Results regarding patronizing feeling experienced by the user with AGP activation in different situations

The acceptance of the AGP advice without feeling patronized is comparatively high in the highway and rural road situations, which cannot be perceived by the driver early and possess unexpected character ("HR Highway jam", "RR Construction site behind a curve on a straight segment", "RR Construction site behind a right curve"). Also the situation "HR Stagnant traffic" is considered by the test subjects as potentially critical, and therefore AGP advice is perceived as applicable.

23 out of the 30 test subjects see the potential benefits of the proposed concept and would be ready to activate it on the highway roads, and 21 on rural roads.

Detailed explanation of the driver's acceptance can be found in [7].

4 Summary

The study presents an investigation of a multimodal human-machine interface (HMI) of an anticipatory driver assistance system. The HMI of the system consists of visual indicators displayed in the digital instrument cluster and a discrete impulse of the active gas pedal (AGP). The assistance recognizes the upcoming driving situation, informs the driver, and suggests the driving action, which execution assures significant reduction in fuel consumption.

The experiment was performed at the fixed-base driving simulator located at the Lehrstuhl für Ergonomie, Technische Universität München. 30 test subjects took part in the experiment; their average age was 37 years. The twelve included deceleration situations systematically differ in their criticality and the duration of the suggested coasting phases. In the fixed-base simulator experiments, the quantitative values for some of the driving measures may differ from those of real drives. However, a clear tendency can be derived.

In 50% of all cases the participants step of the gas pedal for the first time within 1.2 seconds after the first information. In 80% of all cases the reaction takes place within 2 seconds. Throughout the entire drive, the test subjects are able to reduce fuel consumption on average by 7.5% with activated assistance. This occurs due to the fast reaction and longer coasting phases which are being proposed by the assistance system, and, as a general rule, are reflected in driver's actions.

The results regarding AGP appeal to the test subjects in deceleration situations are following: eight test subjects do not show any interest in the currently proposed concept of AGP assistance (two of whom clearly state their dissatisfaction), 21 of the 30 participants assess the concept as "good" or "very good", and one test subject cannot decide. Overall, it can be stated that test subjects perceive haptic assistance in form of discrete impulses of the accelerator in deceleration situations rather as a warning, than as information indications for a fuel saving strategy.

Up to 80% of the test subjects prefer activation of an AGP on rural and highway roads, especially in the situations which cannot be perceived at larger distances, which are considered to be of rare occurrence, and which can demand extreme decelerations if the driver is not informed in advance. Such situations include construction sites on rural roads, stagnant traffic and jams on highways. Situations with slower moving vehicles are excluded from the list where AGP assistance would be welcomed. It can be concluded, that the haptic impulse of the gas pedal is subjectively felt as a warning regarding approaching deceleration situations, rather than an advice to coast a vehicle for a gain in fuel consumption. Low interest for AGP is observed in the investigated urban situations.

References

1. Busch, F.: Car-to-X im Verkehrswesen, munich network Tagung, München (2007)
2. Härri, J., Hartenstein, H., Torrent Moreno, M., Schmidt-Eisenlohr, F., Killat, M., Mittag, J., Tillert, T.: Car-to-X Communication Simulations: Tools, Methodology, Performance Results. In: Network on Wheels (NoW) final Workshop, Daimler AG, Ulm (2008)
3. Donges, E.: Ein regelungstechnisches Zwei-Ebenen-Modell des menschlichen Lenkverhaltens im Kraftfahrzeug. Zeitschrift für Verkehrssicherheit 24, 98–112 (1978)
4. Nestler, S., Duschl, M., Popiv, D., Rakic, M., Klinker, G.: Concept for Visualizing Concealed Objects to Improve the Driver's Anticipation. In: Proc. 17th World Congress on Ergonomics IEA, Beijing, China (August 2009)
5. Samper, K., Kuhn, K.-P.: Reduktion des Kraftstoffverbrauchs durch ein vorausschauendes Assistenzsystem. Düsseldorf, aus: VDI Berichte 1613, VDI Verlag (2001)
6. Popiv, D., Rommerskirchen, C., Bengler, K., Duschl, M., Rakic, M.: Effects of assistance of anticipatory driving during deceleration phases. In: Proc. European Conference on Human Centered Design for Intelligent Transport Systems, Berlin (2010)
7. Hajek, H.: Untersuchung des Einflusses von einem aktiven Gaspedal auf die Fahrerreaktion und die Akzeptanz zur Unterstützung des vorausschauenden Fahrerns. Lehrstuhl für Ergonomie, Technische Universität München, unpublished Diploma Thesis (2010)
8. Rommerskirchen, C.: Zeitliche Anforderungen zur Unterstützung von vorausschauenden Fahren unter Optimierung von Effizienz und Akzeptanz, Lehrstuhl für Ergonomie. Technische Universität München, unpublished Diploma Thesis (2009)
9. SILAB, http://www.wivw.de/ProdukteDienstleistungen/SILAB/index.php.de/
10. Heißing, B.: Fahrwerkshandbuch. Vieweg und Teubner, Germany (2008)

Real User Experience of ICT Devices among Elderly People

Ayako Hashizume[1], Toshimasa Yamanaka[1], and Masaaki Kurosu[2]

[1] Graduate School of Comprehensive Human Science, University of Tsukuba, Japan
hashi-aya@kansei.tsukuba.ac.jp, tyam@geijutsu.tsukuba.ac.jp
[2] The Open University of Japan, Japan
masaakikurosu@spa.nifty.com

Abstract. As a representative device of ICT-related devices and systems, authors selected the mobile phone for our research. Authors started their analysis from the questionnaire research roughly focusing on the use of mobile phone. As a result, it was revealed that the elderly people have low literacy for using the mobile phone compared to the young people. Furthermore, the elderly people living in urban area have higher level of mobile phone literacy, while those who are living in rural area tend to rely on others, especially young people. In order to analyze the difference between urban are and rural area and between young people and elderly people, authors then conducted a field survey adopting the contextual inquiry and analyzed the data by applying M-GTA, then summarized the information as a category relationship diagram. In the diagram, such factors as the motivation for using mobile phone, the active involvement to the communication and the mobile phone literacy were regarded as the principal components.

Keywords: user experience, usability, mobile phone, elderly people, M-GTA, literacy.

1 Introduction

In Japan, the progress of aging is very rapid and since 2005 Japan has become the most aged society in the world. Whereas the progress of ICT is smoothly going upward

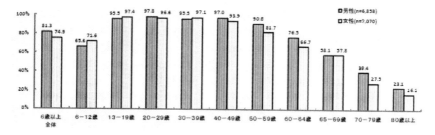

Fig. 1. Usage ratio of internet for each age group (male: grey, female: white). Cited from the report of Ministry of Public Management, Home Affairs, Posts and Telecommunications (2010).

M. Kurosu (Ed.): Human Centered Design, HCII 2011, LNCS 6776, pp. 227–234, 2011.
© Springer-Verlag Berlin Heidelberg 2011

recently, it was found that the elderly people are low in their usage rate of internet, for example, compared to young thru middle-aged people as is shown in Fig. 1.

As the aging progress, various factors shown in Fig. 2 contribute to the change of social relationship and communication.

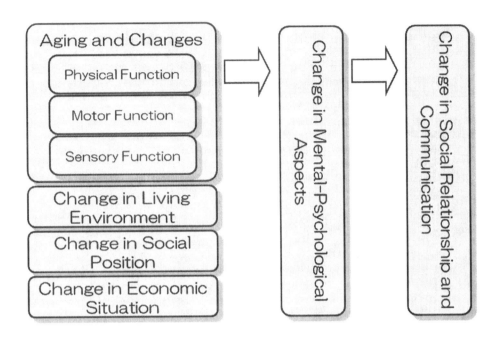

Fig. 2. Changes that accompany with aging

In addition to the age group difference, there is also a regional difference. Mobile phones and computers are much used in such urban areas as Tokyo, Kanagawa, Chiba, and Saitama while they are not much used in such areas as Okinawa, Hokkaido and Aomori.

Based on the conceptual analysis, Fig. 3 was described that shows how changes in society may affect the life of elderly people.

Because the lack of communication and the lack of information seemed to be two of the key factors for the negative aspects in the life of elderly people, authors decided to focus on how they can become able to use ICT-related device and system, especially the mobile phone with an expectation that the facilitation of such device and system may decrease the negative aspects in their life.

As for the research method, authors adopted the mixed methods research (Creswell 2003) that combines the quantitative approach such as the questionnaire and the qualitative approach such as the ethnographic research.

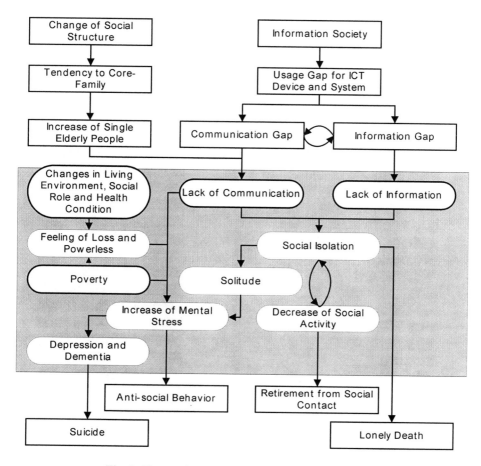

Fig. 3. Changes in society and the life of elderly people

2 Study 1: Questionnaire Research (Quantitative Approach)

A questionnaire research was conducted for clarifying how the use of the ICT-related device and system, especially the mobile phone, differs between young people and elderly people and between urban are and rural area.

2.1 Method

The questionnaire contained such questions as human-relationship, use of high-tech devices, mobile phone, various means for communication, communication means depending on situation, etc.

Informants were 77 male (20's 42, 60-70's 25 in urban area, 60-70's 10 in rural area) and 61 female (20's 26, 60-70's 25 in urban area, 60-70's 10 in rural area).

2.2 Result

Fig. 4 shows the comparison of young people and elderly people on how they know about the function of mobile phone. It is clear that there is a distinct difference between the young people and elderly people.

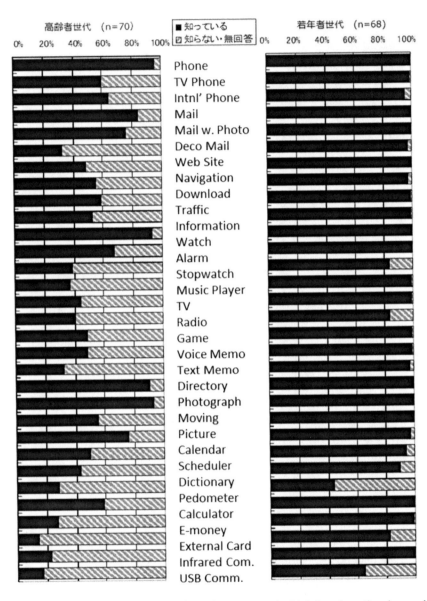

Fig. 4. Comparison of elderly people (left) and young people (right) on how they know about the functionality of mobile phone

Fig. 5 shows the comparison of senior people living in urban area and rural area on how they know about the function of mobile phone. It is evident that elderly people living in urban area are more acquainted with various functions of mobile phone.

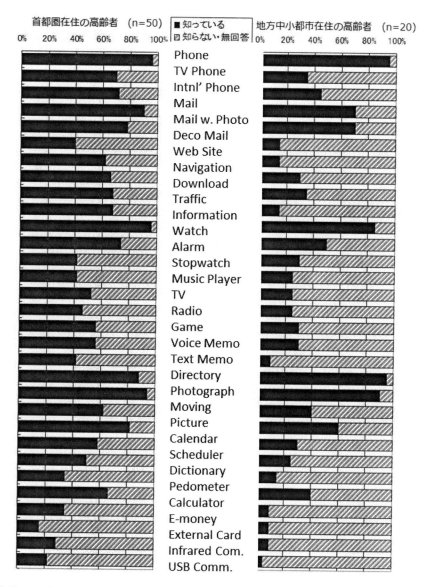

Fig. 5. Comparison of urban area (left) and rural area (right) on how elderly people know about the functionality of mobile phone

3 Study 2: Ethnographic Research (Qualitative Approach)

Although the result of study 1 brought us an outline on the use of mobile phone, individual analysis was not possible to see how individualistic factors may affect the use of the mobile phone among senior people. This is the reason why authors adopted the qualitative approach in our 2nd study.

3.1 Method

The focus of this 2nd study is described in Fig. 6.

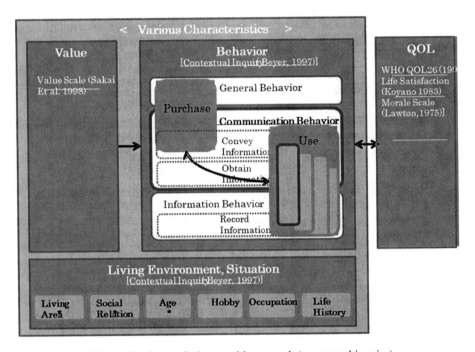

Fig. 6. The focus of ethnographic research (contextual inquiry)

Informants were 18 male (4 20's and 6 60-70's in urban area and4 20's and 4 60-70's) and 18 female (the same).

Research questions included following items: 0. life history, social relationship, hobby, job career, use of the mobile phone and other ICT-related device and system, 1. Value based on the notion of Spranger, 2. Possession and usage of ICT-related device including mobile phone, 3. Use of device depending on the situation, 4. Criteria for information search and communication, 5. QOL and satisfaction, 6. PGC Morale Scale, 7. Detailed information about the mobile phone and other device.

3.2 Result

The interview (contextual inquiry) data was first transcribed and then analyzed by M-GTA (Kinoshita 1999, 2005) using MAXQDA software. We finally draw the category relationship diagram as shown in Fig. 7.

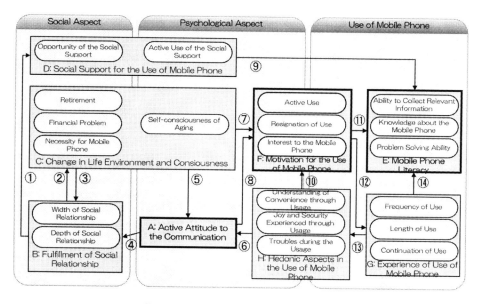

Fig. 7. Category Relationship Diagram

Motivation	High H**		Low L**	
Literacy	High HH*	Low HL*	High LH*	Low LL*
High **H	**HHH** YC1 | YR1 YC4 | YR2 YC7 | YR3 | YR7	**HLH**	**LHH** YC5 YC6	**LLH**
	SC1 | SR3 SC3 | SR8		SC5 SC9	SC7 | SR2 SC10 | SR4
Middle **M	**HHM** YC2 | YR4 YC3 | YR5 YC8 | YR6	**HLM**	**LHM** YR8	**LLM**
	SC2 | SR1 SC6 | SR7		SC4 SC8 SC12	SC11 | SR5 | SR6
Low **L	HHL	HLL	LHL	LLL

Active Attitude to the Communication

Fig. 8. Plot of Informants based on important 3 dimensions

In Fig. 7, the motivation, the literacy (for the mobile phone), the active attitude to the communication were regarded fundamental dimensions in terms of the use of mobile phone among elderly people. Hence, we formed the table as Fig. 8 and plotted the informants (as shown as YC1 or YR1 etc).

4   Conclusion

Findings based on the data analysis are as follows:

1. In order to increase the mobile phone literacy among the elderly people, the motivation and the real experience of usage are quite important.
2. If an adequate social support is given to the elderly people and the problem they may be facing could be solved, the mobile phone literacy will be increased.
3. Factors that facilitate the motivation for the use of mobile phone include the active attitude to the communication and the positive hedonic experience during the usage.
4. On the contrary, factors that demoralize the use of mobile phone include the change of life environment due to the retirement, the decrease of necessity of use of the mobile phone, the decrease of motivation to the active communication due to the self consciousness of the aging, etc.
5. But if the retirement may be regarded positively as the attainment of freedom, the elderly people may have the active motivation. And because they may have more free time than when they were working, the interest to the mobile phone can be increased.
6. Furthermore, changes in the life situation and the life consciousness will influence the social relationship in general. This fulfillment of the social relationship is related to the existence of the opportunity to get the social support and the acceptance of such support.

References

Beyer, H., Holtzblatt, K.: Contextual design: defining customer-centered systems. Morgan Kaufmann Publishers Inc., San Francisco (1997)

Creswell, J.W.: Research Design –Qualitative, Quantitative, and Mixed Methods Approach, 2nd edn. Sage Publications Inc., Thousand Oaks (2003)

Kinoshita, Y.: Grounded Theory Approach (in Japanese) Kobundo 1999 (2005)

Ministry of Public Management, Home Affairs, Posts and Telecommunications A Report on the Use of Communication System in 2009 (2010)

Spranger, E.: Lebensformen, Max Niemeyer (1921)

WHO ICF (International Classification of Functioning, Disability and Health) (2001)

Human Affordance as Life-Log
for Environmental Simulations

Masayuki Ihara[1], Minoru Kobayashi[1], and Takahiro Yoshioka[2]

[1] NTT Cyber Solutions Laboratories, NTT Corporation
1-1, Hikari-no-oka, Yokosuka, 2390847 Japan
{ihara,minoru}@acm.org
[2] Graduate School of Information Science and Electrical Engineering,
Kyushu University, 744 Motooka, Nishi-ku, Fukuoka, 8190395, Japan
takahiro.yoshioka@inf.kyushu-u.ac.jp

Abstract. This paper presents the design principle of establishing environmental simulation systems on the "human affordance" collected as user life-log. We envisage that combining life-log applications with a consideration of cognitive science will yield better life-log utilization. Research questions in this study are how to collect life-logs without user resistance to exposing the logs and how we can continuously utilize the latest life-logs. Our answer to the first question is to transform the recorded data to the extent that the user willingly accepts the automatic release of his/her life log. Our answer to the second question is to employ the affordance theory in cognitive science.

Keywords: Life-log, human affordance, transformation, environmental simulation.

1 Introduction

A variety of life-logs can be recorded in our daily life such as location data or device operation history. These life-logs can be used to create recommendations based on an inference of user preferences or activities. Such recommendations are a user-feedback type of service. We note that the life-logs from many users in the world could be used for environmental simulations, which will play an important role in visualizing future problems with the earth and the identification of solutions (Fig. 1). In order to achieve such simulations, it is important to well design the system, which must integrate user devices, networks and simulation servers. To determine the design principle for these system, this study addresses the research questions of how to collect life-logs without user resistance to exposing the logs and how we can continuously utilize the latest life-logs. Our answer to the first question is to include a privacy control function on the user's own device and to transform the recorded data to the extent that the user has no resistance to the automatic release of his/her life-log. Our answer to the second question is to employ the affordance theory in cognitive science. In this paper, we propose the design principle of establishing environmental simulation systems that use "human affordance" collected as user life-log. We envisage that combining life-log applications with a consideration of cognitive science will enhance life-log utilization.

M. Kurosu (Ed.): Human Centered Design, HCII 2011, LNCS 6776, pp. 235–242, 2011.
© Springer-Verlag Berlin Heidelberg 2011

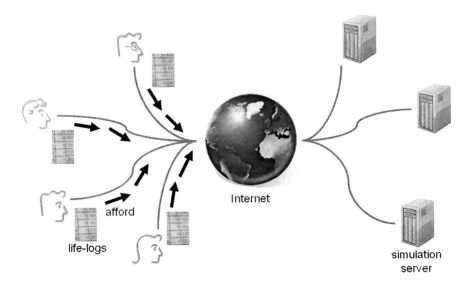

Fig. 1. A simulation system using life-logs from many users in the world

2 Research Background

Life-log technologies can be categorized into three types; collecting, processing and presenting. Sensors can be regarded as being covered by collecting technology. One example is logging the remote controller operations of an appliances [1]. Data mining for recommendation and user feedback interface are covered by processing and presenting technologies, respectively. The effective combination of those three life-log technologies is demonstrated by recent location-based services (LBS) using GPS (Global Positioning System) or Wi-Fi. In particular, a large amount of life-logs from people in many areas is regarded as collective intelligence and is being used for several kinds of simulation [2][7]. Many kinds of life-log will be used for LBSs in the future while identification number and position being the dominant data.

Environmental simulations play an important role in visualizing future problems with the earth and the attempt to find solutions. To raise the accuracy and effectiveness of the simulations, more simulation systems must process life-log data. Note that the same life-log data will be used by several simulation systems. For example, air conditioner logs can be used for both simulating the power consumption in a city and global warming. All such data can be collected and made accessible through the Internet. This means that it might be possible for future simulations to use real-time data collected from all over the world.

For future environmental simulations based on many kinds of life-log, this paper proposes the design principle of establishing environmental simulation systems that use "human affordance" collected as user life-log. We envisage that combining life-log applications with a consideration of cognitive science will yield better life-log utilization and that our design principle will stimulate the research field of human-computer interaction.

3 Problems with Life-Log Application

3.1 Continuous Life-Log Utilization

In order to continuously utilize the latest life-logs, we employ the "affordance" theory of cognitive science. In life-log-based simulations, there is a risk that user life-logs might not be sent or that their transmission will be intermittent. Ideally, life-logs should be sent to a server automatically but with the user's preferred exposure setting. To do this, we employ the affordance theory. For continuous life-log utilization, system design should not force users to perform additional operations to record logs.

3.2 User Resistance to Releasing Life-Logs

Users resist the release of their life-logs. A questionnaire answered by 1,104 experimental subjects revealed that 40% to 50% of them did not want to release their life-logs regardless of the kind; information search history, purchase history, geographical movement data etc. In order for users to accept the release of their activity data, their recorded life-logs should be stored on the user's own device such as a cell phone, and be sent to a server following commonly agreed principles. Another way to gain the user's acceptance of data release is log transformation. Transformation is needed because users are worried about sending negative data (such electricity use) to a third party such as a simulation system.

4 Proposed Design Principle

We propose the design principle of establishing environmental simulation systems that use "human affordance" collected as user life-log.

4.1 Affordance Theory

The term of affordance comes from the perceptual psychologist Gibson, who provided an ecological alternative to cognitive approaches[3][4]. His theory is that *the*

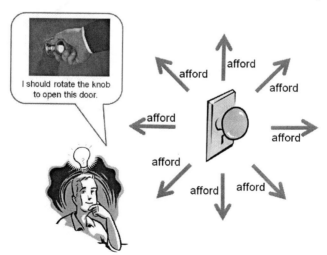

Fig. 2. Affordance theory: The attribute of the knob is afforded by its round shape

affordances of the environment are what it offers the animal, what it provides or furnishes, either for good or ill. The concept of affordance is popular in the field of user interface design as it provides a means of enhancing usability[6]. Fig. 2 shows the affordance of door knob. Human can understand how to operate it from its round shape.

4.2   Human Affordance

So as to utilize the cognition viewpoint for engineering, we extend the concept of affordance to cover human activities [5]. Human affordance is afforded from humans, not artifacts. In Gibson's affordance theory, information about an artifact is afforded to the ambient space through its "look" no matter whether a human perceives the artifact or not. As shown in Fig. 3, in the same way information about a human is afforded to the simulation server as the automatically gathered life-logs.

One advantage of human affordance is the focus it places on human factors, which will yield user-centered design of life-log-based simulation systems. Many kinds of simulations can be conducted anytime using desired data taken from the user life-logs. This is because many computers running simulation programs will be connected to the network and the life-logs collected from many users will always be ready to be used. This simulation flexibility in program choice, data range, and execution timing will lead to advanced simulations.

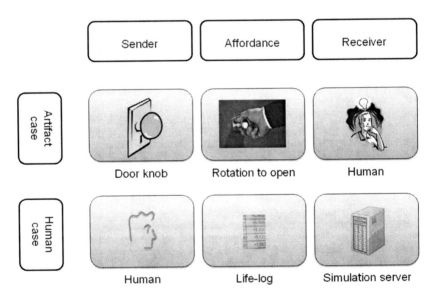

Fig. 3. A metaphor to understand human affordance

4.3   Encouraging the Release of Log Data

The underlying reason why users resist the exposure of their own life-logs is their fear of creating a negative impression. They do not want to afford information about their activities which would give others negative impression on themselves such as the overuse of electricity data. This hesitancy can be overcome by transforming the data to eliminate the negative aspects. For example, if the user runs the air conditioner continuously for one month, the user's device sends only the average value of the other months. When the user alters the device usage pattern, switching the air conditioner off when not needed, the real air conditioner log is sent as environmental contribution data.

5 Experiments

We are investigating the degree of which life-logs should be transformed to create a positive feeling and thus secure release of the data. In this section, we introduce the experiments we conducted to examine user resistance to release of life-logs recorded on a cell phone device.

5.1 Experimental Design

The aim of this experiment is to determine the level of user resistance to transformed life-log data and to know what kind of logs trigger the strongest resistance. A questionnaire survey was conducted by 30 subjects in their 20s to 50s. We gave each subject the following instruction, question and answer choices.

Instruction:
   Please assume that you live in a society where the logs on your cell phone such as device identification number, location and battery power were collected by a service provider for urban planning (i.e. environmental protection). Note that the device identification number will be used only for information feedback to you and is never used for any other objective.

Question:
   How do you evaluate your resistance to the release of your logs in the following three exposure cases?
      Exposure case:
      A: Release as raw data with short cycle (e.g. every 5 seconds)
      B: Release as raw data with long cycle (e.g. every 1 hour)
      C: Release as transformed data with some cycle (e.g. every 5 minutes)

Answer:
   1. I agree because the release will lead to my convenience or contribute to society.
   2. I accept if the released data will be kept anonymous.
   3. I think that there is no choice though I am worried.
   4. I do not want to release data.

Note that the answer 4 indicates the highest level of resistance. The cell phone logs that subjects should evaluate in this experiment are a device identification number, location (latitude and longitude), whether talking on the phone or not, and battery level. Each subject was instructed additionally in the case of transformation (Case C).

The location is transformed to a name of city or town. Whether talking or not is transformed to a total time of talking. The battery level is transformed into the amount of change in battery level.

5.2 Results

The questionnaire results are shown in Fig. 4 to Fig. 7. The horizontal axis is the degree of transformation. The vertical axis is the percentage of subjects who responded with each resistance level.

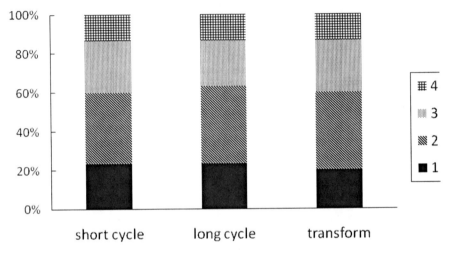

Fig. 4. Result for Device ID

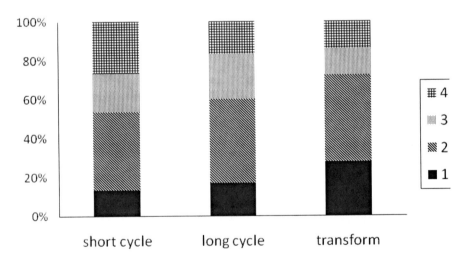

Fig. 5. Result for location

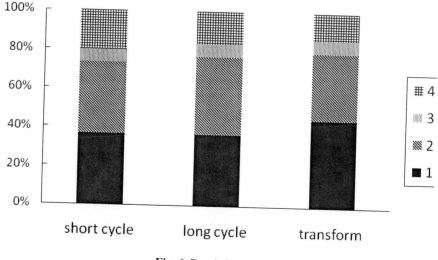

Fig. 6. Result for phone

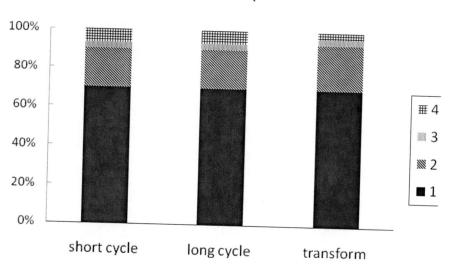

Fig. 7. Result for battery

6   Discussion

Figs. 4 to 7 prove that people strongly resist the release of location data and less concerned about battery data regardless of the level of transformation. As shown in Figs. 5 and 6, the level of resistance to the release of location data and phone use do not depend on the level of transformation. The decrease in resistance is not based on any specific transformation method but on the transformation concept itself. Future studies should examine specific transformation methods so as to eliminate negative factors in user life-logs and thus encourage user acceptance. For example, the transformation that protects the environmental credentials of the user will be accepted because it provides the user with the impression of being a good steward of the environment. The first step is to encourage the acceptance of releasing life-logs. Such transformation will be more powerful when combined with sensing and information feedback technologies, for example, the EnergyLife [8] and the DEHEMS [9].

Another interesting point in this experiment is the low resistance to exposing the battery data (See Fig. 7). Although the amount of battery use on each cell phone device is very small, the total amount of batteries in use all over the world is too large to ignore. Given that the target of this study is the environmental simulation system of global scale, information about cell phone battery consumption is of significant interest.

7 Conclusion

This paper presented the design principle for establishing environmental simulation systems. We employed the concept of human affordance for continuous life-log utilization, and conducted a questionnaire survey to investigate the potential of life-log transformation in reducing the resistance of users to exposing their life-logs. Future work will include a study on effective life-log transformation methods and the development of an environmental simulation system that utilizes life-logs afforded from many users.

References

[1] Abe, M., Morinishi, Y., Maeda, A., Aoki, M., Inagaki, H.: A Life Log Collector Integrated with a Remote-Controller for Enabling User Centric Services. IEEE Transactions on Consumer Electronics 55(1), 295–302 (2009)

[2] Cuff, D., Hansen, M., Kang, J.: Urban Sensing: Out of the Woods. Communications of the ACM 51(3), 24–33 (2008)

[3] Gibson, J.: The Senses Considered as Perceptual Systems. Allen and Unwin, Ltd., London (1966)

[4] Gibson, J.: The Ecological Approach to Visual Perception. Houghton Mifflin, New York (1979)

[5] Ihara, M., Kobayashi, M., Sakai, Y.: Human Affordance. International Journal of Web Based Communities 5(2), 255–272 (2009)

[6] Norman, D.A.: The Psychology of Everyday Things. Basic Books, New York (1988)

[7] PlaceEngine, http://www.placeengine.com/

[8] Jacucci, G., Spagnolli, A., Gamberini, L., Chalambalakis, A., Björksog, C., Bertoncini, M., Torstensson, C., Monti, P.: Designing Effective feedback of Electricity Consumption for Mobile User Interfaces. PsychNology Journal 7(3), 265–289 (2009)

[9] Sundramoorthy, V., Liu, Q., Cooper, G., Linge, N., Cooper, J.: DEHEMS: A user-driven domestic energy monitoring system. In: Internet of Things 2010 Conference, pp. 1–8 (2010)

Guideline Development of Mobile Phone Design for Preventing Child Abuse

JoonBin Im, Eui-Chul Jung, and JungYoon Yang

Dept. of Human Environment & Design, Yonsei University
50 Yonsei-Ro, SeoDaeMun-Gu, Seoul 120-749, Korea
joon99@nate.com, jech@yonsei.ac.kr, jyoon08@naver.com

Abstract. This paper is studied to develop a guideline of mobile phone design that protects a child from crimes. It is studied that the criminal commits crimes through the three steps of 1) approach, 2) attract, and 3) plunder – movement. Through this research, it was found that the child's recognition of the criminal situation could prevent the child from crimes. Because of limitations of child's cognition, this study is conducted to provide diverse solutions that could notify abnormal signs to the guardians using a mobile phone, even when the child has no awareness of the situation. In addition, the patterns of the child's usual route and walking speed are logged for detecting the criminal situation. This feature enables the guardians to intervene the crime. A design guideline for developing hardware and applications of a mobile phone is suggested through case analysis and expert interviews.

Keywords: Cracking Crime, Anti-Crime Design, Child Abuse, Mobile Phone Design, Design Guideline.

1 Introduction

Recently, four major violent crimes are increasing every year. Among those crimes, crimes against a child are showing gradual increase [1]. Although government and prosecutor's office has announced they would take strong and rigid action against child abuse, crimes continued. It is pointed out that the crime rate still increases even with the new legislations and revised policy because all these efforts focus on the follow up measures rather than prevention. Therefore, the goal of this research is to propose hardware and App. design guideline using a mobile phone that can prevent crimes against a child. To propose the guideline, this research was conducted with four phases. In the Phase 1, characteristics of a child and child abuse cases, existing prevention systems, and existing mobile phones were analyzed to understand the present state of related researches, and to find opportunities. In the Phase 2, initial guidelines were proposed based on the possibilities of prevention in each step of child crime processes by analyzing criminal cases and composing scenarios. In the Phase 3, experts' interviews were conducted in order to evaluate the scenarios and guidelines. In the final phase, a mobile phone design guideline for prevention focusing on hardware and App. was summarized.

M. Kurosu (Ed.): Human Centered Design, HCII 2011, LNCS 6776, pp. 243–252, 2011.
© Springer-Verlag Berlin Heidelberg 2011

2 Theoretical Background

This chapter analyzes children's performance and circumstantial judgment through literature, and studies preventing methods through understanding the procedure and characteristics of precedents of crimes against child for the past seven years.

2.1 Characteristics of a Child

Child refers to an age of 6 to 12, and during this period child learns similar tasks that he or she has to perform once they become an adult [2]. Child development is often discussed within the perspective of developmental psychology, and defined in biological and psychological perspectives, and believed to have characteristics of dependency, continuous development, susceptibleness, demand, and adaptability [3]. Jean Piaget's "concrete operations stage" which is third stages of four stages of development applies to 7-11 year old children and believed that children at this stage can perform simple operation, and logical reasoning replaces intuitive thinking which means a child can operate a mobile phone. However, how well a child can operate a mobile phone in case of emergency need to be carefully examined.

Table 1. Mobile phone design in accordance with characteristics of children

Items	Contents
General Characteristics	-should be able to get attention from same age group -should seek for a way to educate children about utilization in case of emergency through game -should prepare for malfunction caused by shock -should educate children to avoid unintended operation out of curiosity -child should be familiarized with operation
Physical Characteristics	-design should consider physical development of childhood -design should consider safety of a child so he/she does not get injured from unintended operation due to physical development
Cognitive Characteristics	-cognition of age of 6, 9, and 12 is different and should have flexible structure or icon design for respective ages. -not only a button but also a variety of methods need to be considered to respond in case of emergency -circumstantial judgment and discernment may insufficient compare to an adult

Therefore, followings need to be considered when designing a mobile phone for kids. Although child at concrete operations stage can perform operation that requires quite complicated and logical thinking, it must be designed so a child can operate quickly in case of emergency. Table 1 shows list under consideration in accordance with characteristics of children.

2.2 Characteristics of Child Abuse

Based on the analysis of the precedents, what marks the child abuse is that most of the crimes involve sexual motivation, although some leads into murder of a child. In order for a crime to occur there must be three prerequisites to be complete, i.e. 1) crime will, 2) victim, 3) crime scene. Once these are satisfied, there are five phases to commit a crime; crime prerequisite – approaching – luring and kidnapping – movement – committing crime. The intention of classification of the crime is to extract the Design Specification of the cell phone for children that can respond to each phases of the crime and prevent it beforehand. First of all, it is found that the prevention of the first phase, crime prerequisite, is already been conducted in West in various ways. Nevertheless, the criminals will still approach to the victims, wearing electronic Tagging. Criminals would lure the victims using feigned identity or asking for directions or capture the victims by force and move to the certain spot and commit the crime. It is a fact that preschool, crime scene, and the criminal's house are located within the 300m in 'Kim Kil-Tae case', and it shows that lots of crimes are committed on the circulation of a child. Also a study shows that the luring point and the crime scene is only 30 min. away i.e. within 2km.

2.3 Prevention System and Current Technologies

To have better understanding of current technologies, domestic and foreign prevention system were compared and analyzed. Domestic system includes child safety guards, Amber alert, Electronic Tagging, posting the identities of sexual offenders. Foreign system includes isolation of the sexual offenders (U.S), Amber alert, posting sexual abuse record/identities (U.S), surgical or drug castration of sexual offenders (U.S.), and genetic database (U.K.). It shows that the domestic system shows more favor to the rights of a criminal over the victims or the many and unspecified possible victims' right. The followings are the current technologies used for crime prevention; SOS feature of mobile phone, LBS (Location Based Service), Electronic Tagging. These technologies cannot cognize criminals during first phase (approaching) and cannot be utilized. However, The followings are the current technologies used for crime prevention; SOS feature of mobile phone, LBS (Location Based Service), Electronic Tagging. These technologies were useful for second (kidnapping) and third phase (movement), however, these cannot cognize criminals during first phase (approaching) and cannot be utilized.

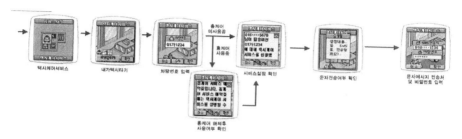

Fig. 1. Example of setting up the security company ADT's service

The reason why it is only useful for kidnapping and not luring for second phase is that child has to intentionally operate the system. SOS feature had reliability issue due to frequent unintended operation. LBS is effective system, however, as seen from Figure 1, security company ADT service is not easy for a child to operate. Electronic Tagging is being enforced under new legislation in 2007, however, only inspector receives signal from Electronic Tag and child cannot respond to it.

2.4   The Current Design of Mobile Phone for Kids

The evaluation of the existing mobile phones for kids (Verizon-Migo, Kyocera-Mamorino(for kids), Cingula-Firefly, willcom-Nico, au-A5520SA, Hop-on –Chitter chatter, willcom-papipo, Gigabite-Sergeant Keroro, Dmodo-M900) were conducted to analyze features that are needed to be considered in this study.

Table 2. Useful current technologies to prevent crimes against children

Categories	functions	Contents
Button	Specified Key	Easy operation for various ages and child with disabilities.
	Emergency Button	Informs the current situation and self location to the preregistered person in emergency
	Track Me	Can inform self location at any time
Power	Battery Cover	When the battery cover is forcibly removed, warning sign will be sent out to preregistered number
	Powerless Positioning Signal	Can notify one's location without any power in certain length of time
Lamp	Lamp	The lights will be turned on based on the setting.
Emergency	Strap	If a child pulls the strap, it will sound the alarm and transmits a text message to the preregistered number.
App.	GPS	Parents can confirm their child's location any time and will be notified when the child approach to a specific location.
	DB	Parents can protect their child from harmful contents and decorate the mobile phone using DB for kids.

The precedents analysis shows that it is not easy for a child to control the device in case of emergency, nor protecting a child through third party involvement seems an easy task. However, some of the features of the devices were useful such as A5520SA from AU which informs parents when a child approach to a specific location, and informs the child's location every certain length of time even if the power is off, as well as Chitterchatter's wearing method found to be useful. Table 2 summarizes useful features. The users have needs for normalization and avoid admitting that they are special [4]. Thus, mobile phones for kids currently available have very small LCDs or none at all; nevertheless, designs for both specified-key mode and 10-key mode would be viable by parents setting up the key-type screens, based on the Full Touch Technology, adjusted for their children's age-specific cognition level.

3 Design Concept of Mobile Phone for Preventing Child Crime

3.1 Preparing Scenario for Design Development

This paragraph summarizes 'characteristics of crime phases' and 'basic setting' for developing a scenario. Then more in-depth measures were analyzed by categorizing four types of crime phases and condition.

Characteristics and Procedures of Child Crimes. As described before, crimes against children involves crime prerequisites – approaching – luring and kidnapping – movement – commit a crime. The following Table 3 shows set up time and place for scenario development based on analysis of crimes. Based on Table 3, it is possible to make a scenario of a 35 year old man with two previous convictions (or less than twice), kidnapping a 11 year old girl near a local elementary school and bring her to his house and commit a crime. The scenario was compared with two precedents from district court and an article to verify the validity of it and they matched in many ways. It shows that the perpetrators stroll around the victim's course.

Table 3. Setting up for scenario development

Category	Detail	Setting	2km within crime scene	2km within luring point	2km within perpetrator's house
Perpetrator	Age/ gender	35 year old / Man	-	-	-
	With previous conviction	More than one previous conviction	-	-	-
	Without previous conviction	-	-	-	-
	Perpetrator's house	-	58%	57%	-
Victim	Age/gender	11 year old /girl	-	-	-
	Victim's house	-	72%	76%	50%
	Victim's school	-	74%	75%	50%
Crime	Time	15:00	-	-	-
	Date	Weekend	-	-	-
	Place	Playground, road, park	96%	-	58%
	Crime scene	Perpetrator's house	-	-	57%

It is concluded that if child's routine pattern is recorded and if parents can detect any abnormal activity such as different routine pattern or sudden acceleration of a speed, parent will be informed in case of emergency of their child, even if a child was kidnapped forcibly and not able to make any signals.

Categories of Scenario for Child Crimes. The scenario was written based on crime categorization of Table 4 and was used to support reasons for necessary mobile phone features.

Table 4. Setting up four categories

Criminal records	methods	cognition	categories
Less than two previous conviction	Luring	does not perceive danger	category 1
	kidnapping	recognize danger	category 2
More than two previous conviction	Luring	does not perceive danger	category 3
	kidnapping	recognize danger	category 4

Category 1: without Electronic Tagging and luring. Category1 shows case of a criminal who has less than two previous convictions approaching a victim of 6-12 year old girl. This is categorized as the worst case for the perpetrator can hide intention and child can never be aware of threat until right before a crime. Child's mobile phone detects change of speed and breaking the course and notifies guardians, and guardians may access to remotely controlled camera and microphone to check and respond to the situation.

Category 2: without Electronic Tagging and kidnapping. Category 2 shows case of threatening and forcible kidnapping by a criminal with same type of previous conviction. If a child cognizes the situation from threatening of a criminal, he/she can press emergency button and notify to third party. However, when child is being kidnapped forcibly or too nervous to control the device, the mobile phone will automatically notify the guardians about the change of speed and breaking course

Category 3: with Electronic Tagging and luring. Category 3 shows crimes committed by 40 year old man with more than one previous conviction, wearing Electronic Tagging. Proposed mobile phone receives signals from Electronic Tagging, however, does not inform the child to avoid the human rights violation of the ex-convict and also, it is not a crime to be nearby. This information will be informed to guardians and they may determine the level of threat through remotely controlled camera and microphone and sound the alarm or report to the police.

Category 4: with Electronic Tagging and kidnapping. Category 4 is the same scenario with category 3 which is criminal wearing Electronic Tagging and kidnapping a child by threatening. In such case, child recognizes a danger and may take immediate action to notify a third party by pressing emergency button. Even if child fails to press button, he/she still can get help because their guardians are being notified already by receiving signals from Electronic Tagging using mobile phone.

3.2 Drawing Mobile Phone Design and Demanding App.

As stated previously, prevention approach may different depends on whether a child recognize threat or not, and to have better understanding of this matter, mobile phone design and demanding app. service i.e. feature and applying the feature to the each crime phases is summarized at Table 5.

Table 5. (Part of) Studying effectiveness of demanding feature to each crime phases

Type	Mobile phone for kids	Phase 1	Phase 2 kidnapping	Phase 2 luring	Phase 3	Phase 4
hardware	Button for parent	-	O	V	O	O
	Transmitting preregistered short message	-	O	O	O	-
	Remotely controlled camera to watch their child	O	O	O	O	O
	Remotely controlled microphone to listen	O	O	O	O	O
	Remotely controlled high quality speaker to sound the alarm	O	O	O	O	O
App. Service	Should know child's moving speed	V	O	O	O	-
	Transmits signal when battery is detached	O	O	O	O	O
	Notify guardians (parents) when abnormal activities are detected	O	O	O	O	O
	Notify guardians (parents) when power is off	O	O	O	O	O

O: very effective   V: effective

4 Evaluation of Design Concept

4.1 Interview with the Experts

The interviewees were composed of four experts; a child and crime expert, two of design experts, and an engineering expert. Each of them has practiced more than ten years. The interview examines validity of the scenario, validity of the function and

Table 6. (Part of) Analyzing the interview with the experts

Elements		Suggestions
Supplement of the scenario		Victims may be hit with a blunt weapon or drugged and become unconscious.
Hardware Design	Remote controlled camera	The purpose is for parents to remotely check the situation once they receive emergency signal and it is suggested to add intercept function.
	Preventing mobile phone extortion/ emergency button	-Proposed to send emergency signal through pressing a button or pulling a strap by a child in case of emergency. -Suggested extra device (ex. watch, bracelet, belt, buckle, etc) is needed in case of extortion or other cases that make operation impossible.
App. Service Design	Mobile phone design and fun element	-Design experts' suggestion: hardware does not have to be look immature. To encourage kids to utilize the mobile phone more frequently, it was advised to add more fun element, thus using a larger screen was suggested.

service, and solution for problems of proposed function and service through eight questions specific to each expert. Additional suggestions were included based on the comments from the experts and Table 6 summarizes the result of the analysis.

4.2 Proposed Mobile Phone Design Guideline

Guideline for Hardware Design. The characteristics of guideline for hardware design include notification by a child through emergency button and a strap and remotely controlled camera and microphone for guardians to check the situation. It uses screen touch type with age specific designed for 6-9 and 9-12 respectively and can be set to Hot-key or 10-key type. Table 7 summarizes parts of the guideline.

Table 7. (Parts of) Guideline for Hardware Design

Categories	Types		contents
LCD-Screen	With	-	Informs a child how to respond through various information
	Type	Window	Provide information through text or illustration
		Full touch	Button type(hotkey & 10key) can be jointly used and age-specific usage is possible
Button	Number of button	6-9 year old	Using illustration over button Hot-key type
		9-12 year old	Uses numeric key and should be limited with call, end, select menu, and number button. (includes emergency button)
	Type of button	Hot key	Specified key type makes easy to operate for a child under age of 9.
		Emergency key or strap	Can use specified key in addition to keypad to supplement weakness of 10 key type
Emergency function	Emergency strap		Has risk of unintended operation, but can be quickly operated.
	Track me button		Child can inform his/her location to guardians when it is hard to make phone call
Battery	Battery cover	Enclosed (integrated)/ Screw down type	Prevents force power down
	Latent battery		Has enough power to transmits signal of one's position for a certain length of time even after the battery is removed
Ubiquitous	Camera / Mic.	Emergency	Remotely collects data about the environment
	Sound	Siren	Third party can remotely sound the alarm

Guideline for App. Service Design. The guideline for App. service design includes screen layout, text, LBS, Electronic Tagging, SMS, education, fun, and etc and Table 8 summarizes part of it. Screen layout has proposed while considering age specific interest. The size and number of text is designed in corresponding to lightweight and smaller hardware design that considers child's body size. LBS positions child's location and records time-based moving speed and location and determines level of threat it any abnormal activities are detected and notify guardians using recorded child's pattern.

Table 8. (Part of) Guideline for App. service design

Categories	Features	Characteristics
Screen layout	Selective layout is possible.	Cognition and operational skill of 6-9 and 9-11 year old should be considered respectively for age specific design.
Text	Maximum number of characters within screen.	Should be able to give information with minimum number of letters; Korean 4-6, English 9 letters (ex. EMERGENCY)
Electronic Tagging	Receives signal from Electronic Tag.	Receives signal once Electronic Tag and a child fall within certain distance.
Emergency service	-	Informs when child breaks out of route. Informs when detects change of moving speed.
SMS	Transmits preregistered message in case of emergency.	Child can transmit text message to his/her guardians with only one touch.

5   Conclusion

It was found that measures of the third party intervention are very essential because a victim is inferior physically and mentally. Thus, the key features in the mobile phone design guideline are as follows. The mobile phone can induce the third party intervention automatically even under the situations that a child does not recognize a threat and/or that a child cannot ask for help. In terms of hardware feature, the mobile phone can trace child's positions in the usual situation and maintain the child's route pattern and speed in database. Based on these data, it will monitor the presence of any abnormal movement. When it detects an abnormal movement, the App. service in the cell phone sends a signal to child's guardians. The third party remotely manipulates the camera and microphone of the phone to determine the safety of a child and can ask protecting the child safety. In addition, the mobile phone can identify the electronic tagging signal of an ex-convict and informs the child's guardians that there an ex-convict around a child, which inherently prevents crimes. The significance of this study lies on the attempt of preventing crimes against children in the early stage in context awareness ways using mobile phones. For further study, the guideline should be refined by evaluating prototypes developed using the guideline in various aspects.

References

1. The Legal Research and Training Institute: White Paper on Crime (2005-2008)
2. Laura e. Berk: Infants & Children (in Korean) (trans. Nang Ja Park) Jung-min, Seoul (2007)
3. Han, S.S., Song, J.M.: Child Welfare, Chang Ji, Seoul, p. 12 (2003)
4. Jo, S.A.: M.S. Dissertation: Research on Designing Customized Menu for Users in Mobile Phone. Kyung Sung Univ., Korea (2007)

The Impact of Robots Language Form
on People's Perception of Robots

Yunkyung Kim, Sonya S. Kwak, and Myung-suk Kim

Dept. of Industrial Design, KAIST
335 Gwahak-ro, Yuseong-gu, Daejeon, Republic of Korea
{yunkim,sonakwak,mskim}@kaist.ac.kr

Abstract. Robots in people's daily life have social relationships with human. This study investigated how the expression of social relationship in human communication is applied to human-robot relationship. We expressed two axes of social relationship through robots' verbal language. In a 2 (address: calling participants' name vs. not calling participants' name) x 2 (speech style: honorific vs. familiar) between-participants experiment (N=60), participants experienced one of four types of the robot and evaluate the robot's friendliness and dominance. Participants rated robots friendlier when it called their name than when it didn't call their name. In the case of robots' dominance, there was no significant difference in whether the robot called participants' name as well as the robot's forms of language. Based on the experiment results, we discussed the use of a social relationship concept for designing robots' dialogue.

Keywords: Human-Robot Interaction, Robot dialogue, Interpersonal traits, Social relationship.

1 Introduction

As robots appear and exist in daily life, it as social agents having their own identities rather than just autonomous products has a social relationship with human [1]. Social relationships are identified by each individual's social identity, status, mutual interpersonal closeness, etc. [2]. Among these factors, intimacy and status which represent horizontal and vertical axis of social relationship contribute to interpersonal attraction which is related to how much we like, love, dislike, or hate someone [3]. Therefore concepts of social relationships need to be applied and explored to the relationship between human and robot.

In this study, we set social relationships between human and robot by forms of robots' language and explored difference of people's perception of the robot through measuring robots' interpersonal traits (friendliness and dominance) perceived by people. According to Kiesler and Goetz (2002), dialogue affects people's perception more than the appearance of the robot [4]. As such, we used forms of language in dialogue because this is an effective way to represent social relationships between humans and robots.

M. Kurosu (Ed.): Human Centered Design, HCII 2011, LNCS 6776, pp. 253–261, 2011.
© Springer-Verlag Berlin Heidelberg 2011

2   Related Works

In human-robot interaction research, researchers mainly consider three primary types of dialogue: low-level (pre-linguistic), non-verbal, and natural language [5]. Among these, it is expected that people will have natural language interactions with robots in the near-future [6]. Natural language interactions with robots will be needed for people to understand instructions given by the robots such as robot receptionists [7], serving robots [8], teaching assistant robots [9], museum guide robots [10], and so on. Natural language is also required for socially interactive robots, because it helps the robots be engaging with people [11]. Therefore, we applied forms of interaction in human communication on the interaction between human and socially interactive robots.

There are various factors of an interaction changed by social relationships between humans. In verbal communication, it can be addresses, speech styles, ways of speaking, etc., while nonverbal communication includes posture, eye contact, smiling and gestures [12]. Forms of language, including types of address, are one of the critical ways to represent social relationships between two people [13]. In this study, we explored the effects of calling name and selection of speech style on people's perception of robots.

2.1   Name and Social Relationships

Introducing, perceiving and calling someone's name occur when people make a relationship at the beginning of a meeting and keep other's attention to them. The meaning of calling name has been studied in a play to investigate the relationships between characters. Lee (2005) describes that an unnamed person is a stranger and nonexistence, that is to say one person of numerous nameless people [14]. Introduced and called someone's name indicates that existence of him/her is realized by others and he/she forms relationships with others.

2.2   Speech Styles and Social Relationships

In linguistics, an honorific speech grammatically encodes and represents the relative social status of the participants of the conversation [15]. Honorific speech style is the most common speech style and is commonly used between strangers or between superiors and subordinates, while familiar speech style is typically used when the addressee is below the speaker in age or social rank [16]. In addition, sentence-final speech styles of honorific and familiar forms are certainly divided in Korean language [17].

When one person calls another's name, the linguistic forms of address is governed by the relation between the speaker and his addressee [13]. If addressee is in an intimate distance from a speaker, such as close classmates, elder brothers, boy or girl friends, lovers, then a speaker calls addressee with first name, FN [4]. In contrast, a speaker uses title and last name, TLN, or don't call the name to addressee distantly related, such as teachers, strange fellow workers, neighbors, governors, strange waiters. The forms of address are reasonably well described by a single binary contrast in English as well as Korean language: FN (First Name) or TLN (Title + Last Name). FN is typically used when people talk to another with familiar speech style, while people who speak honorifically call another by TLN.

3   Method

We used a 2 (address: addressing participants' name vs. not addressing participants' name) x 2 (speech style: honorific vs. familiar) between-participants experiment design. All participants experienced one of four types of the robot.

We were interested in how people's perception of robots is different according to horizontal and vertical of social relationships between human and robots. Because Kim and her colleagues (2010) found that people allow the robot to come into their personal space more when the robot call names of people [18], we believed that participants would perceive interpersonal traits of the robot which calls their name more positively than the robot which don't call their name. The survey results also showed that people accept the robot which didn't call their name more than the robot which called their name when the robot used honorific speech style. As such, we anticipated that the effect of calling participants' name would vary by the speech styles.

This analysis led to the following research hypotheses:

H1. Friendliness of the robot's traits will vary with whether the robot calls participants' name.

H2. Dominance of the robot's traits will vary with whether the robot uses honorific or familiar speech style.

3.1   Participants

We recruited participants who are in their twenties (Male: 28, Female: 32). Korean students from an engineering college who is familiar with technology participated in the experiment. Because people might have poor mental representations of those with whom they have little experience [19], it may hard to make judgment confidently about robots.

3.2   Materials

We used the Nettoro robot which is a Mechanoid type rather than a Humanoid or Android type because most participants have never seen humanlike robot and should adapt to the robot easily to hold a natural conversation with it. The Nettoro robot is a cleaning and an information robot that can navigate the space autonomously. The robot's words were recorded in advance through text-to-speech (TTS) program with a mechanical voice. During the experiment, the robot's words for a conversation with participants were controlled using the Wizard of OZ technique (WOZ) [20].

3.3   Procedures

At the beginning of the experiment, participants were welcomed and asked to take a seat in front of the robot. An experimenter explained the experiment and notified that this robot can detect participants' voice and have a simple conversation with people, while complicated sentence is hard to recognize for it. After that, participant was asked to wait a minute and an experimenter went to the hidden area to operate the robot and the conversation, as shown in Fig. 1. After the conversation, the researcher asked participants to answer the questionnaire about the robot's interpersonal traits.

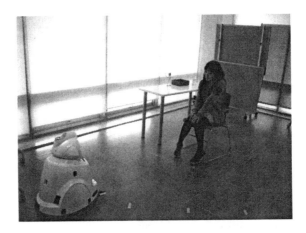

Fig. 1. One participant talking with a robot in testing room

3.4 Experimental Manipulations

One of the independent variables *Speech style* had two levels: honorific vs. familiar speech style. To make participants realize the speech style used by the robot, the robot said *"I'll use honorific (familiar) speech style because I'm a robot"* in the early part of the conversation.

In the case of another independent variable *whether the robot calls participants' name or not*, the robots ask three questions regarding participants' name such as *"What is your name?"*, *"What means your name?"*, and *"Who named you?"*, when it called participants' name. In addition, during the experiment, the robot called participants' name seven times to let participants know that it knows and calls their name.

The overall contents of the robot's words were neutral including instruction of the experiment to avoid an effect of it on participants' perception of the robot.

3.5 Measures

According to McCrae and Costa (1989), most critical traits to interpersonal relationships are dominance and friendliness based on Wiggin's circumplex [21]. For measuring interpersonal traits of the robot, revised interpersonal adjective scales (IAS-R) which includes 8 adjectives in each of the eight octant scales *PA: Assured-Dominant, BC: Arrogant-Calculating, DE: Cold-Hearted, FG: Aloof-Introverted, HI: Unassured-Submissive, JK: Unassuming-Ingenuous, LM: Warm-Agreeable,* and *NO: Gregarious-Extraverted* was used, as shown in Fig. 2. In this experiment, 64 Korean adjectives which are verified in psychology were used, because all participants were Korean [22]. Participants were presented with a list of single adjectives (e.g., "Extraverted") and asked to assess each adjective on a seven-point Likert scale ranging from "extremely inaccurate" to "extremely accurate".

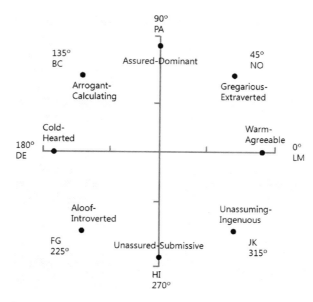

Fig. 2. Circumplex structure of Revised Interpersonal Adjective Scales (IAS-R; Wiggins, Trapnell, & Phillips, 1988)

For deriving aggregate measures of dominance and friendliness from the IAS-R, the following formulae [23] were used.

$$Dominance = PA - HI + .707(NO + BC - FG - JK) \qquad (1)$$

$$Friendliness = LM - DE + .707(NO - BC - FG + JK) \qquad (2)$$

(PA, BC, ..., and NO in these formulae are average scores of each of its items.)

4 Results

All statistical analyses were conducted using analysis of variance (ANOVA) with address (calling participants' name or not) and speech style (honorific or familiar speech style) as independent variables.

4.1 Friendliness

As predicted by H1, a significant effect of whether the robot called participants' name on friendliness of the robot was found, $F(1, 58) = 8.53$, $p < .01$, as described in Fig. 3. Participants feel the robot more friendly when it called their name, $M = 2.64$, $SD = 3.92$, than when it didn't call their name, $M = 0.02$, $SD = 2.96$. When the robot used honorific speech style, however, there was no significant difference of robots' friendliness rated by participants between whether participants were called by the robot or not, $F(1, 28) = 0.20$, $p = .66$. When participants had a conversation with the

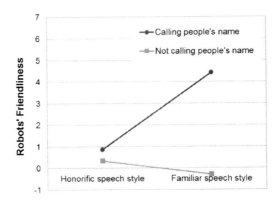

Fig. 3. Robots' friendliness perceived by participants according to whether the robot called participants' name and robot's speech style

robot which used familiar speech style, they felt the robot which called their name, $M = 4.43$, $SD = 3.49$, friendlier than the robot which did not, $M = -0.28$, $SD = 3.11$, $F(1, 28) = 15.24$, $p < .01$.

4.2 Dominance

In the case of dominance, there was no significant difference in whether the robot called participants' name, $F(1, 58) = 1.48$, $p = .23$, as well as robots' speech styles, $F(1, 58) = 1.21$, $p = .28$. Therefore, H2 was not supported.

5 Discussion

5.1 Summary and Interpretation of Results

Consistent with H1, people rated the robot as being friendlier when their name was called by the robot, while the robot which did not call was rated less friendly. During the experiment, the robot not only called participants' names but also asked their name, the meaning of it, and the person who named them which the participants revealed personally relevant information. It was the kind of self-disclosure which has traditionally been considered an important component and index of intimacy between people [24, 25, 26].

Regarding the friendliness, the results also showed that calling participants' name had no effect on robots' friendliness perceived by participants, when the robot used honorific speech style. Even if a robot called people by their names, the score of friendliness of the robot using honorific speech style was similar to that of the robot which did not address people by name. Because an honorific speech style is usually used by products, such as cash machines or automatic service machines, letting a robot know people's name might be just a way to check who will use the machine. On the other hand, if robots call people' name using a familiar speech style, it might be perceived not only as intelligent products but also as close objects which people can have social relationships with.

Related to the robots' dominance, no independent variables had an effect on it and all types of the robot were rated as being dominant.

5.2 Implications for HRI

From the results of this study, it may be argued that a robot's speech should be designed to be friendly and acceptable, because people perceive a robot communicating verbally as a dominant entity.

If a robot calls people by their name, it may be perceived as a social entity that can make close relationships with people rather than only a product. A robot would not be a social agent like a human with only verbal communication ability. Being a social agent may require having some factors which influence on forming social relationships, such as self-disclosure and being named. In the case of personal service robots which mostly interact with particular individuals, therefore, robot designers should make the robot call users' name to make users feel friendliness from the robot and make a social relationship with it. In particular, people's name should be naturally introduced by themselves as one of the self-disclosure process. In addition, if the robot talks to people with familiar speech style, people may accept the robot as if the robot is a close friend of them.

If a robot doesn't use or know people's names, it should use an honorific speech style because it would be just one of many products having no relationship with specific individuals as if two strangers newly meet. Therefore, a robot having a conversation using an honorific form of language would be more acceptable, if it is interacting with unspecified individuals in a public space, such as museum guide robots or information robots. In addition, results of this study suggest that inducing users' self-disclosure is not essential interaction factor to a robot in the public space, because asking and calling people by name have no serious effect on people's perception of robots.

5.3 Limitation

There are several limitations in this experiment. First, participants of this experiment were limited in their twenties. People's perception of addresses and speech style may be different in different ages. Children may be used to be called with familiar speech style than adults, while adults were usually called by others with honorific speech style. Therefore the results of this experiment are hard to be applied on all ages. Second, most contents of conversation in this experiment were consist of instructions for the experiment and then participants might perceive the robot dominant. Interactions with less dominant contents may produce different results. Third, the experiment was conducted in a short period of time. We expect that we can identify appropriate time to change robot's speech style from honorific to familiar one in a long term experiment, because as relationships between humans and robots are gradually developed. Further studies would handle this point of view.

6 Conclusion

Interacting with people is the most important ability for robots, as robots emerge in people's daily lives not only as assistant workers but as social partners. In particular, verbal communication has been regarded as the basic ability of robots. The objective of this study is to examine the effect of representing ways of social relationships between human and robots on people's perception of robots through forms of robots' language. The results indicate that it is enough to make people perceive robots friendly with only by calling people's name. When robots use honorific speech style, however, whether robots call people's name have no effect on people's perception of robots. These findings suggest a way of enhancing positive human-robot interaction through designing a robot's dialogue.

References

1. Nass, C., Steuer, J.S., Tauber, E.: Computers are social actors. In: SIGCHI Conference on Human Factors in Computing Systems: Celebrating Interdependence, pp. 72–77. ACM Press, New York (1994)
2. Schmitt, M.H.: Near and Far: A re-formulation of the social distance concept. Sociology and Social Research 57(1), 85–97 (1972)
3. Berscheid, E., Walster, E.H., Hatfield, E.: Interpersonal Attraction. Addison-Wesley, MA (1969)
4. Kiesler, S., Goetz, J.: Mental Models and Cooperation with Robotic Assistants. In: Human Factors in Computing Systems, pp. 576–577. IEEE Press, New York (2002)
5. Fong, T., Nourbakhsh, I., Dautenhahn, K.: A survey of socially interactive robots. Robotics and Autonomous Systems 42(3-4), 143–166 (2003)
6. Kriz, S., Anderson, G., Gregory Trafton, J.: Robot-directed speech: using language to assess first-time users' conceptualizations of a robot. In: The 5th ACM/IEEE International Conference on Human-Robot Interaction (HRI 2010), pp. 267–274 (2010)
7. Gockley, R., Bruce, A., Forlizzi, J., Michalowski, M., Mundell, A., Rosenthal, S., Sellner, B., Simmons, R., Snipes, K., Schultz, A.C., Wang, J.: Designing robots for long-term social interaction. In: Intelligent Robots and Systems (IROS 2005), pp. 1338–1343 (2005)
8. Lee, M.K., Kiesler, S., Forlizzi, J., Srinivasa, S., Rybski, P.: Gracefully mitigating breakdowns in robotic services. In: The 5th ACM/IEEE International Conference on Human-Robot Interaction (HRI 2010), pp. 203–210 (2010)
9. Kanda, T., Hirano, T., Eaton, D., Ishiguro, H.: Interactive Robots as Social Partners and Peer Tutors for Children: A Field Trial. Human-Computer Interaction 19, 61–84 (2004)
10. Nourbakhsh, I., Bobenage, J., Grange, S., Lutz, R., Meyer, R., Soto, A.: An affective mobile robot educator with a full-time job. Artificial Intelligence 114(1-2), 95–124 (1999)
11. Sidner, C., Lee, C.: Robots as laboratory hosts. Interactions 12(2), 16–24 (2005)
12. Rosenfeld, H.M.: Approval-seeking and approval-inducing functions of verbal and nonverbal responses in the dyad. J. Personality and Social Psychology 4(6), 597–605 (1966)
13. Brown, R., Ford, M.: Address in American English. J. Abnormal and Social Psychology 62(2), 375–385 (1961)
14. Lee, S.: Le nom et le theatre moderne dans Les amants du Metro. d'etudes de la culture francaise et des arts en France 13, 1–17 (2005)
15. Brown, P., Levinson, S.C.: Politeness: Some Universals in Language Use. Cambridge University Press, Cambridge (1987)
16. Lee, I., Robert Ramsey, S.: The Korean Language. State University of New York Press (2000)
17. Brown, L.: The honorifics systems of Korean language learners. SOAS-AKS Working Papers in Korean Studies, Dept. of Japan & Korea. London Univ. (2008)

18. Kim, Y., Kwak, S.S., Kim, M.: Effects of Social Relationships on People's Acceptance of Robots: Using Forms of Language by Robots. In: The 7th International Conference on Ubiquitous Robots and Ambient Intelligence (URAI 2010), pp. 85–88 (2010)
19. Gill, M.J., Swann, W.B., Silvera, D.H.: On the genesis of confidence. J. Personality and Social Psychology 75, 1101–1114 (1998)
20. Gould, J.D., Conti, J., Hovanyecz, T.: Composing letters with a simulated listening typewriter. Communications of the ACM 26(4), 295–308 (1983)
21. McCrae, R., Costa, P.T.: The structure of interpersonal traits: Wiggin's circumplex and the five factor model. J. Personality and Social Psychology 56(5), 586–595 (1989)
22. Yun, J.: Development and Validation of Korean Interpersonal Adjective Scales (KIAS). MS.diss., Dept. of Counseling Psychology, Catholic Univ., Seoul, Republic of Korea (2003)
23. Wiggins, J.S., Trapnell, P., Phillips, N.: Psychometric and Geometric Characteristics of the Revised Interpersonal Adjective Scales (IAS-R). Multivariate Behavioral Research 23(4), 517–530 (1988)
24. Altman, I., Taylor, D.A.: Socialpenetration: The development of interpersonal relationships. Holt, Rinehart & Winston, New York (1973)
25. Derlega, V.J., Metts, S., Petronio, S., Margulis, S.T.: Selfdisclosure. Sage, Newbury Park (1993)
26. Jourard, S.M.: Self-disclosure: An experimental analysis of the transparent self. Wiley, New York (1971)

Adjustable Interactive Rings for iDTV: First Results of an Experiment with End-Users

Leonardo Cunha de Miranda, Heiko Horst Hornung,
and M. Cecília C. Baranauskas

Institute of Computing, University of Campinas (UNICAMP)
13083-852, Campinas, SP, Brazil
professor@leonardocunha.com.br, heix@gmx.com,
cecilia@ic.unicamp.br

Abstract. Based on previous results of our research in the field of physical artifacts for interaction with Interactive Digital Television (iDTV) we developed a new digital device we named Adjustable Interactive Rings (AIRs). This work presents a quantitative analysis of an experiment conducted with twelve end-users in order to investigate the interaction of users with the hardware prototype of AIRs for iDTV. The experiment results presented in this paper indicate a positive acceptance of our solution and a good learning curve with respect to the interaction language of this physical artifact of interaction.

Keywords: interactive digital television, interaction design, user experiment, quantitative analysis, gesture-based interaction, human-computer interaction.

1 Introduction

Literature indicates that the remote control, the main physical artifact of interaction with the television system, is not appropriate for mediating the interaction between users and applications of Interactive Digital Television (iDTV), especially in a scenario of diversity of user profiles.

Based on results we already obtained conducting research in the domain of physical artifacts of interaction for iDTV with a Human-Computer Interaction (HCI) perspective [4,5,6,7,8,9,10], this paper presents quantitative results of an experiment with end-users aiming at the evaluation of the interaction enabled by the hardware prototype of the Adjustable Interactive Rings (AIRs). AIRs for iDTV is a technology resulting from a research project [11] conducted to develop a hardware alternative for the context of iDTV use, overcoming the many barriers of interaction that current remote controls present, as discussed by various authors, e.g. [1,2,3,5,12].

This paper is organized as follows: Section 2 presents the Adjustable Interactive Rings for iDTV; Section 3 describes the scenario and the methodology of the experiment; Section 4 presents the activity results and discusses these results under a quantitative perspective; Section 5 presents concluding remarks and indicates future work.

M. Kurosu (Ed.): Human Centered Design, HCII 2011, LNCS 6776, pp. 262–271, 2011.
© Springer-Verlag Berlin Heidelberg 2011

2   Adjustable Interactive Rings for iDTV

Based on the understanding that a more direct interaction with iDTV requires the physical artifact of interaction to be more transparent to the user, i.e., that the focus should be at the interactive applications interface and not at the artifact itself, we designed and developed a patent-pending physical artifact of interaction for iDTV called Adjustable Interactive Rings (AIRs).

The design solution was informed by the principles of Universal Design [14] and developed in a participatory approach [13] with representatives of the target audience. The resulting physical artifact of interaction, as specified and implemented in [6], is accessible, adjustable, ergonomic, and ambidextrous. It also enables flexible use and has a simple gesture-based interaction language for use by all people, to the greatest extent possible. This new digital artifact has two parts: i) the physical artifacts that enable interaction between the users and the television system, i.e., a set of three different AIRs; and ii) the device responsible for the capture of gestures made by users that are using the AIRs. This capturing device is called Receiver Interfacing Module (RIM).

This new digital device does not use textual input or output, to provide increased accessibility and enable a wider range of potential users, since a significant portion of the target audience of our research can be characterized by different levels of visual impairment and low literacy. The gesture-based interaction language has been shown easy to learn and allows the AIRs use in arbitrary fingers or not in the fingers at all, in order to maximize its use by people with different levels of motor impairment. Each AIR has only one button. Fig. 1 presents a user using the hardware prototype of AIRs for iDTV.

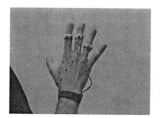

Fig. 1. User using the hardware prototype of AIRs for iDTV

3   Research Scenario and Methodology

The AIRs for iDTV experiment aimed at evaluating the interaction enabled by the hardware prototype of this new digital artifact. Activities started in July 2010 and were recorded in audio and video, with prior consent from users, to facilitate the *a posteriori* analysis of this activity.

Users were grouped in six different pairs, identified below as P1, P2, ..., P5, P6. The initial idea was that for each pair, only one member would in fact use the AIRs to interact with a simulated application of iDTV. Thus, the user not using the AIRs could observe and optionally help the partner in performing the tasks. The decision of who would use the AIRs was made by the users without the intervention of the facilitator of this experiment, the first author of this paper.

The persons who participated in this experiment originally came from different regions of Brazil – South, Southeast, North and Northeast – and showed different levels of experience with the use of technology, ranging from computer instructors to people who had never used a computer before and have trouble using ATMs, mobile phones or other digital artifacts. The 12 participating users are identified by U1, U2, ..., U11, U12. Table 1 shows details of these users.

Table 1. Ethnographic data regarding the participants

User	Pair	Age	Gender	Schooling Level	Occupation
U1	P1	23	M	Bachelor's degree	Graduate student
U2	P1	24	M	Master's degree	Graduate student
U3	P2	29	M	Master's degree	Graduate student
U4	P2	35	F	Master's degree	Graduate student
U5	P3	48	M	Bachelor's degree	Economist
U6	P3	48	F	Doctorate degree	Researcher
U7	P4	22	F	Bachelor's degree	IT professional
U8	P4	23	M	Ongoing bachelor's degree	Undergraduate student
U9	P5	57	F	High school	Housekeeper
U10	P5	23	M	Bachelor's degree	Informatics instructor
U11	P6	58	F	Bachelor's degree	Handcrafts woman
U12	P6	56	F	Some high school	Elderly nurse

During the experiment, the six pairs were divided into two groups: i) pairs without prior knowledge about the solution of AIRs for iDTV; and ii) pairs with basic knowledge about the AIRs. Before conducting the actual experiment, the pairs in group (ii) were invited to attend a presentation about the AIRs for iDTV, where the facilitator demonstrated the gesture-based interaction language by showing examples of different AIR functions using non-functional AIR mock-ups. The intent of this approach was to test if a previous exposure to the interaction language had a positive effect on task performance. The pairs who attended this explanation were P1, P4 and P6.

The following materials were used during the conduction of the experiment:

- hardware prototypes of AIRs and RIM;
- mock-ups of AIRs;
- one laptop with RGB output;
- one datashow;
- one table to support the RIM;
- one digital camcorder;
- one stopwatch;
- twelve forms of personal data;
- twelve feedback forms;
- twelve forms of the Self Assessment Manikin (SAM);
- six forms of the observer.

The following six (T)asks were performed during the experiment and presented to (ii) group in this sequence:

- <u>T1</u>: Turn on the television;
- <u>T2</u>: Enter name or nickname;
- <u>T3</u>: Move three geometric figures to their correct positions in the interface one at a time;
- <u>T4</u>: Switch from channel 2 to channel 12; in Campinas (Sao Paulo, Brazil) where the experiment was conducted, these correspond to the channels Rede Record and Rede Globo/EPTV;
- <u>T5</u>: Adjust the volume;
- <u>T6</u>: Turn off the television.
- During the conduction of the experiment, three researchers were present executing the following functions:
- <u>Facilitator</u>: Responsible for conducting the experiment and audio capture;
- <u>Observer</u>: Its function was to fill out the evaluation of tasks, including the marking of time that each user took to accomplish tasks;
- <u>Cameraman</u>: Responsible for the filming the user interaction with the prototype of AIRs for iDTV.

4   Experiment Results

After a brief explanation about how the users would conduct the tasks of the experiment, group (ii), i.e. P1, P4 and P6, assisted a presentation of the gesture-based interaction language with AIRs mock-ups. After this explanation, the second moment included the accomplishment of tasks by users. This explanation was necessary so that all users were aware of what they would do. However, it is noteworthy that the explanation did not teach users how to use the AIRs, except P1, P4 and P6 who, before the experiment, watched a simulation of the interaction language of AIRs through mock-ups. Thereafter, the actual experiment started (cf. Fig. 2).

Fig. 2. Experiment moments of AIRs for iDTV

4.1   Quantitative Results

Table 2 shows the time each user required to perform the tasks. As mentioned in Section 3, the initial idea was that only one user of each pair would use the prototype. However, during the activity, all users expressed their wish to use this new digital artifact, hence Table 2 presents the performance of all twelve users.

Table 2. Users vs. Tasks (times in seconds)

User/Task	T1	T2	T3	T4	T5	T6
U1	2.00	79.55	31.79	17.75	9.62	2.00
U2	2.00	108.85	19.42	24.54	43.25	5.28
U3	13.62	16.07	9.06	60.08	18.19	18.81
U4	264.63	36.37	19.56	3.41	4.67	22.03
U5	2.00	113.56	29.50	83.42	103.12	2.00
U6	70.13	381.84	27.82	38.03	55.50	2.00
U7	2.00	101.78	10.07	n/a	n/a	2.00
U8	10.37	79.50	84.07	46.72	4.72	18.04
U9	17.08	82.75	105.06	8.28	2.34	16.88
U10 (1st)	111.81	166.49	93.28	34.44	88.41	4.09
U10 (2nd)	2.35	35.56	42.56	54.53	54.81	2.00
U11	4.83	691.32	80.18	76.66	109.84	16.93
U12	25.16	640.88	n/a	n/a	n/a	n/a
Average	**40.61**	**194.96**	**46.03**	**40.71**	**44.95**	**9.34**

Fig. 3 presents a chart that summarizes the data of Table 2. Regarding the exact two seconds that some users required to carry out T1 or T6, according to the interaction language, in order to turn the TV on or off, the button of a certain AIR had to be pressed for 2s. Thus, two seconds for T1 or T6 mean optimal task performance.

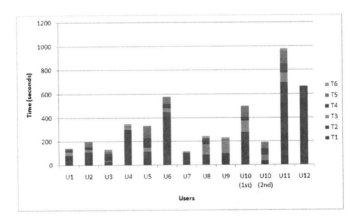

Fig. 3. Time required by each user to perform the tasks

4.2   Quantitative Analysis

The following is an analysis of this experiment using the quantitative data (i.e. task performance time) as a point of reference for our considerations about user interaction with the prototype. In the case of P1, U2 was the first user to accomplish the tasks, followed by U1. As can be seen in Fig. 4, U1 performed most tasks more efficiently than the U2. Regarding the worse performance of U1 in the realization of T3, we observed a minor problem of the prototype hardware during task performance regarding the accuracy of the electronic sensor that captures the AIR movement.

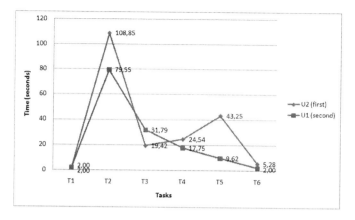

Fig. 4. Time P1 required accomplishing the tasks

In the case of P2, U4 was the first user to accomplish tasks and U3 was second. As can be seen in Fig. 5, U3 performed most tasks more efficiently than U4, except for T4 and T5. However, analyzing the video and data collected by the observer, we realized that U4 completed these two tasks "by chance", i.e. without actually knowing the necessary gestures and without being able to explain how she did it. Moreover, the performance of U3 regarding T4 was hampered by the low accuracy of the electronic sensor that captures the movements of the AIRs.

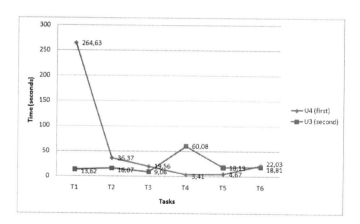

Fig. 5. Time P2 required to accomplish the tasks

In the case of the P3, U6 was the first user to participate in the experiment, followed by U5. As can be seen in Fig. 6, U5 performed most tasks more efficiently than U6. Similarly to the interaction of P2, in the case of this pair, U6 performed T4 and T5 more efficiently than U5. However, reviewing the video of the experiment we perceived that U6 did not knew how to accomplish these tasks. Furthermore, U5's task performance was less efficient due to this user's desire to make sure to absolutely understand how to perform the task. U5 only finalized the tasks after having completely understood which gestures were required to perform the respective tasks.

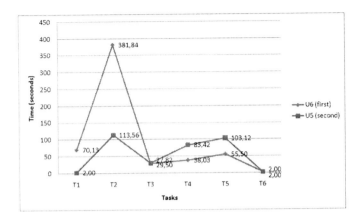

Fig. 6. Time P3 required to accomplish the tasks

In the case of P4, U8 was the first user to perform the tasks, and U7 the second. As can be seen in Fig. 7, U7 performed most tasks more efficiently than the U8, except for T2 since U7 found it difficult to align the sensor electronics from the AIRs with the RIM. U7 completed T4 and T5, but the respective times were not recorded since the facilitator had to intervene due to a hardware malfunction.

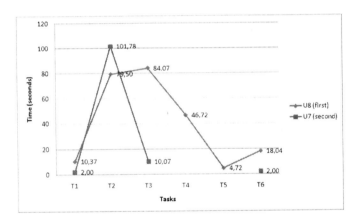

Fig. 7. Time P4 required to accomplish the tasks

In the case of P5, U10 was the first user to participate in the experiment, followed by U9. However, differently from what occurred with the other teams, after U9 having performed the tasks, U10 asked to use the AIRs once more again, and the facilitator authorized the interaction. As can be seen in Fig. 8, U9 performed most tasks more efficiently than U10 during the first iteration. Moreover, it is noteworthy that the second interaction of U10 was more efficient than the first iteration.

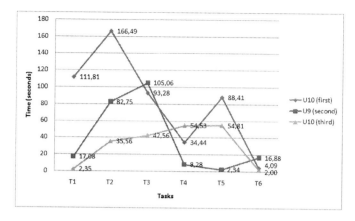

Fig. 8. Time P5 required to accomplish the tasks

In the case of the P6, U12 was the first user to accomplish tasks, followed by U11. As can be seen in Fig. 9, U12 completed only the first two tasks. After that, she experienced a discomfort in the arms and decided not to proceed with the experiment. Perhaps discomfort is related to the long time interval of approximately 640s during which the user remained with the hand facing the television to conclude T2. We understand that T1 was executed more efficiently by U11 than by U12, due to U11 having observed U12 performing the first experiment.

Differently from the other five pairs, during the performance of T2, U11 and U12 required a very long time to complete this task – more than three times the average task performance time – because of the difficulty of these users to align the RIM's electronic sensor that captures the movements of the AIRs. An important fact, which may explain this difficulty is, that U11 and U12 are not familiar with the use of technology in their daily lives.

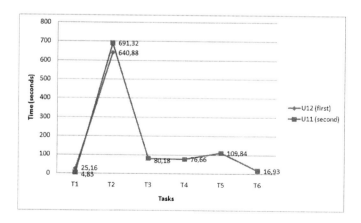

Fig. 9. Time P6 required to accomplish the tasks

Based on the six pairs' measured task performance times, we observe a tendency of the second user to perform tasks more efficiently than the first one. We understand that this is due to rapidly learning the AIR interaction language by merely observing others using the AIRs. This can be corroborated by a comment by U10 during the interaction of U9. U10 said: "You were much faster" and U9 replied: "But I saw you using [the AIRs]".

We conjecture that in some of the cases in which the time to perform a particular task by the second user was less efficient than that of the first, this was due to problems related to the accuracy of the electronic sensor that captures the movement of the AIRs, and not due to a lack of understanding of the interaction language. Although some of the measured task completion times appear to be high, 10 of 12 users completed all tasks. Another positive aspect of the experiment is that all users and not only one of each pair expressed their desire to use the AIRs, which indicates a positive acceptance of this new digital artifact.

Finally, we conclude that the learning curve of the interaction language of AIRs for iDTV was satisfactory, regardless of whether or not users had a prior knowledge of the AIR interaction language. It is worth noting that, like for every new technology or form of interaction, there is always a need for prior training and adaptation to the pattern of interaction. As presented and discussed in this section, adapting to the interaction language of the AIRs for iDTV showed to be a fast process.

5   Conclusion

This paper presented the first results of an experiment conducted with end users to evaluate the interaction enabled by the hardware prototype of Adjustable Interactive Rings. In this experiment participants were asked to perform six tasks regarding the interaction with a simulated iDTV application using AIRs as physical artifact of interaction.

This experiment allowed us to investigate the use of the solution directly with the target audience, including users with little or no knowledge about how to use the AIRs for interacting with iDTV applications. The feedback on how these users perceive and make use of this new digital artifact indicates that users can learn to use the AIRs after only few interactions.

The dissemination of the results of our experiment enables other researchers to work with AIRs, for example customizing the device for employing it in areas other than iDTV, e.g. immersive virtual environments, electronic games, etc. As future work we propose to carry out a qualitative analysis of this experiment focusing on the identification of affective and emotional states of users while using the AIRs for iDTV.

Acknowledgments. This research was partially funded by the National Council for Scientific and Technological Development – CNPq (grants #141489/2008-1 and #141058/2010-2), by the Brazilian Federal Agency for Support and Evaluation of Graduate Education – CAPES (grants #01-P-09881/2007 and #01-P-08503/2008), and by the State of São Paulo Research Foundation – FAPESP (grant #2010/11004-9).

References

1. Berglund, A.: Augmenting the Remote Control: Studies in Complex Information Navigation for Digital TV. Doctoral Thesis, Linköping University (2004)
2. Cesar, P., Bulterman, D.C.A., Jansen, A.J.: Usages of the Secondary Screen in an Interactive Television Environment: Control, Enrich, Share, and Transfer Television Content. In: Tscheligi, M., Obrist, M., Lugmayr, A. (eds.) EuroITV 2008. LNCS, vol. 5066, pp. 168–177. Springer, Heidelberg (2008)
3. Cesar, P., Chorianopoulos, K., Jensen, J.F.: Social Television and User Interaction. ACM Computers in Entertainment 6(1), 1–10 (2008)
4. Miranda, L.C., Baranauskas, M.C.C.: Anéis Interativos Ajustáveis: Uma Proposta de Artefato Físico de Interação para a TVDI. In: 4th Latin American Conference on Human-Computer Interaction, pp. 16–23. UABC, Ensenada (2009)
5. Miranda, L.C., Hayashi, E.C.S., Baranauskas, M.C.C.: Identifying Interaction Barriers in the Use of Remote Controls. In: 7th Latin American Web Congress, pp. 97–104. IEEE Computer Society, Los Alamitos (2009)
6. Miranda, L.C., Hornung, H.H., Baranauskas, M.C.C.: Adjustable Interactive Rings for iDTV. IEEE Transactions on Consumer Electronics 56(3), 1988–1996 (2010)
7. Miranda, L.C., Hornung, H.H., Baranauskas, M.C.C.: Prospecting a New Physical Artifact of Interaction for iDTV: Results of Participatory Practices. In: Design, User Experience, and Usability. LNCS, pp. 1–10. Springer, Heidelberg (2011)
8. Miranda, L.C., Hornung, H.H., Baranauskas, M.C.C.: Prospecting a Gesture Based Interaction Model for iDTV. In: IADIS International Conference on Interfaces and Human Computer Interaction, pp. 19–26. IADIS Press, Lisbon (2009)
9. Miranda, L.C., Piccolo, L.S.G., Baranauskas, M.C.C.: Artefatos Físicos de Interação com a TVDI: Desafios e Diretrizes para o Cenário Brasileiro. In: VIII Brazilian Symposium on Human Factors in Computing Systems, pp. 60–69. SBC, Porto Alegre (2008)
10. Miranda, L.C., Piccolo, L.S.G., Baranauskas, M.C.C.: Uma Proposta de Taxonomia e Recomendação de Utilização de Artefatos Físicos de Interação com a TVDI. In: Workshop on Perspectives, Challenges and Opportunities for Human-Computer Interaction in Latin America, pp. 1–14 (2007)
11. Miranda, L.C.: Artefatos e Linguagens de Interação com Sistemas Digitais Contemporâneos: Os Anéis Interativos Ajustáveis para a Televisão Digital Interativa. Doctoral Thesis, University of Campinas (2010)
12. Nielsen, J.: Remote Control Anarchy, http://www.useit.com/alertbox/20040607.html
13. Schuler, D., Namioka, A.: Participatory Design: Principles and Practices. Lawrence Erlbaum Associates, Inc., Hillsdale (1993)
14. Story, M.F.: Maximizing Usability: The Principles of Universal Design. Assistive Technology 10(1), 4–12 (1998)

Part III

Human Centered Design in Health and Rehabilitation Applications

Cognitive Prostheses: Findings from Attempts to Model Some Aspects of Cognition

Norman Alm[1], Arlene Astell[2], Gary Gowans[3], Maggie Ellis[2], Richard Dye[1], Phillip Vaughan[3], and Philippa Riley[1]

[1] School of Computing, University of Dundee, Scotland, UK
[2] School of Psychology, University of St Andrews, Scotland, UK
[3] School of Design, Duncan of Jordanstone College of Art and Design,
University of Dundee, Scotland, UK
{nalm,rdye,philippariley}@computing.dundee.ac.uk,
{aja3,mpe2}@st-andrews.ac.uk,
{g.m.gowans,p.b.vaughan}@dundee.ac.uk

Abstract. Improvements in the power and portability of computing systems have made possible the field of cognitive prostheses, which attempts to make up for cognitive impairment by to some degree modeling cognitive processes in software. Research on interfacing directly with the brain is at a very early stage. However, in research into dementia care, a number of non-invasive research prototypes have been developed to support people with dementia in specific areas of functioning, such as carrying out everyday activities, holding a conversation, being entertained, and being creative. Findings from the individual projects which may have general applicability are highlighted.

Keywords: Cognitive prostheses, assistive technology, dementia.

1 Internal and External Cognitive Prostheses

Cognitive impairment can be from congenital or developmental causes, or can arise from illness or injury. This creates difficulties for the person in coping with everyday life. Improvements in the power and portability of computing systems have led to the development of the field of cognitive prostheses, which attempts to make up for cognitive impairment by to some degree modeling cognitive processes in software.

Developing systems which can interface directly with the brain is at a very early stage. Cochlear implants have been developed to replace the function of the cochlea for profoundly deaf people. Although the results are much cruder than a sound signal processed in the normal way, they do represent a successfully working physical interface between technology and nerve cells [1]. Attempts are being made to similarly develop a range of implants which can restore cognitive function to individuals with brain tissue loss due to injury, disease, or stroke by performing the function of the damaged tissue with integrated circuits. An example of current work is developing a prosthetic for the treatment of hippocampus damage, including that caused by Alzheimer's disease [2]. An implant is currently being tested in rats.

M. Kurosu (Ed.): Human Centered Design, HCII 2011, LNCS 6776, pp. 275–284, 2011.

Having implanted cognitive prostheses may be quite a long way off, but it may be possible to provide a significant amount of help for people with cognitive deficits by focussing on the activities which their impairment is affecting, and developing external systems which prompt and support them through these specific activities. In this way a system can to some degree take over cognitive functions that a person is having problems with.

One inspiration for this work is the realisation that many computer applications which assist all of us in our daily tasks could actually be considered extensions of our cognitive processes. This includes the cognitive assistance offered to us by pocket calculators, spreadsheets, and financial decision support systems.

An example of people with cognitive difficulties who can benefit from computers without any specialist software is people with learning difficulties. In many cases a standard word processor provides better motivation and better access to writing than a pen and paper. The result is always neat and mistakes can be undone very easily. Computers can make the text in emails, documents and web pages more accessible through the use of text-to-speech. Creating multimedia using modern computers is becoming increasingly easier. It can allow people with cognitive difficulties to bring together video, pictures, sounds and symbols into a single package that communicates their needs, wishes or achievements [3].

The use of computers in this way relates to another new field of enquiry: distributed cognition. Distributed cognition is a psychological theory incorporating findings from sociology and cognitive science. It is a study that makes clear the deep inter-relationship between individuals, artefacts and the environment. Distributed cognition sees human knowledge and cognition as not being confined to the individual. Instead, it is distributed by placing memories, facts, or knowledge on the objects, individuals, and tools in our environment [4]. As Salomon puts it, "People think in conjunction and partnership with others and with the help of culturally provided tools and implements" [5].

2 Dementia

Cognitive deficits can be produced by a variety of causes, but by far the most widespread and rapidly growing cause of this problem is dementia in older people. Dementia is the loss of cognitive abilities, particularly the use of working (short-term) memory, usually as a result of Alzheimer's disease or stroke. Dementia occurs primarily in older people, and while it does not affect all of them, its rate of occurrence rises steeply from about 1 in 5 of people in their 80s to 1 in 3 of those in their 90s [6,7]. As the world's population balance shifts towards the older end of the spectrum, the incidence of dementia will continue to increase [8]. The effects of dementia can be quite devastating for the person and their family, and pose significant challenges to professional carers. There is currently no way to stop or reverse the physical causes of dementia. Until such help is found, there will be a need to develop social and technological supports for people with dementia and their carers. This paper outlines a number of research projects with this aim.

2.1 Daily Activities

People with dementia find it increasingly difficult to carry out daily activities, which puts a great strain on family carers, and eventually can result in referral to residential care. Most people with dementia, and their families, prefer that they remain in their own homes as long as possible. Work is ongoing to find ways in which technology could assist by providing appropriate and acceptable support to relieve some of the burden on carers.

The Institute for Cognitive Prosthetics in Pennsylvania, USA, has been working for twenty years on developing ways to support people with cognitive difficulties in daily activities through technology [9].

The group has found that cognitive assistive technology requires substantial customization for each patient, and that there are islands of deficits in seas of ability and vice versa. Activities that are objectively simple can be complex for the user, but encouragingly, brain plasticity has been seen as a result of intensive use of cognitive prostheses. Also cognitively impaired patients will develop creative unanticipated uses for new systems. And, to put technology in its proper place, just because a computer can perform a task more accurately doesn't mean that the patient should not be allowed the satisfaction of performing the task on their own.

A system under development at the University of Toronto aims to develop non-invasive ways of monitoring the activities of a person with dementia in their home, in order to deliver timely and relevant prompts to help them with daily activities [10]. The system uses computer vision and so does not need any equipment attached to the person. This of course makes it more difficult to interpret what is happening and a great deal of computational intelligence is employed. The person's movements must be tracked, and then a determination made of what they are attempting to do. Then a decision must be made as to how best to prompt them. For tracking, the system makes use of a Bayesian system, which is a way of calculating probabilities based on developing knowledge gained. It then uses a Markov process to decide what the user is doing and how to prompt them. As well as its possible practical applications, this work represents an interesting contribution both to computer vision modeling of human activities and decision making under uncertainty.

The first task the project has tackled is hand washing in the bathroom. The initial results have been successful. Further work for the team will include trying to find ways to assist with tooth brushing and then the far more challenging task of helping with using the toilet. Other technologies to be considered are using sound as well as visual input (e.g. whether the water is running), and speech recognition to detect what the user is saying during the process.

A prototype comprehensive home support system for people with dementia has been developed through a European funded collaboration [11]. The aim of the project was to develop a touchscreen based support system which collated sensor information from around the house and assisted the user by supplying a wide range of prompts and supports, including a daily agenda, pop-up reminders, a locator for lost items, and step-by-step instructions for daily activities. Navigation help outdoors was provided by means of a linked handheld device. Door alarms were in place as well as automatic control of lighting and easy control of household appliances. Music and other entertainment was provided with easy to use controls.

The system performed well in tests. The team felt that particular strengths of this project were the involvement of potential users throughout the design process, and the multidisciplinary nature of the effort.

The Laboratory for Assisted Cognition Environments at Rochester University in New York is working on a number of memory and problem solving aids that help an individual perform the tasks of day-to-day life [12,13]. This interdisciplinary project combines computer science research in artificial intelligence and ubiquitous computing with clinical research on patient care.

The systems were based on multisensor (video and RFID) sequences, all of which were synchronized together. The scenarios of activities in daily living which were supported were walking around the indoor space, sitting and watching TV, preparing to use and then storing a kitchen utensil, preparing cereal, and drinking water. The experimental approach was based on calibration-free multi-view approaches which did not require cumbersome and sensitive camera calibration procedures.

At the University of York in England work has been done on assisting people with dementia to use a cooker. People with dementia have problems carrying out multi-step tasks such as are involved in cooking. Using a cooker of course carries with it significant possible risks to the safety of a person with cognitive difficulties. One problem with intelligent systems built to prompt people through a set of tasks is that the prompts used by these systems are likely to be viewed as novel. As people with dementia are known to have difficulties with novelty this could be a problem. An experiment was performed by this group to determine how to prompt people with dementia about what knob controls what burner on a cooker. A cue using a fluorescent visual path to call attention to the connection between the knob and the hot plate was found to provide comparable or better results than more conventional alternatives. It is concluded that design in this area does not need to be constrained by the need to avoid novelty as long as the design is well carried out. The experiment is also of interest because of the way that it was embedded in a natural cooking task suitable for people of varied cognitive capacity [14].

A number of projects have experimented with using on-screen avatars to prompt the users of cognitive support systems. There have been concerns about the acceptability and the effectiveness of avatars, particularly for older users. Some work has indicated that computer-generated 3D faces can be perceived quite negatively by people with dementia, whereas traditional 2D cartoon representations can work better [15]. At the Austrian Research Centers in Vienna some recent work has involved avatars based on photographs of faces, which are then animated [16]. The assumption is that these faces will find easier acceptance than cartoon-like or less realistic avatars. The photorealistic avatars were shown to elicit a high degree of attention holding and a positive reaction from people with mild dementia.

2.2 Conversation

One of the most devastating effects of dementia is the deprivation of the ability to communicate. Without working memory, conversation becomes impossible. For many people the basic interactional structure can remain to an extent, so greetings, farewells and other forms of ritualised speech are still possible, but the rest can be repetitive or apparently meaningless to a listener. As a result people with dementia can become socially isolated and deprived of the range and variety of social interactions that characterise everyday life for unimpaired people. This can have a profound effect on the person's sense of wellbeing, and put severe strains on family and carers.

Although short-term memories are increasingly not available in dementia, longer term memories can be relatively well preserved. This is because long-term memories are stored in the brain in a different way from working memory. With dementia, the difficulty is to find a way to prompt these long term memories. This can be done by a family member who knows the person's history well, but it can be hard work, and does not make a for a relaxed and natural interchange, with both participants contributing equally.

There may be a role here for technology to provide more stimulating and complex prompting and supporting, freeing both parties to enjoy a conversation. A group at Dundee and St Andrews Universities in Scotland have developed and evaluated a system called CIRCA, which performed this role. CIRCA consisted of a hypermedia structure with reminiscence material as content, accessible via a touchscreen. The system relieved the carer or relative of the task of continually supporting the person with dementia in a conversation. Instead they could join with them in exploring and enjoying the multimedia material, which then had the effect of regularly triggering long-term memories. The person with dementia could then relate their story or recollection [17].

The first prototype system presented the users with a choice of three reminiscence themes and three media types, drawn from approximately 10 video clips, 30 music clips, and 230 photographs. The material was chosen after consultation with about forty people with dementia and their families and carers, to determine what sort of material would be the most engaging and stimulating of conversation. Fig. 1 shows the system in use.

After an iterative development of the system, for a final formal evaluation, communicative interactions were set up between people with dementia and a carer, with and without the system being used. Since CIRCA used reminiscence contents to stimulate conversation, the control condition used standard physical reminiscence material to prompt and facilitate conversation.

Given the difficulties people with dementia may have in communicating their opinions, objective measures for such aspects of the interaction as engagement, enjoyment, and the degree to which a satisfying interaction is taking place were devised. Sessions were video recorded and then coded. A set of coding techniques was devised to describe both verbal and nonverbal behaviour that allowed focusing on (i) the people with dementia, (ii) the carers and (iii) the relationship between the two. In particular, an attempt was made to determine if people with dementia could be supported to take the lead more in conversations, rather than the contents and course of the interactions being directed by the carers. This would have a beneficial effect on the quality of life of people with dementia as the provision of a positive interaction, at whatever level a person with dementia understands it, can be considered a successful intervention [18]. In addition, facilitating staff to engage in successful reminiscence activities has been shown to have a positive impact on their attitudes towards the people they work with that continues beyond the activity sessions [19].

Fig. 1. The CIRCA system in use

The CIRCA system proved successful in facilitating communication and in particular was able to give the person with dementia an increased control over the conversation, because control was shared via both participants interacting with the computer system. Staff were uniformly positive about using the system, and in some cases reported that they had learned more about the person in twenty minutes with the system than they had ever known before.

A group in Japan has developed a system which facilitated a conversation, based on reminiscence materials, which took place between a person with dementia and a family member or volunteer over the internet, thus giving support to people who might otherwise be isolated in their own homes. The system provided a two-way videophone plus the ability to easily share photographs and video clips between the participants. Trials were conducted first with people with dementia in a hospital setting, to determine if they could understand and make use of a videophone connection. Subsequent field trials of the system in more realistic settings showed that people with dementia could communicate with therapists by videophone and that the reminiscence sessions over the network were generally as successful for individuals with dementia as face-to-face reminiscence [20].

At the University of Toronto work has been done on developing multimedia reminiscence DVDs for people with dementia and their families. These were 40 minute DVD presentations made from material supplied by the family. In contrast with the CIRCA prototype this project made use of material personal to each individual. The idea was to give them and their families a pleasant and stimulating experience, and to prompt conversation based on the person's long term memories. Responses were positive, and the long-term aim of the group is to contribute to the restoration to some degree of the 'personhood' of the individual with dementia [21,22].

2.3 Interactive Entertainment

The ability to entertain oneself is an important facet of a full life. As well as a quality of life issue, a very practical problem is that people with dementia currently need uninterrupted attention from carers all day, which can lead to exhaustion in relatives, and can make paid carers retreat from an overwhelming sense of demand into just providing the basics of physical care.

The team in Scotland who developed the CIRCA system have investigated ways in which an interactive entertainment system for people with dementia could engage them and then support and prompt them in such a way so that they would be able to use the system unaided. The first issues addressed were what sort of content for an interactive system would be appropriate and engaging for older people and what kinds of prompts would be necessary to keep the person engaged and enjoying using the system [23].

The activities developed were: games of skill, creative activities, and virtual experiences. Fig. 2 shows one of the games. The object is to get the ball past the goalie. Touching the ball launches it. Physics is taken account of, so that the ball's

trajectory depends on where it is touched. Most of the users engaged well with the activities. Much was learned about how to structure the interaction to avoid confusion or boredom. Activities with a clear and always present goal worked best. Activities which were familiar, not surprisingly, also worked well. Activities which were less successful involved occasional pauses when it was not clear what to do next, or just what was happening.

Fig. 2. A game of Beat the Goalie in the LIM system

Devising ways to replacing a helper by having the system itself prompt the user was a important part of this project. The team experimented with increasingly intrusive levels of prompting, so that the minimum level necessary to still ensure success could be determined.

Prompts could be provided, in increasing order of obtrusiveness, by:

1. An interface which was simple and easy to figure out
2. Visual reminders (such as an onscreen button flashing)
3. Text boxes which could pop up with instructions or suggestions
4. Spoken messages to the user
5. An onscreen avatar to deliver instructions and suggestions to the user

Having tried the first four prompting methods, it was found that a well-designed interface, along with occasional text-box prompts worked well. The spoken prompts with synthetic speech did not work well. The synthetic speech was quite understandable to people without dementia, but participants with dementia simply ignored the speech, as if nothing had been said. This interesting outcome relates to work being done on difficulties older people in general and people with dementia is particular may have with understanding synthetic speech. Visual prompts were better than verbal prompts in that they were better at keeping the initiative with the user. Users have enjoyed even the unfinished prototypes and engaged with them for extended periods without continual support from another person.

The conclusions of this project were that a successful interactive experience should have the following characteristics:

1. In engaging, attractive and colourful interface, which promises enjoyment
2. It is always obvious what to do next, either because of the way the interface behaves, or because of a specific prompt
3. An element of challenge and skill mastery to the experience works well (i.e. there is a 'point' to it, even for people with little working memory)
4. Providing continual feedback provided on the user's performance - encouragement when they are not succeeding at a task, and praise when they do succeed.

2.4 Being Creative

Looking beyond entertainment, as important as that is, a further challenge would be to devise technology which could help a person with dementia to carry out a satisfying creative activity. A simple definition of computer-supported creativity would be an activity which is directed by the user, within limits provided by the system, which produces a result which is innovative, individual and, ideally, aesthetically pleasing.

A number of studies have shown the potential of people with dementia to be creative. Activities such as music and painting can positively influence the sense of well-being of people with dementia, with those participating showing improvements in self-esteem and mood and decreased agitation. All of these approaches have involved a great deal of planning and hard work on the part of carers to support the person with dementia in being creative. A project at Dundee University in Scotland has involved working on ways to use technology to take over some of the support needed by the person with dementia in order to carry out creative activities. A system called ExPress Play was developed to support failure-free musical composition for people with dementia. The system was intended to enable people with dementia to participate in active music making regardless of prior musical experience. Using the system always had a successful outcome. Express Play was accessed via a touch screen, and used a method of producing music which means it always sounded 'musical' [24]. Twenty-five people with dementia participated in evaluations, each taking part in three sessions spaced out over a number of days. All the sessions were videotaped. A tool was developed to continuously track and record all the selections, movements, and timings during a session, making it possible to replay the sound and display generated during each session for analysis.

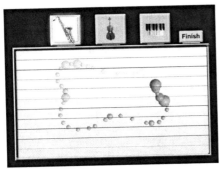

Fig. 3. The ExPress Play interface showing the 'bubble trail' left as the user drags their finger across the screen, creating music

Each session was intended to last ten minutes. However it was decided not to stop participants abruptly if they were deeply engaged in play, but rather stop them at the next appropriate time after ten minutes. In fact, more participants played for longer than ten minutes than expected.

A helpful feature of the system was a display of a trail of 'bubbles' that appeared as the user dragged a finger across the screen, creating the music. The bubbles faded with time so as not to fill the screen entirely, but stayed on long enough to create the effect of a visual memory of where the finger had been. See Fig. 3.

Users of the system all produced unique musical output. The tracking data showed that each person had a completely individual pattern of use of the system. The different trails of finger movement clearly showed some individuals favouring staccato dabs at points in a row, some making points all over the screen, others opting for continuous lines horizontally, vertically diagonally, or in swoops. None of the participants played in the same way as the others. The project thus fulfilled its aim of assisting the users to be creative in an engaging and satisfying way.

3 Conclusion

There are a number of interesting general findings from this work which may have a more general application to future attempts to provide effective cognitive prostheses.

Methodology. All of the projects involved potential users throughout the design process, as difficult as this is given the nature of dementia. In some cases novel methods of measuring the effectiveness of the prototypes were developed [17]. The projects all were multidisciplinary in their approach. Given the complex nature of cognitive deficits and their effect on everyday life, this is an important feature to ensure that the results of the work are usable and desirable for intended beneficiaries.

Hardware. Recent work in this field has moved away from sensors which require to be carried by or attached to the person, and instead rely on using sophisticated computer vision algorithms and to track users' movements and to analyse the behaviour being observed [10,12,13].

Human Computer Interaction. Systems which require human-computer interaction for people with dementia need be sure it is always obvious to the user what to do next, either because of the way the interface behaves, or because of a specific prompt [23]. It is important, because of the working (short-term) memory problems to provide users with continual feedback [23]. It may be possible to devise ways having a simple stylised visual representation of short-term memory, such as in the music making system [24]. Efforts should be made when using prompt either to have them obvious in their intent, or familiar in design. Unobvious novel prompts are likely to fail [10]. With good design, however, even novel technologies can be introduced, if it is obvious how to use them [14,20,23,24]. Interestingly, even cognitively impaired patients will develop creative unanticipated uses for systems designed for another purpose, and such creativity is to be encouraged [9].

Impact on Users. Improving the ability of people with cognitive deficits to function better will have a positive effect on staff attitudes towards the people they work with that continues beyond any particular activity session [19]. Ultimately such systems can contribute to ambitious goal of restoring the 'personhood' of the individual with dementia [21, 22].

References

[1] Waltzman, S., Roland, J.T.: Cochlear implants. Thieme, New York (2006)
[2] Berger, T.W., Ahuja, A., Courellis, S., Deadwyler, S., Erinjippurath, G., Gerhardt, G.: Restoring lost cognitive function. IEEE Engineering in Medicine and Biology Magazine 24(5), 30–44 (2005)
[3] Better Learning Through Technology, http://www.bltt.org
[4] Hutchins, E.: Cognition in the Wild. MIT Press, Cambridge (1995)
[5] Salomon, G.: Distributed cognitions: Psychological and educational considerations. Cambridge University Press, Cambridge (1997)
[6] Jorm, A., Korten, A., Henderson, A.: The prevalence of dementia: a quantitative integration of the literature. Acta Psychiatrica Scandinavica 76, 465–479 (1987)
[7] Knapp, M.: Report into the Prevalence and Cost of Dementia prepared by the Personal Social Services Research Unit (PSSRU) at the London School of Economics and the Institute of Psychiatry at King's College London, Alzheimer's Society, London (2007)

[8] Foresight Panel: The Age Shift: Priorities for Action. Report of the Foresight Ageing Population Panel, p. 11. Office of Science and Technology / Department of Trade and Industry, London (2000)

[9] Cole, E.: Patient-centered design as a research strategy for cognitive prosthetics: lessons learned from working with patients and clinicians for two decades. In: Proceedings of CHI 2006 Workshop on Designing Technology for People with Cognitive Impairments, Montreal, Canada (2006)

[10] Mihailidis, A., Boger, J., Candido, M., Hoey, J.: The COACH prompting system to assist older adults with dementia through handwashing: An efficacy study. BMC Geriatrics 8(28) (2008)

[11] Meilland, F., Reinersmann, A., Bergvall-Kareborn, B., Craig, D., Moelaert, F., Mulvenna, M., Nugent, C., Scully, T., Bengtsson, J., Dröes, R.: COGKNOW: Development of an ICT device to support people with dementia. Journal on Information Technology in Healthcare 5(5), 324–334 (2007)

[12] Kautz, H., Harman, C., Modayil, J., Levinson, R., Halper, D.: Integrating cueing and sensing in a portable device. In: Univ. of Washington Institute on Aging Conf.: Supportive Technology and Design for Healthy Aging. Univ. of Washington Press, Seattle (2008)

[13] Messing, R., Pal, C., Kautz, H.: Activity recognition using the velocity histories of tracked keypoints. In: 12th IEEE International Conference on Computer Vision. IEEE Kyoto, Los Alamitos (2009)

[14] Wherton, J., Monk, A.: Technological opportunities for supporting people with dementia who are living at home. Inter. Journal of Human-Computer Studies 66(8), 571–586 (2008)

[15] Alm, N., Dobinson, L., Massie, P., Hewines, I., Arnott, J.: Cognitive prostheses for elderly people. In: Proc. of IEEE Systems Man & Cybernetics Conf., pp. 806–810. IEEE, Tucson (2001)

[16] Morandell, M.M., Hochgatterer, A., Fagel, S., Wassertheurer, S.: Avatars in assistive homes for the elderly: A user-friendly way of interaction? In: Holzinger, A. (ed.) USAB 2008. LNCS, vol. 5298, pp. 391–402. Springer, Heidelberg (2008)

[17] Alm, N., Dye, R., Gowans, G., Campbell, J., Astell, A., Ellis, M.: A communication support system for older people with dementia. IEEE Computer 40(5), 35–41 (2007)

[18] Baines, S., Saxby, P., Ehlert, K.: Reality orientation and reminiscence therapy: a controlled cross over study of elderly confused people. British Jour. of Psychiatry 151, 222–231 (1987)

[19] Woods, R.: Psychological therapies in dementia. In: Woods, R. (ed.) Psychological Problems of Ageing. Wiley, Chichester (1999)

[20] Kuwahara, N., Abe, S., Yasuda, K., Kuwabara, K.: Networked reminiscence therapy for individuals with dementia by using photo and video sharing. In: Proceedings of ASSETS 2006, pp. 125–132. ACM, New York (2006)

[21] Damianakis, T., Crete-Nishihata, M., Smith, K., Baecker, R., Marzialia, E.: The psychosocial impacts of multimedia biographies on persons with cognitive impairments. The Gerontologist 50(1), 23–35 (2010)

[22] Cohene, T., Baecker, R.M., Marziali, E., Mindy, S.: Memories of a life: a design case study for Alzheimer's disease. In: Lazar, J. (ed.) Universal Usability, pp. 357–387. John Wiley & Sons, London (2007)

[23] Alm, N., Astell, A., Gowans, G., Dye, R., Ellis, M., Vaughan, P., Riley, P.: Engaging multimedia leisure for people with dementia. Gerontechnology 8(4), 236–246 (2009)

[24] Riley, P.J., Alm, N., Newell, A.F.: An interactive tool to support musical creativity in people with dementia. Journal of Computers in Human Behaviour 25, 599–608 (2009)

Management of Weight-Loss: Patients' and Healthcare Professionals' Requirements for an E-health System for Patients

Anita Das[1], Arild Faxvaag[1], and Dag Svanæs[2]

[1] Norwegian University of Science and Technology (NTNU),
Department of Neuromedicine, Medisinsk Teknisk Forskningssenter,
NO-7491 Trondheim, Norway
[2] Department of Computer and Information Science, Sem Sælandsvei 7-9,
NO-7491 Trondheim, Norway
{Anita.Das,Arild.Faxvaag}@ntnu.no, Dag.Svanæs@idi.ntnu.no

Abstract. An increasing number of patients with overweight undergo weight-reduction treatment. However, many people experience challenges with long-term maintenance and are in risk of weight-regain. Currently there is no unique solution that ensures long-term maintenance of lost weight. Several studies have explored the effectiveness of web-based and e-health interventions, on improving the outcomes of weight-management. The results are unclear. This paper describes requirements for e-health solutions for weight-loss patients. Our findings suggest that such solutions need to be developed in collaboration with both patients and healthcare professionals to ensure that they are in line with medical treatment in addition to taking consideration to the behavioral aspects of using such systems.

Keywords: Design, E-health, Healthcare, Obesity, User involvement.

1   Introduction

Involving multiple stakeholders in the design process is challenging due to the required time and investment. Within the healthcare domain, e-health systems typically have multiple end-user groups with widespread backgrounds and interests [1]. Patient-centered e-health solutions are patient focused, but are not always in conjunction with disease-management programs, in which healthcare professionals have a central role. For such systems it is therefore important to include the perspective of all stakeholders, both different patient groups and the relevant healthcare professionals.

Until lately, obesity have been managed within the primary care, but due to the increased prevalence of severe obesity, the demand for interventions such as surgical interventions and lifestyle programs offered by the specialist care is rising. People that undergo such treatment require lifelong lifestyle modification with focus on dietary habits and physical activity. With the treatments taking place within the specialist care, patients increasingly need to conduct self-monitoring activities in their home environment with little follow-up by healthcare professionals. E-health solutions hold

M. Kurosu (Ed.): Human Centered Design, HCII 2011, LNCS 6776, pp. 285–294, 2011.
© Springer-Verlag Berlin Heidelberg 2011

the potential to support patients after initial weight-loss, to help establish, support and maintain lifestyle changes. Successful long-term maintenance is associated with self-care management and self-monitoring [2]. However, conducting such activities are labor intensive, and compliance is often difficult [2]. Hence, we need a better understanding of the experienced challenges after treatment, about aspects influencing upon non-compliance, and how self-management can be promoted by the use of e-health systems in weight-loss patients. To be able to design a clinical e-health solution for this patient group, a first step is to gain knowledge and understanding of the challenges they experience, and further investigate how the behavior change process can be promoted by the use of e-health solutions. In this study we have involved patients and healthcare professionals in a participatory design process of a clinical e-health system, to elicit the multiple user groups' requirements and perspectives towards such a system.

2 Background

The prevalence of obesity in the western countries has increased the last decades [3]. Obesity is associated with increased morbidity and mortality, and is a risk factor for diabetes, cardiovascular problems, hypertension, cancer illnesses, osteoarthritis as well as other health problems of psychosocial characters [3]. Increasingly, obesity is being recognized as a chronic disease itself, requiring health interventions. Weight-loss has beneficial effects in co-morbidities and long-term survival, and can be achieved through lifestyle intervention, bariatric surgery or pharmacotherapy [4]. However, long-term maintenance of lost weight is difficult, and studies show that conventional treatment (incl. lifestyle modification programs and pharmaceutical agents) is relatively ineffective in a long-term perspective [5]. As for today, surgical interventions are shown to be the most effective, and produces substantial initial weight-loss in the great majority of patients [5]. However, studies imply that weight-loss of bariatric surgery is temporary, and that many patients regain weight after a while [5,6]. Long-term weight maintenance is therefore a challenge regardless of initial weight-reduction treatment, as many experience weight regain after a period of time.

Weight-reduction programs are resource demanding for the individual patient as well as for the healthcare services considering the time, economical costs and emotional investment it requires. Currently there is no unique solution that ensures long-term maintenance of lost weight [7]. Several studies has explored the effectiveness of web-based and e-health interventions, on improving the outcomes in the area of weight-management, physical activity and dietary intake with unclear results [7].

Since its infancy in the 1980s [8], the perspectives and techniques of Participatory Design (PD) have become part of the state-of-the-art in systems development. As exemplified by Druin's work with children [9], certain user groups require modification and adaptations to the existing PD methodologies. We will here report from a Participatory Design project with another non-standard user group: Obesity Patients.

3 Methods

During 2009 we conducted a qualitative study involving a series of four participatory design workshops including patients and healthcare professionals. The inclusion criteria for the patients were that they (a) had completed a weight-reduction program offered at the local hospital, (b) were age 18 or above, and (c) had basic proficiency in Norwegian language. The study got approval from the regional Ethics Committee (Central Norway, Trondheim), and all participants provided written consent when enrolling to the study.

3.1 Workshops

We conducted separate workshops with the multiple groups, and two facilitators from the research team had the roles as moderators during each workshop. Healthcare professionals were included in the first workshop, followed by two workshops that included patients that had undergone weight loss treatment either through conventional treatment (lifestyle therapy) or bariatric surgery.

The workshops consisted of design tasks, semi-structured interviews and group discussion. In advance, we had clarified the topics that were to be discussed, and we used open-ended questions that were followed by probing questions that clarified the participant's responses. The objectives of the workshops were to systematically gather information, ideas and perspectives from the multiple user groups.

Finally, we conducted a last workshop where we included selected participants from the previous three workshops. The purpose of the last workshop was to present our results, and to validate the findings. Each workshop lasted for 3.5 hours, and the whole session was video-and audio recorded.

3.2 Analysis

The recordings from the four workshops were transcribed verbatim. The data was analyzed qualitatively using a grounded theory approach [10]. The transcripts were coded before these were grouped together, and themes were identified.

4 Results

The aim of this study was to describe patients and healthcare professionals requirements of an e-health system to promote self-care management after weight-loss, to explore differences between these perspectives, and to assess the implications this may have for further e-health system development.

4.1 Participants

In total 20 people participated in the workshops, 12 people that had undergone weight-loss treatment at the local hospital, whereas six people had gone through bariatric surgery, and six had attended conventional treatment. Eight healthcare professionals attended the workshops, and their professional background was from nursing, medicine and clinical nutrition. Demographics are presented in table 1. To validate our findings from the first three workshops, we invited all the participants to

Table 1. Demographics

Participants	Female	Male	Total
Professionals	6	2	8
Lifestyle group	5	1	6
Surgery group	4	2	6
Total	15	5	20

a final workshop. Not all were able to attend this workshop due to personal practicalities, and the last workshop consisted of 10 participants. Five were from the healthcare personnel group, two from the conventional therapy group, and three from the bariatric surgery group.

4.2 User Requirements

All the groups suggested a secure-web-portal solution that would be accessible through the Internet. The benefit of gathering several features in one portal was the main reason. The possibility to send news feed and reminders from the web-portal to the users mobile telephone was proposed as an extra feature to motivate use of such a system. Ethical and privacy issues were discussed, and secure access that requires username and password for all users was pointed out to be important, as access to the system would only be give to patients treated at the hospital. Patients that had undergone weight reduction treatment would be the primary user, and healthcare professionals would have the role as moderators and facilitators.

The findings from the workshops identified that the multiple user groups have rather similar requirements when it comes to the features that they emphasize as important in a clinical e-health system for weight-loss patients. Table 2 gives an overview of the suggested features to be included in the system. However, the multiple groups have differing perspectives and rationales towards such a system.

Table 2. Required features in a clinical e-health system after weight-reduction treatment

Requirement	Healthcare Professionals	Patients
Information	X	X
Articles		X
Links	X	X
Discussion Forum	X	X
Private Communication	X	X
Buddy System		X
Self-management tools:	X	X
- Diary	X	X
- Notes		X
- Calendar	X	
- Diet plan	X	X
- Exercise plan	X	X
- Clinical Measures	X	
- Reminders on sms	X	X

Several broad themes were identified, and the participants emphasized patient education, communication and disease management as the most crucial ones. These issues will be further elaborated from the multiple perspectives.

4.3 The Healthcare Professionals Perspective

Healthcare professionals have the responsibility to provide patient education. According to the healthcare personnel group they use much time on providing and repeating the same information to the same patients several times, but they experience that the patients claim not have received this information, and that they are non-compliant. As a result, the healthcare professionals had started documentation of provided information, and experienced that this was the reality. As one of the professionals expressed:

> "Not all, but let us say that 70% (of the patients) asks questions about things they already have received information about."

The professionals were particularly worried about the patients that undergo bariatric surgery, due to the nutritional and metabolic problems that these patients may experience if non-compliant to recommended post-surgery regime. Such surgery is an intervention to help patients loose weight and involves a decrease of the size of the stomach. The necessity of preventing undesirable repercussions of the treatment is crucial. The patients need to undergo major changes considering eating habits, and they are dependent on taking lifelong vitamin supplements daily to prevent developing nutritional deficiencies over time [5]. The surgery alone is not the solution to weight-loss, and they need to follow guidelines for diet, exercise and lifestyle changes to prevent long-term weight-regain.

The healthcare professional group suggested to provide more elaborate patient information in the e-health system, and had hopes that this could support patient education. Insecurity among the patients was another issue that was discussed in relation to patient education. Particularly this was observed among the bariatric surgery patients, where lack of information regarding food, diet and nutrition was a recurring topic. Positive patient outcomes, confidence and prevention of future malnutrition were the rationales that the professionals emphasized.

Enhanced communication between healthcare professionals and patients, but also among patients was highlighted as important issues that could be promoted in a future e-health system. To serve different communication purposes, an online discussion forum where the patients could discuss and share experiences was suggested, as well as private communication for one-to-one dialogues. The healthcare professionals experienced that patients seek contact with others in the same situation, and are positive towards that patients can learn from each other. However, they had experienced patients that had been mislead due to wrong information they had found online and in chat-forums. If possible, they would like to be there for their patients, as one of the healthcare professionals put it:

> "The first thing is how important it can be if some of use are moderators (for the forum). There are enough online forums for these people, but these are not forums where experts (professionals) are present. That is where we can contribute - as a source to knowledge, as a source to correction."

The healthcare professionals were aware of the fact that some patients would like more frequent consultations, and professional guidance and advice in some periods after treatment. Particularly they experienced that during the first phase after treatment, the number of telephones to the clinic increased, and sometimes a simple question could turn out to become a long lasting telephone consultation. With the time pressure-and workload, some emphasized the potential of using electronic medias for information, and one-to-one communication channels, that could provide efficacy benefits for the clinic. As of today, electronic communication with the use of e-mail and chat rooms are established ways of communication. However, within the specialist healthcare in Norway this is not an established reality. The reasons are complex, and may be explained by the legal issues that limit healthcare professionals to use electronic communication channels due to the security aspects. Another factor is the economical model behind the financing of the healthcare system in Norway, where face-to-face consultations and telephone calls gives benefits, rather than electronic communication that still does not exist as an option within the billing system.

Patients self-care management and disease management was emphasized as crucial by the healthcare professionals. An e-health system that could promote self-care management activities, and where the patient can organize and structure their day were heavy arguments for implementing such a system closely connected to the clinic. The professionals experienced that several patients lacked structure regarding eating and exercise. They suggested that reminders about eating could be of help, and that the patients could receive these reminders on their mobile telephone, as most people carry their phone wherever they go. Such reminders were also proposed to apply to vitamin pills and doctors appointments. Another self-management activity they currently recommended for their patients was to write diaries. They observed that when patients wrote a diet record or diary, this raised awareness about diet and diet pattern. However, according to the healthcare professionals, many patients were non-complaint towards diary writing, but they would advocate implementing a diary feature in the system anyway. The possibility of for instance sharing the diet record with the dietitian (for feedback) could be a motivational factor for starting to use such a tool.

4.4   The Patients Perspective

The patient groups expressed that in contrast to the healthcare professionals, they experienced that their information needs were not addressed sufficiently in the current clinical practice. An e-health system could promote information access, providing patient information retrieval whenever they needed it. Tailored information according to the treatment was underlined as important. Patients that were operated emphasized the need for elaborated information regarding the postoperative phase, about expected side effects of the operation, and about food and nutrition, as many experience food intolerance due to the surgery. Some could tell that the information they had received on paper had been misplaced, and after that the information was gone. In lack of information from the hospital, some had searched for information online, with the result of becoming even more confused. The Internet provides amounts of information, and the overload makes it difficult for patients to filter what to trust, and what to ignore. A place with validated information that they knew was correct would make the information retrieval process easier. One of the patients that had undergone surgery searched online, and ended up in a obesity forum with the following experience:

*"That page... if you write something there, then you get an answer
that belongs nowhere. Then you get a sensible answer, and then you
have three others that criticize the sensible answer. And then you
are back where you started, what is correct?"*

The patients in the conventional therapy group did not express the same need for
information regarding food and nutrition as the other patient group. These participants
had gone through a lifestyle therapy program involving a residential intermittent
program at a Rehabilitation Centre. The basis for this treatment is dynamic group
based psychotherapy, but in addition the patients take part in a structured Physical
Activity (PA) program daily (2 sessions of group PA+ one individual/day), and in a
nutritional education program (about estimation of energy-balance, food, healthy
cooking and eating etc). The ultimate aim of the treatment is to empower the patients
so they can be in charge of their own lifestyle change. However, these patients
emphasized that they needed information that clearly stated guidelines for diet and
physical activity according to their physical health. Several of these participants had
knowledge about healthy food, but had difficulties estimating the amount of food to
eat, and to establish healthy lifestyle habits. Further they emphasized communication
with healthcare professionals and other patients, for both evidence-based information
that professionals holds, in addition to the experiential knowledge that patients have.
Several of the participants in the conventional therapy group had reduced significant
amount of weight, but had regained weight after a while. Using food as coping
mechanisms, and lack of structure in daily life were issues that were mentioned as the
causative factors. A woman in this group put it as follows:

*"I eat because I like it, not because I necessarily am hungry. Food
is, I probably eat in response to emotions – many people do that...So
if I am happy - I eat, if I am bored - I eat, if I am exited - I eat."*

Regarding Disease Management, suggested tools to support self-care management
were reminders on sms, diary writing, meal plans, PA-plans and commitment to
others. One of the patients experienced that commitment was a motivational factor to
exercise:

*"For me, it helps with commitments, and for instance such as
workout partners. And then it can be the fact that the exercise group
has limited number of places. So when you get place, then you just
have to show up."*

Lack of structure was mentioned as a challenge in both patient groups. The
majority of the patients described their daily life before treatment as unstructured
regarding food, diet and exercise, and many described poor and inconsistent eating
patterns. Changing their lifestyle was a challenge for many. As one woman said:

*"I am structured when I am at work, then everything goes according
to the clock in my mind. It is during the weekends (the challenges
come), because then I am free."*

The patients believed that the disease management process could be facilitated by
the use of a clinical e-health system. A system with several features gathered in one

portal was the main argument for wanting to use such a system. Validated information by professionals, and that professionals can act as moderators and correct invalid statements or misinformation, were highly rated requirements. In contrast to commercial web-portals and websites, the expertise that professionals hold about the treatment of severe obesity is invaluable for the patients going through such treatment. Several patients could tell about little understanding about their condition from their surroundings, and underlined that professionals' with knowledge about their particular condition was extremely important. General information about weight-loss and dieting becomes irrelevant, as these people experience their situation as more complex than only reducing weight.

4.5 The Perspectives Compared

The main differences between the rationales of the patients and the healthcare professionals can be summed up as show in table 3.

Table 3. Multiple rationales

The Professionals' Rationales	Requirements	The Patients' Rationales
Patient education	Information	Self-care
Efficacy	Communication	Support, guidance
Prevent side-effects	Self-management tools	Social network
Patient self-care		Health outcomes

Our findings imply that a clinical e-health system can promote validated information delivery and retrieval, enhanced communication, and self-care management tools for the patients. Even though the multiple groups' requirements were similar, the rationales differ.

5   Discussion and Conclusion

Traditionally healthcare treatment has taken place in hospitals and healthcare institutions. Lately there has been a paradigm shift, where treatment also takes place in the home environment of the individual, particularly within chronic care. An understanding of this environment is crucial when implementing new technology. Patients hold on unique experiential knowledge, and they provide insight to their daily life, and about what their challenges and needs are.

Patients that have undergone weight-reduction treatment due to severe obesity need to implement lifelong lifestyle modification, and are dependent on implementing self-care and disease management. Our findings indicate that a clinical e-health system for this patients group would benefit from active involvement of healthcare professionals, even though the patients would be the primary users. Commercial e-health systems are probably more economical, but provide less professional contact [11]. The patients emphasize access to validated health information and communication with healthcare professionals. They require a social network with patients for social support and experiential knowledge, and their main rationales are improved health outcomes and weight-management. The healthcare professionals emphasize patient education, prevention of side-effects of the treatment (bariatric surgery patients), and enhanced patient self-care management. Their rationales are efficiency of clinical practice, patient health outcomes, and improved quality of care.

In this project we also determined that we were dealing with two distinct user groups: Those undergoing surgery and those going through conventional lifestyle therapy. The difference was not mainly concerning medical or social characteristic, but simply the fact they had different needs. Having undergone weight-loss surgery leads to specific needs that are irrelevant for the other group. Mixing the two would only lead to confusion, and underlines the importance of involving multiple user groups during the development process. Professional knowledge and the experience that clinicians have gained over years are irreplaceable, as they have evidence-based knowledge about a whole group, and not only about one individual patient. That a clinical e-health system contain information that is in line with medical treatment is crucial. Our findings imply that when designing such a system, the perspectives of both healthcare professionals and patients need to be addressed, particularly when the system is to act in the continuation of the medical treatment offered by the specialist care.

Multiple stakeholders have different backgrounds, interests and expectations towards a system. In this study we found that the multiple groups' required features for a clinical e-health system for weight-loss patients are quite similar, but the multiple groups had different perspectives and rationales. These are important in a system development process, and imply that inclusion of multiple user groups may provide an added value when it comes to input about what to prioritize for implementation. The multiple user perspectives and requirements complement each other, and provide valuable input for system design. Our findings imply that development from only one perspective may contribute to a system that lacks important content and functionality.

Acknowledgments. We thank the participants for sharing their time, valuable experience and great ideas.

References

1. Årsand, E., Demiris, G.: User-centered methods for designing patient-centric self-help tools. Informatics for Health and Social Care 9 33(3), 158–169 (2008)
2. Tsai, C.C., Lee, G., Raab, F., Norman, G.J., Sohn, T., Griswold, G.G., Patrick, K.: Usability and Feasibility of PmEB: A mobile Phone Application for Monitoring Real Time Caloric Balance. Mobile Networks and Applications 12, 173–184 (2007)
3. World Health Organization, http://www.who.int/en/
4. Fujioka, K.: Management of Obesity as a Chronic Disease: Nonpharmacologic, Pharmacologic, and Surgical Options. Obesity Research 10, 116–123 (2002)
5. Karlsson, J., Taft, C., Rydén, A., Sjöström, L., Sullivan, M.: Ten-Year trends in health-related quality of life after surgical and conventional treatment for severe obesity: the SOS intervention study. International Journal of Obesity 31, 1248–1261 (2005)
6. Fujioka, K.: Follow-up of Nutritional and Metabolic problems After Bariatric Surgery. Diabetes Care 28(2), 481–484 (2005)

7. Neve, M., Morgan, P.J., Jones, P.R., Collins, C.E.: Effectiveness of a web-based interventions in achieving weight loss and weight-loss maintenance in overweight and obese adults: a systematic review with meta-analysis. Obesity 11, 306–321 (2009)
8. Ehn, P.: Work-Oriented Design of Computer Artifacts. Lawrence Erlbaum Associates, Inc., Mahwah (1991)
9. Guha, M.L., Druin, A., Chipman, G., Fails, J.A., Simms, S., Farber, A.: Working with young children as technology design partners. Communications of the ACM 48(1), 39–42 (2005)
10. Glaser, B.G., Strauss, A.L.: The discovery of grounded theory: strategies for qualitative research. Aldine, Chicago (1967)
11. Gold, B.C., Burke, S., Pintauro, S., Buzzell, P., Harvey-Berino, J.: Weight Loss on the Web: A Pilot study Comparing a Structured Behavioral Intervention to a Commercial Program

A Study on the Visibility of
the Light Emitting Braille Block

Hiromi Fukuda[1], Noriaki Kuwahara[1], Takao Suzuki[2],
and Kazunari Morimoto[1]

[1] Graduate School of Science and Technology, Kyoto Institute of Technology,
Kyoto, 612-8585, Japan
[2] Tanabe Co., Ltd, Kyoto, 602-8164, Japan
hiromi@fukudanet.com

Abstract. About 60% of the visually impaired people are low-vision. Light-emitting textured paving block by using LED is developed to support the mobility of such visually impaired because such block is considered to be effective for notifying specific places such as the entrance, the exit, and so on to the weak eyesight people in the night. This block uses the innovative lighting mechanism by which extremely long-life light emission by using the battery is enabled. So far, there is no report on the visibility of the LED by using this mechanism. Therefore, in this paper, we report the result of the preliminary experiment for evaluating the visibility, and discuss on the trade-off between the visibility and the electric consumption.

Keywords: Low-vision, Light-emitting textured paving block, LED, Visibility.

1 Introduction

There are about 310,000 visually impaired in Japan. Visually impaired people are not necessarily total blindness, because about 60% of them are weak eyesight. For the persons who have visual impairment, walking guide is general requirement. To fulfill the requirement, many high-technology solutions including wearable devices [1].

The Braille blocks had invented originally by Mr. Miyake who lived in Okayama Prefecture in 1965. Two years later, 230 blocks were paved around a school for the blind in Okayama. After that, the blocks were spreading everywhere. Though there are several arguments that these blocks are not useful for wheelchair users, they became already necessary infrastructure for the blind persons living in Japan.

Figure 1 is an example of the visually impaired people's' view. Therefore, the light-emitting textured paving block ensures the security of the night walk of the people with such weak eyesight, and is used to notifies the specific places such as the entrance, the exit and so on (Figure 2).

M. Kurosu (Ed.): Human Centered Design, HCII 2011, LNCS 6776, pp. 295–303, 2011.
© Springer-Verlag Berlin Heidelberg 2011

The block shown in Figure 1 uses the LED as the light source, and the innovative lighting mechanism enables it to emit light for extremely long period of time by using the battery. This technology will be assumed what is applied to various public signatures in future (Figure 3).

There is also a report on programmable light-emitting Braille blocks by using LED [2]. These blocks have each ID number and its luminance can be controlled in eight levels. Most remarkable feature is that the program which controls illuminating timing and the brightness is rewritable. By that feature, these blocks became an emergency guide system by changing the animation patterns.

However, there is no evaluation report on the visibility of such light-emitting Braille block for the people with weak eyesight. Therefore, we conducted the preliminary experiment for evaluating the visibility. In this paper, we report the result of the preliminary experiment for evaluating the visibility, and discuss on the trade-off between the visibility and the electric consumption. And there are already many studies of the walk support of weak sight person [3]- [5].

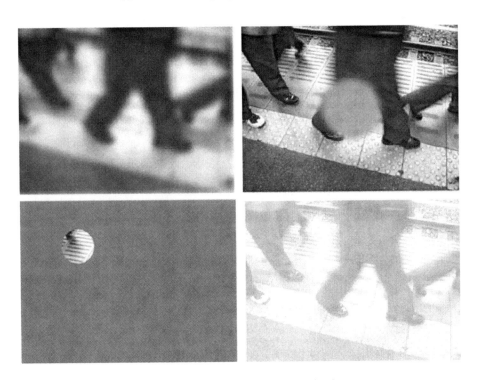

Fig. 1. Example of weak sight person's view

Fig. 2. Example of light-emitting textured paving block by using LED

Fig. 3. Example of public sign that shows no smoking area

2   Light-Emitting Textured Paving Block by Using LED

Figure 4 shows the lighting mechanism of LED that enables the long-life light emission by using the battery. A is installed in the places such as road surfaces catching the strong vibration. Therefore A is driving by a battery to decrease risks such as the disconnection. It is well-known that the brightness of the LED depends on the steep of the rising edge of the electric current in the LED (Broca-Sulzer phenomenon) [6].

In order to generate such electric current, the tantalum condenser is used because of its low internal electric resistance. The high-speed switching transistor enables to charge and discharge it in an extremely short time. By making the interval of this cycle shorter than the critical fusion frequency of human, people perceive only one blight blink rather than several dark blinks. We can control the brightness and the electric consumption by

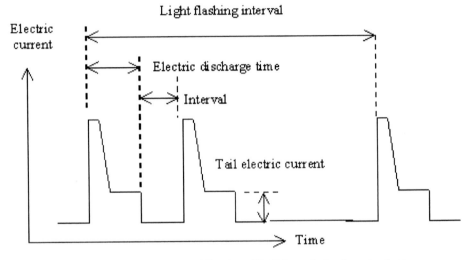

Fig. 4. Lighting mechanism enabling long-life light emission by using battery

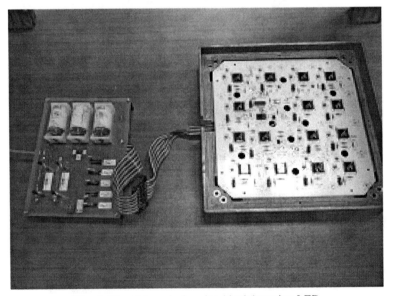

Fig. 5. Circuit board of paving block by using LED

Fig. 6. Appearance of flashing of block in laboratory

changing the interval of the blinking and the number of the blinks, but so far, it remains uncertain which combination of these parameters are the best in terms of both the visibility and the electric consumption. Figure 5 shows the circuit board which we used for this evaluation, and Figure 6 is appearance of blocks' flashing. The number of times of the light flashing is once twice three times from the left.

3 Experiment 1

3.1 Method

In this experiment, I controlled this parameter and experimented on evaluation of the visibility by the physically unimpaired person. The subjects of this experiment were nine students of Kyoto Institute of Technology. And an experiment place is a classroom of Kyoto Institute of Technology, and all of them are normal eyesight.

The experiment was conducted in a room surrounded in a blackout curtain. The block has 25 LEDs, but we covered LEDs except the center one. Subjects could observe the blinks of the LED at the center of the block. The fluorescent lamp of 100W was set above the LED. We gradually lowered the height of the lamp position. Accordingly, the surface of the block became brighter, and perceiving the blinks of the LED became difficult. When subjects could not perceive the blinks of the LED, we measured the illumination of the block surface as the visibility of light-emitting Braille block. (Figure 7).

The subject stood at 90cm apart from the block. Three patterns of the interval of the charge and discharge were tested: 2.5msec, 7.5msec, and 25msec, and three patterns of the number of times were tested: one, two, and three. On each combination of the interval and the number of time of charge and discharge, we measured the illumination mentioned above three times per subject.

Fig. 7. The machine parts which used for an experiment by the physically unimpaired person

3.2 Result

Figure 8 shows the experimental result. When the interval was 2.5 msec, the luminance of the block surface when subjects could not perceive the LED blinks increased according to the number of time of charge and discharge. There was the significant difference between each result on the number of time in this interval.

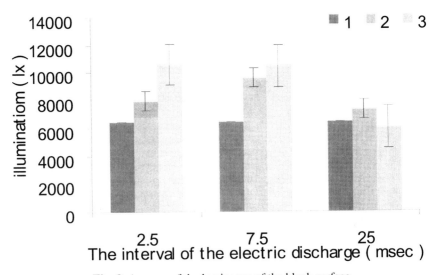

Fig. 8. Average of the luminance of the block surface

However, when the interval was 7.5 msec, there was no significant difference between two and three times of charge and discharge. Furthermore, when the interval was 25msec, there was no significant difference between one, two, and three time. Therefore, in order to gain sufficient visivility even in the daytime outdoor shade (about 10,000 lx), the best trade-off point is to charge and discharge electricity twice at 7.5msec interval.

4 Experiment 2

4.1 Method

In this experiment, the method was the same as Experiment 1, but the subjects were weak sight persons. The number of the subjects was twenty, but There were the subjects who had a difficulty to find the blinking of the LED due to the severe tunnel vision. Also, there were the subjects who could not perceive the blinking of the LED even in the lowest illumination condition of our experimental setup. Therefore, only nine subjects could complete this experiment.

4.2 Result

Figure 9 shows the experimental result. There was no significant difference between the results of the number of time of charge and discharge in every case of the interval of the electric discharge. It also shows that weak sight person could not perceive the LED blinking at about one quarter of the luminance in comparison with the normal vision person even though they could find the position of the blinking LED. Almost all subjects developed severe tunnel vision. They lost not only the most of the visual field, but also the sensitivity of their retina due to the disease. Furthermore, their cornea, lens, and vitreum might have become opacified.

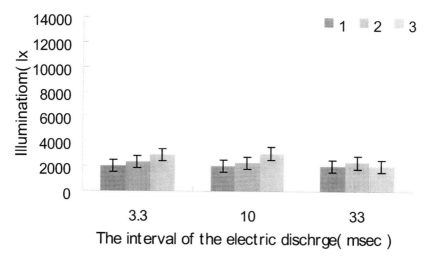

Fig. 9. Average of the luminance of the block surface

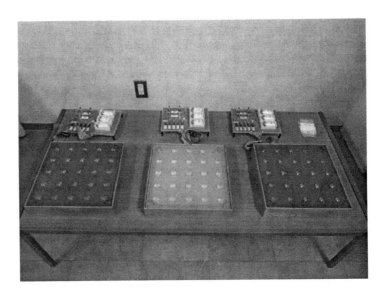

Fig. 10. Flashing of block of various colors

5   Conclusion and Future Works

According to the experimental results of the subjects with normal vision, increasing the number of times of charge and discharge increased the visibility of the LED when the interval was 2.5msec and 7.5msec. On the other hand, when the interval was 25msec, increasing the number of times of charge and discharge had no effect for increasing the visibility. Furthermore, when the interval was 7.5msec, there was no significant difference between two and three times of charge and discharge. In terms of the trade-off between the visibility and the electric consumption, setting the interval to 7.5msec, and setting the number of times to two is the best in this preliminary experiment.

On the other hand, the result of the weak sight subjects showed no significant difference between every combination of the parameters, although it indicated the same tendency as that of people with normal vision.

As for the future works, we would like to evaluate the visibility of the LED under the condition where the parameters are controlled more flexible, because we tested very limited combination of the parameters due to the hardware constraint. We are developing the new board for this purpose. Also, we would like to examine the effect of the block color to the visibility by using the light-emitting Braille blocks of other colors as shown in Figure 10.

Acknowledgments. This research was supported by Adaptable and Seamless Technology Transfer Program through Target-driven R&D, JST.

References

1. Kim, C., Song, B.: Design of a wearable walking-guide system for the blind. In: Proceedings of the 1st International Convention on Rehabilitation Engineering & Assistive Technology (2007)
2. Kobayashi, M., Katoh, H.: Development and installation of programmable light-emitting braille blocks. In: Miesenberger, K., Klaus, J., Zagler, W., Karshmer, A. (eds.) ICCHP 2010. LNCS, vol. 6180, pp. 271–274. Springer, Heidelberg (2010), doi:10.1007/978-3-642-14100-3_40
3. Takai, C., Ishida, H.: Visibility of Tactile Indicators: Study on safety for vision-impaired pedestrians in public space Part 1. J. Mater. Sci.: Memoirs pro-Architectural Institute of Japan Plan 6, 153–158 (1999)
4. Takai, C., Ishida, H.: Visibility of Tactile Indicators: A Way of Recognition Improvement of Tactile Indicators: Study on safety for vision-impaired pedestrians in public space Part 2. J. Mater. Sci.: Memoirs Pro-Architectural Institute of Japan Plan 5, 141–148 (2000)
5. Oka, M., Kano, T.: Basic Study on the Effectiveness of Light Emitting Curbstones of Nighttime Walking For Persons With Low-Vision. J. Mater. Sci.: Memoirs pro-Architectural Institute of Japan Plan 8, 1707–1713 (2008)
6. Morita, K., Abe, M., Motomura, H., Jinno, M.: Luminous Improvement by Pulsed LED Using Psychometric Effect. The Institute of Electronics, Information and Communication Engineers 108(227), 35–40 (2008)

Knowledge Based Design of User Interface for Operating an Assistive Robot

Rebekah Hazlett-Knudsen[1], Melissa A. Smith[2], and Aman Behal[2]

[1] University of Central Florida, School of Social Work, Orlando, Florida
[2] University of Central Florida, NanoScience Technology Center, Orlando
rhazlett@mail.ucf.edu, {Florida,abehal}@mail.ucf.edu

Abstract. In this paper, the research objective is to develop and implement a procedure for integration of user preferences and abilities into the Graphical User Interface (GUI) of a Wheelchair Mounted Robotic Arm (WMRA) to be operated by users with Traumatic Spinal Cord Injury (TSCI).

Keywords: Graphic user interface (GUI), robotic assistive devices, MANUS robotics system, heuristics, traumatic spinal cord injury.

1 Introduction

Traumatic spinal cord injury (TSCI) affects nearly 262,000 individuals within the United States [1]. The level of injury, associated paralysis and resulting limits in upper and lower body function differentially impair individuals' abilities to perform activities of daily living (ADLs). To respond to these challenges, researchers seek new and enhanced interface options to assist TSCI individuals and their caregivers with ADL tasks through the use of assistive devices.

This work focuses on interaction with robotic assistant devices that contribute to social reintegration through enhancing function and cutting caregiver cost and reliance, thereby promoting a sense of rejuvenated self-esteem in the impaired individual. TSCI users in wheelchairs need to perform pick-and-place tasks in an unstructured workspace whose volume is generally larger than that of any fixed-base robot's reachable workspace. Therefore, wheelchair mounted robot assistants (WMRAs) – which are lightweight, highly maneuverable robotic arms with standard grippers that attach to a wheelchair – have been developed to increase functional independence in a variety of tasks and environments for wheelchair-bound individuals with upper extremity debilities.

The MANUS Assistive Robotic Manipulator (ARM) is a WMRA which has consistently been adopted by users and researchers as the robotic assistant of choice (e.g., see [2], [3], and [4]). The commercial ARM system is controlled by a joystick or a keypad controller that requires much training and memorization to be used practically. It is difficult, if not impossible, for use by users with higher-level TSCI. Researchers in UCF's Assistive Robotics Laboratory developed a preliminary PC based graphical user interface (GUI) that can be operated using a mouse, a track ball and jelly switch, a head tracker, or voice recognition. This paper focuses on a system

M. Kurosu (Ed.): Human Centered Design, HCII 2011, LNCS 6776, pp. 304–312, 2011.
© Springer-Verlag Berlin Heidelberg 2011

procedure for selecting relevant interface components and their ideal layout so they are accessible to the target group of users with TSCI.

2 GUI Design

A graphical user interface (GUI) is an interface based on graphics, such as icons, pictures and menus, instead of pure text. Present in everything from websites to automated teller machines to mobile phones, GUIs are a present in various areas of technology, with their applications subject to continuous expansion. The MANUS robotics system is specifically designed to assist patients with limited upper-body movement, which makes the native joystick controller a limiting factor for many users. In developing a graphical user interface that could be controlled via touch screen, track ball, jelly switch or voice recognition, the accessibility of the MANUS assistive robot to its intended population is greatly increased.

3 Development of a Component-Based Ranking System

To work towards a GUI design with integrated user preferences, the researchers designed a component-based ranking system to analyze and rank specific criteria needed within the various graphical user interface designs. A component-based ranking system is comprised of all the aspects deemed necessary for the ideal system ranked in order of priority and importance. The ideal GUI design would possess the maximum possible 'score' for each of the components of the system. The creation of the component-based ranking system for the development of the MANUS GUI was accomplished by integrating focus group and participant feedback regarding heuristic means and preferences with the guiding principles of Nielsen's computer (web-based) heuristic information.

3.1 Heuristics

Heuristics refer to the experience based methods used by an individual to problem-solve and integrate knowledge. Nielson's [5] general heuristics for web-based interface design includes ten usability principles for web interface design, including visibility of system status, user control and freedom, efficiency of use, minimalist design, error prevention and error recovery. These principles when combined with heuristics, or means of learning, are important to the effective design of any user interface and are not limited to website design. Combining Nielson's principles for web-based interface with the heuristic patterns identified in focus group results, a ranking and evaluation system for GUI design was developed.

3.2 Focus Group

The researchers conducted a focus group in which four TSCI patients of different age, ethnicity and injury level participated. The main purpose of the focus group was to explore and identify the heuristic means by which users integrate new assistive technology for use with activities of daily living (ADLs). Researchers were interested in identifying the techniques persons with TSCI use to self-educate. Of particular use to the development of the GUI, the focus group identified discrete features of a robotic assist device seen as preferable by users.

Focus group analysis was conducted using Nvivo 8 qualitative software [6]; a qualitative research software which allows the researcher to identify and aggregate patterns in a source such as a focus group transcript. The application of Nvivo involves the creation of a coding system based upon thematic questions and sensitizing concepts. Sensitizing concepts [7] are constructs and organizing ideas that guide the researchers coding and identification of patterns. The analysis process resulted in preliminary information on learning heuristics, training preferences, social learning, and interface preferences. Three main methods of participant problem solving emerged and included: adaptation, seeking instruction, and trial and error. Preferences for interface with a robotic assist device fell into eight areas, including simplicity, affordability, reliability, accessibility, customizable, speed, responsiveness, and accuracy. The most frequent references were to simplicity, affordability and reliability. This indicated a need for TSCI individuals to have a robotic assist device which they can easily access and on which they can depend. Less frequent, yet important references were also made to customizability, speed, responsiveness, and accuracy. Future focus groups should further inform learning heuristics and interface preferences. More details about the focus group can be found in [8].

3.3 Component-Based Ranking System

By combining the information gained from the focus group and referring to Nielson's interface heuristics, a component-based ranking system was developed. The six-main components found necessary, as identified by the focus group as preferences, for the interface are safety, simplicity, responsiveness, accuracy, reliability, and customizability. For each user preference, GUI design implications were identified by the researchers in order to rank a prototype interface in each category. The different preferences were given a percentage rating based on their overall relative importance to the system, based on feedback from the focus group and user comments during usage. Table 1 shows the relative importance of each component within the system via the maximum percentage value for each component. The following formula was used for score calculation:

$$SCORE = \sum_i \text{Importance of Factor } i * \text{Efficacy of Factor } i \text{ in Design}$$

Safety was considered the most important feature for any of the graphical user interfaces designed, due to the essential need for user safety in interactions with the MANUS assistive robot. GUI design safety was assumed to be at its maximum as the safety of the system is due more to the mechanical components of the robot than it is to the arrangement of the GUI. The overall safety of the system would not change greatly due to changes in the button arrangement within the GUI.

Table 1. Component-based ranking system by component

Component	Description	Maximum Percentage Value
Safety	-user activated stop function -prevents user from impossible actions	30%
Simplicity	-minimum number of buttons and screens necessary for functional use of system -function of buttons clear to user -dialogue contains relevant information -interface layout optimum for functional usage -interface accessible to both healthy and disabled users	25%
Responsiveness	-interface acknowledges user entries immediately -interface keeps users informed of system status	20%
Accuracy	-interfaces offers feedback/suggestions when expected (user-created) errors occur -interface responds the same way every time user does the same actions	10%
Reliability	-design has minimal system resets and unexpected errors -interface behaves in expected manner – minimal errors, unfamiliar functions/messages	10%
Customizability	-design has relevant customizable features	5%
Total		100%

The component-based ranking system was utilized to assess the value of adding a feature to the proposed GUI designs. For example, adding a stop button to the interface decreased the simplicity score due to the addition of another button, but increased the safety score by allowing the user to have more control over the system; since safety has a higher percentage ranking than simplicity; inclusion of the button was found to be beneficial. The goal with each GUI design was to attain the highest maximum percentage value of each component.

4   GUI Designs

The initial GUI design (GUI design 1), as shown in figure 1, was scored using the component-based ranking system to establish a baseline score. GUI design 1 represents the initial interface designed by the researchers for use with the MANUS assistive robot. Using the component-based ranking system this initial design was shown to lack in all areas except reliability. Based on these areas of concern, other GUI designs were developed.

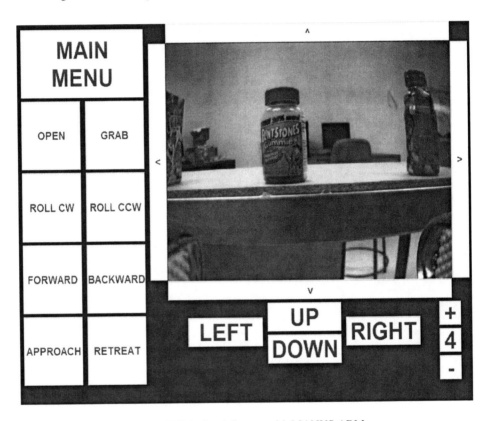

Fig. 1. GUI design 1 for use with MANUS ARM

Key areas of concern with GUI design 1 included: the lack of clear button functionality (simplicity), no representation of system status (responsiveness), no confirmation given by system when object selected by user (accuracy), and no interface customizations available to user (customizability). Overall, GUI design 1 received a component score of 56 percent. Table 2 summarizes component concerns.

Table 2. GUI design 1 areas of concern by component

Component	Concern
Simplicity	Button functions not as clear as they should be
Responsiveness	No representation of system status
Accuracy	No confirmation given by system when object selected by user
Reliability	Interface is overall very reliable
Customizability	No interface customizations available to user

*Design received a score of 56 percent.

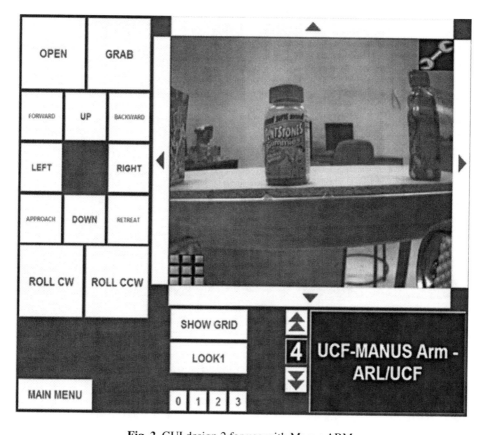

Fig. 2. GUI design 2 for use with Manus ARM

In GUI design 2, shown above in figure 2, a status box was added to give the current system status and feedback to the user (responsiveness). Buttons were rearranged to keep the most used buttons in close proximity (simplicity) and position buttons were added at the bottom to allow the user to send the robot to customized preset positions (customizability, simplicity). Finally, a 'look' button was added which allowing users to choose text-only buttons, image-only buttons, or text + image buttons. GUI design 2 received a component score of 85 percent, an increase of 29% from GUI design 1.

In spite of the increase in component-based score further corrections were applied to the GUI in the areas of simplicity, responsiveness, accuracy, reliability, and customizability adjustments.

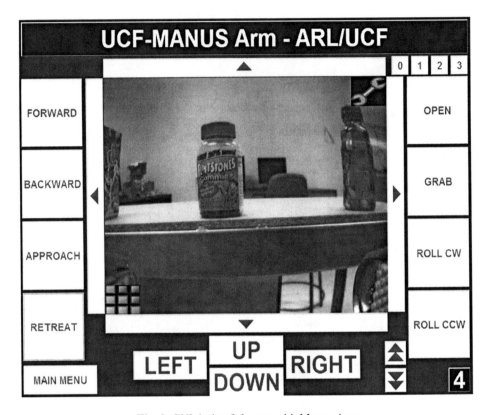

Fig. 3. GUI design 3 for use with Manus Arm

GUI design 3, shown in the image above, was arranged differently than GUI design 2, with the buttons now surrounding the image from the camera. This configuration allowed for the buttons to be directly located on the screen in the corresponding directions (i.e. the pan-up button at the top of the screen). Changes made in design three focused on improving simplicity and responsiveness. To increase the component-based ranking scores the dialog box was moved to the top of the screen, stretching the size over the width of the screen. Buttons were moved and enlarged for ease of use. GUI design 3 received a component score of 85 percent.

4.1 Final GUI Design

The final GUI design (seen in Figure 4) ranked the highest on the component-based ranking system with a score of 94 out of a total possible 100 points. This ranking was due to the three factors which were not present in earlier versions of the GUI, specifically 1) an easily visible dialog box in the bottom right corner, 2) the most

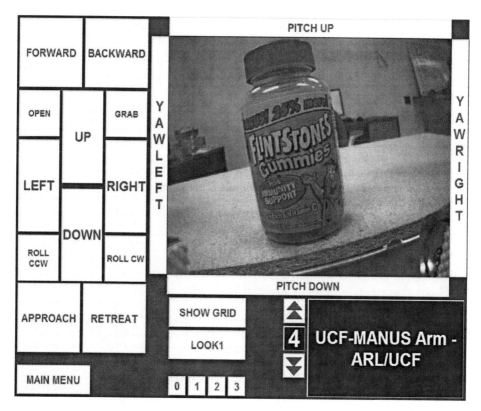

Fig. 4. Final GUI design, scored 94/100 on the component based ranking system

frequently used buttons displayed in a larger size and 3) an undo button, which, when selected by the user, returned the robot to the last "safe" position. The undo button appears in the lower-right corner of the camera-view screen after a button has been selected by the user. This improvement in the safety component score earned the final GUI design an increase of four percentage points from GUI design 3 and an increase of 19 percentage points from the initial design. This substantial increase allowed the final design to receive the highest component score of the four designs.

5 Conclusions

The researchers created and tested four GUI prototypes. Following three GUI design iterations, the final GUI design resulted in the highest component-based ranking score. Due to an easily visible dialog box in the bottom right corner, buttons in a larger size and an undo button allowing users to correct error-causing actions the final GUI design demonstrated a clear score and functionality increase from previous designs. All four GUI designs demonstrated consistent levels of safety, accuracy, reliability, and customizability. The final design maximized the integration of simplicity and responsiveness functions.

Acknowledgments. The researchers would like to thank the Orlando Health Rehabilitation Institute which is a part of Orlando Health, Inc. along with the participants of the focus group and user study.

References

1. Spinal Cord Injury Statistical Center. Spinal cord injury facts and figures at a glance. University of Alabama, Birmingham, Alabama. Retrieved on (September 2, 2010), from `https://www.nscisc.uab.edu/public_content/pdf/Facts%20and%20Figures%20at%20a%20Glance%202010.pdf`
2. Kwee, H., Quaedackers, J., van de, B.E., Theeuwen, L., Speth, L.: Adapting the control of the MANUS manipulator for persons with cerebral palsy: an exploratory study. Technology and Disability 14, 31–42 (2002)
3. Tsui, K., Yanco, H.: Simplifying Wheelchair Mounted Robotic Arm Control with a Visual Interface. In: Proceedings of the AAAI Spring Symposium on Multidisciplinary Collaboration for Socially Assistive Robots (March 2007)
4. Tijsma, H., Liefhebber, F., Herder, J.: Evaluation of New User Interface Features for the Manus Robot Arm. In: Proceedings of IEEE International Conference on Rehabilitation Robotics, pp. 258–263 (2005)
5. Nielson, J.: Ten Usability Heuristics (2005) (retrieved on September 3, 2010) from, `http://www.useit.com/papers/heuristic/heuristic_list.html`
6. International, Q.S.R.: Nvivo 8, main webpage (2008), `http://www.qsrinternational.com`
7. Holloway, I.: Basic concepts of qualitative research. Blackwell Science, Oxford (1997)
8. Hazlett, R., Kim, D.-J., Godfrey, H., Bricout, J., Behal, A.: Exploring Learning Heuristics for Adopting New Technology to Assist with Activities of Daily Living (ADL): Results of Qualitative Analysis using Nvivo 8 Software. In: 2010 Florida Conference on Recent Advances in Robotics, Jacksonville, FL (May 20-21, 2010)

Holistic Prosthetic Approaches to the Hearing Handicapped People: Communication Tools in Various Situations

Kazuyuki Kanda[1] and Tsutomu Kimura[2]

[1] Chukyo University, Japan
[2] Toyota National College of Technology, Japan
kanda@lets.chukyo-u.ac.jp, kim@toyota-ct.ac.jp

Abstract. We difine our term holistic prosthesis, and stated why we coined new word and its background. We showed some example of holistic use of prosthetic manufactures. A new idea is proposed for supporting the hearing handicapped people from our point of view. We introduced an experiment at a museum as an example of informational support in public.

Keywords: holism, prosthesis, public, hearing handicapped.

1   Introduction

A man always creates the way of using a tool in a way which the manufacturer could not anticipate in advance. For Example, a simple stick can be used as a rod for hanging, fishing, or whipping, or as a pole for propping up the tree or flag, or as a stake, or even for chopstick, or for magic wand. He always creates another way different from an original use of the tool. A matchstick quiz is an example of non-original use of a match to light a smoke.

Fig. 1. Match stick game

A child always finds a new way of using a tool for a new toy. A simple block could be a car, a truck, a house, etc. anything he imagines in his playing. This is because of his imaginations.

M. Kurosu (Ed.): Human Centered Design, HCII 2011, LNCS 6776, pp. 313–320, 2011.
© Springer-Verlag Berlin Heidelberg 2011

2 An Invention of Telephone

An inventor is sometimes tormented by a guilty conscience. Alfred Nobel left his will to found Nobel Prize for invention of dynamite to be used as a weapon. Alexander Graham Bell, the inventor of the telephone had a guilty conscience to his invention, because hie original purpose was to invent a hearing aid. His father, grandfather and brother had all been associated with work on speech. His father, Alexander Melville Bell was known as teacher of speech in deaf education and he created a system named Visible Speech which is composed of symbols that show the position and movement of the throat, tongue and lips when a man produces a speech sound.

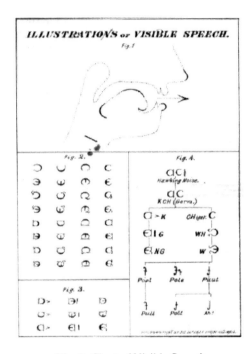

Fig. 2. Chart of Visible Speech

Both Graham's mother and his wife were deaf and his research was led to invent an hearing aid. Accidentally he invented a telephone which caused the deaf people disadvantageous to tele-communication. He set up the Volta Laboratory Association with money from the Volta Prize. He set aside his share of the profits from his invention as an endowment of the Volta Fund. He became the first president of the American Association to Promote the Teaching of Speech of the Deaf and the association became Alexander Graham Bell Association for the Deaf and Hard of Hearing now.

An invention of telephone was inconvenient for communication of the deaf who uses sign language and finger spelling. It caused them a lot of trouble on the contrary to Bell's aboriginal intention.

After a hundred years, the invention of an e-mail on a cellular phone helps the hearing handicapped people. The inventors might not expect that.

3   Definition of Holistic Prosthesis

The term "holistic prosthesis" was a coinage by The Holistic Prosthetic Research Center, Kyoto Institute of Technology, Japan. We would like to introduce the concept and definition of our idea below.

3.1   Holism

The word "holistic" is an adjective form of 'holism' and it is widely used in many areas of sciences, such as in anthropology, business, ecology, economics, philosophy, sociology, psychology, theology, neurology, architecture, education, medicine, etc. (*wikipedia*: http://en.wikipedia.org/wiki/Holism).

The general principle of holism was concisely summarized by Aristotle in his Metaphysics that "the whole is different from the sum of its parts". He thought that 1 + 1 is not simply 2 in the human actions. The counterpart of holism is called reductionism saying that a complex system can be explained by reduction to its fundamental parts. Rene Descartes stated that non-human animals could be reductively explained as automata in *De Homines* 1662. Therefore he was called a reductionist.

If we read his famous book *Discourse on the Method and Principles of Philosophy* carefully, he had a holistic idea in general and he was always not in reductionism:

"but if there were machines bearing the image of our bodies, and capable of imitating our actions" as far as it is morally possible, there would still remain two most certain tests whereby *"to know that they were not therefore really men"*. Of these the first is that *they could never use words or other signs arranged in such a manner as is competent to us in order to declare our thoughts to others"* (Part V).

The second test is, that although such machines might execute many things with equal or perhaps greater perfection than any of us, they would, without doubt, fail in certain others from which it could be discovered that *"they did not act from knowledge, but solely from the disposition of their organs"* (ibid.).

These lines seem to show his reductive concept to the organs, but the next lines below suggest the difference between animals and humans. Nor does this inability arise from want of organs: for we observe that magpies and parrots can utter words like ourselves, and are yet unable to speak as we do, that is, so as to show that they understand what they say; in place of which *"men born deaf and dumb"*, and thus not less, but rather more than the brutes, destitute of the organs which others use in speaking, *"are in the habit of spontaneously inventing certain signs by which they discover their thoughts to those who, being usually in their company, have leisure to learn their language"* (ibid.).

His idea is reductive as far as we see the animals or machines, but holistic when he considered a human who has mind and thought.

3.2 Prosthesis

The term 'prosthesis' is most frequently used in the field of medicine, meaning an artificial device that replaces a missing body part, like limb, or sometimes heart valve, eye, etc. It seems that prosthesis is realization of reductionism and our term "holistic prosthesis" looks like a contradiction in this sense.

If we stand on the manufacture's viewpoint, it is yes, it is a contradiction. But if we stand on the user's one, we say it is no. The manufacturer always uses his brains to create a tool to replace a missing body and to substitute the original function of the body part. All he is interested in is a mechanism of the body part. On the contrary, the user of the prosthetic tool might use it in a different way from the manufacturer's intention.

In a recent TV drama in Japan, we saw the murderer used a prosthetic arm as a weapon to hit the victim's head to kill him. No one but the detective could find it because other officers tried to find a hammer or some other blunt instrument. They could not imagine that a prosthetic arm could be used as a deadly weapon, being preoccupied to the idea that a prosthetic arm was used only for medical use.

A wheelchair was originally invented as a prosthetic tool for moving but it is now widely used a sporting tool for racing, yachting, playing tennis or basket ball, or some are designed as a furniture in a living room, which are made of all woods.

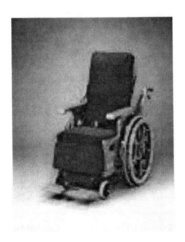

Fig. 3. Wooden Wheelchair

A maker of this chair changed his mind to create a new tool. He says "we want to be a partner to convey our mind of gentleness and heartfulness, not as a simple manufacturer of a tool".

Here we propose to expand the sense of the term prosthesis to mean by that it is a supporting system not only replacing the missing part but creating a new usability of human action. When we think a modern robotic prosthesis for physically handicapped people, a robot suit, for example, functions more than a normal human body parts with more power and no fatigue.

Fig. 4. Robot Suites

4 Prosthesis for Hearing Disabled People

A hearing aid is an only prosthetic tool at present for the hearing impaired person. However it does not work for many of them. Graham Bell's idea for inventing a hearing aid was not successful as far as he stack on the concept that hearing through the hearing aid and producing a speech were the best for the deaf. Actually the hearing aid is not a replace of an ear organ. It is just a set of small microphone and ear phone and it works for disorders in an external part of the ear. It does not work for nervous system nor brain, that is, not for sensori-neural hearing loss, which most of the deaf has.

All born-deaf persons spontaneously learn and acquire sign language. However, those elder people who lost hearing after being aged cannot learn sign language easily and they often do not recognize and accept to be deaf in their mind. They usually think that they have some disorder in their ears and they are different from deaf persons. Even if a hearing aid does not work very well, it is better than nothing for them. Those hearing aids now available in the market are rather small. Many of them are so-called ear mold type which hide themselves to the other people's eye. The elderly hearing disabled people would prefer to wear a hearing aid if it is a modern and fashionable one as an accessory of a designer's brand like Christian Dior or Chanel. Or like an SONY music player or iPod, because they do not need hide that it is a hearing aid. Moreover they are proud to wear a new fashion.

In the notion of holistic prosthesis, a proper tool for the hearing disable people has a variety. One is the fashionable hearing aid above. Another is a signing robot which can replace an sign interpreter in a special situation in which an interpreter must be avoided such as in an X-ray laboratory, in a private medical consultation or in a secrete judicial court. It is not a simple replace of a certain missing body part but it supports communication by signing.

Signing of a human is composed of manual movement of fingers and hands, movement of fists and arms, postures of a body torso and facial expressions. There are no special body parts for signing, like a mouth for eating or a nose for breathing. It is a sum of all the moving of those organs. If we are ignorant for the meaning of each movement, we do not understand the meaning of the signing. As Descartes said in the above, sign language is a human invention by the deaf and it is a holistic prosthetic tool for communication, not specified to certain human body.

5 QR Code System for the Public Exhibitions

Here we show an example of holistic prosthetic idea for the hearing handicapped people. Kanda et al. 2006, Kimura et al. 2008 and Kimura 2011 had an experiment of presentational system for the hearing handicapped people at public facilities as a museum, using QR codes.

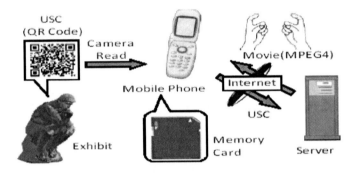

Fig. 5. QR Code Presentation System

The QR code is patched on the exhibition. The user catches it on his cellular phone. The corresponding text message or signing is idicated on its screen. We had an experimental presentation at Toyota Science Museum at Toyota City, Japan in 2009. We collected the comments and estimates of our system and usability from the deaf users at real situation.

All of them gave us a good reputation and suggestions, such a cost, signing speed, and future use.

Fig. 6. Experiment

The result of the display is shown below.

Fig. 7. Display on the Cellular Phone

6 Conclusion

The user of the above us a hint of another use of this system. If there is a Chinese deaf visitor, and if Chinese sign presentation is available, it must be helpful. Now we think of multilingual presentational system by Chinese text message, Korean text message, Japanese sign language, Chinese sign language, Korean sign language, and international sign language, if we could get translations of them. There would be no space to show all the languages at the same time on the exhibitions. In such a case, QR code is small enough to patch on the limited area.

As our experiment shows, the user always gives a hint to the manufacturer to develop newer and more comfortable usability. The concept of holistic prosthesis is significant to both the user and the manufacturer. They must be always open minded to the use of a tool and a human activity.

Acknowledgment. This research is partially supported by the Grants-in-Aid for Scientific Research of Japanese Ministry of Education, Basic Research (A) in 2010, titled "Morphemic Dictionary for Sign Language and its Applications", #20242009 , Representative Researcher, Kazuyuki Kanda, PhD.

References

1. Kimura, T., Hara, D., Kanda, K., Morimoto, K.: Development and Evaluation of Japanese Sign Language-Japanese Electronic Dictionary. Japanese Journal of Sign Linguistics 17, 11–27 (2008)
2. Kanda, K., Kimura, T., Hara, D.: A proposal of the universal sign code. In: Miesenberger, K., Klaus, J., Zagler, W.L., Karshmer, A.I. (eds.) ICCHP 2006. LNCS, vol. 4061, pp. 595–598. Springer, Heidelberg (2006)
3. Kimura, T., Katoh, M., Hayashi, A., Kanda, K., Hara, D., Morimoto, K.: Application System of the Universal Sign Code - Development of the Portable Sign Presenter -. In: Miesenberger, K., Klaus, J., Zagler, W.L., Karshmer, A.I. (eds.) ICCHP 2008. LNCS, vol. 5105, pp. 678–681. Springer, Heidelberg (2008)
4. Kanda, K. (ed.): A Basic Course of Sign Linguistics. Fukumura Publishing, Co. Ltd. (2009)
5. Kimura, T., Hara, D., Kanda, K., Morimoti, K.: Expansion of the System of JSL-Japanese Electronic Dictionary-An Evaluation for the Compound Research System. In: Proceedings of the 14th International Conference on Human-Computer Interaction (2011) (in print)

Human-Centered Design in the Care of Immobile Patients

Thomas Läubli[1], Roger Gassert[1], and Masaru Nakaseko[2]

[1] Swiss Federal Institute of Technology, Zurich, Switzerland
[2] Kyoto Institute of Technology, Japan
tlaeubli@ethz.ch

Abstract. Nurses frequently suffer from low back pain, but oppose against using mechanical lifting devices. It was found that the nurses' reluctance to use technical aids may be due to several drawbacks of currently used lifting devices in patient care: 1) the lifting maneuver is controlled through a control device located at a distant position form the patient (e.g. fixed to the supporting structure). 2) Conventional lifting devices are position controlled and operate at a low velocity. 3) The lifting device holds the entire weight of the patient, while the nurse performs translational movements. Therefore existing technological solutions were studied and novel ways were explored of achieving intuitive interaction, e.g. through the use of force and position sensors and shared control strategies. The initial results of our task analysis suggest that both the handicapped/ immobile person and the nurse may be supported by intelligent assistive lifting devices.

Keywords: nurses, lifting device, intuitive interaction.

1 Introduction

Musculoskeletal injuries comprise the largest proportion of total injuries in all types of nursing activities. Musculoskeletal injuries occur in both acute care and nursing homes, with care aids presenting twice the risk of registered nurses [1]. Several countries use a so-called no lifting policy to improve the care of immobile patients and limit excessive physical loads of nurses [2]. Nurses frequently suffer from low back pain, but strongly oppose against using mechanical lifting devices. It is hypothesized that enhancing such devices by ergonomically designed interaction control modalities would result in more intuitive use and help to overcome the nurses' reluctance. Similar "lifting and balancing" devices have been proposed for automotive production environments to minimize operator strain and are seeing increasing application in body unweighting systems.

2 Aim

To perform a user-oriented analysis of typical lifting and bed to wheelchair transfer tasks, to explore and develop potentially useful solutions that better fit the control of patient lifting devices to the needs of nurses and patients, e.g. by an intelligent assistive device which partially unweights the patient while the nurse initiates and controls the lifting and/or transfer movement in a collaborative manner while holding and guiding the patient in a natural manner.

M. Kurosu (Ed.): Human Centered Design, HCII 2011, LNCS 6776, pp. 321–326, 2011.
© Springer-Verlag Berlin Heidelberg 2011

3 Methods

Two computer-based questionnaire studies were used to analyze the origin of back pain at work. The principal provider of occupational health services for Swiss hospitals carried out a questionnaire study among supervisors as well as among hospital nurses to identify the perceived physical strain during work, especially during patient care tasks. A similar study was started among Japanese hospital nurses.

A representative study among Swiss employees, the Swiss data from the Fourth European Working Conditions Survey (847 employees, response rate 32%) [3], was used to gain more insight into the complex relationships between musculoskeletal disorders and physical loads at work, organizational factors, work satisfaction, as well as lack of recovery [4] [5]. The answers to questions on risk factors for work-related musculoskeletal disorders (n=67) were dichotomized in such a way that for each variable, the twenty percent with the most unfavorable working conditions were considered to be at risk. Of these variables, 29 showed significant bivariate relationships with work-related musculoskeletal disorders and were entered into a logistic regression model with backward elimination.

In a limited number of hospital nurses recordings of heart rate, arm accelerations (by actimeter), and electromyography (EMG) of the trapezius muscle are underway and will help to discriminate between static and dynamic musculoskeletal strain at work. Further, a task analysis was performed based on expert ratings, video-analysis and nurses' comments.

With the aim of alleviating musculoskeletal strain during patient transfer and overcoming nurses' reluctance to use technical aids, existing technological solutions were studied and novel ways of achieving intuitive interaction, e.g. through the use of force and position sensors and shared control strategies, are being explored.

4 Results

532 nurses answered the emailed questionnaire. Nurses showed an increased risk (relative risk=1.4; p<0.01) to suffer from low back pain compared to a random sample of Swiss women (n=4505). Lifting/ transferring immobile patients indeed are the most strenuous tasks for nurses (fig 1).

The analysis of the Swiss data set on working conditions and work-related disorders revealed for musculoskeletal disorders that eight variables had a significant predictive effect in the logistic regression model (Odds Ratios and 95 confidence intervals):

- Poor fit of working hours with family or social   3.4 (2.0–5.6)
  commitments
- Carrying or moving heavy loads or persons   2.9 (1.7–4.9)
- Exposed to vibrations from hand tools, machinery, etc   2.8 (1.8–4.5)
- Frequent, disruptive, unforeseen tasks   2.3 (1.5–3.4)
- Not very / not at all satisfied with working conditions   2.0 (1.2–3.5)
- Rare / no assistance from superiors   1.7 (1.1–2.7)
- Pace of work dependent on numerical performance targets 1.7 (1.1–2.5)
- Not free to decide when to take holidays or days off   1.6 (1.1–2.3)

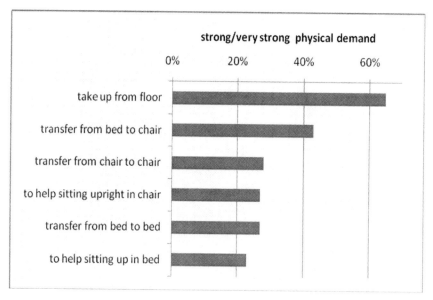

Fig. 1. The physically most demanding tasks in handling immobile patients. Percentage of nurses indicating strong or very strong exertion (n=532).

The analysis of the activity (based on accelerator data) from Japanese hospital nurses showed that a high incidence of continuous high activity was observed especially during night shifts. An example is shown in Fig. 2.

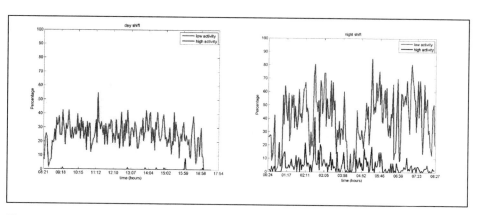

Fig. 2. Registration of work activities (illustrative example) during two 8-hour work shifts in a Japanese hospital nurse. Legend: percentage of low (upper, red curve) and high physical activity (lower, blue curve) during a day (left) and night shift (right).

Often nurses use the so-called kinesthetic approach to optimally activate a patient with every maneuver needed. The kinesthetic approach tries to activate patients as much as feasible, nurses are supposed to give the needed advice but not more physical support than is necessary. Such a close collaboration between nurse and patient is well illustrated in Fig. 3.

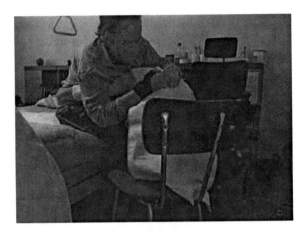

Fig. 3. By using kinesthetic techniques (as can be seen on fig 3) the nurse partly protects her back and activates the patient by giving him precise indications on how to contribute as much as possible with his own force

If patient and nurse have a good personal relationship, such a physical and mental collaboration between patient and nurse can help to improve the patient's capacities and supports the healing process. In contrast to this situation, the use of a traditional lifting device tends to impede the physical contact between patient and nurse. The nurse may perceive such a situation as an artificial or emotional barrier between patient and nurse, and therefore avoids such technical aids.

Fig. 4. Use of a traditional electric lifting device in the patient's bedroom. Note the difficulty to simultaneously control the lifting machine and provide support/contact to the patient.

Figure 4 shows several limitations of lifting devices currently used in patient care which motivate the development of more intuitive, collaborative systems: 1) the lifting maneuver is controlled through a control device located at a distant position form the patient (e.g. fixed to the supporting structure, or hanging on a cable attached to the frame). 2) Conventional lifting devices are position controlled (i.e. stiff) and operate at a low velocity. 3) The lifting device holds the entire weight of the patient, while the nurse performs translational movements. These factors result in inconvenient and non-intuitive use and unnecessarily high cognitive load. This interaction could be improved by employing principles of intelligent assist devices that are already finding increasing application in industrial environments [6]. The user interface could be simplified by achieving an interaction in which touching/pushing the patient directly activates the mechanical assistance with appropriate speed and direction. This could be achieved using passive, spring-balanced, patient unweighting systems, or active, force-controlled systems that unweigh the patient based on the input from force sensors placed close to the patient's body (e.g. integrated into the harness). This would make the interaction more intuitive and result in true collaboration between nurse and lifting mechanism.

5 Discussion

Swiss and Japanese nurses showed an increased prevalence of low back pain, as it is well-known in other countries [2]. Since the risk factors of work-related musculoskeletal disorders include both physical and work-organizational factors and since multiplicative models are highly predictive, all different aspects of the workload must be taken into account when developing preventive measures. Especially time pressure and work organizational factors must be accounted for when designing lifting devices for nurses. Societal and legal aspects, but also intuitive use with short learning curve strongly influence the provision and actual use of lifting devices in hospitals as well as in the home care setting. The initial results of our task analysis suggest that both the handicapped/ immobile person and the nurse may be supported by intelligent assistive lifting devices.

6 Conclusions

Transfer tasks of immobile patients represent a high physical load for nurses, and are a major source for musculoskeletal disorders. As existing lifting devices provide a non-intuitive operator interface, nurses are often reluctant to use them. The human-machine interaction should therefore be designed in such a way that sensors which can register the intention of the nurse can be attached near the body of the patient, so that they can be used to intuitively guide direction and speed of movement trajectories imposed by the nurse while the assistive device unweights the patient, resulting in an ergonomic, collaborative approach.

References

1. Alamgir, H., Cvitkovich, Y., Yu, S., Yassi, A.: Work-related injury among direct care occupations in British Columbia. Canada. Occup. Environ. Med. 64(11), 769–775 (2007)
2. Hignett, S., Fray, M., Rossi, M.A., Tamminen-Peter, L., Hermann, S., Lomi, C., Dockrell, S., Johnsson, C.: Implementation of the Manual Handling Directive in the healthcare industry in the European Union for patient handling tasks. International Journal of Industrial Ergonomics 37(5), 415–423 (2007)

3. Krieger, R., Graf, M.: Arbeit und Gesundheit - Zusammenfassung der Ergebnisse der Schweizerischen Gesundheitsbefragung 2007. SECO, Arbeitsbedingungen (2009), http://www.seco.admin.ch/dokumentation/publikation/00008/0002 2/02415/index.html?lang=de (retrieved form January 25, 2011)
4. Läubli, T., Müller, C.: Arbeitsbedingungen und Erkrankungen des Bewegungsapparates – geschätzte Fallzahlen und Kosten für die Schweiz. Die Volkswirtschaft 82, 22–25 (2009)
5. Läubli, T., Müller, C.: Conditions de travail et maladies de l'appareil locomoteur. Revue de politique économique 82, 22–25 (2009)
6. Colgate, J.E., Peshkin, M., Klostermeyer, S.H.: Intelligent assist devices in industrial applications: a review. In: Proceedings of the 2003 IEEE/RSJ International Conference on Intelligent Robots and Systems (IROS), vol. 3, pp. 2516–2521 (2003)

Electronic Medication Reminder for Older Adults

Yi-Lin Lo, Chang-Franw Lee, and Wang-Chin Tsai

Graduate School of Design, National Yunlin University of Science and Technology
123, University Road Section 3, Touliu, Yunlin, 64002, Taiwan, R.O.C.
G9730804@yuntech.edu.tw

Abstract. As the numbers of the elderly people is increasing rapidly, it is important and urgent to design appropriate products for older adults. Because of physical and mental function decline, the elderly need to take multiple drugs, they often occurring medication non-compliance behavior, seriously affecting the health of the elderly. Forget to take medicine is one the most frequency problem, so there were several products designed for reminder, such as electronic pillbox. In recent years, with the development of smart phone, some software was also designed for medication reminder. For the lifestyle and electronic products using experience, not all of the elderly use cell phone in Taiwan, and they do not operate any other function except making a phone call. To discusses whether the software suitable to the elderly was the purpose of this study.

This is an exploratory study about electronic medication reminder in Taiwan, there were total 30 volunteers join the project, included 15 older adults; through interviews and a questionnaire survey with the elderly, try to gather difficulties and needs from the elderly when using an electronic medication reminder. The results showed correlations between the interface-complexity and preference of older adults; they would rather choose electronic pillbox than smart phone because it is easier setting and more "approachable". Through the study results, hope the findings will help clarify the direction of further research and to develop more suitable for the elderly on the operating trends.

Keywords: Older Adults, Medication Compliance, Pillbox, Electronic Medication Reminder.

1 Introduction

The problem of aging population is the trend all over the world [1] [2]. It has become an aging society in Taiwan since 1993. If the elderly people can take care themselves well in the daily living, it will be able to lower the care burden among younger generation; If we can improve the medication compliance, it will promote the physical and mental health of the senior citizens and assist them to live independently [3].

With the aging process is often accompanied by the occurrence of chronic diseases, there are more than half of the older people suffered from different diseases in Taiwan [4]. Because of physical and mental function decline, the elderly need to take multiple drugs, they often occurring medication non-compliance behavior, seriously affecting the health of the elderly. The behavior not only affects the disease, but also seriously

M. Kurosu (Ed.): Human Centered Design, HCII 2011, LNCS 6776, pp. 327–334, 2011.

endangers their health and security [5][6]. Compliance can be defined as the extent to which a patient's behavior corresponds to the physician's therapeutic recommendations. Improvement of the medication compliance would increase cost-effectiveness. Studies have demonstrated the prevalence of poor adherence (/compliance) across all types of regimens and diseases, including life threatening illnesses; for this reason, often caused by medication non-compliance with more discussion and concern [7].

The non-compliance behavior of the elderly often caused from forgotten or misunderstanding[8]; a review of the literature indicates that forgotten is the most frequent problem on older adults[9], whether "forget to take medicine" or "forget taking medicine", these problems were hazardous to health for the elderly; so there were several products designed for reminder, such as electronic pillbox.

In recent years, with the development of electronic communication products, some software was also designed for medication reminder. Not only notice the medication time, some software also could connect to a comprehensive medication database, or search a list of local personal physicians with office contacts [10]. These powerful features are developed in general NB, PDA, or smart phone, change the act and lifestyle on people. However, the previous studies were targeted on younger adults with computer or other small screen device usage, but few data are available on how different variables vary with electronic reminder products using performance and comprehension for older adults. There are much more information to be presented on the electronic reminder products, however, there is no accurate way to be understood by all users. We would like to analyze the layout and usability of existing products by interviews through the concept of universal design.

As the numbers of the elderly people is increasing rapidly, it is important and urgent to design appropriate products for older adults. Universal design is a process intended to promote the development of products or environments that can be used effectively by all without adaptation or stigmatization [11]. To put it simply, describes universal design as "design for people of all ages and abilities" [12]. For the lifestyle and the different electronic products using experience, not all of the elderly use cell phone in Taiwan, or they do not operate any other function except making a phone call. Although the smart phone had bring more possibility of medication reminder to people, it should be put more concern on the technological generation gap, to make the products close to the elderly's living habits, and easier to be learned and used. There has thus far been relatively little research into the area; the study try to primarily explore the issue about needs and difficulties for older adults on using electronic medication reminder products.

2   Methods

For the purpose, the study includes three aspects: Operating practice, observation and interview to the elderly are carried out to primarily explore the issue about the electronic medication reminders products. Their response measures included operating performance (speed and accuracy) and subjective satisfaction assessment. The study was a 3 x 2 factor between-subjects design. The first factor was cell phone (three levels: Phone call services, reminder software, and built-in alarm) and the second factor was pillbox (two levels: electronic reminder and no reminder); show as Table.1. Based on the factorial design, there are totally six articles composed of designated levels for the study.

Table 1. The study factors

Personal Database →			Pillbox	
			Electronic	NoReminder
	Smart Phone	Service	A	B
		Software	C	D
		Alarm	E	F

→ Subjective Satisfaction Assessment

2.1 Participants

There were total 30 volunteers join the project, included 15 older adults. All participants were fluent in Chinese as their first language and educated to at least secondary/high school level. The 7 females and 8 males were between the ages of 65 and 81 years (mean = 70.3, SD = 3.69). They were required to have at least 20/25 visual acuity with corrective lenses and to be without physical or mental problems. They were also requested not to stay up late, alcoholic drinks and any other substance that might possibly affect the test results. All older subjects had experience of taking medicine over 5 years. A small gift was given to participants as payment for taking part.

2.2 Equipment

In this study, there are 3 major equipments used in operating practice: 1. An iPhone 3GS (http://www.apple.com/) with the software of pillboxer (charged) and iPills (free), The touch screen could allow the participants use the finger to control the options. 2. An electronic pillbox, with a LCD panel and three buttons for setting remind alarm; and 3. A set of rainbow pillbox, three grids a day, totally seven days a set; show as Figure.1. Both pillboxes are made of transparent plastic.

Fig. 1. The study Equipment

2.3 Materials

Pillboxer (charged): Search through a database of more than 11,000 FDA-approved medications. Notification engine let's people know when ones medications should be taken. Visual "pill box" icons immediately let users know what medicines have taken during the week.

iPills (free): Start by entering the medications that needs to be taken along with the intervals for each, it will then present people with a daily pillbox showing the medicines one need to take for the day. Every time after taking the medicine the user needs to tap the corresponding pills on the screen. When it comes to appearance, users can further customize the way pills look with different colors and shapes. The application also lets user keep an overall record of the medications he have taken, or missed.

Language: Under the premise of maintaining the normal operation of the software, take English (the original language, Figure.2) as Interface plate; the researchers provide translation and interpretation.

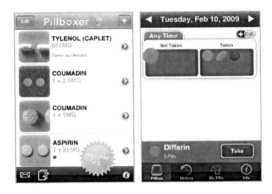

Fig. 2. The study materials of Pillboxer and iPills

2.4   Operating Performance and Subjective Satisfaction

Researchers would explain the functions and operating steps to the participants before the operating practice. All participants must complete the three setup tasks, including smart phone, electronic pillbox and phone alarm; Time-consuming and the behavior of the process were recorded as the reference for the follow-up analysis.

2.5   Experiment Procedure

A standard laboratory desk and chair were provided for experimentation. The experiment environment was standardized. Prior to the experiment each participant was instructed about the purpose and procedure of the study. Participants were asked hold the iPhone (or electronic pillbox) and to read the information from a comfortable position and were told that they could bring the iPhone (or electronic pillbox) closer to the face if necessary. Each participant should operate the process completely to setting the medication time on 13:45, then remove the medicine and cease the alarm.

After completed the process, the participants were requested to answer the subjective satisfaction on what they thought of the different electronic reminder and pillbox by 5 point Likert scale. An additional brief semi-structured interview was carried out on the elderly. For the analysis of the data, this study applied the statistical analysis by utilizing the Windows SPSS Statistics 13 Program.

3 Results

All participants completed the operating practice and subjective satisfaction survey; through interviews and a questionnaire survey with the elderly, they expressed the difficulties and needs on using the electronic medication reminders.

3.1 Operating Performance

Operating performance was measured by setting speed and accuracy. Figure.3. shows the trends of the two groups. For the elderly, the operating performance of existing smart phone reminder was not as good as the electronic pillbox; however, the performance was not affected significantly for the general adults. The operating performance of older adults is: electronic pillbox> setting alarm> software reminder, and so does the general adults, but the difference among the products are not Obvious. Over all, the results indicate that the interface complexity has a positive effect on operation performance; however, this part yielded limited information about product design. Therefore, further questionnaires and interviews to be executed, the results show as the next section.

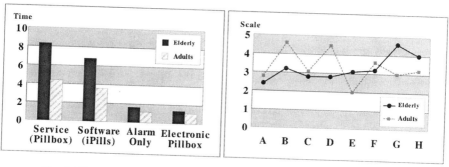

Fig. 3. The Operating Performance (L) and The Subjective Satisfaction (R)

3.2 Subjective Satisfaction Results

Similar as the results of operational performance, the elderly were most satisfied with the electronic pillbox; the results of the satisfaction questionnaire were mixed, for the smartphone generally have higher adult satisfaction, but the elderly has the opposite.
The following is a summary of two interviews with the participants.

The Older Adults: The electronic pillbox could storage medicine and offer reminds at the same time, but the other two products could not. Despite the quite popularity of mobile phones in Taiwan, there are still a certain percentage of the elderly usually do not carry the mobile phone with them; so the medication reminder software might not exact fit the current older adults. Furthermore, the Flat-panel touch screen is too small, and the no-button design is relative unfamiliar to the elderly. Some advantages can attract the elderly try using the smartphone, such as the large medications information database and the medication identification photos; the prerequisite is to simplify operations.

The General Adults: The size of electronic pillbox is still too large to take away, for the participants, they would rather take a pillbox + smartphone, but do not take a "ringing box" with them, the ringing pillbox will remind other people how many times he has to take medicine. Second, the smartphone has a more beautiful interface, it also integrate more functions; to people whom carry mobile phones at any time, this product would conform more needs in more situation. Not all of the advantages are attributed to smartphone, existing software are in English, the steps are not simple enough to eliminate the language barrier; consideration of economic factors is another reason.

4 Discussions

The operating performance of smart phone was not as good as the electronic pillbox; consider of the reasons can be attributed to the interface type: The flat-panel touch screen is a new operation technique for the elderly, diversity options for the decision-making process is also a challenge; despite the two products are equal to the auditory feedback, but in the sense of touch, the elderly are more accustomed to a specific button or particles, can serve as a tactile aid, although smart phones have vibration feedback, the experience of the elderly is relatively new; the visually compensate for this disadvantage, compared to the limited LCD screen of electric pillbox, a larger area of smartphone with bright colors provide a better operating environment for the elderly. This inference should also be supported from the interviews.

The interview results showed that the language affect the user's operational behavior, thus have an influence on the interface preference. This study take two software as materials, which are both of English; this may suggest some effects of subjective satisfaction, another partial explanation for this may lie in the suppose that under an unfamiliar language, the interface essence could be tested more directly.

The elderly did not completely reject the use of smart phones, if the motivation stronger or the setting process easier, the elderly will be more willing to learn it; in addition, part of the interview reveal that some elderly do not often use the cell phone, not to mention carry with them. Even they heard the reminded on cell phone, but forgot to take the pillbox out of the house. Another interview has shown a similar reason: "electronic pillbox could storage medicine, but the cell phone could not." This is a direct and interesting answers, and also shows the demands of the elderly is so simple.

Comprehensive questionnaire and interview results, three of these finding are worth summarizing:

Smartphone with Medication Reminder Software: In the past ten years the science and technology for the creation and use of smartphone have progressed tremendously; we have seen a shift in patterns of user-behavior. In spite of the smartphone has some essentially problem (i.e. it cannot carry medicines, and economic considerations) there are four points support the trends: 1.Variety of interface make a easier and more beautiful reading experience; 2. Integration of multiple functions; 3. More complete and scalable information database; 4. Greater emphasis on privacy.

Electronic Medication Reminded Pillbox: It seems a kind of transitionary products, before a better form of the medication reminder is developed, or before the "new elderly" whom are familiar with the electronic products are getting older; the contemporary older adults would choose these products which are easier operate or closer to their lifestyle. The advantage are integrating two functions, the elderly could receive the reminder message then take medicine right away.

The Non-Reminder Pillbox: The pillbox only used for medication storage. Because of the simple function, the product could be designed as different forms, or be designed from different material; so it received less controversy.

5 Conclusions

The results showed correlations between the interface-complexity and preference of older adults; they would rather choose electronic pillbox than smart phone because it is easier setting and more "approachable". This interviews brought us a positive inference, improving operational processes and links medications to the cell phone, will help the elderly to accept the new form or new product of reminders. For the lifestyle and electronic products using experience, not all of the elderly use cell phone in Taiwan, and they do not operate any other function except making a phone call. With the design closer to the elderly, so can we reduce the learning difficulties of the elderly. Through the study results, hope the findings will help clarify the direction of further research and to develop more suitable for the elderly on the operating trends.

This method of investigation is not without problems, we readily acknowledged that our research is exploratory and that there are not precise with the statistical mode. While this study has its limitations, it can serve as a basis for further study in electronic medication reminders for the elderly.

References

1. Sterns, A.A.: Curriculum Design and Program to Train Older Adults to Use Personal Digital Assistants. The Gerontologist 45, 828–834 (2005)
2. Fisk, A.D., Rogers, W.A., Charness, N., Czaja, S.J., Sharit, J.: Designing for older adults: principles and creative human factors approaches. CRC press, U.S.A (2004)
3. Gravely, E.A., Oseasohn, C.S.: Multiple drug regimens: Medication compliance among veterans 65 years older. Research in Nursing & Health 14, 51–58 (1991)
4. Lamy, P.P.: The elderly, communications, and compliance. Pharm Times 58, 33 (1992)

5. Sterns, A.A.: Curriculum Design and Program to Train Older Adults to Use Personal Digital Assistants. The Gerontologist 45, 828–834 (2005)
6. Neely, E., Patrick, M.: Problems of aged persons taking medicine at home. Nursing Research 17(1), 52–55 (1968)
7. Huang, L.H.: Medication-taking behavior of the elderly. Kaohsiung of Journal Medical Science 12, 423–433 (1996)
8. Conrad, P.: The Meaning of Medications: Another Look at Compliance. Sociology of health and illness- critical perspectives, 4 edn. (2005)
9. Lo, Y.L., Lee, C.F.: A Study of the Drug Bag Design. In: The 3rd Conference of International Association of Societies of Design Research 2009/10/18-24, IASDR 2009, Seoul (2009)
10. Hamilton, G.A.: Measuring adherence in a hypertension clinical trial. European Journal of Cardiovascular Nursing 2, 219–228 (2003)
11. Beecher, V., Paquet, V.: Survey instrument for the universal design of consumer products. Applied Ergonomics 36(3), 363–372 (2005)
12. Tuljapurkar, S., Li, N., Boe, C.: A universal pattern of mortality decline in the G7 countries. Nature 405(6788), 789–792 (2000)

Development of a Wearable Airbag for Preventing Fall Related Injuries

Toshiyo Tamura[1], Takumi Yoshimura[2], Masaki Sekine[1], and Mitsuo Uchida[3]

[1] Department of Biomedical Engineering, Chiba University Graduate School of Engineering,
33, Yayoi-Inageku, Chiba 263-8522 Japan
[2] Department of Biomedical Engineering,
Tokyo Metropolitan College of Industrial Technology
[3] Prop Co., Tokyo Japan
tamurat@faculty.chiba-u.jp

Abstract. We have developed a wearable airbag that incorporates a fall detection system that uses both acceleration and angular velocity signals to trigger inflation of the airbag. The fall detection algorithm was devised using a thresholding technique with the signals of an accelerometer and gyro sensor. The thresholds of acceleration less than ± 3 m/s^2 and the integral of angular velocity exceed 0.52 rad/s were used. Five young healthy subjects mimicked falls, and their signals of acceleration and angular velocity were monitored. Then, we developed a fall detection algorithm that could detect signals 300 ms before the fall. This signal was used as a trigger to inflate the airbag to a capacity of 2.4 L. The system has been manufactured but the accuracy was not 100% of operation. In this study we have improved fall detection algorithm to operate correctly in daily life.

1 Introduction

Falls are a serious problem for elderly people as well as people who work at tall environment. One-third to one-half of the population aged 65 years and over have experienced falls. Half of the elderly people who fall do so repeatedly. Falls are a complex phenomenon, suggesting present disease and predicting future disability. They are caused by interactions between the environment and dynamic balance, which is determined by the quality of sensory input, central processing, and motor responses. Falls are the leading cause of injury in older adults and the leading cause of accidental death in those over age 85.

Even a fall that does not result in injury can have serious consequences. Psychological trauma and fear of falling can produce a downward spiral of self-imposed reduced activity, leading to a loss of strength, flexibility, and mobility, thereby increasing the risk of future falls and injuries.

Fall detection and prevention are important issues for the elderly population. From an engineering perspective, two main fall detection technologies have been proposed: one uses motion sensors, such as accelerometers and gyro sensors [1-4], and the other uses image processing with a camera installed in the room [5,6]. The thresholding technique is used to determine the direction and strength of a fall [3]. In addition, a gyroscope-based fall detection sensor array can be used for image processing [4].

M. Kurosu (Ed.): Human Centered Design, HCII 2011, LNCS 6776, pp. 335–339, 2011.

Another study installed a small digital video camera in the ceiling and classified falling patterns [5]. We sought to develop a means to reduce or prevent injuries associated with falls. Our method measures the acceleration before and during falls, and then inflates an airbag [7]. In previous study, all falls were detected, but the airbag was activated occasionally during daily activities. Therefore, in this study, we developed an improved fall detection algorithm that uses both acceleration and angular velocity. Then, we tested a prototype airbag system.

2   Objective and Methods

2.1   Apparatus

The key issues regarding a wearable airbag are: 1) It must be able to detect falls while the wearer is standing or walking; 2) It must protect the head and thighs; 3) It must be small, lightweight, and simple to wear; and 4) It must be activated only during falls and not during daily activities.

Based on these considerations, our system consists of a tri-axial accelerometer and a tri-axial angular velocity sensor and an airbag. Figure 1 shows a block diagram of the device.

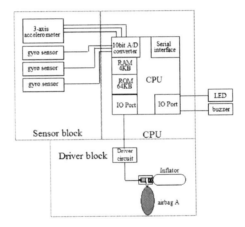

Fig. 1. Block diagram of the wearable airbag system

To monitor the acceleration and angular velocity signals, we designed a system with low power consumption. The system must be small, light, and able to be worn without discomfort.

The monitor measures 50 × 56 × 18 mm and weighs 50 g. The system was designed to operate without any complex setting. A tri-axial acceleration and a tri-axial gyro-sensor (Gyrocube 3A Oni-23503, O-NAVI, USA) were used to measure the subject's movement. The acceleration and angular velocity waveforms were converted into digital data with 16 bit resolution. The received data were transferred to the CPU and then analyzed with the fall detection algorithm.

The airbag system consists of an inflatable airbag, a battery, a gas cartridge, sensors to determine acceleration and angular velocity, a trigger mechanism to release the gas, and an inflator to inflate the gas. Airbags protect both the head and the hips shown n Figure 2. Before inflation, one bag is folded on the back; this bag covers the head and neck after inflating. This airbag measures 470 × 330 mm and has a volume of about 10 liters. The airbag protecting the hips is folded inside a pouch; this bag covers the hips and thighs when inflated. It measures 250 × 450 mm and has a volume of about 10 liters. Each airbag has an independent inflator and cartridge. The two airbags weigh 1.1 kg and are worn like a jacket. When the user falls, the sensor detects this and sends the trigger signal to automatically release gas from the cartridge to inflate the airbag and protect the user. Gunpowder is used to release the gas. When the triggering signal is generated, a 3-V signal is transmitted to cause ignition. Then, the gunpowder explodes, making a small hole in the gas cartridge. The inflator is made from an aluminum block big enough to avoid accidents when the gunpowder explodes. Each gas cartridge weighs about 160 g and measures 125 × 25 mm. The gunpowder weighs 100 mg. Our invention is superior to available devices because of its automatic deployment, compact size, light weight, ease of use, and reusability.

Fig. 2. Photograph of the airbag system

The system operates for about 200 hours with an A3 alkaline battery. The operating switch is built into a belt, and a magnetic switch inside the buckle works when the buckle is closed.

2.2 Fall Detection Algorithm

The main assumption in the fall detection algorithm is that the wearer is in free fall. The acceleration signal during the fall is similar to one that occurs during free fall, i.e., the acceleration is zero. Considering human posture, the position of the senor, and the signal-to-noise ratio of a low acceleration value, we assumed that a prevention acceleration value would be below ±3 m/s^2. In addition, after a preliminary study of angular velocity, we added the stipulation that an angular velocity of less than 0.52 rad/s did not indicate a fall. Therefore, a fall occurred when the acceleration was less than ±3 m/s^2 and the angular velocity exceeded 0.52 rad/s.

We have tried moving average period to compare the accuracy of detecting falls

2.3 Experimental Setup

The prototype system was tested by five young, healthy subjects (mean age 19.6 ± 1.3 years, weight 62.8± 9.6 kg, and height 172.9 ± 3.2 cm) who mimicked falls to the front, back, and sides while wearing the device without airbags. To prevent injuries, the subjects fell on a double mattress. In addition, we evaluated the effectiveness of the algorithm during misleading activities such as vertical jumping.

This trial was approved by the ethics committee of Chiba University, Graduate School of Engineering. Written informed consent was obtained from each subject.

3   Results

We determined that an acceleration below ±3 m/s^2 indicated free-fall conditions and that the angular velocity exceeded 0.52 rad (30 degrees)/s while falling.

Table 1 shows the relationship between the fall detection and activities.

Table 1. Relationship between activities and fall detection with different moving average time

Fall detection with averaging time of	0.5	1	1.5	2
Vertical jump	3	1	0	0
Long jump 1m	2	0	0	0
Long jump 1.5 m	1	0	0	0
Jumping from 45 cm	0	0	0	0
Jumping from to cm	0	0	0	0
Jumping backwards	4	0	0	0
Falling Backwards	1	4	5	5

The moving averaging time above 1.5 s shows good results for fall detection out of five each five trial.

4   Discussion

We determined that a fall occurred when the triaxial acceleration was below ±3 m/s^2 and the angular velocity exceeded 0.52 rad/s. The key issues in fall injury prevention are the time it takes for the airbag to inflate and incorrect inflation caused by the algorithm.

Any fall detection system must be very reliable. Our proposed moving averaging time was 1.5 s and had an accuracy of 100%. In our previous study, errors occurred when the free-fall condition appeared before the angular velocity exceeded 0.52 rad/s. This resulted from a discrepancy in the detection times for acceleration and angular velocity. In this study, we proposed moving average to avoid misleading of fall detection. The subject responded to the fall in a defensive manner by bending his/her knee and falling on the knee. Consequently, the body turned less, and the angular velocity was lower. If the subject had not assumed a defensive posture, the angular velocity would have been large enough to detect.

New algorithm was able to discriminate between a real fall and similar acceleration and angular velocity signals resulting from events in daily life.

5 Conclusion

We developed a jacket to protect the head and hips when a person falls by automatically inflating an airbag that absorbs the shock of the fall and reduces the impact on the human body. The heart of this life jacket is a sensor that detects falls. The key characteristics of this fall sensor are the new fall-sensing algorithm, which uses a triaxial, single-unit accelerometer and angular velocity sensor, the jacket's compact design, and its battery-powered operation, which make it readily portable. For a reusable airbag, we need to develop a sophisticated electromagnetic valve. The use of this fall sensor and an airbag-equipped life jacket may also save lives and reduce injuries from falls at construction sites and other locations

Acknowledgement. We thank the Prop Company for producing the prototype airbag.

References

1. Hwang, J.Y., Jang, Y.W., Kim, H.C.: Development of novel algorithm and real-time monitoring ambulatory system using Bluetooth module for fall detection in the elderly. In: Proc. Annu. Intern. Conf. IEEE Eng. Med. Biol. Soc., vol. 26(3), pp. 2204–2207 (2004)
2. Diaz, A., Prado, M., Roa, L.M., Reina-Tosina, J., Sanchez, G.: Preliminary evaluation of a full-time falling monitor for the elderly. In: Proc. Annu. Intern. Conf. IEEE Eng. Med. Biol. Soc., vol. 26(3), pp. 2180–2183 (2004)
3. Bourke, A.K., O'Brien, J.V., Lyons, G.M.: Evaluation of a threshold-based tri-axial accelerometer fall detection algorithm. Gait & Posture 26(2), 194–199 (2007)
4. Bourke, A.K., Lyons, G.M.: A threshold-based fall-detection algorithm using a bi-axial gyroscope sensor. Medical Engineering & Physics 30(1), 84–90 (2008)
5. Nait-Charif, H., McKenna, S.J.: Activity summarisation and fall detection in a supportive home environment. In: Proceedings of the 17th International Conference on Pattern Recognition, ICPR 2004, vol. 4, pp. 323–326 (2004)
6. Lee, T., Mihailidis, A.: An intelligent emergency response system: preliminary development and testing of automated fall detection. J. Telemed Telecare 11, 19–194 (2005)
7. Tamura, T., Yoshimura, T., Sekine, M., Uchida, M.: A wearable airbag to prevent fall injuries. IEEE Transactions on Information Technology in Biomedicine, 910–914 (2009)

Semantic-Conditioned Peripheral Vision Acuity Fading Awareness (PVAFA)

Ming-Chia Wang and Manlai You

National Yunlin University of Science and Technology, College of Design,
123 University Road, Section 3, Douliu, Yunlin 64002, Taiwan
stmk@ms54.hinet.net

Abstract. This is a pilot study report that explores one of the factors that influence one's awareness of the extent of vision acuity other than biological reasons. Semantic factor is chosen to put to test to match the tests' linguistic nature of words reading. Look-then-answer style of self-report method is adopted to better reflect this experiment's goal of understanding how one "consciously knows" his or her quality of vision at that moment of words reading. By comparisons of fixating and gazing at a two-character segment of a reading line set in forms of Chinese and Korean characters of right-reading and wrong-reading versions, it can be checked to see how semantic factors influence one's *Peripheral Vision Acuity Fading Awareness* (PVAFA). Results show the tendencies that partially support semantic-conditioned interpretations that: (1) the better a reading line's semantic meaning understood, e.g., native Chinese readers gaze at Chinese characters, the more peripheral visions smeared than gazing at Korean characters; (2) the harder the lexical information can be identified, i.e., gazing at wrong-reading characters (in this case, upside-down typesetting), the lenient the PVAFA effect to occur. A follow-up discussion stresses how semantic factors mingle with vision acuity awareness in a lab set-up is worthy further hypothesized to probe its broader implications on visual form perception in both real world situations and human-computer interacted environments.

Keywords: foveal visions, vision acuity, visual form perception, visual logics.

1 Introduction

Foveal vision is an area or region of the retina within which an image has its sharpest resolution and thus shows the best clarity. Because sharp vision is restricted to a much smaller region as mentioned, foveal vision subtends a very small angle of view, while impressions in the parafovea (the region surrounding the fovea) are somehow less distinct [1]. The span of material clearly seen at each fixation point is stringently limited due to the narrow field of foveal vision.

In general, one gazes longer at interesting or puzzling things and shorter at mundane or simple things. It is during this fixation period that one "sees" a feature before moving on to another feature [1]. Thus, the fixation of a gazing is not only about biophysical vision acuity, but also a matter of psychological visual understanding.

M. Kurosu (Ed.): Human Centered Design, HCII 2011, LNCS 6776, pp. 340–347, 2011.
© Springer-Verlag Berlin Heidelberg 2011

Even for biological reasons, there are different kinds of factors that singly or confoundedly affect vision acuity. According to Bursill's findings, under hot conditions, peripheral signals at greater eccentricities were more likely to be missed through a "funnel" of the field of awareness [2]. Sanders suggests that the functional vision fields may be elastic or sensitive to the overall information-procession demands [3].

These studies certainly suggest that the useful field of view is sensitive to overall task demands, and most agree that increases in foveal load should hurt performance more and more at increasing eccentric locations [4]. But it is difficult to ascertain whether the effect was cognitive in nature or due to the visual complexity of the foveal information [5].

Early foveal vision researches mostly started from reading material tests. In addition to pure biologics-driven theories, those reading-test studies reveal factors causing vision acuity fading other than mere biological reasons. There is also a large body of evidence suggesting that higher lever linguistic factors affecting fixation duration [6]. Some studies even have shown that semantic information is extracted up to 6-8 characters to the right of the fixation location [7]. And, lexical information may induce more on foveal load that prevents attentiveness to disperse into peripheral vision zone and beyond.

Our understanding is that word reading is a state of visual perception, which in one way draws data from the sensation of physical input through biological cannels, and in the other induces his or her previously stored concepts to participate in the current visualization process. Semantic factor is one of the most obvious ones that impose a person's foveal load of word reading and that in turn delineate the extent peripheral vision acuity fading awareness (PVAFA) can be dispersed to.

We tested our conjecture by conducting a series of test-run experiments in forms of cross comparison on Chinese and Korean characters presented in both right-reading and wrong-reading fashions to see how semantic factor influences the peripheral vision acuity fading awareness (PVAFA), and to see whether any further formal and larger-scale experiment should be carried forward for a long-shot expectation of theorizing a form-content(meaning) interactive model of general visual form perception theory.

2   Methods

2.1   Subjects

Opportunity sampling is used to recruit testees for the investigation. Though Opportunity sampling might produce biased sample, it is still adequate when investigating processes that are assumed to work in a similar way in all "normal" individuals, especially for a cognitive experiment like this one. A total of 7 Ph.D. students and 5 master students and 14 undergraduate students from the College of Design of YUNTECH served as testees. A number of them wore spectacles, but all demonstrated the capability to read and the sufficiency to recognize and understand all the Chinese characters that have been printed on A4 photocopy papers which were viewed at the normal reading distance in the tests. None of these testees speaks Korean language and no one recognizes or understands Korean characters either.

2.2 Design

Repeated measure design (also called related measures or within group design) is adopted not only for the benefit of small number of needed participants it requires, but also for its consistence in terms of participants' variables.

2.3 Apparatus

Three 4-ft fluorescent tubes in a ceiling fixture illuminated the testing room without Lucite covered. A standard light-gray color office desk (720mm high) and a matched-color height-adjustable office chair, which can be adjusted to match every testee's hip height was used to sit him or her. No other testee was presented while the tests were conducted except one taster who sat next to the testee's right side giving instructions and administered the tests. Subject would be instructed to sit on the chair with his or her back fully straightened up and bellbottom lightly pressed to the desk's edge panel.

2.4 Stimuli

A 210mm x 297mm regular white A4 photocopy paper was used for each test. All the printed characters were horizontally typeset of 5mm height in a one-line fashion in the middle and across the paper surface with blank margin of 15mm to both edges. Two kinds of language characters—Chinese and Korean were used. Only one side of each paper sheet was printed with black ink, one with Chinese characters, one with Korean.

2.5 Procedure

We conducted four rounds of tests. The first test round included two separate non-comparison tests. Each of the rest three test rounds included a pair of in-tandem comparison tests. Before all the tests took place, the tester introduced testees what they were going to do with what material, and all were told to answer ONLY what the tester would ask and nothing else. Each subject was then instructed to sit and rest his or her both hands on the tabletop with palms facing downward. After the tester helped adjusting the testee's head in a position paralleling his or her face with the desktop surface to a distance about 350mm in between.

In the first round of tests, two tests were conducted with a short intermission in between: First, a right-reading Chinese characters (CR) test sheet was laid on the table surface between the testee's two resting palms. Subject was then asked to look at the printed line and was guided to concentrate on the target area—a two-character segment locates right in the middle of the printed line which was underlined by the tester with an inkless ballpoint pen tip.

While telling the testee to keep on gazing at the targeted segment for about 3 seconds, the tester started to ask the testee if he or she had the awareness of fading vision acuity on both extending sides of the targeted area by answering: "Yes", "No" or "Not sure". Once the answer given, the testee was told to stop gazing and look away from the test sheet. The test sheet was then replaced by a right-reading Korean characters (KR) test sheet in the same position. The subject was then told to get back to the testing position and postured as instructed before and repeated the same procedure as he or she did in the first test. Once the same question was asked and answered, a 60-second intermission was announced to conclude the first round tests.

For the second round tests, the same right-reading Chinese characters (CR) test sheet was used and the same question was asked ("Have you the awareness of fading vision acuity on extending sides of the character line besides the targeted area's characters?"). Once the question answered, the test sheet was turned upside down to form a wrong-reading Chinese characters (CW) line, and the testee was immediately asked to gaze as before and compare previous "right-reading (CR)" line with then current "wrong-reading (CW)" one by saying which had him or her "stronger" peripheral vision acuity fading awareness (PVAFA) or "not sure (NS)".

After one more 60-second break, in the third round tests, both "right-reading" and "wrong-reading" test sheets were tested in the same fashion as the second round tests did. Only this time with Korean characters lines printed instead which can be simply coded as KR and KW respectively.

The fourth round tests were conducted in identical procedures as previous rounds did, but with a straight comparisons on Chinese characters line with Korean characters line and on "right-reading" versions only.

3 Results

Four rounds of tests were conducted and five sets of congruent answers were collected. All 26 testees took tests. Two responses of theirs were eliminated due to failures to meet the minimal requirements of stable reading postures and adequate verbal responses. The overall responses, in terms of peripheral vision acuity fading awareness (PVAFA), are presented as follows:

Test round 1

Test 1: 18 testees reported "Yes" to "right-reading Chinese characters" for PVAFA effect. 2 testees answered "No". 4 testees answered "Not sure (NS)".(See figure 1).

Test 2: 12 testees reported "Yes" to "right-reading Korean characters" for PVAFA effect. 8 testees answered "No". 4 testees answered "Not sure (NS)" (See figure 2).

Test round 2

11 testees reported that "right-reading Chinese characters (CR)" induced stronger PVAFA effect than "wrong-reading Chinese characters (CW)" did. 7 testees gave opposite answers and 6 testees responded "Not sure (NS)" (See figure 3).

Test round 3

5 testees reported that "right-reading Korean characters (KR)" induced stronger PVAFA effect than "wrong-reading Korean characters (KW)" did. 7 testees gave opposite answers and 12 testees responded "Not sure (NS)" (See figure 4).

Test round 4

15 testees reported that "right-reading Chinese characters (CR)" induced stronger PVAFA effect than "right-reading Korean characters (KR)" did. 6 testees gave opposite answers and 3 testees responded "Not sure (NS)" (See figure 5).

Yes ████████████████ 18
No ☐ 2
NS ▨▨▨ 4

Fig. 1. Right-reading Chinese characters test sheet and the frequency of each answer

Yes ██████████ 12
No ☐ 8
NS ▨▨▨ 4

Fig. 2. Right-reading Korean characters test sheet and the frequency of each answer

CR ██████████ 11
CW ☐ 7
NS ▨▨▨ 6

Fig. 3. Right-reading Chinese characters (CR) vs. wrong-reading Chinese characters (CW) test sheets and the frequency of each answer

Fig. 4. Right-reading Korean characters (KR) vs. wrong-reading Korean characters (KW) test sheets and the frequency of each answer

Fig. 5. Right-reading Chinese characters (CR) vs. right-reading Korean characters (KR) test sheets and the frequency of each answer

4 Discussion

According to the test results—a tendency of stronger induced PVAFA effects on "Chinese" over "Korean" (figure 1 and 2, 5) characters lines and "right-reading" over "wrong-reading" of Chinese characters lines (figure 3), initial conclusions can be drawn to support most of our conjecture that a reading material's semantic factors do influence the range a foveal vision to be able to encompass, and that in turn cause peripheral visions to lose their degree of acuity.

On the other hand, the lenient impact of PVAFA effect on Korean "right-reading" materials (Figure 2) and smaller differences between "right-reading" and "wrong-reading" Korean characters (Figure 4), further strengthen semantic-conditioned hypothesis by demonstrating that the harder the characters' lexical message to be detected, the weaker the impact of fading peripheral vision acuity to be aware of.

Yet, due to the experiments' test-run nature and the tentative arrangement of apparatus, the whole procedure did have encountered several uncontrollable situations. Some of them are foreseeable and are anticipated, and some are newly discovered. All of which are invaluable to be incorporated into the consideration in our future formal experiments.

In addition to the aforementioned shortcomings, the too small number of valid testees (total of 24) is too bare-boned to meet the minimum requirements of being a representative sample. Also, a counter-balance technique should be considered to minimize the elusive testee's expectancy effects.

Finally, since this is a "self report" kind of test, how to construct a more precise verbal wording to better direct the viewers to truly stick to the "target zone" would also greatly decide the overall outcomes to be valid and reliable or else.

5 Conclusion

This is the first paper in a series that attempts to explore two levels of investigation: First, to probe factors that might condition the awareness of fading peripheral vision acuity other than biological reasons, and the second, to bring those findings to a higher level investigation about visual logics in visual form perception. Just as Engle mentioned that as the shape and/or size similarity among targets increased, search performance deteriorates [8], our propositions can push the prospects of vision acuity study beyond specific factor domain identification and point to the understanding of a higher rank of structural aspects of visual logics and to untangle some theoretical dimensions about a person's visual form perception around his or her daily life situations and around human-computer interacted environments.

Our current study focuses on the initial level of the investigation starting with a single variable—semantic factor of reading material. We carried out these bare-bone experiments to see whether an elusive phenomenon with confound factors like peripheral vision acuity fading awareness can be explained with simple yet elegant concepts or not.

The overall tendency toward the semantic-conditioned phenomenon not only suggests that the study of peripheral vision acuity fading awareness (PVAFA) needs more full-fledge experiments, it is also encouraging that one-more-notch-up research target on issues of visual logics that are capable of dealing with more abstract concepts and deeper regularities of visualization deserves further exploration.

References

1. Solso, R.L.: Cognition and Visual Arts. The MIT Press, MA (1994)
2. Bursill, A.E.: The restriction of peripheral vision during exposure to hot and humid conditions. Quarterly Journal of Experimental Psychology, 10–11 (1958)
3. Sanders, A.F.: Some aspects of the selective process in the functional visual field. Egonomics 13, 101–129 (1970)
4. Williams, L.J.: Forveal load affects the functional field of view. Human Performance 2(1), 1–28 (1989)

5. Craik & Lockart, F.M., Lockhart, R.S.: Levels of processing: A Framework for memory research. Journal of Verbal Learning and Verbal Behavior 11, 671–684 (1972)
6. Findlay, J.M., Walker, R.: A model of saccade generation based on parallel processing and competitive inhibition. Behavioral and Brain Sciences 22, 661–721 (1999)
7. McConkie, G.W., Rayner, K.: The span of the effective stimulus during a fixation in reading. Perception & Psychophysics 17, 578–586 (1975)
8. Engel, F.L.: Visual conspicuity and selective background interference in eccentric vision. Vision Research 14, 459–471 (1974)

Part IV

Human Centered Design in Work, Business and Education

Culturally Situated Design Tools: Animated Support Tools for Mathematics

Albanie T. Bolton[1,2] and Cheryl D. Seals[1,3]

[1] Auburn University, Auburn, AL 36849
[2] 741 Plummer Rd. #408, Huntsville, AL 35806
[3] Shelby Center for Engineering Technology, Suite 3101M, Auburn, AL 36849
{atb0010,sealscd}@auburn.edu

Abstract. Culturally Situated Design Tools (CSDTs) are web-based software applications that allow students to create simulations of cultural arts: Native American beadwork, African American cornrow hairstyles, urban graffiti, and so forth; using these underlying mathematical principles. CSDTs are the rationale of creating a set of culturally designed games utilizes gaming as a teaching tool to attract and instruct students with familiar methods and environments. The focus of this study is on Ron Eglash and others research on the indigenous design of various cultures using computer game simulations to teach math and computer science in the classroom sector. This study will review the development and evaluation of CSDTs, and discuss how various activities attempt to navigate through the potential dangers and rewards of this potent hybrid of information technology (CSDTs), traditional culture and individual creativity.

Keywords: Culturally Situated Design Tools (CSDTs), educational gaming, ethnomathematics, mathematics, culture, computing.

1 Introduction

Diversity does not refer only to ethnicity or race. [It refers to] differences in social class, family culture, geographic, religious backgrounds, and learning styles which are all reflected in our classrooms as important components of diversity [2]. Culturally responsive instruction addresses the specific interests, concerns, and experiences of students in the classroom. Therefore, teaching math and computer science in culturally responsive ways means using students' own habits, experiences, and cultural references to connect to real-world experiences with numbers, shapes, patterns, chance, and measurement. Successful methods for learning, calculating, memorizing and communicating about math actually differ quite a lot across cultures. This is where the term Culturally Situated Design Tools come into play.

Culturally Situated Design Tools allow students and teachers to explore mathematics and computer science with depth and care, using cultural artifacts from specific times, places, and cultures. Ethnomathematics is the study of mathematical ideas and practices situated in their cultural context. The Culturally Situated Design Tools website provides free standards-based lessons and interactive "applets" that help students and teachers explore the mathematics and knowledge systems using

M. Kurosu (Ed.): Human Centered Design, HCII 2011, LNCS 6776, pp. 351–359, 2011.
© Springer-Verlag Berlin Heidelberg 2011

ethnomathematics in areas such as African, African American, Youth Subculture, Native American, and Latino. The supporting materials for the CSDTs include lesson plans and evaluation instruments to ensure they are integrated into the curriculum through state and national standards. Based in K-12 schools with significant numbers of African-American, Latino, and Native American students (current locations include Alaska, California, Idaho, Illinois, Michigan, New York, and Utah), preliminary evaluations indicate statistically significant increase in both math achievement and attitudes toward technology-based careers.

The CSDT simulation software and teaching materials are copyrighted to Ron Eglash and Rensselaer Polytechnic Institute. The software is provided on the web by Dr. Ron Eglash at *http://www.rpi.edu/~eglash/csdt.html* [3], a professor at Rensselaer Polytechnic Institute and the author of *African Fractals: Modern Computing and Indigenous Design*. When instructors have a sense of what issues motivate and are interesting to their students they may want to search various ways to teach more culturally effective. Using this site is a very effective way of doing this. Although the use of mathematics and computing is universal, math is not culturally neutral. This is why there is research being done on how to incorporate math and computer science into gaming.

Ron Eglash sees that there is not enough diversity in these fields and what can individuals do to get the attention of students into these areas. Therefore, he constructed these tools on the basis of researching in the areas of ethnomathematics, mathematics, computing, and educational gaming. The sole inspiration of this project is based on Dr. Ron Eglash's research.

The project we focused on out of the CSDT Series is the Break Dancer tool (Figure 1), located on the CSDT website. The Break Dancer tool is a software-simulated game that teaches 3-dimensional space or solid geometry. Real-world objects exist in 3-dimensions. For example, a cuboid, or a box, is described by three parameters, length, breadth, and height. Corresponding to that, each point in the Cartesian space has three coordinates - x, y, and z [1]. In this project, the Break Dancer tool integrates the youth subculture, promotes physical activity, and makes it into an educational game.

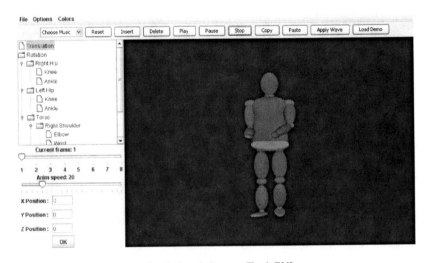

Fig. 1. Break Dancer Tool GUI

The project aims to provide a source for increasing teaching aide in schools. In addition, develop a gaming convention that will hold the student's attention in the field of mathematics. This field scares most students and keeps them away from Computing and IT jobs because this is required in the coursework. If we can help instill in students that math can be fun and innovative, this will bring more high school graduates entering college with a higher confidence level in math.

2 Using CSDTs as Teaching Aids

Ways of saying games reflect a cultural value is that games are social contexts for cultural learning. This means that games are one place where the values of a society are embodied and passed on. Although games clearly do reflect cultural values and ideologies, they do not merely play a passive role. Games also help to instill or fortify a culture's values system. Seeing games as social contexts for cultural learning acknowledges how games replicate, reproduce, and sometimes transform cultural beliefs and principles. This way of looking at games forms the basis of this schema Games as Cultural Rhetoric [6].

Games contribute to the development of knowledge by having a positive effect on the atmosphere in the class which produces a better mental attitude towards math in the pupils. Educational games provide a unique opportunity for integrating the cognitive, affective and social aspects of learning [3]. Each CSDT topic comprises a number of resources that enable teachers to integrate the topics into standards-based math instruction. Resources for each topic include:

1. A section on cultural background and history
2. A tutorial on the math topic and its connection to cultural artifacts and systems of knowledge
3. Software (applets) that enable teachers and students to simulate the development of these artifacts
4. Links to extensive teaching materials—including lesson plans, pre- and post-tests, and samples of student work from a wide variety instructional settings.

Many game features, combined and designed effectively into educational gaming, could teach many things in an engaging and motivating manner. Games could be used for the expansion of cognitive abilities, as well as a platform for developing new or practicing existing skills in the context of real world goals, rules, and situations. Games could also be used to teach old subjects in new ways.

There are skills that are hypothesized to help with the use of games and simulations: higher order skills, practical skills training, high performance situations, rarely used skills, developing expertise and team building.

3 Break Dancer Tutorial Development

The tutorial development was done to improve the current tutorial at hand. The current tutorial (Figure 2) needed more information added to get a better understanding for what was being taught.

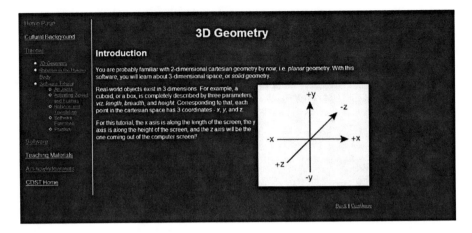

Fig. 2. Online Break Dancer Tutorial

The current tutorial is in the following outline:

- 3D Geometry
- Rotation in the human body
- Software Tutorial

    - All Joints
    - Adjusting Speeds and Frames
    - Rotation and Translation
    - Software Functions
    - Pointers

In doing the revised tutorial, we did extensive research on the Break Dancer tool and we contacted math teachers to get an understanding of what would be needed to compile a lesson plan. The previous version of Break Dancer's tutorial only highlighted a few of the topics that were discussed in the tutorial.

In the new tutorial, we talked about every topic in extensive detail, also including examples with each explanation (Figure 3).

The revised tutorial outline was changed to the following:

- Introduction to 3D Geometry & Space
- Tait-Bryan Angles
- Cartesian Coordinates in 3D Space
- Transformations (Translations & Rotations)
- Sine Function

The revised lesson plans should be equally if not more beneficial to the students and the teachers. The tool we used to design the new tutorial is PowerPoint and Captivate. We also created lessons plans and mini tests. We did data analysis on the tutorial, lesson plans and mini tests. The results are expressed in the next section.

Introduction to 3D Geometry & Space

You are probably familiar with 2-dimensional Cartesian geometry by now, i.e. *planar* geometry. With this software, you will learn about 3-dimensional space, or *solid* geometry.

Three-dimensional space is a geometric model of the Physical universe in which we live. The three dimensions are commonly called length, width, and depth (or height), although any three mutually Perpendicular directions can serve as the three dimensions.

Real-world objects exist in 3 dimensions. For example, a cuboid, or a box, is completely described by three parameters, *length, breadth,* and *height.* Corresponding to that, each point in the Cartesian space has 3 coordinates *x, y,* and *z*.

For this tutorial, the x axis is along the length of the screen, the y axis is along the height of the screen, and the z axis will be the one coming out of the computer screen!

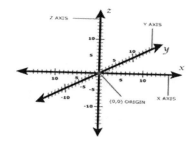

Tait-Bryan Angles

Yaw, pitch, and roll, also known as **Tait–Bryan angles**, named after Peter Guthrie Tait and George Bryan, are a specific kind of Euler angles very often used in aerospace applications to define the relative orientation of a vehicle respect a reference frame. The three angles specified in this formulation are defined as the roll angle, pitch angle, and yaw angle.

These angles are particularly seen when looking at the rotation of an object in 3D space. The rotations can be split into three parts. This will be further discussed under the Transformations heading.

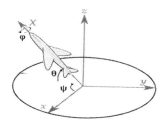

Fig. 3. Revised Break Dancer Tutorial

4 Data Analysis

4.1 Data from the Usability Testing of Tutorial

The questionnaire used for the testing of the tutorial contained four questions that focused on the teachability and learnability.

There were ten participants who were teachers and not previously exposed to the CSDTs. For most cases, we would normally have about 30 participants for a full analysis. Due to the response level, we went with an approach of opportunistic sampling. It consists of taking the sample from people who are available at the time the study is carried out and fit it to the criteria you are looking to test. We felt that it

Table 1. Usability Testing Current Tutorial

	Average General User Satisfaction
Terrible ------------------Wonderful	3.5
Frustrating----------------Satisfying	3.8
Difficult--------------------------Easy	3.5
Boring----------------------------Fun	2.0
Strongly Agree = 5, Agree =4, Neutral = 3, Disagree = 2, Strongly Disagree = 1	

was adequate in the sense that it was the population we wanted and the participants supplied very detailed information. The result showed that the participants particularly like the tool and how the tutorial conveyed the information to the students and teachers.

Table 1 represents data taken on the current tutorial that is on the site. Overall, people did not like the tutorial that was initially used and commented that it need drastic improvement. The participants noted that the current tutorial did not convey enough information to the user and would be complex to teach lesson from what was in the tutorial.

Table 2. Usability Testing of Revised Tutorial

	Average General User Satisfaction
Terrible ------------------Wonderful	4.5
Frustrating----------------Satisfying	4.8
Difficult--------------------------Easy	4.5
Boring----------------------------Fun	5.0
Strongly Agree = 5, Agree =4, Neutral = 3, Disagree = 2, Strongly Disagree = 1	

A moderate percentage of the participants answered that the tutorial was fun and overall satisfying in comparison to the older tutorial. The Anova results for each level listed in tables 3, 4, and 5. Our initial null hypothesis was done at a level of significance of 3.0. By looking at the P-Value we see that there is are differences between the two results.

Table 3. Anova Comparison Results – Terrible & Wonderful

Anova: Single Factor

SUMMARY

Groups	Count	Sum	Average	Variance
Column 1 (table 1)	10	35	3.5	0.277778
Column 2 (table 2)	10	45	4.5	0.277778

ANOVA

Source of Variation	SS	df	MS	F	P-value	F crit
Between Groups	5	1	5	18	0.00049	4.413873
Within Groups	5	18	0.277778			
Total	10	19				

Table 4. Anova Comparison Results – Difficult & Easy

Anova: Single Factor

SUMMARY

Groups	Count	Sum	Average	Variance
Column 1(table 1)	10	35	3.5	0.277778
Column 2 (table 2)	10	45	4.5	0.277778

ANOVA

Source of Variation	SS	df	MS	F	P-value	F crit
Between Groups	5	1	5	18	0.00049	4.413873
Within Groups	5	18	0.277778			
Total	10	19				

From the results, we see that the participants found that this tutorial conveyed the appropriate information to the teacher and student. Therefore, the overall revised tutorial makes it easier for the student to learn the suggested lesson with the Break Dancer tool.

Table 5. Anova Comparison Results – Boring & Fun

Anova: Single Factor

SUMMARY

Groups	Count	Sum	Average	Variance
Column 1 (table 1)	10	20	2	0.222222
Column 2 (table 2)	10	50	5	0

ANOVA

Source of Variation	SS	df	MS	F	P-value	F crit
Between Groups	45	1	45	405	8.65E-14	4.413873
Within Groups	2	18	0.111111			
Total	47	19				

4.2 Suggestions from Data Collected

The main suggestions that were collected from the data were to incorporate more lesson plans and to include more tests. In addition, one suggested that the lesson should be taught in a 3D lab to allow for more of an appropriate atmosphere for the Break Dancer tool. The participants did think that what we have now is very nice and would recommend that any teacher use this tool in their classroom. The participants found the Break Dancer tool fun and innovative. In addition, a great way to grab the students attention and teach a math lesson at the same time.

5 Future Work

The future work for this project is basically focused on implementing the software in Unity 3D for future use with Wii. That is our final ultimate goal for this project. There is work that needs to be done from evaluation viewpoint, including qualitative and long-term evaluation. One hypothesis is that a math game that engages students will motivate them in other Computer Science and IT course and encourage them to pursue these careers in college. By measuring student grades in the math pre-class and post-class, this could determine if the game engaged students sufficiently to improve their performance in later classes. We would also like to do a more rigorous evaluation over a six to nine week time span with some validated instrument to look more into our data analysis portion. We propose that we do a mini ethnography and sit in the classroom with the teacher to see exactly how effective the tool can actually be in a classroom setting.

In addition to implementing Break Dancer in Unity 3D, we need to find a way to implement the drag and drop interface. This idea was not implemented and will continue to be worked on for a future project. As stated earlier, the problem encountered is that the current interface only has two panels, and if you want to implement a scripting drag and drop interface, you need three panels. Therefore, we concluded that this task will be for a future project and that by then the initial designer will have directed us in the path that the group in the path that needs to be taken.

Other possible future work could be to explore other opportunities for game enhanced versions of other math courses, with the intention to motivate and improve the education of possible undergraduate students.

6 Conclusion

We implemented a better tutorial, lesson plan, and mini tests to help teach the lesson better and grab the user's attention. This factor will help to make the Break Dancer tool a better instrument for teachers as well as students. The CSDTs offer an exciting convergence of both pedagogical and cultural advantages. Computer games are very popular among children and adolescents. In this respect, they could be exploited by educational software designers to render educational software more attractive and motivating [5]. Unlike many other ethnomathematics examples, we can modify the interface to allow a close fit to the math curriculum, which makes it easy for teachers to incorporate into their class. At the same time, their ability to move between virtual and physical implementations allows use in the arts; and their historical connections provide teaching opportunities in history and social science. Most importantly, they allow for a flexible, creative space in which students can reconfigure their relations between culture, mathematics, and technology.

Research could help make games more attractive to different types of uses, and address the differences in the types of games that appeal to either sex. Females tend to be more attracted to games that involve relationship building than do males, who tend to prefer action games. Educational games also need to be culturally sensitive.

Acknowledgments. I would like to extend a heartfelt thanks to Dr. Cheryl Seals for being a great mentor during this research. A special thanks to Dr. Ron Eglash, for your support and contributions to the subject matter. I would also like to thank Rahul Potgan, Ravali Gondi, Human Computer Interaction Laboratory (Auburn University) and STARS Alliance members for their contributions in this project. All ideas expressed are through research and solely opinions/conclusions of the authors.

References

1. Culturally Situated Design Tools, http://www.rpi.edu/~eglash/csdt.html
2. Math and culture, http://www.commcorp.org/dys/pdf/DYSMath-Math Culture
3. Mathematical Games as an Aid to teaching Mathematics, http://www.bbc.co.uk/dna/h2g2/A798221
4. Ramer, W.: What Makes a Game Good. In The Games Journal (2000), http://www.thegamesjournal.com/articles/WhatMakesaGame.shtml
5. Salen, K., Zimmerman, E.: Rules of Play Game Design and Fundamentals. Games As Cultural Rhetoric. Massachusetts Institute of Technology, 515–534 (2004)
6. Virvou, M., Katsionis, G., Manos, K.: Combining software games with education: Evaluation of its educational effectiveness. In: Educational Technology & Society, pp. 54–65 (2005), http://www.ifets.info/journals/8_2/5.pdf
7. Claypool, K., Claypool, M.: Teaching software engineering through game design. In: Proceeding of the 10th Annual SIGCSE Conference on Innovation and Technology in Computer Science Education, ITiCSE 2005, Caparica, Protugal, June 27-29, pp. 123–127. ACM Press, New York (2005)

Towards a Paperless Air Traffic Control Tower

Tanja Bos[1], Marian Schuver–van Blanken[2], and Hans Huisman[2]

[1] National Aerospace Laboratory NLR, Anthony Fokkerweg 2, 1059 CM,
Amsterdam, The Netherlands
[2] Air Traffic Control, LVNL, The Netherlands
Tanja.Bos@nlr.nl, {M.Schuver-vanBlanken,H.Huisman}@lvnl.nl

Abstract. A prototype of an Electronic Flight Strip (EFS) system for air traffic controllers in the tower was developed in a participatory design process with rapid prototyping. The process which included five intermediate part task evaluations resulted in a prototype in which the existing working methods could be maintained. During a whole task evaluation of the EFS system in a tower simulator the usability of the EFS system was evaluated as well as the impact of the EFS system on strip hand-over, the controllers' mental picture and head-down time. It revealed that controllers were able to handle peak traffic with EFS after just 20 minutes of familiarization. Furthermore, the hand-over of traffic with EFSs was better supported according to the controllers. Nevertheless incoming strips were left unnoticed longer with EFSs and head-down time increased. For these reasons the support of the controllers' mental picture was rated slightly lower with EFSs. With small improvements and more familiarization the concept is considered ready for implementation.

Keywords: Air traffic control, Tower, Electronic Flight Strips (EFSs), Design, Evaluation, Simulation, Usability.

1 Introduction

The objective of this research was to prototype an Electronic Flight Strip (EFS) system user-interface for the air traffic control tower at Schiphol Airport Amsterdam and to evaluate it in a human-in-the-loop exercise in NLR's tower research simulator with a realistic traffic sample. This prototyping and human-in-the-loop evaluation was required by ATC the Netherlands to obtain system and implementation requirements for an EFS system.

In Air Traffic Control (ATC) towers, EFS systems are currently being enrolled to replace paper flight strips [1], [4], [5], [6]. Flight strips are mainly used:

1. to present flight information to the controllers;
2. to allow the controller to administer his instructions;
3. to build and maintain a mental picture of the aircraft under control; and
4. to support the handover of flights between controllers.

Each strip represents one aircraft or other traffic on the airport surface. Paper strips are a representation of the flight or traffic information available in the central ATC system. After the strips have been printed, controllers administer the progressing

M. Kurosu (Ed.): Human Centered Design, HCII 2011, LNCS 6776, pp. 360–368, 2011.

traffic situation by making notes on the paper strips. These handwritten notes contain valuable operational information, not only for the controller himself/herself, but also for other controllers, in or outside the tower. Digitization of the strips allows the controllers to enter instructions in the central ATC system making the information available for more controllers and other parties such as airport personnel or airlines. In addition, the paper strips are put in holders requiring extra handling and creating a noisy environment in the tower.

This article describes the design process of a digital strip system to replace the existing paper strips conducted by a multi-disciplinary team. The design process (Fig. 1) consisted of: detailed investigation of working methods with paper strips, iterative prototyping of the defined system functions with intermediate part-task evaluations with controllers and finally whole task evaluation in the NLR's tower research simulator with the final prototype.

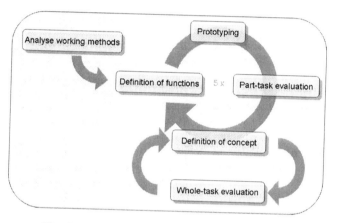

Fig. 1. Schematic representation of design process

2 Design Process and Part Task Prototyping

The first step in the design process was the analysis of existing working methods in the tower. For the fairly complex and busy airport of Schiphol Amsterdam it became apparent that the function of paper flight strips was much broader than administering the information of individual flights. Especially ground controllers - who have most aircraft under their responsibility simultaneously - use the organization of their strips to plan the traffic and to indicate future actions or conflicts very intensively. It appeared that the way in which paper strips were placed in the strip bay or holder were indications of the traffic situation. For example, paper flight strips are not always placed in the strip bay, not completely in the holder or are turned upside down. These are indications of some event that needed to have their equivalent in the prototype of the EFS system. In addition, most standard controller actions (e.g. provide push-back and taxi instructions) were indicated by moving the strip to a next strip bay and were to be represented in the prototype.

In the design process, five design workshops were held with a multi-disciplinary team consisting of representatives of Human Factors, systems, safety and procedures and of each controller function: clearance delivery, startup controllers, ground controllers, runway controllers and assistants. In each workshop, part of the functionality was discussed that was to be prototyped, for evaluation in the next session. This resulted in an incremental prototyping process where existing functionality was improved and new functionality was added to the prototype and described in its specification after each session.

The resulting prototype consisted of an interaction concept allowing to drag strips and to move them by selection and choosing a new location on a touch display operated with an electronic pen.

The approach taken in the design of the EFS prototype was to adapt the system to the current way of working. This implied that all movements of strips were initiated by the controllers and the layout of information on the strip and of the placement of strip in bays were maintained. An important starting point for the hardware was that all positions should be generic allowing for interchangeability of positions independent of the role of the controller.

Because of the requirement to have six strip bays adjacent to each other to accommodate for all different statuses and taxiways of aircraft under control of ground control a decision was made to use two 21" WACOM displays per working position, to be operated using a single electronic pen (Fig. 2). The two displays were placed against each other and on top a transparent plastic sheet was placed to bridge the gap between the two displays. In this way strips could be dragged from one display to the other without lifting the pen.

To adopt the current working method, all changes on strips and movements and handovers were to be initiated by the controllers.

Fig. 2. Ground controller using the EFS system in the simulation

Strips could be moved in three different ways, by:

1. dragging the strip, by placing the pen on a strip and dragging the pen over the screen to the place where the strip was to be placed to;
2. ticking the strip and the intended position; or
3. pushing strips vertically in a bay, in which case a number of strips could be pushed up or down.

Strips were handed over on initiation by the controller. The sending controller would place the strip in a pending bay, which is visible to the sending and the receiving controllers and in which the strip was presented smaller and with merely the most important information. The receiving controller had to move the strip out of the pending bay in order to take the strip and corresponding flight under control.

For the transfer between two ground controllers it was desirable to place strips in the other ground controller's active field. To meet this requirement a picture of the other working positions could be accessed in which the strip could be placed. At the receiving controller's working position the strip would then appear in grey, and would be presented in regular colors after the acknowledgement (touching) of the strip.

The strip system allowed for making changes, for example change a runway or a gate through menus. Also annotations on the strip could be made as if writing with a pen on a paper strip. The latter was used for example by the runway controller to indicate a clearance for line-up and take-off.

3 Whole Task Evaluation

In a limited simulation set-up in NLR's ATC tower simulator with a 360 degrees field of view, the use of EFS was compared to the use of paper flight strips for the most time-critical controller positions in the tower. The main objective was to provide insight on the following questions:

- Does the system adequately support hand-over between positions?
- What is the consequence for head-down time of the controllers with the introduction of EFS?
- Does the system adequately support the creation and retention of a mental representation of the traffic?

Due to limited size of the evaluation it was not expected that enough statistical information would be delivered and that controllers would be familiar with the system enough for a firm answer to these questions, but to get an indication.

3.1 Evaluation Setup

The controller working positions of interest in this study were the ground controller (GC) and the runway controller (RC). These positions were considered most critical and moreover the GC has the highest number of strips (and flights) under control at a given time, with the most different statuses. This position was considered the most complex concerning strip usage.

In the simulation, two generic tower working positions were equipped with the EFS prototype allowing for the evaluation of two working positions simultaneously and the interactions between these positions. Two set-ups were evaluated: two GC working positions adjacent to one another and one GC adjacent to one RC. Two crews, each consisting of two GCs and one RC participated in the simulation, each for one day. Each crew evaluated each set-up in two conditions: once with de prototype and once with paper strips. Runs lasted each 45 minutes.

One crew consisted of controllers that had been involved in the development process of the prototype. The other crew was completely new to the system.

A training run at the beginning of the day was to familiarize the crew with the system, of 45 minutes. The GC and the RC position were available. The GCs took over from one another and it was practiced until all controllers felt confident to use the system.

Table 1. Experimental Design

Run	Day 1	Day 2
Familiarization EFS	GC, RC	GC, RC
Paper strips	GC1, GC2	GC, RC
EFS	GC1, GC2	GC, RC
Paper strips	GC, RC	GC1, GC2
EFS	GC, RC	GC1, GC2

3.2 Scenarios

Two traffic samples of 45 minutes were developed. They simulated an inbound traffic peak that gradually changed to an outbound traffic peak. During the inbound peak two parallel runways (18R and 18C) were used for landing and one (24) for departure. In the transitional phase runway 18C was closed and 18L was taken into use for landing. The traffic sample for the GC-GC exercise included around 64 flights, for the GC-RC exercise around 42. Events were included such as a need for de-icing, occupied gates for incoming traffic, an aircraft that had to return to the gate, the need for aircraft to cross an active runway. Also two strips of inbound traffic appear in the wrong sequence and two aircraft of the same type and airline that were taxiing to the runway called the RC in the wrong sequence, to see whether the controller would notice this in both the exercises with paper strips and EFS.

3.3 Measurements

For the evaluation, questionnaires were to be completed by the controllers, debriefing sessions were held and analysis of simulator loggings and video recordings were made.

A questionnaire was to be completed by the controllers after each run with EFS. Several aspects were to be rated for the EFS system in comparison to the paper strips. The ratings were made on a seven-point scale for which the tick box in the one end indicated "much better", the other side "much worse" and middle indicated 'the same'.

The aspects to be rated were about the primary functions of strips (e.g. support the mental picture of the traffic), the consequences of working with EFS (e.g. workload, efficiency) and the usability.

Hand over. The hand-over of strips from one controller to the other was assessed subjectively through controller comments during debriefing and in the post-experiment questionnaire. In addition, the time was assessed from the moment the strips were handed over until the receiving controller integrated the strip in the strip bays. For the EFS this was done by logging the time from the moment the strip appears on the working position until the time the controller first selects the strip.

Mental representation. In order to measure the creation and retention of the mental representation of traffic, the controllers' reactions to certain events that were introduced in the scenario were observed. For example there were two aircraft of the same type and company taxiing behind one another, of which the second would call the RC for a line-up clearance before the first one. It was observed whether the controllers noticed this and reacted likewise. In addition, controller judgments in questionnaire and debriefing were analyzed.

Head down time. The percentage of time that the controllers were looking down was assessed. During the busiest moment in the scenario, a ten minute sample of the video recordings was taken and analyzed. This sample started after 25 minutes of simulation time for each run. A stopwatch function was used to time the head-down moments.

Table 2. Means of measurement per objective

Aspect	Observation	Logging	Questionnaire	Debriefing
Transfer between positions	Timing of video (paper)	Automatic logging (EFS)	X	X
Head-down time	Timing of video		X	X
Mental representation	Observation of events		X	X

4 Results

Even though the number of runs was limited and the familiarization time was short, the evaluation yielded some interesting results.

4.1 Use of the EFS

Controllers were observed to work with the EFS system quite easily. Nevertheless GCs and RCs mentioned that it was quite different to work with the system and use the pen for all movements and changes of the strips. It requires more visual attention than moving the paper strips. During the movement of paper strips the controllers would often look to the outside view.

The use of EFS resulted in a much lower level of noise in the tower simulator and controllers remained seated in their position for each hand-over.

For the main research questions the results are reported in the following.

4.2 Hand-Over

The support of the hand-over between positions was rated higher with EFS than with paper strips according to the controllers. The reasons mentioned were that it was quieter (paper strips placed in metal holders make noise), it was noticed that it costs less time and did not result in a temporary loss of the mental picture, because the controller can stay on the working position.

Nevertheless, strips were left unattended longer with electronic than with the papers strips. With EFS were left pending for 38 second on average, with paper strips only seconds, too short to measure accurately with a stopwatch. All GC had occurrences where strips were left pending for more than 100 seconds. One controller mentioned to use the pending strip bay also as the 'passive bay' and left the strips there longer for this reason, but other controllers did not notice incoming strips for quite a while. The silent appearance of strips was identified as a potential cause as well as that the location of the pending bays on the sides of the displays were outside the scanning field of the controllers.

For hand over of strips from GC to GC, where the controller opens a picture of the other controllers strip display, also strips were left unnoticed longer with the EFS.

In the simulation this did not cause any potential dangerous situations. However in hand-overs of active, moving traffic (e.g. from RC to GC and from GC to GC) it is undesirable that aircraft are left unattended for too long.

The task load at a position seems to be of influence to this effect.

A sound for incoming strips could be a solution but it is also expected that more familiarization with the system could help.

4.3   Mental Picture of Traffic

The events of changing runway configuration, de-icing, occupied gates, return to gate and crossing an active runway were handled well by both GCs and RCs. Also the events of aircraft calling in the wrong sequence or strips appearing in the wrong sequence were noticed and well responded to by the controllers.

Nevertheless, controllers rated building up and maintaining a mental picture of the traffic lower with the EFS prototype, on average 0.5 point below the paper strips (out of three). A first reason identified, was the strips remaining unnoticed longer, and are therefore not integrated in the overall picture and controllers are 'surprised' by a call

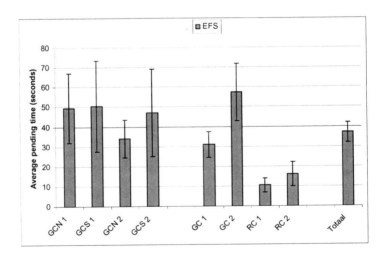

Fig. 3. Average time strips were pending for each position (*GCNorth, GCSouth, GC* and *RC*) on day *1* and 2 and 95% confidence interval

from a flight. Secondly, the introduction of an extra strip bay was a novelty that contributed to the unfamiliarity of working with the new EFS system, and the limited simulation time did not familiarize the controller enough to get used to this.

4.4 Head-Down Time

The time controllers were looking down to the displays and strips was assessed on the basis of the video recordings of the simulation. Time samples of ten minutes were analyzed. The time interval was taken between the 25th and the 35th minute in the simulation. The accuracy of the measurement was estimated around 1% since several repeated assessments revealed a maximum difference of five seconds.

The percentages of head-down time were very high, with paper strips as well as with EFS. On average a head-down time of 74% was measured with paper strips and 79% with EFS. In simulations the head-down times measured are higher than in reality, in [3] head-down times of 80 % were measured. The lower resolution of the outside view in the simulator causes a certain simulator effect.

For the RC the task load was quite low, therefore the RC head-down times are not considered as representative. For the first GC-GC run on the first day the workload was too high, requiring the traffic sample to be altered before the comparable run with EFS. This did not allow to compare these two runs. Comparison between paper strips and EFS with and without the GCN 1, GCS 1 and RC 2 data was significant using a paired T-test (t=2.99, n=8, p=0.05 and t=2.27, n=5, p=0.05). Thus the use of the EFS prototype increased the head-down time in comparison to the use of paper strips in the simulation. Controllers also mentioned that using EFS simply required more visual attention than paper strips. Controllers mentioned that unfamiliarity with the system was a factor and they had the feeling that more practice would decrease the head-down time. On the other hand it was mentioned that annotation on paper strips that also require inputs in the central system (which was not seen in the intervals measured) would require a lot more head-down time than with EFS.

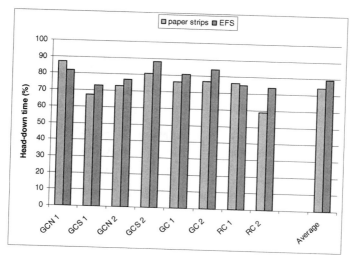

Fig. 4. Head-down times in percentages for the different positions (*GCN, GCS, GC* and *RC*) on day 1 and 2, for time samples of ten minutes starting from minute 25

5 Discussion of Results

The iterative participatory design process resulted in a prototype that proved to be fit for handling peak traffic after just 20 minutes of familiarization by experienced controllers.

The limited simulation set-up yielded several potential improvements and requirements for implementation. The controllers had the opinion that the hand-over of strips was better supported by the EFS system, because it reduced the noise in the working environment, and required less time and effort. Controllers can remain in their position which allows them to maintain their mental picture. Nevertheless, the hand-over of the paper strip was earlier noticed by the receiver than with EFS. After prolonged use, the controllers may adapt their scanning pattern on the new working position. The delay in noticing the pending strips is caused by their silent appearance in the peripheral viewing angle of the controllers. This can be mitigated by a sound to announce the appearance of a strip, especially for active traffic or as [3] argues that essential information should be presented within 15 degrees field of view.

Furthermore, working with the EFS system resulted in an increase of head-down time. Moving the strips requires more visual attention than paper strips. This may decrease when controllers are more familiar with the system, but it obstructs to a certain extent in building and maintaining a mental representation of the traffic.

This simulation focused on the criticalities of implementing the EFS system. It provided enough confidence that the system with small improvements is ready for implementation. For future research are of interest if the head-down time decreases after more prolonged usage of the system and/or how this concept of interaction compares to a concept where the first most likely input is supported by a single tick, transferring the strip to a next location or to a next controller.

References

1. Durso, F.T.: Spinning paper into glass: transforming flight progress strips, F.T. Durso, C.A. Manning. Human Factors and Aviation Safety 2(1), 1–31 (2002)
2. Hilburn, B.: Head-down Time in Aerodrome Operations: A Scope Study, Technical report, Center for Human Performance Research, Den Haag, The Netherlands (2004)
3. Hilburn, B.: Head-Down Time in ATC Tower Operations: Real Time, Simulations Results, Technical report, Center for Human Performance Research, Den Haag, The Netherlands (2004)
4. Pavet, D., et al.: Use of Paper Strips by Tower Air Traffic Controllers and Promises Offered by New Design Techniques on User Interface, USA/Europe R&D seminar ATM 2001, Santa-Fe (2001)
5. Truitt, T.R.: Implementing electronic flight data in airport traffic control towers (2005)
6. Truitt, T.R.: Electronic Flight Data in Airport Traffic Control Towers: Literature Review, Federal Aviation Administration (2006)

A Design-Supporting Tool for Implementing the Learning-Based Approach: Accommodating Users' Domain Knowledge into Design Processes

Jung-Min Choi[1] and Keiichi Sato[2]

[1] Faculty of Crafts and Design, Seoul National University,
599 Gwanak-ro, Gwanak-gu, Seoul 151-742, Korea
[2] Institute of Design, Illinois Institute of Technology,
350 N LaSalle St., Chicago, IL, 60610, USA
jmchoi@snu.ac.kr, sato@id.iit.edu

Abstract. In the current interactive product/system design, while users' acquisition of sufficient knowledge for operating a product or system is increasingly considered important, their acquisition of problem-solving knowledge in the task domain has largely been disregarded. Without enough domain knowledge, users will bot be able to learn how to creatively adjust their product us to produce satisfactory results and better experiences. This research aims to develop a methodology for designing interactive products/systems that can support users' development of domain knowledge through interaction. This new approach to user-product interaction is named the Learning-Based Approach (LBA). Based on the previous theoretical and empirical studies, this paper proposes some mechanisms for implementing the LBA. Then, a computer-based tool is developed in order to support designers' more effective and efficient application of the LBA mechanisms in design processes.

Keywords: users' domain knowledge, Learning-Based Approach, design-supporting system, interaction design methodology.

1 Introduction

Interactive technologies embedded in products and systems have drastically reduced people's efforts in solving problems and achieving goals in everyday practices. In order to improve the quality of interaction with products and systems, designers and researchers has mainly been concerned with how to help users solve problems and achieve goals more easily and efficiently. In the current design process, while users' acquisition of sufficient knowledge for operating a product or system (e.g. how to operate a coffee machine) is increasingly considered important, their acquisition of problem-solving knowledge in the task domain (e.g. how to make delicious coffee) has largely been disregarded. As a result, domain problem-solving knowledge has been hidden behind step-by-step product operational procedures, detached from actual user experiences.

However, people's primary purpose of using a product is not proficient product operation itself, but the acquisition of satisfactory results and experiences through

M. Kurosu (Ed.): Human Centered Design, HCII 2011, LNCS 6776, pp. 369–378, 2011.
© Springer-Verlag Berlin Heidelberg 2011

interaction. In the authors' previous research, it was found that without sufficient knowledge of problem-solving mechanisms in the task domain (i.e. domain knowledge), users were not able to creatively adjust their product use in order to achieve satisfactory results and generate better experiences [1]. Therefore, when designers seek to support users' interaction with a product, it is not sufficient to consider how to help them obtain sufficient knowledge of operating the product. Rather, designers should also consider how to support users' acquisition of their own problem-solving knowledge in the domain.

The present research aims to develop a methodology for designing interactive products and systems that can help users actively develop their domain problem-solving knowledge through interaction. This new approach to user-product interaction is named the Learning-Based Approach (LBA) [1]. Based on the authors' previous theoretical and empirical studies, this particular paper proposes some mechanisms for implementing the LBA. Then, a computer-based tool is developed by employing the mechanisms in order to support designers' more effective and efficient application of the mechanisms in design practices.

By supporting users' acquisition of domain knowledge – i.e. by adopting the LBA – designers can provide users with more robust and fundamental way for assisting their goal achievement through interaction. In addition, for designers, this new approach provides not only a new understanding of user-system interaction but also applicable methods and tool with which they can effectively develop products that can facilitate users' development of domain knowledge.

2   Learning-Based Approach (LBA) to Interaction

2.1   Perspectives on Users' Knowledge Acquisition in Interaction

In order to achieve satisfactory results while using a system, it is necessary for users to access and apply existing knowledge and obtain new knowledge through interaction. Current approaches to system design tend to consider these kinds of user knowledge development processes mainly as skill and/or knowledge development processes for better *operating* a product/system (e.g.,[2] [3]). Consequently, most designers' primary concern regarding supporting users' knowledge development focuses only on how to help users gain knowledge and skills for operating a product as quickly as possible (e.g., [4] [5]). By obtaining only proficient skills for operating a particular product, however, users may have difficulty transferring the obtained knowledge when they are using a new product or solving different types of problems.

Some researchers, such as [3], argue that product/system designs should support users' generation of appropriate mental models (part of knowledge) because successful user-product interaction depends on whether users can build up proper mental models and apply existing mental models to newly encountered problem-solving situations. However, proper mental models of how to operate a product/system do not completely resolve the problem of users' problem-solving capability limited within the functioning of the machine. Furthermore, proficient product operations may not be users' primary purpose of product use.

Some more recent research, particularly in activity theory approaches, provides an alternative viewpoint by explaining the relationship between users and products from a more fundamental viewpoint (e.g.,[6] [7]). Those approaches consider products/ systems of as tools that mediate inherent mutual knowledge development between a user and an object of activity. In other words, interactive products are viewed as knowledge-mediating tools that help users attain their object of activity.

From this viewpoint, in order to fully support users' object-oriented activities, it is not sufficient for designers to consider how to support users' development of skills and knowledge for operating a product (tool). Rather, they should also consider how to help users obtain knowledge needed for fulfilling the object of activity domain.

2.2 Designing Interaction as a Learning Process

By thinking of interactive products as knowledge-mediating tools and user-product interaction as knowledge development processes, one can say that supporting users' knowledge development (i.e. learning) in interaction can be an effective design strategy for improving interaction quality.

This new approach to user-product interaction is named the Learning-Based Approach (LBA). The LBA emphasizes the significance of users' domain knowledge in interaction; when users have enough knowledge about problem-solving mechanisms in the task domain, they will be able to figure out how to achieve a goal by creatively adjusting their product use to meet their variable needs. Therefore, designers should consider how to help users develop their own domain knowledge through interaction.

In order to empirically investigate the effects of users' operation and domain knowledge, the authors previously conducted observational user studies [1]. The studies compared four different groups' problem-solving behaviors, which were organized according to types of knowledge (domain vs. operation knowledge) and their levels of knowledge (novice vs. expert). From the analysis results, it was clear that in order to achieve quality results in using a product, people need to acquire and effectively apply domain knowledge. For example, participants who were accustomed to operating a product (e.g. coffee maker), but did not have knowledge of the task domain (coffee-making principles), had difficulty creatively adjusting their knowledge in certain unexpected situations. Unlike the group, participants who had enough knowledge of both operation and the task domain were able to flexibly modify their behaviors in different situations.

2.3 Some Terminology: T*ask Domain* and D*omain Knowledge*

Task Domain. A task domain is defined here as the domain *in* which the user is working, but it is also concerned with a user's *reasons for* conducting an activity. The term task domain in this research may seem similar to the notion of domain that is often used in the field of knowledge-based system development, and yet, there are some important distinctions to be made. In the system development field, some researchers emphasize that designers need to understand the "task domain" in order to develop a software system that can better fulfill users' needs when performing tasks in the domain. For example, [8] argues that software designers should analyze the domain by considering "where and for what the system design is used." The knowledge-based system development field works to provide users with specific knowledge necessary for particular problem-solving. While agreeing with this overall frame, the present research is based on some different viewpoints. This research argues that designers should not think of users' problem-solving methods as fixed or entirely predictable, and that designers should account for users' dynamic needs when developing products and systems. Therefore, while knowledge-based system development often holds that a task domain is stable, this present research focuses on the notion of the task domain as constructive and changing. In other words, a task domain is differently *constructed* based on an individual user's different needs in variable situations, and can also be continuously modified according to the user's changing contexts and needs.

Domain Knowledge. Domain knowledge refers to knowledge necessary for solving problems in the task domain. It is about the kinds of factors that shape the quality of results, and about how to adjust the factors in order to achieve a desired quality of results. It includes conceptual, hierarchical, sequential, and cause-effect knowledge. There is also a distinction between *domain knowledge* as it is used here and the same term used in knowledge-based system development. Knowledge-based system designers are concerned with *designers'* understandings of domain knowledge structures, which allow them to develop systems that *solve* the problems users encounter. By formalizing domain knowledge structures and using the structures to pre-configure *solutions* for problems in the domain, knowledge-based systems usually aim to develop systems that provide users with a set of domain knowledge *pre-selected to solve users' problems*. The present research, on the other hand, focuses on how to support *users' development of problem-solving abilities*.

3   Mechanisms for Implementing the LBA in Design Processes

In order to help designers apply the concept of LBA in their design processes, this research proposes some mechanisms for implementing the LBA. The basic mechanisms for integrating the LBA in product/system designs are depicted in Fig. 1. The LBA mechanisms are divided into two main processes. The first part is for defining *what kinds of* domain knowledge should be represented for users during task performance. This involves such sub-processes as structuring domain knowledge, analyzing tasks, and mapping between domain knowledge and tasks. The second part describes the processes for defining *how to deliver* specific domain knowledge by appropriately representing this knowledge according to users' contextual conditions. This part involves sub-processes such as defining contexts, defining representational factors, and mapping between contexts and representational methods.

Fig. 1 also shows the boundaries of this research, to what extent the present research addresses system design processes and what kinds of external information or mechanisms should be input in order to complete a full system design process. In other words, the LBA methodology is not intended to constitute self-sufficient and independent design processes in and of itself; rather, it is integrated in regular design processes as a new and important constituent. At a more detailed level, the LBA-implementing mechanisms involve the following steps.

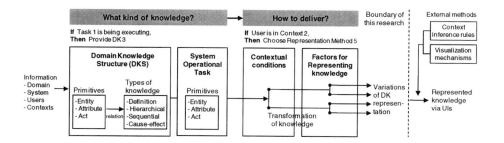

Fig. 1. Mechanisms for implementing the LBA in design processes

Understanding Domain Knowledge Structure (DKS). In the first phase, designers should understand and represent the DKS in a way that a product/system can recognize and process it. In this phase, designers need to define what kinds of particular domain knowledge should be delivered during users' interaction with the product/system. In order to represent domain knowledge, several sets of elements are defined.

1. Ontological elements of knowledge representation: act, entity, attribute, and relation
2. Elements for DKS description: goals, actions, entities and procedures
3. Types of knowledge: definition, hierarchical, sequential, and cause-effect knowledge – by making relations among the DKS elements

The present research describes domain knowledge structure (DKS) by adapting the format of task knowledge structure (TKS), which was proposed by [9]. The purpose of TKS is to represent the structure of knowledge involved in users' task performance. By using a similar format to TKS, the DKS aims to describe the structure of domain knowledge. Fig. 2 describes how domain knowledge can be described differently from operation knowledge in the same format. Although this research adopts the format of [9], the intention of the representation is somewhat different. TKS is described for developing a system that can support users' task execution in using the system, based on *designers' (systems')* understanding of users' task knowledge. The present research, on the other hands, aims to enhance *users'* development of domain problem-solving abilities.

Analysis of System Operational Tasks. In the second phase of the LBA mechanisms, users' system operational tasks are analyzed in order for designers to consider delivering domain knowledge that can support particular stages of system operations. In the system design field, task analysis is typically used to identify users' tasks in their goal-oriented behaviors. Since users need to obtain particular domain knowledge to achieve goals, and because goals can be attained by conducting a series of tasks when using a product, designers should also analyze users' tasks involved in operating the target product.

Types of knowledge	Elements of representation	Domain Knowledge Structure	vs.	Operation Knowledge Structure

Hierarchical knowledge	⌈Goal
	⌊Subgoals
	Actions
	Entities
		Take. Photo (**Quality : Satisfactory**)
		Adjust . Exposure control
		  Control. Shutter speed controller (Speed) OR
		  Control. Aperture controller (Opening degree) OR
		  Control. ISO sensitivity controller (Sensitivity)
				Take. Photo
				Adjust . Exposure control
				  Control. Shutter speed menu (Speed) OR
				  Control. Aperture menu (F degree) OR
				  Control. ISO menu (ISO)

Sequential knowledge	Subgoal
	Subgoal
	Procedure
	(Sequential)
	
	Subgoal
		Take. Photo (Brightness : Brighter)
		Change. Shutter speed controller (Speed)
		  {Select. Shutter speed controller THEN
		  Select. Shutter speed controller (**Speed : value**)}
		
		OR
		Change. ISO sensitivity controller (Sensitivity))
		  {Select. ISO sensitivity controller THEN
		  Select. ISO sensitivity controller (**Sensitivity : value**)}
		OR …
				Take. Photo (Brightness : Brighter)
				Change. Shutter speed menu (Speed)
				  {Select. Function control button THEN
				  Select. Shutter speed menu THEN
				  Select. Shutter speed menu (**Speed : number**)}
				
				OR
				Change. ISO menu (Sensitivity)
				  {Select. ISO menu THEN
				  Select. ISO menu (**Sensitivity : number**)}
				OR ….

Cause-effect knowledge	Procedure (Conditional)	IF **Increase**. Aperture controller (**Opening degree : value**) THEN Increase. Photo (Brightness) IF Increase. ISO sensitivity controller (**Sensitivity : value**) THEN Increase. Photo (Noise)		IF **Decrease**. Aperture menu (**F degree : number**) THEN Increase. Photo (Brightness) IF Increase. ISO menu (**ISO : number**) THEN Increase. Photo (Noise)
Hierarchical knowledge	Goal structure	Goal1 (Goal3, Goal4) Goal2 (Goal1, Goal4, Goal5)		Goal1 (Goal2, Goal3, Goal4) Goal2 (Goal3, Goal5)
Definition knowledge	Entities	Definition (Shutter speed controller) Definition (Aperture controller)		Definition (Shutter speed menu) Definition (Aperture menu)
Focus :		How Domain-Entities can be manipulated		How Operation-Entities can be manipulated
		Focus of DK-oriented Learning-Based Approach		Focus of conventional OK-oriented approaches

Fig. 2. Comparison the representation of domain knowledge-oriented representation with operation knowledge-oriented representation

Mapping between Tasks and Domain Knowledge. The essence of the mapping mechanism in this phase is based on the idea that a system should help users be aware of what *domain entities (defined in DKS)* they are actually manipulating, how their manipulations relate to the resulting quality, and to reach this awareness while they are controlling a particular *operation entity (defined in task analysis)*. In other words, by using a LBA-applied product, users will be able to continuously learn what kinds of *domain problem-solving mechanisms* they are actually manipulating while they are controlling *the mechanisms of interface elements* to achieve their goals. The consequences of these processes can be described as such:

"If a user is executing task 1, then domain knowledge 3 and 5 should be provided."

Delivering Domain Knowledge through User Interface Designs. The next consideration should be choosing the most effective methods for representing specific domain knowledge objects in a way that can facilitate users' acquisition of the knowledge through interaction. Since the appropriateness of the representation methods of particular knowledge is thought to depend on whether the representation method is matched to the user's contexts, designers need to define particular types of users' contextual conditions. Then, designers should also create variations of representational methods by categorizing factors for representing knowledge and defining the properties for each factor so that a system can infer appropriate representational methods according to recognized contexts. The results of mapping between contextual conditions and representational methods can be described as such:

"If a user is in context 2, then domain knowledge 3 will be represented in method 5."

Developing Prototype Specifications. By selectively employing some part of the design information produced in the prior phases, designers will be able to develop prototypes that reflect particular concerns of the project.

4 Development of Domain Knowledge-Based Tool (DKT)

In order to help designers apply the LBA mechanisms through the design processes, this research develops a computerized design-supporting tool, the DKT (Domain Knowledge-based design Tool).

4.1 Overview of the DKT

The DKT is developed to incorporate the concept of the LBA into product and system design processes. By using the computational tool, designers will be able to apply the LBA mechanisms to their practices more efficiently and effectively. The DKT provides step-by-step functional modules, each of which corresponds to the specific step for implementing the LBA mechanisms described above. The DKT consists of separate but interrelated five-phase functional modules, as shown in Fig. 3: *DK Structuring, Task Analyzing, Task-DK Mapping, Knowledge Delivering,* and *Prototype Specification.*

Based on the conceptual model shown in Fig. 3, basic components of the tool and the information flow amongst the components are defined and illustrated in a block diagram, as shown in Fig. 4. *Database layer.* Database components in the database layer are classified into two groups: 1) a primitive database group and 2) a composite database group. Database components in the primitive database group are designed to store foundational data such as *DK Entity, Operational Entity, Context, and Factor for Representation.* These types of data cannot be further deconstructed. On the other hand, database components in composite database groups are designed to accumulate data that are generated using the components in the primitive database group and/or other components in the composite database groups.

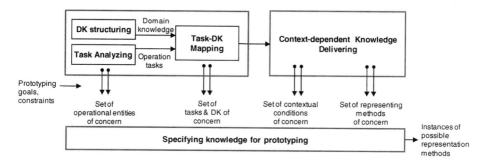

Fig. 3. Conceptual model of the Domain Knowledge-based Design Tool (DKT)

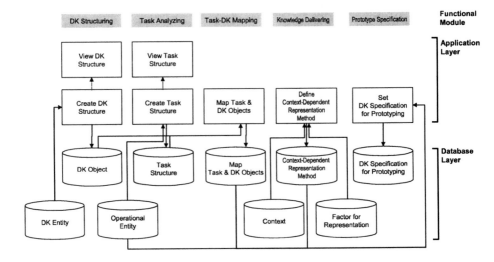

Fig. 4. System architecture of the DKT

4.2   Functional Modules of DKT

In order to more effectively explain the details of using functional modules of the DKT, consistent examples drawn from the task domain of photo-taking are used throughout all main features of the tool.

1) *DK (domain knowledge) Structuring.* Fig. 5 shows the full screen of the *DK Structuring* module. The *DK Structuring* module consists of two main features: – in the upper area of the screen – and composing DK objects based on the defined entities – in the lower area of the screen.

2) *Task Analyzing.* This module allows designers to define operational entities of a particular product and the structure of users' tasks that are related to using the entities.

3) *Task-DK Mapping.* This module enables designers to specify particular DK corresponding to each operational task. As shown in Fig. 6, the functions of this module are divided into the *Composed Task* section (left) and the *Relevant Domain Knowledge Object (DKO) for Task* section (right).

4) *Knowledge Delivering.* In this module, designers can generate variations of context-dependent knowledge representation methods. This consists of three sub-modules: *Creating Context, Creating Factors for Representation,* and *Defining Context-Dependent Knowledge Representation Method.*

5) *Prototype Specification.* The *Prototype Specifier* allows designers to easily recognize available information elements in the pull-down lists so that they can select specific sets of information necessary for prototyping.

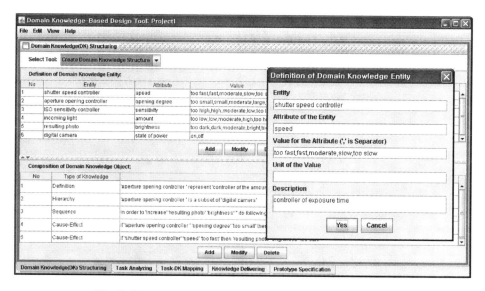

Fig. 5. *Domain Knowledge (DK) Structuring* module of DKT

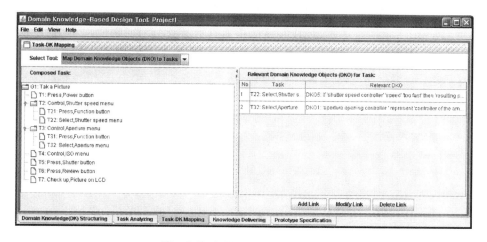

Fig. 6. *Task-DK Mapping* module

5 Conclusions

The Learning-Based Approach (LBA) to interactive product and system design is proposed to provide designers with a new viewpoint of users' knowledge development in interaction, as well as applicable methods and tools for developing a product from this perspective. The LBA emphasizes the significance of users' domain knowledge in using a product/system; when users have enough knowledge about problem-solving mechanisms in the task domain, they will be able to figure out how to achieve a goal by creatively manipulating their interaction behaviors to meet their variable needs. Therefore, system designers should consider how to help users develop their own domain knowledge through interaction. By obtaining sufficient domain knowledge through the interaction with a system, users will be able to attain high quality results, richer user experiences, and overall higher levels of satisfaction. In this paper, mechanisms for implementing the LBA are proposed, including 1) describing domain knowledge structure in a way that a product/system can recognize and process it, 2) analyzing users' tasks for operating a product, 3) mapping domain knowledge onto relevant particular tasks, 4) delivering domain knowledge to users via user interfaces by flexibly representing domain knowledge according to users' contextual conditions, and 5) developing prototype specifications by selectively employing some part of the design information produced in the prior phases. In order to help designers incorporate the mechanisms in design processes, this research develops a computerized design-supporting tool. The future research will introduce the prototype development of a case product. Through the prototyping process, the user interface outputs that are drawn on the LBA mechanisms will be demonstrated and evaluated.

References

1. Choi, J., Sato, K.: Interaction as Learning Process: Incorporating Domain knowledge into System Use. In: 5th NordiCHI, pp. 73–82. ACM Press, New York (2008)
2. Card, S.K., Moran, T.P., Newell, A.: The Psychology of Human-Computer Interaction. Lawrence Erlbaum Associates, Hillsdale (1983)
3. Norman, D.A.: The Design of Everyday Things. Basic Books, New York (1988)
4. Carroll, J.: The Nurnberg Funnell: Designing Minimalist Instruction for Practical Computer Skill. MIT Press, Boston (1990)
5. Nielsen, J.: A Meta-Model for Interacting with Computers. Interacting with Computers 2, 147–160 (1990)
6. Engeström, Y.: Activity Theory and Individual and Social Transformation. In: Engeström, Y., Miettinen, R., Punamäki, P. (eds.) Perspectives on Activity Theory. Cambridge University Press, Cambridge (1999)
7. Kaptelinin, V.: Activity Theory: Implications for Human-Computer Interaction. In: Nardi, B.A. (ed.) Context and Consciousness: Activity Theory and Human-Computer Interaction, 3rd edn. MIT Press, Boston (2001)
8. Wu, Y.: What else should an HCI pattern language include? In: Pattern Languages for Interaction Design: Building Momentum, CHI 2000, The Hague, The Netherlands (2000)
9. Johnson, P.: Human-Computer Interaction: Psychology, Task Analysis and Software Engineering. McGraw Hill, London (1992)

A Methodical Approach for Developing Valid Human Performance Models of Flight Deck Operations

Brian F. Gore[1], Becky L. Hooey[1], Nancy Haan[2], Deborah L. Bakowski[1], and Eric Mahlstedt[1]

[1] San Jose State University at NASA Ames Research Center
MS 262-4, P.O. Box 1, Moffett Field, CA 94035-0001
{Brian.F.Gore,Becky.L.Hooey,Debi.Bakowski,
Eric.Mahlstedt}@nasa.gov
[2] Dell Services Federal Government
MS 262-4, P.O. Box 1, Moffett Field, CA 94035-0001
Nancy.Johnson@NASA.gov

Abstract. Validation is critically important when human performance models are used to predict the effect of future system designs on human performance. A model of flight deck operations was validated using a rigorous, iterative, model validation process. The process included the validation of model inputs (task trace and model input parameters), process models (workload, perception, and visual attention) and model outputs of human performance measures (including workload and visual attention). This model will be used to evaluate proposed changes to flight deck technologies and pilot procedures in the NextGen Closely Spaced Parallel Operations concept.

1 Introduction

The National Airspace System (NAS) in the United States is currently being redesigned because it is anticipated that the current air traffic control (ATC) system will not be able to manage the predicted two to three times growth in air traffic in the NAS [1]. The goal of the Next Generation Air Transportation System (NextGen) is to increase the capacity, safety, efficiency, and security of air transportation operations [1]. However, in doing so, it is expected that the data available to pilots on the flight deck (e.g., weather, wake, traffic trajectory projections, etc.) will be increased substantially in order to support more precise and closely coordinated operations. If not designed with consideration of the human operators' capabilities, these NextGen concepts could leave pilots, and thus the entire aviation system, vulnerable to error.

Human Performance Models (HPMs) have been shown to play a role in all phases of the concept development, refinement, and deployment process of next generation systems [2,3]. HPMs can be used to develop and evaluate new technologies, operational procedures, and the allocation of roles and responsibilities among human operators and automation. HPMs hold the most promise when they are used early in the system design, or system redesign, process and when used iteratively with human in the loop (HITL) simulation output [3,4]. However, before HPMs can be successfully implemented to evaluate how NextGen concepts will impact pilot

M. Kurosu (Ed.): Human Centered Design, HCII 2011, LNCS 6776, pp. 379–388, 2011.

performance, baseline models of current-day pilot performance must first be developed and validated.

The objective of this research effort was to develop and validate a baseline HPM of current-day pilot performance. This model will then be used to evaluate proposed changes to flight deck technologies and pilot roles and responsibilities in NextGen Closely Spaced Parallel Operations (CSPO) concepts. Model outputs, including pilot workload and visual attention will be used to draw conclusions regarding the requirements necessary to support NextGen concepts and predict human performance effects, identify safety vulnerabilities, and recommend mitigations.

1.1   Man-machine Integration Design and Analysis System (MIDAS)

The Man-machine Integration Design and Analysis System (MIDAS) is a dynamic, integrated human performance modeling environment that facilitates the design, visualization, and computational evaluation of complex man-machine systems [5]. MIDAS symbolically represents many mechanisms that underlie and cause human behavior. Figure 1 illustrates the model's organization and flow of information among the model's components.

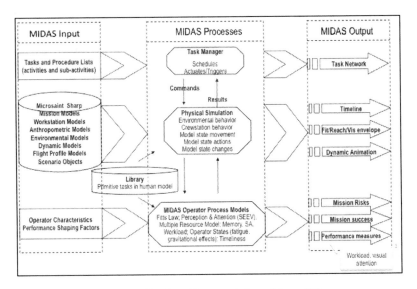

Fig. 1. MIDAS architecture (adapted from [2])

MIDAS inputs (Fig. 1, left column) include the operators' task and procedures, the operational environment (e.g., flight profiles, scenario objects and events, cockpit layout etc), and operator characteristics (e.g., operator expertise, and fatigue).

The *MIDAS processes* (Fig. 1, middle column) are comprised of a task manager model that schedules tasks, definitions of the state of models within the physical simulation, a library of "basic" human primitive models that represent behaviors required for all activities, and cognitive models such as operator perception, visual attention, and workload. These basic process models have been extensively validated.

For instance, MIDAS' attention-guiding model operates according to the SEEV model [6], an extensively validated model that estimates the probability of attending, P(A), to an area of interest in visual space, as a linear weighted combination of the four components - salience, effort, expectancy, and value. The SEEV model has been integrated into MIDAS [7] and drives the operators' visual attention.

Visual perception in MIDAS depends on the amount of time the observer dwells on an object and the perceptibility of the observed object. The perception model computes the perceptibility of each object that falls into the operator's field of view based on properties of the observed object, the visual angle of the object and environmental factors. In MIDAS, perception is a three-stage, time-based model (undetected, detected, comprehended) for objects inside the workstation (e.g., an aircraft cockpit) and a four-stage, time-based perception model (undetected, detected, recognized, identified) for objects outside the workstation (e.g., taxiway signs on an airport surface) [8]. Information then passes into a three-stage memory store [9] that degrades according to empirically-driven memory decay rates [10].

The cognitive models interact with a series of validated anthropometric models that call a number of validated motor movement models [11]. For a description of the MIDAS processes and empirical models, the reader is directed to [5, 12].

The *MIDAS output model* (Fig. 1, right column) generates a runtime display of the task network, timeline, fit, reach, and visibility envelopes, a dynamic animation of the operator carrying out his/her tasks within the environment, and mission performance measures such as workload and visual fixations.

2 Modeling Flight Deck Operations

2.1 Developing a Model of Current Day Approach and Land Operations

The objective was to develop a high-fidelity model of two-pilot (pilot flying, PF, and pilot-not-flying, PNF) commercial transport operations, with ATC tasks and procedures modeled at a lower level of fidelity, but at a level sufficient to represent the interactions between pilots and ATC. The model was based on a scenario in which pilots flew an area navigation (RNAV) approach into Dallas Fort Worth (DFW) with current-day Boeing-777 equipage (see Figure 2). The scenario began with the aircraft at an altitude of 10,000' and 30nm from the runway threshold. The cloud ceiling was 800', with a decision height of 650' at which point the modeled pilots disconnected the autopilot and manually hand-flew the aircraft to touchdown.

The RNAV model was based on cognitive task analyses of flight tasks [13,14] and cognitive walkthroughs with a commercial pilot and ATC. This process generated a comprehensive set of tasks in each of the following major phases of flight: Descent, Approach, and Land. During descent (10,000' to 4,000'), the PF controls the aircraft autopilot using the MCP and the PNF is primarily responsible for radio communications, checklists, and crosschecking. During approach (4,000' to 650'), the crew configures for the aircraft for landing by progressively lowering flaps and then the landing gear. At the Final Approach Fix (FAF), the PNF radios Tower Control, to obtain landing clearance. In the Land Phase (650' to touchdown (TD)) the crew prepares to land the aircraft. After obtaining a visual identification of the runway, the PF disconnects the autopilot and flies the aircraft to touchdown on the runway.

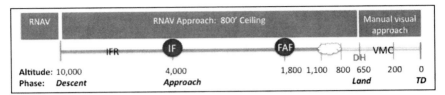

Notes: DH = Decision Height; FAF = Final Approach Fix; IF = Initial Fix; IFR = Instrument Flight Rules; RNAV = Area Navigation; TD = touchdown; VMC = Visual Meteorological Conditions.

Fig. 2. Baseline RNAV model of approach and land

The task model is composed of major procedures that are then broken down into a set of task primitives at a fine-grained level of fidelity. For example, the task of pressing a button on the MCP is translated into the following sequence of behavioral primitives: *reach, push and release, return arm*. These are then translated into MIDAS' Micro Saint Sharp task network structure. The model was composed of over 970 tasks including environment parameters and flight crew or ATC tasks.

Verifying the Model. To verify that the model was implemented error-free, a new task analysis was reverse-engineered from the model output of task begin and end times and this was compared to the original task analysis. The reverse engineering process culminated in a list of pilot tasks and associated task times and sequence. This was evaluated by an independent pilot, not involved in the initial model development process, for accuracy and completeness.

2.2   Validating the Model of Flight Deck Operations

Model validity can be considered from many different perspectives [4,5] including evaluation of model inputs and model outputs. Model inputs refer to the task trace and model input parameters such as task times and task loads, whereas model outputs refer to operator performance measures, such as visual fixations, workload, or situation awareness. HPMs of complex operations should be evaluated using multiple measures that address varying levels of fidelity. Relying only on output validation, (also referred to as results validation) as the sole measure as is frequently the case is insufficient because there is no guarantee that the model represents the operators' tasks or cognitive processes accurately [2,3,15]. Indeed, it is possible that one could make parameter manipulations until the model output fits the data, while misrepresenting the sequence or order of tasks, the workload associated with carrying out the individual tasks, or the way the operator processes the information from the environment [3]. In this case, the model does not validly represent the pilot's tasks and may lead to invalid conclusions when the model is extended to new scenarios, tasks, or environments.

The MIDAS approach for model validation is presented in Figure 3. This methodical approach is multi-dimensional using multiple variables at varying levels of resolution. The validation process involves three validation components: validation of the inputs, validation of the process models, and validation of the model outputs. The input and output components are scenario-specific while the architecture process models are general and not specific to the domain application model (i.e. perception is perception no matter if the task is driving or flying). Note that validation of the process models themselves, including workload management, perception, and visual attention are important aspects of the validation progression. These process models within MIDAS have been previously validated (as discussed in Sect 1.1) and are held constant across domain applications, so will not be discussed further here.

The application model of flight deck operations was validated by a thorough evaluation of *model inputs*, including the task trace and model input parameters required for the workload and visual attention models and *model outputs* including workload and visual fixation (percent dwell time; PDT). The process was iterative, in that the model was refined based on the input validation process and the output were validated by comparing the refined model to empirical HITL data.

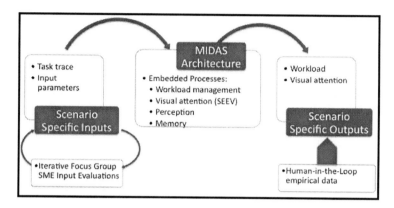

Fig. 3. Comprehensive model validation approach

2.3 Input Validation

Our validation of the model inputs included two aspects. First, a formal validation of the task trace was conducted to determine the extent to which the modeled tasks represent the pilots' actual tasks. Second, a formal analysis was conducted to determine the validity of the model input parameters of workload assigned to the basic task primitives. MIDAS uses behavioral primitives that contain workload estimates based on the Task Analysis and WorkLoad (TAWL) index [16]. These values are based on inputs from military rotorcraft pilots, and have not previously been validated by commercial pilots, for the task of conducting approach and landing tasks in fixed-wing aircraft.

Method. Two full-day focus group sessions were conducted to evaluate the validity of the model inputs. Each session was composed of four pilots. The eight pilots (six Captains and 2 First Officers) were current commercial pilots of glass-cockpit aircraft (M=1,317 flight hours), with RNAV-approach experience. Using a scenario-based format, the focus group pilots conducted a cognitive walkthrough of a typical approach-and-land scenario, starting at 10,000' and continuing to touchdown. Pilots were asked to consider the tasks required of the PF and the PNF, and the nature of communications between ATC and the pilots.

The PF, PNF and ATC tasks as modeled in MIDAS were presented on a worksheet and each pilot was asked to independently review the task and identify any tasks that were assigned to the incorrect operator, occurred in the incorrect sequence, or at an incorrect altitude or navigation marker, and identify tasks that were missing. Upon completion of the worksheet, the pilots discussed their evaluations in a semi-structured round-table format and the source of any discrepancies among pilots was identified and resolved. Differences among tasks were attributed to differences due to aircraft type, airline, pilot technique, and airport/airspace procedures.

Next, the pilots were trained to estimate task workload along five dimensions (Visual, Auditory, Cognitive, Speech and Motor) using the 7-point modified TAWL scale with behavioral anchors [16,17]. The pilots were asked to first identify the workload dimensions that were applicable for the given task, and then estimate the workload for each relevant dimension using the 7-point scale. Two categories of behavioral primitives were evaluated: 1) Basic behavioral primitives existing in MIDAS based on the TAWL that were deemed a valid representation for rotorcraft operations, and, 2) RNAV model-specific behavioral primitives, which had not been previously validated. The basic behavioral primitives served as a baseline upon which to evaluate whether the focus group pilots' workload estimates were comparable to those MIDAS behavioral primitives that were based on the previously validated TAWL scale.

Results

Task Trace Validation. Out of 74 pilot procedure tasks in the MIDAS RNAV application model, the focus group pilots identified 12 tasks that should be removed, reordered, or added. In addition, pilot-ATC phraseology was refined to better reflect actual operations. Incorrectly representing the communication length or the information contained within the communication results in misestimates of workload and task time to reach comprehension. The MIDAS input model was modified to reflect these changes.

Input Parameter Validation. For each task, the mean estimated workload for each workload dimension was compared to the MIDAS input parameter, with the constraint that at least six of the eight focus group pilots determined that the dimension was relevant for the task. One sample t-tests were conducted, which compared the mean focus group rating to the MIDAS value. Significant results indicated that the pilots' estimated workload values were significantly different than the MIDAS values. Thirty-nine tasks were rated on the visual, auditory, cognitive, and motor dimensions (as relevant for the task) resulting in 75 ratings.

Three of the behavioral primitives were established, previously validated, MIDAS behavioral primitives. They were *push-and-release, reach object, and say message.* For all three primitives, the focus-group mean ratings did not differ significantly from the existing MIDAS ratings (*push-and-release* - $t_{visual}(7)=4.2$, p>.05, $t_{motor}(7)=12.19$, p>.05; *reach object* – $t_{visual}(7)=.306$, p>.05, $t_{cognitive}(7)=1.17$, p>.05, $t_{motor}(6)=1.37$, p>.05; *say message* - $t_{cognitive}(7)=.877$, p>.05). This is evidence that the focus group pilots were trained sufficiently on the TAWL scale to produce answers in accordance with the TAWL, thus providing confidence in the pilots' ratings for the non-TAWL primitives, as discussed next.

The initial model mapped four of the pilots' tasks (*set speed, set flaps, set gear, and tune radio frequency*) to the *push-and-release* task primitive. However, the focus-group ratings of these input parameters revealed that each possessed unique workload properties that differed significantly from *push-and-release* (*set speed* - $t_{visual}(7)=3.5$, $p<.05$; $t_{cognitive}(6)=1.7$, $p<.05$; *set flaps* - $t_{motor}(7)=4.5$, $p<.05$; *set gear* - $t_{motor}(7)=4.9$, $p<.05$; *tune radio frequency* - $t_{visual}(7)=4.2$, $p<.05$; $t_{motor}(7)=12.1$, $p<.05$;). Unique task primitives were developed for each of these tasks using the mean focus-group ratings.

The baseline RNAV model required three new primitives, not contained in the TAWL, which were specific to approach-and land operations in commercial fixed-wing aircraft. These were: *visually acquire runway; manipulate yoke; manipulate pedals*. Table 1 presents the primitive workload values implemented in MIDAS based on the focus group estimates.

Table 1. Validated MIDAS workload primitives for the RNAV model

Task	Visual	Cognitive-		Motor		
		Spatial	Verbal	Fine	Gross	Voice
Push and Release	3.7	1.2		2.2		
Reach Object	3.7	1.2			2.6	
Say Message			5.3			4.5
Set Speed	3.9	3.0		5.0		
Set Flaps	4.0	2.2			4.3	
Set Gear	4.0	2.0			4.7	
Tune Radio	4.5	2.7		5.4		
Acquire Runway	5	3.7				
Manipulate Yoke	1.2	5.9		1.3	1.3	
Manipulate Pedals		5.1			3.2	

Note: These tasks do not contain an auditory component.

2.4   Output Validation

In the output validation phase, the model outputs of workload and dwell percentage from 10 monte carlo model runs were compared to empirical data from the existing literature. This phase was completed after all of the inputs into the HPM were modified based on the task trace and parameter input analyses described previously.

Method. The baseline RNAV model's predicted workload and percent dwell time (PDT) data were compared to empirical data from independent HITL simulations available in the literature. Statistical correlation tests were conducted to evaluate the goodness-of-fit between the model and HITL data. For all analyses, only the PF data are shown.

A survey of the literature was conducted to identify relevant HITL data sources from commercial pilots flying approach-and-land scenarios in a glass cockpit in either an actual flight test or a high-fidelity flight simulator. One HITL study was identified as a suitable comparison for the workload data [18]. This medium-fidelity HITL simulation was previously conducted for a different model validation effort, and as such was unique in that it provided workload estimates from three commercial pilots using the TAWL scale, for the three phases of flight modeled in the current baseline RNAV model. Three additional HITL studies [19,20,21] were identified as suitable comparisons for the visual fixation data. Each study included commercial pilots, flying Instrument Landing System (ILS) approach-and-land scenarios.

Results

Workload. Figure 4 (left) presents the overall workload as predicted by the MIDAS RNAV model and estimated by the pilots in the HITL simulation [18] for each of three phases of flight (descent, approach, and land). For the model data, overall workload was calculated as the mean of the individual workload channels (Visual, Auditory, Cognitive, and Motor) within each phase of flight from 10 monte carlo runs. The HITL data are the mean of the three subjects' subjective estimates of Overall Workload from a nominal baseline IMC scenario. As can be seen, the HPM and HITL data are positively correlated (r^2=.63). The model data tended to over-predict workload during the landing phase. It should be noted that the HPM simulation was a medium-fidelity simulation, and lacked the high-fidelity representations of the instrumentation and controls, which may have actually lead to an under-estimation of workload by the actual pilots in the HITL simulation.

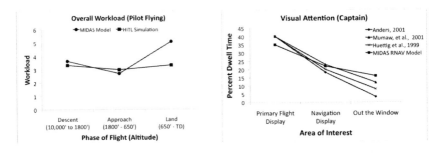

Fig. 4. Output validation: Overall workload (left), visual fixations (right)

Visual Fixations (Percent Dwell Time). Figure 4 (right) presents the model fit of PDT on three areas of interest: Primary Flight Display (PFD), Navigation (Nav) Display, and Out-the-Window (OTW). The model represents the PDT to each display for the entire scenario (10,000' to touchdown) averaged over 10 monte carlo runs. These data were compared to the three separate HITL data sets [19,20,21] from approximately the same phase of flight. There was a strong positive correlation (r^2 = .96) between the RNAV model PDT output and the average of the three HITL studies. This is evidence that both the model inputs and the SEEV process model, which guides visual attention, are related linearly.

3   Discussion and Summary

A HPM of commercial airline pilots conducting approach-and-land procedures was created using the MIDAS software following a methodical development and validation approach. The premise that guided the current work was that model validity is a process, not solely a single value at the conclusion of a model development effort. Valid inputs lead to valid outputs. Conducting only one of these validation processes may lead to invalid models. This is especially true as the complexity of the operational environment and tasks increase.

The pilot focus groups were instrumental in defining valid model inputs. The scenario-based cognitive walkthrough approach captured the context of operations well and enabled the pilots to easily identify tasks that depend on specific phases of flight, and augment the environmental considerations that are used to drive the model's performance.

The workload associated with the behavioral primitives was evaluated with some degree of success. This effort illustrated that MIDAS workload primitives, derived directly from the TAWL, were valid as evaluated by the focus-group pilots. Context-specific workload primitives were modified based on pilot input.

The model output correlated strongly with multiple independent human-in-the-loop simulation studies. These output validation results provide further evidence that the model inputs and the workload and SEEV process model are valid.

In summary, the methodical and comprehensive model validation effort presented in this paper illustrates a candidate process for developing and validating HPMs. This valid current-day RNAV model will next be extended to evaluate the impact of potential NextGen CSPO concepts.

Acknowledgements. The composition of this work was supported by the Federal Aviation Administration (FAA)/NASA Inter Agency Agreement DTFAWA-10-X-80005 Annex 5 (FAA POCs Dr. Tom McCloy, Mr. Dan Hershler; NASA POC Dr. David C. Foyle). The authors would like to thank Connie Socash, Dr. Christopher Wickens, Dr. Andrew M. Gacy, and Mala Gosakan from Alion Science and Technology (MAAD Operations) for their invaluable model development support and all reviewers for their insightful comments.

References

1. JPDO, Joint Planning and Development Office: Concept of operations for the next generation air transportation system. In: Joint Planning and Development Office (Ed.), vol. 3. JPDO, Washington, DC (2009)
2. Gore, B.F.: Chapter 32: Human Performance: Evaluating the Cognitive Aspects. In: Duffy, V. (ed.) Handbook of Digital Human Modeling. Taylor and Francis/CRC Press, NJ (2008)
3. Hooey, B.L., Foyle, D.C.: Advancing the state of the art of human performance models to improve aviation safety. In: Hooey, B.L., Foyle, D.C. (eds.) Human Performance Modeling in Aviation, pp. 321–349. CRC Press/Taylor & Francis, Boca Raton (2008)
4. Gore, B.F.: An emergent behavior model of complex human-system performance: An aviation surface related application. VDI Bericht 1675, 313–328 (2002)
5. Gore, B.F.: Man-machine integration design and analysis system (MIDAS) v5: Augmentations, motivations, and directions for aeronautics applications. In: Cacciabu, P.C., Hjalmdahl, M., Luedtke, A., Riccioli, C. (eds.) Human Modelling in Assisted Transportation. Springer, Heidelberg (2010)
6. Wickens, C.D., McCarley, J.M.: Applied Attention Theory. Taylor and Francis/CRC Press, Boca Raton (2008)

7. Gore, B.F., Hooey, B.L., Wickens, C.D., Scott-Nash, S.: A computational implementation of a human attention guiding mechanism in MIDAS v5. In: Duffy, V.G. (ed.) ICDHM 2009. LNCS, vol. 5620, pp. 237–246. Springer, Heidelberg (2009)
8. Arditi, A., Azueta, S.: Visualization of 2-D and 3-D aspects of human binocular vision. Paper Presented at the Society for Information Display International Symposium (1992)
9. Ericsson, K.A., Kintsch, W.: Long-term working memory. Psychological Review 102(2), 211–245 (1995)
10. Gugerty, L.: Evidence from a partial report task for forgetting in dynamic spatial memory. Human Factors 40(3), 498–508 (1998)
11. Fitts, P.M.: The information capacity of the human motor system in controlling the amplitude of movement. Journal of Experimental Psychology 47, 381–391 (1954)
12. Gore, B.F., Hooey, B.L., Haan, N., Socash, C., Gacy, M., Wickens, C.D., et al.: Evaluating NextGen Closely Spaced Parallel Operations Concepts with Human Performance Models. NASA Ames Research Center, Moffett Field (2011)
13. Keller, J.W., Leiden, K., Small, R.: Cognitive task analysis of commercial jet aircraft pilots during instrument approaches for baseline and synthetic vision displays. In: Foyle, D.C., Goodman, A., Hooey, B.L. (eds.) Aviation Safety Program Conference on Human Performance Modeling of Approach and Landing with Augmented Displays (NASA/CP-2003-212267). NASA Ames Research Center, Moffett Field (2003)
14. Gore, B.F., Hooey, B.L., Salud, E., Wickens, C.D., Sebok, A., Hutchins, S., Koenecke, C., Bzostek, J.: Identification Of Nextgen Air Traffic Control and Pilot Performance Parameters for Human Performance Model Development in the Transitional Airspace. In: NASA Final Report. ROA 2006 NRA # NNX08AE87A, SJSU, San Jose (2009)
15. Campbell, G.E., Bolton, A.E.: HBR validation: interpreting lessons learned from multiple academic disciplines, applied communities, and the AMBR project. In: Gluck, K.A., Pew, R.W. (eds.) Modeling Human Behavior with Integrated Cognitive Architectures: Comparison, Evaluation and Validation, pp. 365–395. Lawrence Erlbaum & Associates, New Jersey (2005)
16. McCracken, J.H., Aldrich, T.B.: Analysis of Selected LHX Mission Functions: Implications for Operator Workload and System Automation Goals (Technical note ASI 479-024-84(b)). Anacapa Sciences, Inc. (1984)
17. Hamilton, D.B., Bierbaum, C.R.: Operator Workload Predictions for the Revised AH-64A Workload Prediction Model: Volume I: Summary Report (AD-254 198). Anacapa Sciences, Inc., Alabama (1992)
18. Hooey, B.L., Foyle, D.C.: Aviation Safety Studies: Taxi Navigation Errors and Synthetic Vision System Operations. In: Hooey, B.L., Foyle, D.C. (eds.) Human Performance Modeling in Aviation, pp. 321–349. CRC Press/Taylor & Francis, Boca Raton (2008)
19. Anders, G.: Pilot's Attention Allocation during Approach and Landing: Eye- and Head-Tracking Research in an A330 Full Flight Simulator. In: Proceedings of the 11th International Symposium on Aviation Psychology, Columbus, OH, USA (2001)
20. Mumaw, R.J., Sarter, N., Wickens, C.D.: Analysis of pilots' monitoring and performance on an automated flight deck. In: Proceedings of the 11th International Symposium on Aviation Psychology. The Ohio State University, Columbus (2001)
21. Hüettig, G., Anders, G., Tautz, A.: Mode Awareness in a modern Glass Cockpit – Attention Allocation to Mode Information. In: Proceedings of the 10th International Symposium on Aviation Psychology, Columbus, OH, USA (1999)

Building Human Profile by Aggregation of Activities
Application to Aeronautics Safety

Laurent Chaudron[1], David Guéron[1], Nicolas Maille[1], and Jean Caussanel[2]

[1] Onera/DCSD & DCSP, Base Aérienne 701, 13661 SALON AIR CEDEX
prenom.nom@onera.fr
[2] LSIS/CODEP – UMR CNRS 6168, Domaine Universitaire de Saint-Jérôme, 13397
MARSEILLE CEDEX 20
jean.caussanel@lsis.org

Abstract. The work related here is devoted to the setup of a methodology regarding the study of polyvalent objects about which our knowledge is incomplete.

It is concerned with the analysis and characterization of flights/flight maneuvers, considered from the standpoint of the involved human operators. The two following issues have been identified: 1) *incompleteness*, which comes from the second-hand nature of the recorded data that describes and situates the pilot's activity, and 2) *variability* of human sensations and reactions, as a result of which identical stimulations may cause different reactions and different observations may correspond to identical sensations/situations.

Our aim is not to close up on the theoretical mechanisms of perception and preference but, based on these mechanisms, to obtain a behavioural model that will be used 1) to characterize observed patterns amongst the various recorded data, 2) to anticipate the patterns to be observed and to relate them to particular flight conditions.

We introduce the three-step process of supervised aggregation, an aggregation driven by experts and expertise, which we successfully put into practice in the case of elementary turns. This process was developed aiming to convey characterizing and predictive power, notwithstanding the incompleteness and variability of observable data.

1 Framework of the Study

From the sets of recorded parameters during these 10 flights, could you come up with a *"mean"* flight ? (Captain D., Mirage squadron — March 22nd, 2005)

The above is a straightforward statement of the goal pursued in the project Activity Modeling and Aggregation of Flight Profiles, 2008-2011. More precisely, by *"mean"* flight, we mean a *representative* flight, that will contain the important and distinctive features of the 10 original flights. This project, in which flights

M. Kurosu (Ed.): Human Centered Design, HCII 2011, LNCS 6776, pp. 389–396, 2011.

are considered according to displayed human activity, attempts a *descriptive* rather than *evaluative* approach.

Because flights, and hence flight recordings, are complex objects, the latter consisting of several figures related to each other in various ways, one cannot generate an aggregate description by simply averaging the data recorded during each individual flight: this data has to be preliminary processed into figures and modes relevantly suited to feed some aggregation procedures.

In this project, we consider flights (the studied objects) from the standpoint of the human activity (mainly that of the pilot, but possibly also that of other crew members and operators), which it originates from. In the analysis that follows, it is essential not to jumble up the data used to depict the piloting activity and this activity itself.

1.1   Issues

While configuring the Supervised Aggregation Process (SAP), we acknowledged the need to reckon with the two important issues examined below.

Incompleteness: The first issue is the incomplete nature of any description, which can either be due to lack of information (some meaningful parameters, such as speed or outdoor temperature, have not been recorded) or to an important parameter being implicitly contained in the recorded data. The first shortcoming cannot be avoided and one must keep constantly aware that meaningful data may be lacking or incomplete. There are two ways to deal with the second shortcoming: 1) first, any implicit parameter can be explicitly added to the description of a flight, 2) second, the aggregation procedure can be selected so as to keep invariant the important parameter. To the benefit of the latter solution, it doesn't require the user to modify the objects to aggregate.

Variability: Concerning processes involving human beings (or more generally living beings), an additional reason for keeping the distinction between the studied objects and the characters signifying them is the variability of the reactions of living organisms.

In [3], Chevallard relates such variability to the learning process, during which knowledge stems from a progressive unification of different representations.

1.2   A Main Tool: Probabilities

The first and one of the main tools developed for the description of polyvalent states and characters, is the science of probabilities. A probability is a real number between 0 and 1 attached to any state of the investigated system, that can be considered as standing for the "chance" for this state to be observed.

Next, one is led to consider different kinds of probabilities, as already advocated by the French economist Augustin Cournot (1801-1877). A descriptive probability, used to depict polyvalent states, is somewhat different from a probability describing ignorance or from one used to depict "randomness" (for an introduction to the connections between randomness and ignorance, see [4]).

Lastly, probabilities may also be used to shade the overly precise descriptions of an uncertain phenomenon that numbers are bound to provide. To this end, several models of imprecise probabilities (Dempster-Shaffer Model, Transferable Belief Model, Capacity Model... see [5] for an overall introduction) have been developed.

To sum up this brief review on probabilities, it appeared to us that they don't possess the precision and univocity that is usually attached to figures, and that seems necessary in order to aggregate a set of objects into an object of the *same kind*. Indeed, in order to achieve this goal, certain choices have to be made that cannot be made without some kind of *meta*-knowledge on the situation. For instance, in order to specify an average speed for a set of maneuvers in which these average speeds are regularly spaced, rather than selecting any specific numeric mean, it might be observed that these speeds all are functionnally related to (a combination of) some other recorded data, such as aircraft's weight and visibility conditions.

Hence, we realize that the outcome of the aggregation of a set of complex objects (to which flights certainly belong) can't end up in a similar object (another "mean" flight), but will result in a set of (functionnal) relations concerning the characters describing the objects, and shared by the latter (a set of *common features*).

Cournot's standpoint over probabilities, highly original for the time, has nowadays attained overall acknowledgement, as can be seen in Barberà's work hereinafter cited.

2   The Supervised Aggregation Process

2.1   Definition

We offer to carry on the three-step SAP, which we developed as an extension of Barberà's preference aggregation process [1]:

1) **decomposition**
2) **maieutics**
3) **reconstruction**

Here, "*supervised*" means that, at each step, the constraints will be set up by expertise: *e.g.* the number and type of characteristics (in the decomposition step), the number of expected rules (maieutic step), as well as the selection and combination of these rules (reconstruction step).

As in Barberà's process, the first step, decomposition, consists in expressing the studied objects using a convenient (from the expert point of view) set of decomposition characters.

It's in the second step, maieutics, that our SAP differs: in Barberà's aggregation process, this step consists in "applying mean operations" to each (or some) of the decomposition character, to end up with similar characters. However, as we indicated above and established earlier in [6], the outcome of the second step should be a meta-object rather than a mere set of means, which would be similar to the characteristics resulting from the first step.

In other words the aggregation of a set of items should lead to a higher-order object that would signify the essential properties shared within the studied set: the aggregation of activities should end up in a "way of doing", the aggregation of flight data files should consist in a set of rules defining the piloting properties shared by these flights.

The last step of the aggregation process, reconstruction, consists in a relevant use of the several rules and relations that sprang out of the maieutic step: i.e. an inference engine, or a sequence of rules defining the sought procedure.

Thus, the SAP allows the design of *specific ways for doing things*, from a set of actual exemplary activities. In the sequel, a static version of the SAP is summarized, before a version of the full dynamic process is introduced.

2.2   Static Aggregation Based on Galois Lattice Induction

In [2], we introduced the Galois Lattice Induction formalism for describing the complete aggregation process of the stimulus/response database generated by a human subject. Such a process enables the construction of a "mean" human activity profile, that we call the "Virtual Subject".

Because only the "stimulus–response" transition is aggregated, this aggregation is said *static*. Its three phases are: 1) decomposition: each stimulus and each response is signified by a vector of characteristics. The set of responses given by the subjects S, each submitted to every stimulus of the set of stimuli St, consitutes the experimental database E, 2) maieutics: this stage consists in running an "association rule process" (a non-supervised rule-induction method) on the database E. A subset of rules, $\Lambda' \subset \Lambda$, whose premises are the properties of the stimulus and whose conclusions are the obtained responses, are then filtered according to the values of their confidence and portance. The output of the maieutic step is a knowledge-base K, i.e. a set of rules relating properties of the environment to observed responses, together with the related subsidiary information (particular cases, portance and confidence...), 3) reconstruction: a classical inference engine runs on K so as to simulate an "aggregated subject".

Formal model for the Static Agregation: let us consider a set of subjects $S = \{S_n\}$, $n \in [1, N]$ that are asked to evaluate a given corpus of phenomena. Each phenomenon is being described through a spectrum of parameters $P = \{p_i\}$, $i \in [1, I]$. The phenomena database is a corpus of events $E = \{E_m\}$, $m \in [1, M]$ (eg: a set of aircraft sounds). Naturally, each E_m is decomposed on the spectrum of parameters $P : E_m \rightarrow (p_{m,1}, \ldots, p_{m,i}, \ldots, p_{m,I})$. The evaluations/reactions of the subjects are measured by a set of evaluation characteristics, $V = \{V_j\}$, $j \in [1, J]$.

The outcome of such an experiment is a set of items defined as follows: given one subject S_n and one event E_m, the final results of the subject's evaluation is the vector

$$\left(\mathbf{v}_{(n,m,1)}, \mathbf{v}_{(n,m,2)}, \ldots, \mathbf{v}_{(n,m,j)}, \ldots, \mathbf{v}_{(n,m,J)}\right) \ .$$

Using the corpus correspondence between the events and their parameters, it is easy to consider that the experiment is a relational database of $N \times M$ items: *(event, subject, event description parameters, subject's evaluations)*.

Formally this *"Experimental Database"* is a set of identifiers *event × subject*, say ES, bearing out a list of values :

$$\left(\mathrm{ES}_{m,n}, \{\mathrm{p}_{m,1}, \ldots, \mathrm{p}_{m,i}, \ldots, \mathrm{p}_{m,I}, \mathbf{v}_{(n,m,1)}, \mathbf{v}_{(n,m,2)}, \ldots, \mathbf{v}_{(n,m,j)}, \ldots, \mathbf{v}_{(n,m,J)}\right) \ .$$

Association rules. Such a database is a natural candidate to be analysed by Galois Lattice theory.

Indeed a large body of literature about *"association rules"* supported by Galois Lattices exists. This theory allows to generate a rule database Λ^* containing all the rules of the template: *(rule identifier, confidence, portance, conclusion, list of premises)*. The confidence c is defined as the proportion of *event × subject* that verify the rule among those that verify all the premises in the *"Experimental DataBase"* (conversely, $1 - c$ is the proportion of counter-examples). The portance, p, is the proportion of such events, though this time taken amongst the whole database (thus $0 < p < c < 1$). Many algorithms exist so as to extensively unearth these rules (and generally considering only those for which $0.5 < c$).

The *"association rules"* approach is known to be the most generic non-supervised rule induction method.

Λ^* is defined as the knowledge-base resulting from the aggregation process (to the c and p thresholds) of the responses given by the subjects $S = \{S_n\}$, $n \in [1, N]$ to the events $E = \{E_m\}$, $m \in [1, M]$.

The Virtual Subject. The final step is to define the *"Virtual Subject"*: a classical theorem prover can run on this Λ^* knowledge-base, so as to define a new functional relation between any new event e (qualified by the list of values of its I parameters) and the derived evaluation values $V_e = (v_i)_{i \in [1,J]}$ proved by the prover ran on Λ^*. We use the fact that each characteristic v_i of the obtained evaluation can then be proved following different proof-chainings to quantify and qualify the rules, with the use of three figures: c_m is the mean confidence, p_m the mean portance (both are quite classical), and pp is the **proof-power** of v_i, *i.e.* the number of proof-chains that result in v_i.

The first application of this induction/deduction loop has been made so as to define a *"Virtual Resident"* (*i.e.* the aggregation of 320 subjects submitted to 100 aircraft sounds). The induction/deduction is programmed in constrained Prolog (*i.e.* also: self-proved programs).

The Static Aggregation methodology is currently studied for application to various kinds of human driven activities, leading to the characterization of interfaces, or to the establishment counciousness characteristics in UAV (Unmaned Aircraft Vehicles) piloting, among others.

However, Galois Lattices are not suitable for cathcing dynamic activities such as the piloting activity itself, for which variational approaches must be used. The purpose of the next section is to sketch out such approaches.

2.3   Supervised Aggregation of Flights: From FDR to Flight Profiles

The first corpus that we used is a set of tactical simulated flights (helicopters) demonstrating a right-or-left turn engaged so as to avoid an obstacle. The second corpus is a set of 20 commercial aircraft flights on which the SAP is blind-tested, as what has to be discovered remains unknown, except from the dates of certain key points.

Here the SAP for flight profiles is developed through a Lagrangian variational methodology: a given flight is decomposed through a set of 83 flight parameters, including time, thus enabling us to consider it as being a trajectory in a state space. The details of the software development for the methodology, that we have implemented in Java, will not be developed in this paper.

The maieutic step consists in considering an N-tuple of pivotal key points given by expertise, $(Kp_1, Kp_2 \ldots)$.

Table 1. Expertise format

Flight	dates			
	Key Point 1	Key Point 2	Key Point 3	...
Flight 1	Kp_1^1	Kp_2^1	Kp_3^1	...
Flight 2	Kp_1^2	Kp_2^2	Kp_3^2	...
...	...	...	...	...

Learning process: selection of the best rule accounting for a keypoint

Description of the learning process. The variationnal method that we implemented consisted in assuming that the observed key points matched the variation (increase or decrease) of some parameter: for each of the 82 parameters (the 83 recorded parameters minus TIME) and for each of the 6 keypoints, we computed the thresholds (cTh and sTh) of the rules of increase and decrease for this parameter, that would account for this keypoint.

The rule best describing "*when and where things are changing in a relevant way*" could then be selected and used to identify the example maneuvers on new flights.

For the first corpus, the rules generated by the SAP are consistent with what was expected from knowledge of flights-mechanics: indeed, the primary effect of a turn is the increase of the aircraft's banking attitude.

The second corpus, that we used more thoroughly, led to the following findings:

- If we obtain several possible rules reporting for the first keypoint, and a bit less for the second, we hardly get any for further keypoints, and none for the last. Moreover, using the thus-obtained rules to find out the latest keypoints appeared hopeless.

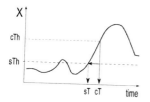 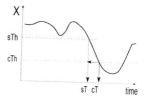

Fig. 1. Identifying the increase and the decrease of a measured parameter, X

- Concerning the first key point (the take-off), the best-evaluated rules were, in order, increases of the lateral wind, the banking attitude, and of two altitude parameters. Among these, only the first is not relevant to a physical phenomenon but due to the fact that the measure of the lateral wind started from the time the airplane's undercarriage had left the ground.
- As for the second keypoint (retraction of the landing gear), the two selected rule were an increase of altitude (which the expert dismissed, as it was due to the coincidence that the landing was retracted at roughly the same heights by the different pilots) and a decrease of incidence (which was validated by the expert as being due to a physical phenomenon related to the studied maneuver).

Further developments. Notwithstanding the fact that the two rules studied are quite simple, they also are basic, meaning that a huge variety of more complicated rules can be obtained from these two rules by suitable combinations.

In order to make our rule-production process more robust and wider-ranged, it therefore appeared important to:

1. implement the possibility of restricting the time-windows when a rule is looked for or is applied,
2. use expert-knowledge to discard some rules (during the phase of rule-identification),
3. develop a wider set of rules, in particular rules to be applied to discrete (non-numeric) parameters,
4. setup a way of combining several elementary rules, over different parameters, to have a complete set of complex rules at our disposal

The reconstruction step. that is currrently being processed, relies on expertise in order to provide results.

3   Conclusion

In this paper we described the main components of a Supervised Aggregation Process (SAP), philosophically grounded on the mathematical tool of probabilities, whose modern acceptation is issued of Cournot's work on the aggregation concept. The SAP model here depicted is an extension of the economical formalism developed by Salvador Barberà in the form of a three-step process: 1) decomposition, 2) maieutics, 3) reconstruction.

The touchstone of the SAP is the production, from a set of elementary measures, of a meta-object, *the rule* that constitutes the structure of the studied field. Its use for the study of human activity has been described, its main application being the semi-automatic aggregation of exemplary flights in a procedure defining them. Java programs are currently under development so as to achieve more relevant results. The application to new flight safety domains is expected soon.

References

1. Barberà, S.: A Theorem on Preference Aggregation. Centre de Referència en Economia Analítica, Barcelona Economics Working Paper Series edn. Working Paper 166 (2006)
2. Chaudron, L., Guéron, D.: Virtual Subject Aggregation based on Galois Lattice Induction. In: 38th Conf. of the European Mathematical Psychology Group, EMPG 2007, September 10-12, University of Luxembourg (2007)
3. Chevallard, Y., Wozniak, F.: Teaching statistics in high-school: mixed mathematics to convey the notion of variability. In: Mercier, A., Margolinas, C. (eds.) Balises pour la didactique des mathématiques. XIIᵉ école d'été de didactique des mathématiques, La pensée sauvage, Grenoble, August 20-29 (2005); French title: Enseigner la statistique au secondaire : des mathématiques mixtes pour penser la variabilité
4. Dubois, D.: Uncertainty: a Unified View. In: IEEE Conference on Cybernetic Systems, Dublin (Ireland) (September 2007) (invited talk)
5. Dubois, D., Prade, H.: Merging vague information. TS. Traitement du signal 11(6), 431–586 (1994); ISSN 0765-0019, French title: La fusion d'informations imprécises
6. Guéron, D., Chaudron, L., Caussanel, J., Maille, N.: Aggregation of activities: discovering flight profiles starting from recorded parameters. In: INFORSID, Atelier ICT (May 27, 2010), French title: Agrégation d'activités, découverte de profils de vol à partir de traces de paramètres

Building a Shared Cross-Cultural Learning Community for Visual Communication Design Education

Takahito Kamihira[1], Miho Aoki[2], and Tomoya Nakano[1]

[1] Senshu University, School of Network Information
2-1-1 Higashi Mita, Tama-ku, Kawasaki, Kanagawa 214-8580, Japan
kamihira@isc.senshu-u.ac.jp, ne190003@senshu-u.jp
[2] University of Alaska Fairbanks, Art Department,
P.O. Box 7755640 Fairbanks, AK 99775 USA
maoki3@alaska.edu

Abstract. This paper discusses a case study of an educational online visual communication design project. The project is to develop an online platform, which facilitates cross-cultural communications and educational experiences for college-level students and educators in the visual communication design field in conjunction with information graphics assignments. The online system developed for this project allows students in visual design courses to share their class assignments and evaluate works posted by the members from other countries. The assignments are designed to encourage students to investigate the cultural differences and roles of images in visual communication design. The pedagogical value of the project is evaluated by analyzing user interview and survey results.

Keywords: Cultural Issues and Usability, Human Centered Design, Visual Communication, Infographics, Learning Community, Design Education.

1 Introduction

The importance of visual design for communication has increased in the globalized society. We are surrounded by large volumes of complex information today, but our time and ability to process the information are limited. Humans have used images as a visual language to describe complex subjects [1]. Visual communication design fields have studied effective ways of communication through images. This includes Information Graphics (Infographics) that uses graphics such as charts and maps to represent information [2]. Well-designed Infographics visualize information and enable people to share knowledge beyond language barriers [3][4].

The role of visual communication design has become more significant in many design fields today. The globalization of the economy has increased the necessity of cross-cultural understanding and communication [5]. For example, international social network websites, user manuals for exported automobiles and government publications in multiple languages must cater to users with various languages and cultural backgrounds.

M. Kurosu (Ed.): Human Centered Design, HCII 2011, LNCS 6776, pp. 397–406, 2011.

1.1 Cross Cultural Education in Visual Design

Experiencing other cultures can be very meaningful for students in the visual design field. Designers today must be aware of users with cultural backgrounds different from their own. In order to understand the diverse users, it's crucial for designers to understand cross-cultural issues. Understanding various cultures leads to flexible and innovative thinking in creative process. There is an increasing necessity for studying visual design in multicultural context, but most of college level foreign culture courses focus on language acquisition or general culture studies, especially in Japan. It's critical for visual design programs today to include multicultural components to prepare their students for successful careers in the globalized world.

1.2 Building Online Learning Community

Social Networking Services (SNSs) have been popular in the last decade on the Internet. More than 60% of the Japanese younger generations have used SNSs in 2010 [6]. Some services gained international popularity and have users worldwide. Building and maintaining SNSs with specialized focuses, such as education, has also become easier, because of affordable open-source SNS systems and API.

An SNS for foreign language education, Lang-8 [7] is one successful example of special-focus SNSs. Participants in this service write diaries in the language they are learning, and others who use the language as their native language correct and comment on the diaries. This system is based on reciprocal relationships among the participants.

There are active discussions on the pedagogical role of communities surrounding individual learners in recent years [8]. The social acceptance and technical advancement of the Internet have made forming reciprocal and practical learning communities drastically easier [9].

1.3 Online Community as Educational Feedback System

The most effective way to test one's communication skill is to attempt to transfer information to others and evaluate the efficiency. The development process of communication skills has many things in common with Human Centered Design Process, which incorporates user feedback in the developing process. Appropriate feedback from the others can reveal previously unrecognized problems, help one to seek solutions and improve one's communication skills [10]. Communication skills are developed through this cyclic process of trial, obtaining feedback, reflection and refinement.

Online communities can function as a supporting system for learning activities in traditional classrooms. In visual design education, the cycle is often incomplete, especially in foundation level courses. Many works produced by students are viewed only by classmates and instructors. Often, students don't have the chance or are reluctant to revisit their works after class critiques. Receiving feedback only from the instructors causes students to feel that they are working only for the grade and many lose interest after receiving a grade.

"... I created the image only because it was a class assignment. However, I felt happy when my image communicated ideas (to students in the US). If I knew that someone would use my design, I probably would have felt differently. I get more motivated and become more willing to create something useful." (interview with a Japanese student, 12/28/2010).

Practical projects, involving real-word applications and users, are very beneficial to students, but they can be overwhelming for beginning level students. Expanding the project beyond the traditional classroom setting also requires financial and time commitments from the instructors and schools. Therefore, we need a new pedagogical model that enables students to share their works and gain feedback beyond the classrooms without burdening instructors, schools and themselves. Forming online communities with design courses from other institutions can be one of the solutions. Peer students from other institutions are excellent sources of feedback. The reciprocal relationship in the community can fill the gap in the learning cycle [Fig. 1].

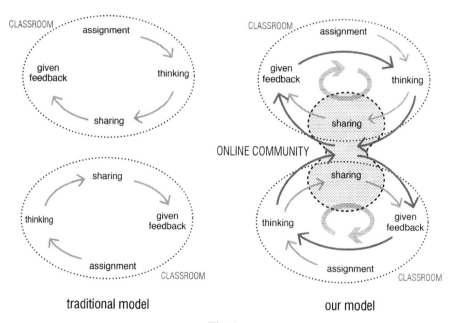

Fig. 1.

2   Project Description

The project is designed to facilitate groups of students and instructors in beginning level graphic design courses in different countries, preferably countries that use different languages. However, the participants have to be able or willing to communicate in one common language. The project consists of two components: classroom and online activities. The classroom component is an assignment which asks them to create an image explaining ideas with minimum use of text. Students are asked to explain a subject that is very common locally but unique to their country or culture. After sharing their images in their classes, students upload the images to the community website. Then students review the works uploaded by their peers abroad and post their interpretations and suggestions. After receiving the initial reviews, students were able to post re-designed works to the site.

In this assignment, students are "designers", and the "users" are the students that live in another country. This helps students to think and express visually and develop the ability to critique their own works. By setting the subject to something unique to their area, the assignment also encourages students to deepen their understanding of their own cultures and country.

The assignment has two design challenges for the students. The first challenge is to convey information visually with a limited amount of text. The second challenge is to select visual grammar and vocabulary that could be understood by the users from different cultures. These challenges are an excellent opportunity to introduce the concept of Infographics and the internationalization issues in visual design.

3 Methods

Two online pilot projects were conducted between two universities, a private university in the area of metropolitan Tokyo in Japan and a state university in the US, located in the interior area of Alaska. The first online project was conducted in April to May 2010 and the second one was from November to December 2010. During and after each project, user interviews and surveys were conducted. We evaluated the pedagogical value of the project by analyzing user experiences.

3.1 In-class Assignments and Reviewing Method

On the Japan side, the assignment was "How to xxx in Japan". Students designed images to explain how to do something very common in Japan but unique for foreign visitors. The students selected from a wide range of subjects, from cooking to religious customs. On the United States side, students were asked to create a graphic temperature scale and explain "How cold is -40°F".

After in-class critique sessions, students uploaded their images to the "Visual Exchange" website. To prevent the reviews from focusing only on initial impressions, we asked students to write their interpretations and analyze the design in the following three points: first impression, interpretations, and suggestions for improvements. After receiving the reviews, it was up to individual students to reply or to revise their images by incorporating the review comments.

3.2 System and User Interface Development

The first test case used a predesigned online social network service. The second case used a customized version of OpenPNE 3.6 (beta), built in php 5.2.14 on Apache 2.0, with a support from the developer of OpenPNE. The system development and interface design incorporated agile and UCD process. We updated the system and design incorporating user feedback during the project period.

The community website had two main pages, the home, which displayed recent overall activities, and the community meeting page, which functioned as a meeting room for the members. More communities can be added for future use. The uploaded images became "topics" under the community page and users added reviews under the uploaded images. Each user comment displayed a profile photo and associated nationality or location by national flags [figure 2].

Fig. 2. A screenshot of "Visual Exchange" website used in Nov. – Dec. 2010 exchange

4 Current Results and Analysis

The first pilot case had 27 participants, and the second one had 38. All posted images received comments including interpretations. This indicates that the images functioned as a communicating tool. Students uploaded 21 images in the first case and 14 images for the second one. The average numbers of reviews and replys posted per image was 5.1 in the first case and 4.6 in the second case.

4.1 Examples of Student Works

Example form Japanese Assignment. This Japanese student design attempted to explain Japanese style futon (sleeping pad) set system and its clearing/airing custom. The student tried to explain how to separate, wash and dry the futon and covers along a timeline. She tried to explain the difference in drying time in summer and winter, but it was not well communicated.

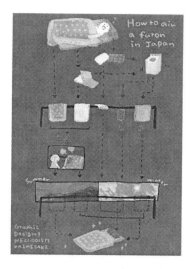

Fig. 3. "How to Air Futon in Japan"

The main confusion students in Alaska had were the seasonal changes of the airing time. One student thought a futon has to be aired from summer until winter while another student thought a half of futon has to be aired in summer and another half in winter. After exchanging ideas, the Japanese student cleaned up the confusing details and indicated the seasons differently.

Example from the US Assignment. This assignment, "How cold is -40°F?" was given to students who live in the interior area of Alaska. Students were given one class to develop their ideas and one week to complete the designs. Figure 4 is one of the images from the assignment.

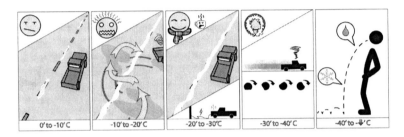

Fig. 4. "It Bounces"

There were six reviews posted by Japanese students for this image. Five reviews were written before the American student posted his explanations. The image attempts to explain the road surface and automobile conditions in various temperature ranges and other phenomena occur in low temperature conditions. The three plates from the left, 0 to -10, -10 to -20, -20 to -30 degrees in Celsius (+32 to +14, +14 to -4, -4 to -22°F) and the plate on the right side, below -40°C (-40°F) were interpreted without major problems. However, no Japanese students understood the plug-in for car heaters featured at the bottom of the central image.

The forth plate, -30 to -40°C (-22 to -40°F), tries to explain "square tires", the phenomenon happen to cars parked outside overnight in cold temperatures. Since this phenomenon was not familiar to Japanese people living in the Tokyo area, it was hard for them to understand.

In the future versions, we would like to build an archive system to keep past student's works and have them available for new participating groups.

4.2   Internationalization and Localization of Design

The reviews posted on the website indicated that students were able to learn issues in the internationalization and localization of visual and interface designs.

Students from the US stated that the right to left order seen in Japanese student works confused them. The right to left order is common in countries that use Japanese, Hebrew and Arabic languages, but not for the countries that use western alphabets. Students learned that the design layout must consider the users' cultural backgrounds.

In the first pilot case, students also learned that some graphic symbols are not universal. A few Japanese students used symbols "o" as good and "x" as prohibited or bad, which is common among Japanese designs, but many American students couldn't understand these symbols, especially in grayscale images.

As already seen in the examples above, objects and phenomena unique to a culture or an area, such as car heater plug-ins, ice fog, snow blowers, shrine gates, futon beating stick and pickled plums were not understood even though there were described visually. *"I've never seen it, so seeing the picture doesn't make me understand what it is." (American student)*

Measurement units were also problems. The student who created the example above, "It Bounces", used Celsius, but other American students used Fahrenheit degrees. Japanese students were not able to understand the temperature intuitively.

4.3 User Feedback

A partially structured interview was conducted with Japanese student participants. The interview was done individually. The following section discusses participants' answers extracted from the interview and its analysis. During the interview, the participants were asked to describe the project's effect on their initial motivation, effect of the feedback on students' view of their works and effect of feedback on students' motivation to revise their works as well as the usability of the website.

Effect of the Project on Students' Initial Motivation. For the Japanese students, this was the first time they communicated with people from abroad on the Internet. One student described the difference between the classroom critiques and the online reviews,

"... My classmates hesitate to say harsh comments, because they are my friends. They don't tell me exactly what they think. On the Visual Exchange site, I didn't know the students in the US. So it was impossible to predict what kind of reviews I would receive. So it was interesting and fun." (12/28/2010, Japanese Student A)

More then a half of students answered that they were curious or eager to communicate with students abroad.

Effect of Peer Feedback on Students' Views on Their Works. Some answers from students suggested that obtaining feedback from people belonging to other cultures can help develop an objective point of view.

"Someone asked me where to put a pickled plum in a rice ball. It's commonsense in Japan, but if you never made it, you don't know that. I realized that there are important things we are not aware of, because we take them for granted. I couldn't imagine that when I was creating the image." (12/28/2010, Japanese Student C)

Many students commented that they were not able to see the assignment project from users' point of view first. The comments similar to the quote above indicate that the online community functions as a feedback system in this cross-cultural context. However, the project needs a scheme for encouraging students to write more productive reviews for works that were already well designed.

"Most of the comments were like, 'everything is nice!' So it's hard to respond. If they pointed out shortcomings of my design, I could have used the comments to improve my work. But I can only say 'thank you' to comments like 'nice!'" (12/28/2010, Japanese Student D)

This is similar to a problem designers and engineers encounter in usability tests. User comments can reveal problematic parts of the designs. However, when there are no obvious problems, many users become unable to suggest a good direction for improvement. Students need to develop the ability to identify strong points of a design and analyze why the design is good. This ability can help students to see their strong points and develop further.

Effect of Feedback on Student Motivation for Revising Their Works. One of the goals of this project was to encourage students to reflect on their initial designs and revise. We asked students how the project affected their motivation to revise their works.

"I didn't want my work to end when I finished creating it. I want to receive more comments. That motivated me to revise my work. It wasn't just about the grade. I just wanted to feel better." (12/28/2010, Japanese Student E).

"If it's an assignment, I try to figure out how much I have to do (to pass). But if I see it as something useful for my study, not an assignment, I would revise it even if it takes time. I would pursue it till it satisfies me." (12/28/2010, Japanese Student F)

5   Current Issues

5.1   Issues in the Project Content and Logistics

During the two pilot projects, the exchange was done in friendly manner. However, cross-cultural communications could cause misunderstandings and conflicts [11]. On the Internet, lack of non-verbal expressions could lead to serious misunderstandings. Therefore, educational projects should be monitored closely by instructors who keep tight communication with each other.

Language barrier had two opposite effects on this project. The language difference motivates students to think and express visually, but it slows down the reviewing process. In both pilot cases, some Japanese students felt reading and writing in English was overwhelming. According to the user survey, more then a half of Japanese participants felt that writing in English was somewhat difficult to very difficult for them and also felt replying to the comments they received was not easy. For the future projects, a reviewing system that relies less on writing needs to be added to increase students' participations.

Scheduling of the assignment and online participation is crucial for keeping students' motivation high. The more time lags between the completion of the initial images and online communications, the lower students' enthusiasm becomes. We need to control the schedule very carefully or need a scheme to reduce the effect of time lags in the future projects.

5.2 Place for Community Building

Building a community requires its members to communicate with each other. Setting an appropriate "space" for the communications is critical to online communities. In these pilot projects, members engaged in communications outside of the main purpose of the projects. Such communications help participants to feel welcome and be more productive member of the community. Some participants exchanged messages about their personal interests and past works. This indicates that there is a strong need for informal communication space. Creating a place for informal exchange accessible to all users is desirable for the future project.

5.3 Interface Design Issues

The interview and web statistics revealed user errors and tendencies, which were hard to predict before the project launch. We were able to make changes quickly during the project period. For example, the interview revealed that many errors in navigation were caused by the two rows of menus placed at the top of the screen. We also changed the list of the uploaded images from text to icon size graphics, but it was not a useful change for the users. Most users were accessing the items from the list of new arrivals. Users were also attracted to images that had higher review numbers. For future versions, we need a new interface design that can direct users to the images with fewer reviews.

6 Conclusion

An online community can be successfully used as educational forum for visual communication design in conjunction with traditional classroom activities. Students participating in Visual Exchange project gained direct cross-cultural experience through exchanging images and reviews. Students were able to learn important issues in today's design field, such as internationalization and localization, communicating visually and cultural awareness. However, simply exchanging images will not yield good educational experience for students. The facilitators have to keep in mind the issues such as the language barrier and scheduling of the assignment. For future development, we hope to develop this project into a platform which can host multiple institutions.

Acknowledgments. We would like to thank the members of Visual Exchange community, Tejimaya Inc., the Senshu University and the University of Alaska Fairbanks. This work was supported by Grants-in-Aid for Scientific Research (Grant Number: 22730702, 2010), from the Ministry of Education, Culture, Sports, Science and Technology of Japan.

References

1. Tufte, E.R.: Envisioning Information. Graphics Press (1990)
2. Wildbur, P., Burke, M.: Information Graphics: Innovative Solutions in Contemporary Design. Thames & Hudson (1999)

3. Neurath, O.: International Picture Language: The First Rules of Isotype. K. Paul, Trench, Trubner & Co., Ltd. (1936)
4. Holmes, N.: Wordless Diagrams, Bloomsbury USA (2005)
5. Aykin, N.: Usability and Internationalization of Information Technology. Lawrence Erlbaum, Mahwah (2005)
6. Global ICT Strategy Bureau, Ministry of Internal Affairs and Communication: Report of Contracted Survey and Research on Social Media Usage (2010), http://www.soumu.go.jp/johotsusintokei/linkdata/h22_05_houkoku.pdf
7. Lang-8, http://lang-8.com/
8. Wenger, E.: Communities of Practice: Learning, Meaning, and Identity. Cambridge University Press, Cambridge (1999)
9. Botsman, R., Rogers, R.: What's Mine is Yours: How Collaborative Consumption is Changing the Way We Live. Harper Business, New York (2010)
10. Schon, D.: The Reflective Practitioner: How Professionals Think In Action. Basic Books, New York (1984)
11. Yamauchi, Y.: An Ethnography on Learning Communities Which Connect a School and Professionals. Japan Journal of Educational Technology 26(4), 299–308 (2003)

Expansion of the System of JSL-Japanese Electronic Dictionary: An Evaluation for the Compound Research System

Tsutomu Kimura[1], Daisuke Hara[2], Kazuyuki Kanda[3], and Kazunari Morimoto[4]

[1] Toyota National College of Technology, 2-1 Eisei-cho, Toyota, Aichi, Japan
[2] Toyota Technological Institute, 2-12-1 Hisakata, Tenpaku-ku, Nagoya, Aichi, Japan
[3] Chukyo University, 101-2, Yagotohonmachi, Showa-ku, Nagoya, Ahchi, Japan
[4] Kyoto Institute of Technology, 1 Hashigami, Matsugasaki, Sakyo-ku, Kyoto, Japan
kim@toyota-ct.ac.jp, daisuke@toyota-ti.ac.jp,
kanda@lets.chukyo-u.ac.jp,
morix@kit.ac.jp

Abstract. We have developed the JSL-Japanese Electronic Dictionary System in which Japanese meaning of a signing was looked in and the corresponding signing video movie was displayed. Our system finds out the target sign through analyzing the phonological components of the sign. We failed to find "e-mail" or "medical doctor" in JSL which are daily used words, because these sings are compounds and the system did not include a compound searching system in it. This paper shows how we developed an enlarged model of the dictionary and result of the evaluation test.

Keywords: sign, Japanese, phoneme, dictionary, compound, database.

1 Introduction

There are many books named sign language dictionary in the market, most of which are Japanese-Japanese Sign Language (JSL below) dictionaries. We find only a few books of JSL-Japanese dictionaries titled *Sign-Japanese Dictionary* [1] and its enlarged version *Enlarged Sign-Japanese Dictionary*[2], and *Practical Sign Dictionary for New Introductory Course of Sign Language*[3]. As for electronic ones, we find only three kinds[4][5][6][7], however, they are not available in the market. The Web version of *Enlarged Sign-Japanese Dictionary*[8] is a pilot version and its content is a part.

In such a situation, a learner can look in the corresponding sign of a Japanese word but not a meaning of a sign. It is equivalent to the situation of a Japanese learner of English without English-Japanese dictionary, even if they have a Japanese-English dictionary.

The four skills of language learning are to be listening, speaking, reading and writing and listening (or 'reading' in sign language) is basic (there is no literal reading and writing in sign language). In sign language learning, sign-Japanese dictionary is not popular and sign language learners are disadvantageous to other language learners at present.

M. Kurosu (Ed.): Human Centered Design, HCII 2011, LNCS 6776, pp. 407–416, 2011.

In the available JSL-Japanese dictionary, we start from choosing a handshape of a sign, and then go to choosing movement or location. That makes us to find a wrong answer if we fail to choose a proper hand. In the printed books, a movement of a sign is showed by a picture which makes us to catch it correctly.

In order to solve the problems, we have developed JSL-Japanese Electronic Dictionary System (EDS below) [9] by which they can look in a meaning of a sign through a phonemic description and capture it correctly using sign movie.

Many signs refer to a single meaning but sometimes there are some combinations of more than two signs means by a single meaning, that is, a compound. We see many of the medical signs and words in e-mail are compounds. Compounds are often used in an actual communication but no EDS includes a compound searching system at present. Therefore we enlarged our system to enable to search a compound effectively and we examined its usability in this paper.

2   Phonemes and Compounds in JSL

A word or a sign is composed of a single or plural syllables, which are the combination of a handshape, a location and a movement or two of those. These elements have no meaning and differentiate a word or a morpheme, so that they are considered to be equal to 'phoneme' of a vocal language in sign linguistics. They are called phoneme in this paper, too.

In Japanese, there are many compounds consisted of plural words, such as *medama-yaki* (fried egg), *tako-yaki* (octopus ball) or *tai-yaki* (fish shaped pancake) and so on. In JSL there also are many examples of compounds such as E-MAIL (KEYBOARD+LETTER) or DOCTOR(PULSE+MAN).

Though they are often used, there was no searching system of compounds in JSL EDS that caused us not to look in the word. We developed it. Here refers the old system below and its enlargement.

3   The Old EDS

Here are the specification of our old EDS [9], and its concept chart in Fig.1 and a sample GUI in Fig.2

1. A user can search a sign intuitively through a phonemic keyword as a handshape, a location or a movement. We offered a specific tab for each phoneme and he can select a phoneme of a sign from any phonemic entry, Handshape, Palm Orientation, Location or Movement.
2. Each time a user select the phoneme, the candidate list of signs which contain the phoneme is indicated at real time. In the initial stage, every phoneme is selected and a full list is indicated. A user deletes the unnecessary phonemes and the candidate sign become limited.
3. If a user clicks a word on the list, a movie of the sign is presented and its phonemic description and a picture of the handshape are presented on the Explanation window as in Fig. 3.

Besides the phoneme, *katakana* (phonemic system of Japanese) is included as a keyword. This enables the EDS as Japanese-JSL dictionary. He can input the *katakana* in the textbox to find a proper sign of completely agreed, partial agreed, head agreed or tail agreed.

This EDS contains the some 2,600 words in *Practical Sign Dictionary at a Glance*[11].

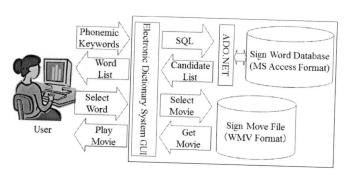

Fig. 1. Conception Chart of the Old EDS

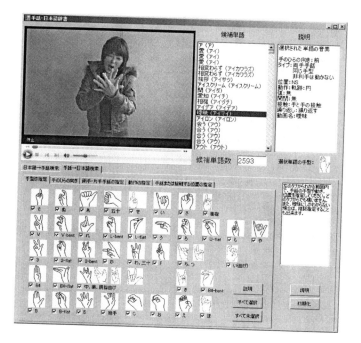

Fig. 2. The GUI of the old EDS

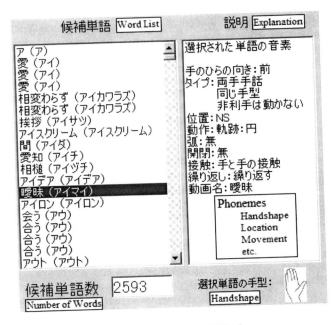

Fig. 3. Explanation of Word

4 Design of Enlarged EDS

4.1 A Guideline for Searching Compound

Here is a guideline of enlargement of searching compound.

1. We provided another database of words which form the compounds to search it in EDS.

2. In JSL-Japanese dictionary, he can search a word through a phoneme selection as he did in the old EDS, and in the next step, he would find a list of compounds using the word.

3. When he looks in Japanese-JSL dictionary, both words and compounds are displayed as a single list. The concept chart is shown below in Fig.4 and our developmental environment in Table 1.

Table 1. Development Environment

OS	Windows XP
Development Language	Visual Basic 2008
Database Format	Office Access 2007
Library	ADO.NET

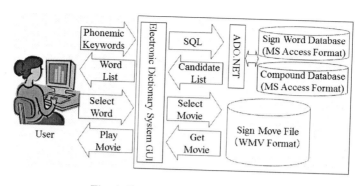

Fig. 4. Conception Chart of the EDS

4.2 Create Database

The database of compounds is created by Access by Microsoft format. The data were adopted from the entries and their descriptions in *Practical Sign Dictionary at a Glance*[11]. The recorded numbers of compounds are about 300 and the name of entry and its consisting words (four at maximum) are the elements.

Moreover, the etymology, synonym, sample sentence, collocation and other information in detail, or explanatory description, reference and others are recorded both in word database and in compound database.

4.3 Revised GUI

We have changed the GUI of the old EDS to display a compound list. In the old EDS as shown in Fig.3, a list of words which include the phonemes selected by a user. We added a tab for compound in the list. When a word of the list was selected, the compounds including the word are displayed on the tab (Fig.5 and 6). For example, when we find MAN after searching, being clicked (Fig.5), the phonemic description of MAN is indicated in the Explanation window and if the sign has some compounds, the list of them including the sign in the Compound Tab. When we click a compound, the corresponding sign movie is displayed and its explanation is presented at the same time. The process is shown in Fig.6.

We can search a compound by any word consisting it. For example as DOCTOR in Fig.6, we can search it by either MAN or PULSE. This makes us to search a compound sign from any information of a sign.

In the old EDS, Japanese phonemic description of a word, its phonemic information and existence of compound were indicated in the Explanation window. We changed it with only Japanese phonemic description and consisting words indicated in the compound searching, because if we display the all phonemic information of the all consisting words, it is entangled and ambiguous to catch up what to what.

The movies of a compound is displayed when clicked, as in the same as in the word searching.

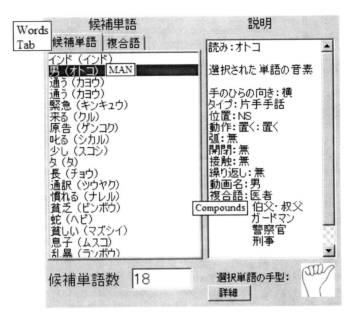

Fig. 5. Compounds Tab

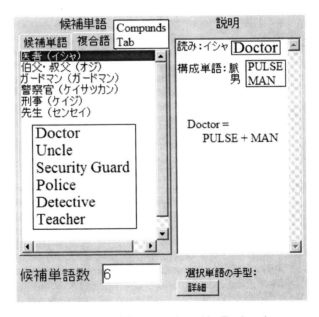

Fig. 6. A List of Compounds and its Explanation

4.4 Using as Japanese-JSL EDS

Our system can be used as Japanese-JSL dictionary, though we explained above how to search a Japanese meaning of a sign through sign phonemes. You can input *katakana* of Japanese for keyword in the text box as in searching a sign word, then a list of compounds is displayed and the compounds are displayed on the list of candidate signs at the same time. A user does not know in advance whether a target word is a single sign or a compound. He might miss it if we separate it into Word Tab and Compound Tab. It also avoids the entangled manipulation.

5 Evaluation and Consideration

5.1 Evaluation Experiment

We performed an evaluation experiment for the enlarged version of our system. The purposes of the experiment are if the use could find a compound, and when he could succeed it, how long he would take a time comparing to when he had searched a simple word. We researched the usability of the system by how the users operated it and by questionnaires sheets.

In the experiment, we showed the user a sign movie and check how long he took to look in the word using our new system.

The user saw the sign movie and found a proper phoneme he thought, then input it into the system, and finally searched out the answer. We recorded his time to get the final answer, the number of the final candidate words, and in the case of compound, which word of the compound he used. We checked three simple words and one compound word. As for time, we judged he failed when he took longer than 6 minutes according to the result of our previous experiment [9]. However in this experiment, almost all the user finished it within 3 minutes. Checking the situations of longer than 3 minutes, he took a wrong search key and no candidate was displayed and he initiated the system and restarted it. That means when the user takes longer than 6 minutes, he would take wrong phoneme.

The subjects were ten boys of 15-16 years old who knew nothing about sign language. The reason why we used them was if there were someone who knows a bit of knowledge of our target compound as *Ao-Mori* (name of the place, literally, Blue Wood), the operation and the time would be significantly different from others.

Table 2. Experimental Results

Word	Info.	Subjects A	B	C	D	E	F	G	H	I	J
BROTHER	Time[sec]	*	65	159	21	131	25	125	181	63	116
	Views	3	1	2	4	2	1	4	2	2	2
	Number of Candidate Words	*	2	32	5	1	5	2	5	5	2
BICYCLE	Time[sec]	124	142	125	113	209	51	178	210	148	145
	Views	2	2	2	4	4	1	5	4	4	4
	Number of Candidate Words	32	37	29	28	17	216	5	99	28	20
BOOK	Time[sec]	189	210	*	123	178	*	270	*	*	206
	Views	1	3	4	4	2	4	5	6	3	2
	Number of Candidate Words	59	42	*	54	22	*	23	*	*	81
Compound Ao-Mori	Time[sec]	356	178	354	138	*	171	*	*	*	*
	Views	3	4	4	8	5	3	9	8	6	4
	Number of Candidate Words	48	24	33	37	*	219	*	*	*	*
	Number of Candidate Compounds	1	3	1	3	*	3	*	*	*	*
	Keyword	Mori	Ao	Mori	Ao	Mori	Ao	Ao	Ao	Ao	Ao

Therefore we gathered the subjects ignorant to sign language and lectured them briefly a phonemic system of sign language, then performed the test. This kept the subjects the same level of literacy for phonemes. Moreover they are well trained to computer literacy and skillful to computer manipulation. In the lecture for sign phonology, they were trained to use the new EDS using some example words and compounds and it assumes that they are well trained to use it. The result is shown in Table 2. The subjects are shown in alphabets. The asterisk means failure in 6 minutes. Consideration of the result is shown below.

5.2 Consideration

1. The comparison of hit rates for simple words and a compound
   The rate of success to find out the target word is; BROTHER: 90%, BICYCLE:100%, BOOK:60% and the average is 80%. As for compound, the average is 50%. We explained the subjects the process of searching compound in advance but many failed. We asked them why.

   - failed to recognize a movie as a compound
   - could recognize it as a compound but could not find out the word boundary
   - took a time to input all the information of the words consisting the compound
   They took a compound for a single word and input all the phonemes to fail.

2. The number of candidate words
   The average number of candidate words is 34.0 and 72.2 for compounds. Subject F did not limit very much since he searched with different approach to other subject. He checked each signing movie when he obtained 200 candidates. Other subjects checked it when they reached to some 30 candidates. But the subject F took about half of the searching time of other subjects, It means it takes less time to check each movie than to think phonemes. It must be more effective for the user who is ignorant to sign language to check each movie than to analyze the phoneme.

   The subjects thought it difficult to input the phonemes of arm movement. There are too many items to input in the arm movement and some are hard to do so intuitively.

3. Searching time
   The subjects were ignorant to sign language in our experiment. Most of those who took a time failed to analyze the phonemes and they restarted to input them. Seeing the case of English learners, it is hard for the very beginners without any learning to use a dictionary. We think it is natural for the user not to deal the dictionary with ease, as in the case of this experiment.

   However, it is important to catch the phoneme when we read signing. We expect the time will be shortened if we improve the system in which it gives an user an advice to check points when he watch a sign movie.

4. Searching compound
   The result of the experiment shows a low rate of compound search. As the subjects said, whether the target signing was a word or a compound was ambiguous and they could not check the compound tab. As in Fig. 5, when they click a word, a list of compound was displayed on the Explanation window, but they did not notice it. They paid attention to movie, not to Explanation window after clicking. That was a part of reason why.

6 Future Problems

1. A List of Compounds

   The result of the experiment shows the low success rate of searching compounds. As some subjects told, one of the reasons of the low rate is that they could not differentiate whether the target signing was a simple word or a compound. The same thing can be said to an English-Japanese dictionary. When we want to look in "over-easy, sunny side-up, fried egg or hot dog", we would not succeed in it until we know whether it is a simple word or a compound. If we would look "hot" or "dog" separately, we could not understand the meaning correctly, even though many dictionaries show it under the title of "hotdog", that is, it is better for us to know that "hot dog" is a compound.

   In order to solve this problem, we focus to the peculiar way to use EDS. A user checks the movie when he looks in a signing, as we explained in the former section. The EDS shows a list of compounds on the lower screen when the movie is displayed which will make him easy to check a compound.

2. Revising GUI for the Sign Language Learners

   Our system was created for the sign researchers at the beginning, and it is true that it is not easy to handle for a sign learner in general. For example, inputting a phoneme, every item can be designated and it is precise. As a result of experiment shows, a user usually inputs a handshape and some movements and checks it through the movie. We would like to put the phonemes in order.

7 Conclusion

We enlarged our EDS to enable to search compounds in JSL-Japanese Dictionary, and estimated it in this paper. It proved that the enlarged version was more effective but the result of the evaluation experiment showed the success rate for compounds were lower than that of simple words. The reason maybe a user cannot differentiate a word for a compound. And it is a problem for a user that it is nuisance for him to input the phoneme. We showed some solutions for it.

The EDS is useful to find a meaning of a sign even if a user remembers it ambiguously, using and checking its movie. It is more advantageous than a printed dictionary.

We will renew and enlarge our EDS in the near future.

Acknowledgment. This research is partially supported by the Grants-in-Aid for Scientific Research of Japanese Ministry of Education, Basic Research (A) in 2010, titled "Morphemic Dictionary for Sign Language and its Applications", #20242009, Representative Researcher, Kazuyuki Kanda, PhD.

References

1. Takemura, S.: Sign-Japanese Dictionary (手話・日本語辞典). Kosaido Publishing (1994)
2. Takemura, S.: Enlarged Sign-Japanese Dictionary (手話・日本語大辞典). Kosaido Publishing (1999)

3. Japanese Federation of the Deaf, Practical Sign Dictionary for New Introductory Course of Sign Language (『新手話教室入門』対応実用手話単語集) (2007)

4. Kanda, K.: Musashi α (ムサシα), the Computerized Dictionary for Japanese Signs, Alpha Media (1996)

5. Language on the Hands, An Introduction (手にことばを・入門編). Fijitsu Palex (1996)

6. Fukuda, Y.: Construction of an Electronic Japanese Sign Language Dictionary. The Japan Journal of Logopedics and Phoniatrics 45(2), 131–138 (2004)

7. Fukuda, Y.: Compilations of the electronic dictionary of Japanese Sign Language (a second edition) and its instruction manual. IEICE Technical Report, WIT 2005-8, pp. 39–44 (2005)

8. Kimura, T., Hara, D., Kanda, K., Morimoto, K.: Development and Evaluation of Japanese Sign Language-Japanese Electronic Dictionary. Japanese Journal of Sign Linguistics 17, 11–27 (2008)

9. Takemura, S.: JSL Japanese Dictionary, Web Version,
   http://www2s.biglobe.ne.jp/kem/test-htm/index.html

10. Kanda, K. (ed.): A Basic Course of Sign Linguistics. Fukumura Publishing, Co. Ltd. (2009)

11. NPO Testing Organization for Proficiency of Sign Language. Practical Sign Dictionary at a Glance (ひと目でわかる実用手話辞典). Shinsei Shuppan (2010)

Co-creation Process of Collaborative Work with Communication Robot

Seita Koike[1], Takeshi Ogino[1], Sari Takamura[1], Tatsushi Miyaji[1], Yuki Miyajima[1], Daishi Kato[2], Koyo Uemura[2], Kazuo Kunieda[2], and Keiji Yamada[2]

[1] Tokyo City University, 3-3-1 Ushikubo Nishi, Tsuzukiku, Yokohama, Japan
[2] NEC C & C Innovation Research Laboratories 8916-47, Takayama, Ikoma, Nara 630-0101, Japan
koike@tcu.ac.jp

Abstract. This paper reports on fieldwork concerning co-creation among primary school students. The students brainstormed ideas and made a play using a robot as a prop. We observed their group decisionmaking process in creating their performance, and concluded that setting up a environment with some "order" is necessary, but freedom must also be preserved for organic decisionmaking. Our results showed that the Japanese children are 1) collaborative decisionmakers (deciding not by majority or by a single leader), 2) avoid conflicts by relying on everyday experience to back their arguments, 3) require outside perspective of a facilitator, and 4) can have equal say if props are used.

Keywords: Robot, co-creation, collaboration, design, communication.

1 Introduction

This paper tries to overcome the traditional dichotomy between freedom and rules-finding an "order" that allows for spontaneous co-creation. By "rules" we refer to rigid laws, whereas "order" is a concept that is more about setting up boxes within which organic decisionmaking is possible. We define co-creation as collaborative work that some people generate ideas and put into practice together.

Often it is thought that the best human decisionmaking must occur through strict rules. Whether in society or politics, rigid rules govern human activity to prevent crime or extreme behavior. However, in our research, we find that children are able to come up with executable, creative ideas with minimum structure or rules.

Our fieldwork examines methods of co-creation among children during a workshop where the children create plays. It is difficult to understand the extent of freedom and order required for children to engage in a collaborative, creative project. We used some order- such as a facilitator, a prop (robot), project phases- and also allowed for freedom- allowing for organic decisionmaking processes, and equal say among children.

M. Kurosu (Ed.): Human Centered Design, HCII 2011, LNCS 6776, pp. 417–424, 2011.
© Springer-Verlag Berlin Heidelberg 2011

2 Research Approach

The workshop was conducted at an elementary school in Tokyo. We organized 2 groups (A, B) of elementary school students, each group consists of about 5 members. The experimental period was about 3 months with 2-hour workshops per months.

We utilized a communication robot called PaPeRo as a prop. The goal of the workshop was to make a short play with them and the robot PaPeRo as actors. The students of our lab Supported each groups as a workshop facilitator. Fieldwork was conducted to assess the activities of students as they gave ideas.

3 The Workshop Process

(1) The elementary students brainstormed on Post-it notes about the theme of their skit. They drew sketches of the stage design.

(2) The groups decide their subjects and plots. They write a script.

    A. Group A decided that PaPeRo is a part-time worker at a convenience store. In the play, PaPeRo fends off a burglar.

    B. Group B decided it would model its skit on "Shoten", a Japanese comedy TV show in Japan. At Shoten, an MC gives a theme to which show guests must answer with a witty story. If the story is not funny, then the guest loses. In their play, PaPeRo acts as MC and the guest roles are performed by the students. What is interesting twist in this play is that the robot is asking questions to the students.

(3) The students program and move the robot according to their script step by step. The facilitators taught the students how to program the robot.

(4) They designed and made the stage using paper and cardboard boxes

(5) On the last day of the workshop, the students presented their slots with PaPeRo in front of other students at the school assembly.

Fig. 1. Brainstorming

Fig. 2. Developing the scenario

Fig. 3. Programing PaPeRo   **Fig. 4.** Presenting their slots with PaPeRo

4   Communication Robot PaPeRo

We use PaPeRoch! with the workshop. PaPeRo is the communication robot developed by NEC. PaPeRo is 385mm hight, 282mm depth, 251mm width, 6.5kg weight. PaPeRo have a function of voice recognition, voice synthesis, image recognition, touch sensor and can move with its own wheel.

We can control PaPeRo by "PaPeRoch!". PaPeRoch! is control software for PaPeRo which based on Scratch. Scratch is developed by the Lifelong Kindergarten Group at the MIT Media Lab. PaPeRoch! is easy to use for elementary students. They can programmed by snap block as a part of function on screen easily.

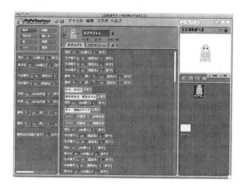

Fig. 5. PaPeRo   **Fig. 6.** Interface of PaPeRoch!

5   Ingredients of Co-creation: Situational Idea Generation, Common Ground, Outside Perspective and Models

5.1   Situational Idea Generation: No Leader, No Voting, But Organic Spontaneous Decision Making

We found that the children generated their ideas spontaneously through a situational process involving a stacking of ideas. A situational process is where a group exchange ideas as they think of them, and the ideas accumulate into a final decision.

There are no predetermined rules. For example, in the following transcript, we see that their ideas eventually converged. Even a student that originally disagreed ended up agreeing. There is no clear logic, but there is some kind of organic synthesis of ideas. Thesis and antithesis are not articulated, allowing for a smoother process towards a conclusion. The children do not feel that one side "wins", so the "losers" are more easily convinced. They can "lose" without feeling that they lost.

Example 1

Evidence of this can be seen below. Student D originally disagrees that PaPeRo (the robot) should not be a part-timer character. Later, however, the same student gives a constructive idea about PaPeRo as a part-timer. The student's opinion is not only reversed, but he is giving a new idea, contributing to the idea he used to object to. In response, Student C encourages Student D, without making any comment about his stance being reversed. Thanks to the follow-up of Student D, Student C feels integrated back into the group. A decision is made without a single leader, without voting, and instead from situational idea creation.

Transcript

Student A: OK! Lets make it a comic skit of PaPeRo!
Student A: Decided!
Student B: Skit, make PaPeRo a part-timer.
Student A: Why do you want him as a part-timer?
Student B, Student C: Sounds good, part-timer!
Student D: Part-timer no! Change!
Student A: Where does PaPeRo work? FamilyMart?
Student C: No, Seven-Eleven is better!
Student A: PaPeRo is eating fried chicken in the shop. PaPeRo gets fired after stealing food.
Student C: So in the skit, PaPeRo becomes a part-timer. Next, we must decide the store, LAWSON or FamilyMart?
Student D: Maybe a customer tells PaPeRo "I want a salmon rice ball", and PaPeRo says "Am I gonna be riceball?"
Student C: Good idea!
Facilitator A: Have you guys decided on the comic skit with PaPeRo?
Student A: OK! OK!

In the situation of fixing content of the play, Student B proposed PaPeRo as a part-timer. Student A and Student D questioned the act of PaPeRo. But they proposed new ideas through their context. Student C declared PaPeRo as a part-timer. Student D opposed the act of a part-timer at first. But he whistle a different tune and proposed "wrapped PaPeRo inside a salmon rice ball" In their discussion nobody only led them, nobody only decide by majority vote, too. They chose their ideas situationally.

Example 2: Stacking ideas

Transcript

Student C: OK, next, let's decide on which convenience store.
All Students: Seven-Eleven.
Student A: Where should it be? Which branch?
Student C: Seven-Eleven's Yokohama branch is good.

Facilitator A: You'd better tell Group B that you decided on this theme.
Student F: We don't have say it.
Student A: We could make a cash register from cardboard...
Facilitator B: Did you decide?
Students: We decided.

In their discussion, they generate ideas by stacking ideas on top of one another. From a general idea, details are stacked up based on other people's ideas. Student C said "convenience store", then all of them said "Seven-Eleven," then the branch and location were also stacked on top of that. From the simple idea of just "convenience store", all the details were provided situationally. A rule does not precede the suggestion, but the suggestion itself creates the rule. The children were never told "Decide on a convenience store, then decide location and branch."

5.2 Common Ground: Every-Day Life Experience

Usually, when workshops are conducted, people from different backgrounds are thrown together. However, in our situation, the workshop members knew each other since they go to the same school. We see that the members used this common ground to their advantage. They generate ideas from their shared everyday-life experience. They fixed cast of the play. Their transcript is as follows.

Transcript
Student E: Are you writing the script? (to Student B)
Student B: That story XXX is no good... (inaudible)
Student E: Not that one, First we have to decide if we "bad students" are gonna be in junior high or high school.
Student B: 9th grade.
Student E: The situation is that 9th graders sneak out of their math class...
Student B: OK.
Student E: Wait, put this down too.
Student E: So he's 9th grade and what does he say?
Student B: He complains at 10am that "the steamed dumpling (made in the convenience store) isn't properly cooked."
Student E: Why10:00 a.m.?
Student B: Because the 9th graders snuck out of their math class...
Student E: It has to be AM?
Student B: It has to be AM. That's why they buy food and go to **Komae.**

They generated the idea of "9th grade students" "snuck out of ath class" and "Komae". No one objects to the idea of Komae because they all are familiar with it. By using a location or ideas they are familiar with, the workshop moves forward smoothly. Ideas generated from their common ground help facilitate a smooth decision making process.

6 Facilitation for Co-creation

The students did not co-create alone. The Facilitators played an important role for co-creation in the workshop. Their outside 3rd person perspective helped the children to reach a conclusion. Whenever the children would get stuck or not know how to continue, the facilitators would step in. They would only step in, however, when they thought it was necessary. The children were given as much freedom as possible, provided that there was some creative activity occurring. A top-down approach was avoided.

Their parts are given as follows.

6.1 Providing an Additional Perspective

In the following transcript we see that while students initially resisted the suggestion of the facilitator to have PaPeRo (robot) play an instrument, after some additional comments, the children accepted the idea. (Student C) The facilitator helped the students recognize an additional function of the robot, but only gave the suggestion as an option, not as a command. Student C agreed without the Facilitator having to make their decision for them. Given a hint by the facilitator, the children are quick to react. They do not require rigid directions.

Transcript
Facilitator: Can PaPeRo play instrument ?
Student H: PaPeRo has no...(inaudible)... so he might be able to play percussion.
Student C: It can't play percussion instrument because has no hands.
Facilitator: So, if you install sound resources into PaPeRo, then you could make it seem like PaPeRo is playing.
Student H: We can't have PaPeRo have an instrument.
Facilitator: Hmm... yeah but you could...
Student C: Yeah, so you're saying we can have him pretend to play...

6.2 Models: Having Equal Say

Usually, when such workshops are conducted, the students who have the loudest voice or have power in the classroom make the remarks. They often end up making most of the decisions as well.

Fig. 7. Model of the stage

One way the facilitator made the process fairer was by introducing a model of the stage made out of paper and cubes of Styrofoam. The children could then discuss the model of the stage by physically moving the cubes around. The students formed a circle around the model and could focus on the stage layout itself, rather than each other. All students had a more equal opportunity since they could all access the model equally.

Again, the facilitator introduced the model, just the "box" or an element of "order", without setting rigid rules that may stifle the students' situational, spontaneous creativity.

7 Conclusion: Designing Environments for Co-creation

In short, the most interesting conclusion from our fieldwork was that there is a delicate balance of freedom and order needed to trigger the right kind of organic, situational decisionmaking process among elementary students.

For co-creation, it is necessary to design co-creative environments that allow for such discussion among the children. Resources for the environment include people, such as the facilitator, as well as props, such as the robot or models. The situation should have some kind of "order" and also should have a facilitator who encourages expansion of the discussion into actual ideas and their applications. Models also help to equalize opportunities for expressing opinions. Setting some kind of goal is important, but should not overbearing. People and props work together to spark motivation for creation.

Additionally, although it is strictly speaking beyond the scope of this research, an interesting observation was that the children treated the "robot" prop not as something strange or foreign, but gave the robot conventional roles, just like humans. Perhaps this suggests some kind of cultural characteristic among Japanese children where they can embed a "foreign" object into daily life situations without feeling it is an alien object. Judging from the ways robots are treated in films in the U.S., for instance, robots are more often considered "foreign" or "unusual". Contrastingly, the Japanese children had no problem giving the robot "human" roles like "part time convenient store worker".

Perhaps the reason the children treat the robot in this way is similar to the reason the children do not divide themselves into clear positions of "for" and "against". The group can include children with different opinions, or even a "robot", into their spontaneous, organic creation. It may be because of the mode of communication which does not clearly set a thesis and antithesis. This finding may be applicable in other situations where situational creation is desired.

Acknowledgments. We deeply appreciate NEC C & C Innovation Research Laboratories as co-researchers and Christopher Gibson as my collaborator.

References

1. Koike, S., Sugawara, M., Yamauchi, S., Kutsukake, Y., Sato, K., Fujita, Y., Osada, J.: Social Robot. In: The Human Computer Interaction CD-ROM 2009, San Diego (July 2009)
2. Koike, S., Osawa, K., Sugawara, M., Suzuki., I., Nishikawa, T., Fujita, Y., Osada, J.: Robot Designing a partner robot by social-technologic network -Ethnography and Design studies for Usability of the robot in kindergarten. In: Proceedings of Japanese Society for The Science of Design, Hiroshima, Japan (May 2008)

3. Koike, S.: Situated-based Robot Design Association for Advancement of Artificial. In: Proceedings of Advanced Robotics, Intelligence - Spring Symposium Multidisciplinary Collaboration for Socially Assistive Robotics, Palo Alto CA USA (March 2007)
4. Sugawara, M., Yasuda, T., Ono, M., Osawa, K., Koike, S., Fujita, Y., Osada, J.: Designing The Robot as Networks. In: Proceedings of Japanese Society for The Science of Design, Shizuoka (June 2007)
5. Osawa, K., Koike, S., Kon, S., Takahashi, M., Fujita, Y., Osada, J.: Robot Design as Social Network. Proceedings of Japanese Society for The Science of Design 2006, Knazawa, Japan (June 2006)

Virtual Office, Community, and Computing (VOCC): Designing an Energy Science Hub Collaboration System

April A. Lewis and Gilbert G. Weigand

Oak Ridge National Laboratory
P.O. Box 2008 Oak Ridge, Tennessee
{Lewisaa,weigand}@ornl.gov

Abstract. [1]The Consortium for Advanced Simulation of Light Water Reactors (CASL) implements a management strategy that imbues physical collocation; community; collaboration; central leadership; multidisciplinary teams executing a single milestones-driven plan; and integrated, co-dependent projects. The CASL-streamlined management structure includes collocation at CASL, use of technology to achieve multidiscipline collaboration, video conferencing for meetings, and a VOCC project that integrates both the latest and emerging technologies to build an extended "virtual one roof." CASL is headquartered at ORNL, where the CASL leadership and a majority of the multidisciplinary, multi-institutional scientists and engineers will be located. Work performed at partner sites will be seamlessly integrated across the consortium on a real-time basis via community and computing (VOCC) capability that integrates both the latest and emerging technologies to build an extended "virtual one-roof" allowing multidisciplinary collaboration among CASL staff at all sites. The paper describes the VOCC collaboration system.

Keywords: User Centered Systems Design, Collaboration, Collaborative Virtual Environments, Collaborative Computing, Human Computer Interaction, Energy Science Hub.

1 Introduction

The United States (U.S.) is currently focused on energy independence and energy efficiency. The Department of Energy (DOE) is making major investments into innovative approaches to improving nuclear reactor energy efficiencies. To realize measurable improvements, DOE needs to optimize its current nuclear power production capability by reducing capital and energy costs per unit, reducing nuclear waste, and enhancing nuclear safety. DOE has funded a consortium of scientists, led by Oak Ridge National Laboratory (ORNL) to develop transformational nuclear computational science models for identifying, understanding and solving nuclear reactor safety and performance issues. These critical virtual reactor models and predictive simulations will not only contribute to extending the life of current reactors, but will be instrumental in supporting future commercial reactor designs.

---

[1] www.casl.gov

M. Kurosu (Ed.): Human Centered Design, HCII 2011, LNCS 6776, pp. 425–434, 2011.
© Springer-Verlag Berlin Heidelberg 2011

To optimally harness the collective conscious of the consortium, ORNL needed to cognitively bring together under "one-virtual-roof" the best nuclear and light water reactor (LWR) scientists, engineers, designers, and industrialists to solve complex alternative energy efficiency designs. The consensus was to construct a physical one-roof space, but extend its reach through a connected virtual collaborative environment. The environment was defined as Virtual Office, Community Computing" and is the first DOE energy hub to use human centric, immersive, and visually analytic design techniques and principles to build a physical work space for the purpose of virtually unifying geographically distributed computational scientist and high-performance computing (HPC) resources.

VOCC is a system of systems integrated in a special way to promote collaboration and critical thinking amongst its users. Critical thinking is presumed to lead to insight and insight to innovation. VOCC's "innovation at the speed of insight" means a more rapid deployment of predictive simulation capability to the LWR industry. A direct consequence of the design is not only reduced current energy costs, but expedited delivery of LWR innovations to industry.

The major components of VOCC include virtual collaboration and design labs and presence and visualization systems. The presence systems provide the primary and essential support for virtual interactions (face-to-face collaboration) amongst teams and consortium members, from scientist to managers. Complimentary collaboration systems provide spaces for ideation and co-creation activities. The Ideation space includes interactive sketching and drawing tools which assist team members in expressing ideas. The space also serves to facilitate stimulation of "collective ideas" and solutions. The visual immersive spaces allow team members to interact with other people, simulation data, and textual information. Lastly, 2d and 3d visual environments permit collective analysis of virtual reactor objects and reactor operation simulations. Defining capabilities contained in these systems was left with project members because we generally feel collaboration is driven by an individual's scientific activities. The CASL collaboration and Ideation Officer (CIO) used the capability to also drive the search for suitable existing technology. The state of technology drove initial collaborative system designs. The three most dominant consortium activities influencing design included agile code development, modeling and simulation, and visual data analysis.

The optimal reactor design strategy exists only when cognitive systems for design are closely coupled with physical engineering design systems inside collaborative virtual spaces. These unique virtual spaces allow researchers to represent holistically diverse components of reactor knowledge in such a way that quick convergence to an understandable solution space is inevitable (*"design at the speed of insight"*). Convergence of the disparate knowledge spaces is facilitated via Collaboration. Collaboration takes place when you can co-locate physically or virtually human dynamics [1] (human perception, perspective, and cognition) and information or data with visually immersive, 3-D design modeling systems. Modeling systems must go beyond static display walls and include ad-hoc, dynamic visual spaces like 3d tele-immersive environments or advanced telepresence communication systems, and interactive touch modeling systems. These devices must be located in or near collaborative spaces so individual team members can be "collectively creative", kick around ideas ('ideate'). Telepresence systems not only serve as advanced spaces for real time visualization, but they are also critical for pervasive collaborative communication amongst creative teams and customers. Collectively these components form a virtual cognitive laboratory representing a collective design conscious of reactor design knowledge and capability. By singularly locating this design knowledge and insight we can guarantee rapid movement of technology and engineering practice into the U.S. energy industry's reactor design centers.

2 Energy Science Hub

In June of 2010 the U.S. Department of Energy (USDOE) launched (CASL), an Energy Innovation Hub. Hubs are new R&D structures created by the USDOE to address the most pressing U.S. energy problems. After reviewing many excellent proposals, the Secretary of Energy identified only three energy innovation focus areas that would be researched under the hub structure 1) building efficiency, 2) fuel from sunlight, and 3) nuclear energy. CASL was selected as the nuclear energy hub. The CASL Hub's end product is a code suite and methodology that industry and regulators will use directly in LWR reactor design and licensing practice, and/or indirectly to justify along with experimental results their proprietary modeling and simulation capabilities.

The Hub research concept will effectively remove current R&D barriers which have historically prevented USDOE from achieving national energy and climate goals. Hubs comprise a highly collaborative team, spanning multiple scientific, engineering, and where appropriate, economics, and public-policy disciplines. Hubs will seek to rapidly drive energy solutions to their fundamental limits. Each Hub will support cross-disciplinary R&D focused on the barriers to transforming its energy technologies into commercially deployable materials, methodologies, and technologies. The ultimate goal of each will be to advance a highly promising area of energy science and technology to the point that the risk level will be low enough for industry to deploy solutions into the marketplace. In the case of the Simulation and Modeling for Nuclear Reactors Hub the USDOE strongly emphasized the "one roof" concept. Co-location of scientists in a single structure was not the only concern for collaboration. It was achieving distributed, effective collaboration using a centrally led "integrated" model of research towards a challenge goal.

2.1 Science Mission and Hub Goals

The Consortium for Advanced Simulation of Light Water Reactors is an exceptionally capable team that will apply existing modeling and simulation capabilities and develop advanced capabilities to create a usable environment for predictive simulation of light water reactors. This environment, designated the Virtual Reactor (VR), will incorporate science-based models, state-of-the-art numerical methods, modern computational science and engineering practices, and uncertainty quantification and validation against data from operating pressurized water reactors. It will couple state-of-the-art fuel performance, neutronics, thermal-hydraulics, and structural models with existing tools for systems and safety analysis and will be designed for implementation on both today's supercomputers and the advanced architecture platforms now under development by the USDOE.

To accomplish this vision for the VR simulation tool, CASL will focus on a set of challenge problems that encompass the key phenomena limiting the performance of pressurized water reactors, with the expectation that much of the capability developed will be applicable to other types of reactors. Broadly, CASL's mission is to develop and apply modeling and simulation capabilities to address three critical areas of performance for nuclear power plants:

1. capital and operating costs per unit energy, which can be reduced by enabling power uprates and lifetime extension for existing nuclear power plants and by increasing the rated powers and lifetimes of new Generation III+ nuclear power plants;
2. nuclear waste volume generated, which can be reduced by enabling higher fuel burnp-ups; and
3. nuclear safety, which can be enhanced by enabling high-fidelity predictive capability for component performance through failure.

3 Hub Innovation Strategies

CASL provides a unique opportunity not only to advance the use of nuclear power in the United States but also to advance the state of distance collaboration in the process – a key element in an increasingly global research society. CASL has a clear commitment to the use of state-of-the-art technology and frequent virtual meetings to enable long distance collaboration.

CASL will undertake a Virtual Office, Community, and Computing (VOCC) Project. The VOCC project will deliver (1) commercially available and when necessary custom web-based virtual office and collaboration technology; (2) advanced telepresence or net-presence technology; and (3) methods and technology for scientific study, analysis, and remote CASL computing on HPC systems. The final deliverable will implement and leverage DOE HPC investments such as Leadership Computing and the National Nuclear Security Administration ASC Program. VOCC's efforts create an integrated multidisciplinary, multi-site collaboration and integration and a singular CASL focus, thus significantly enhancing the opportunity for innovations in nuclear energy.

3.1 A Unified Collaboration Platform

The Virtual Office, Community Computing (VOCC) project is focused on designing a unified collaboration platform and general creative work environment to support the advanced simulation of light water reactors (LWR). The CASL hub concept has two monumental tasks: (1) to cognitively bring together under "one virtual roof" the best LWR scientists, engineers, and industrialists and (2) to create a state-of-the-art scientific collaboration space that not only supports, but also optimizes joint LWR design, fabrication, and assessment. VOCC is the first Department of Energy (DOE) hub project to use human-centric, immersive, and visual analytic design techniques and collaboration principles to build a physical work space for the specific purpose of unifying—virtually—geographically distributed computational scientists and high-performance computing (HPC) resources.

The VOCC platform will be a video-based integrated electronics environment; its primary purpose is to support both CASL's synchronous and asynchronous communications and visualization needs. VOCC will contain the core traditional elements of a collaboration platform like group communication tools, messaging, social networking and computing tools, as well as collaboration tools for modeling and simulation, two-dimensional (2D) visualization, and 3D information manipulation. VOCC must possess virtual collaboration tools that can be accessed by anyone, at any time, from anywhere on any device. Most online virtual tools will be accessible via rich interactive applications (RIAs) on the web. VOCC will leverage existing virtual productivity spaces designed for information sharing such as Wire, Google Sites, SharePoint, Groove, SOSIUS, and Think Free[2].

The VOCC platform is a "system of systems," with each subsystem referred to as a "venue." Each venue has been selected based on a documented CASL user requirement. Once a capability requirement has been established, a technology search is initiated to configure the venue. A candidate set of technologies are identified and an evaluation performed. Technologies are evaluated on many factors, including but not limited to cost (*initial procurement and maintenance/legacy operational costs*), efficiency, mobility, scalability, interoperability, and ease of use. Efficiencies are examined from both a computational and an efficiency point of view, where the latter ranges from sharing of information (data) and knowledge to venue energy consumption.

"Mobility" and "scalability" describe the adaptive nature of a venue to communicate with CASL partners and stakeholders when and where necessary. For synchronous and asynchronous communication, this implies CASL venues scale from room-size voice and video nodes to mobile-size voice and video nodes. Scalability of this nature ensures diverse productive landscapes from venue productivity (groups) to small-scale, handheld smart device (single-user) productivity ("productivity in the palm of your hand",[2]). Interoperability standards for CASL mean that each venue must easily talk to other venues and connectivity is afforded to multiple operating systems (OS), minimally Windows, Mac, and Linux. These types of efficiencies ensure creative collaborative CASL venues are available to users, partners, and stakeholders any time and any place, from any device. Ease of use ensures that no matter how little a user may use a certain collaborative venue or application, he or she can quickly start reusing it in a meaningful way without spending much time retraining (reduced cognitive load on user).

3.2 Human Dynamics and Computing

CASL's vision to create a virtual reactor (VR) for predictive simulations of LWRs necessarily means building a virtual technical team, a community comprising engineers, designers, scientists, researchers, and industry experts. The community

---

[2] Mention of specific commercially available hardware or software does not constitute an endorsement of the items. USDOE does not endorse any commercial products.

members need a virtual-one-roof to engage one another to evolve solutions and approaches to real, difficult, and necessary scientific alternative energy needs.

CASL's one-virtual-roof is VOCC, and it will consist of layered networks, architectures, and communications (including social networks). The networks will provide trusted collaborative computing to enable organizing, distributing, and routing of sensitive data between CASL partners (e.g., intellectual property, export-controlled information, etc.). Cybersecurity is an integral part of CASL. Multiple, redundant resources and tools will be devoted to protecting sensitive digital information inside the sensitive CASL enclave. Additional protocols will be established to ensure there is no transfer of data from sensitive CASL enclaves to CASL open enclaves.

Defining collaboration and collaboration venues and tools is not something for the information technology (IT) group to define on the CASL enterprise level. Such a prescription for enterprise collaboration tools only creates resentment on the part of users because they typically do not truly address users' scientific technology needs. Collaboration is driven at the personal level by individuals. Selection of collaboration tools and venues must be closely coordinated with end user workflows and outcomes. Workers must intuitively see the value of using collaboration tools and they *must* be engaged in their development and build out. However, deployment and secure access to such venues must include organizational information technology resources.

4 VOCC Collaboration Needs Assessment

To understand what CASL's creative collaboration needs are, CASL performed an initial needs assessment. First, a repeatable methodology for information collection was established. We started with the proposal-based concepts of "virtual one-roof" and worked toward reality and eventually to a full initial understanding of stakeholder and user technology needs. We focused on optimizing collaboration in the CASL activity space (speeding up innovation) while simultaneously reducing operating cost (improving CASL's business value). It became quite clear during this assessment that VOCC was not just a milestone in a science proposal; rather, *it is the way CASL would most effectively operate and innovate.*

Next, other key operational considerations were identified that would be critical in establishing the venue-based needs for the assessment process. Among the most important was the need to share data and access on HPC systems such as the Cray Jaguar at ORNL. There are also needs to share intellectual property and tech export control information. Additionally, core computational use cases for evaluation were identified. An initial survey of partner sites was performed to determine the extent of their institutional contributions and limitations (space, bandwidth, firewalls). Great thought and discussion were given to performance measures and how those might be measured in relation to the collaboration venues. Finally, the interoperability issues partner sites might have with selected technologies were considered, along with the infrastructure modifications, if any that would be necessary to accept the technology. VOCC in general does not seek to be prescriptive about collaboration tools or venues, but rather to prepare for being optimally interoperable for the unanticipated use of each organization's resources. It does, however, require that at least one identical form of synchronous communication exist at each core partner site. At this point, it is assumed that the only prescribed venue is immersive telepresence.

4.1 CASL Activities

The needs-based assessment assisted CASL in identifying four primary activities for technology requirements evaluation: (1) Business/Program Management, (2) Agile Code Development, (3) Modeling and Simulation (M&S), and (4) Lectern (Educational)/Research Partnerships. Table 1 lists some of the use cases gathered for each category.

Table 1. Four Main Activities Planned for VOCC

Business/Program Management	Agile code Development	M&S and Analysis	Lectern/Research Partnerships
• B2B partner discussions • Reviews w/sponsors, partners, FA leads, collaborators, etc. • Holistic/Iterative cost estimations for SW development • Technology transfer/sharing • Cooperative agreements, presentation/publication material development • Technical exchange, PowerPoint, MS Office, Primavera, etc. • Sometimes audio only connects	• Daily Stand-ups • SCRUM Sessions • Code assembly and Sketch up • Distributed code development forward & reverse SW engineering • Remote pair programming • Co-authoring SW/user documents and manuals • Distributed configuration management and version control (shared code storage) • Distributed Test Environments • Multiple participants, single interactive session exchange • JAZZ, Eclipse, other IDE's, Rational Asser manager, etc.	• Sharing hi res. Viz. via true telepresence or augmented reality(direct view and manipulation of reactor models) • CRUD – Impact Analysis(Rod vibration, LOCA accidents) • Model validation • Multiple participant, multiple session exchanges • View 3d video objects, share control and edits • Using Access & Para View w/telepresence system • CFD, STAR-CCM, etc.	• Interactive & Immersive classroom instruction • Augmented Reality presentation of guest speakers • 3D learning environment(visual tutors) • Social Software and "shared study & research environments" • Building partnerships, virtual university extensions • Democratizing information access • Expanding human to machine interaction • Simpodium, blackboard, etc.

Business/Program management technology needs to primarily include telepresence or video connectivity and a persistent digital work plan to track document changes and updates. Telepresence would provide a high degree of personal interaction with the senior leadership team (SLT) and focus area (FA) technical team leads. Collaborative document spaces provide a place to share documents involving processes, protocols, and technology exchange agreements. Uses for these document spaces are most needed in the startup phase of the program.

Code development is initially one of CASL's largest user communities in support of the VR. CASL has decided to employ an Agile software development approach to the VR. IBM describes Agile development as a collaborative, incremental, and iterative approach to software development that can produce high-quality software in a cost-effective and timely manner. Unlike traditional software development, Agile development emphasizes flexibility, continuous testing and integration, and rapid delivery of functionality.

The traditional approach to Agile software development is co-location (locating two or more team members in one physical place). Co-location can provide participants immediate coherent visual and auditory feedback from code development team members. The visual feedback is very important, as the majority of collaborative communication and hence understanding is visual [4].

CASL code development teams will need two key technology venues to support agile code development. Scrums are processes for developing such complex software applications as Virtual Reactor Integration (VRI). Scrum teams are relatively small and consist of 5–7 cross-functional team members. Their iterative development life cycle runs every 30 days. They develop code in pairs and as teams. Many coders prefer to design Unified Modeling Language (UML) in an interactive and intuitive touch application space. This requires hardware that is multi-touch enabled and a software development environment designed to emulate traditional interaction like keystrokes and mouse clicks. Code teams have daily Scrum sessions that last approximately 15 minutes. These meetings require telepresence for optimal execution and, minimally, desktop video exchange in remote paired programming instances. Ultimately, augmentation to the telepresence system is desired so that three-dimensional (3D) real-time code modeling/meshing change can be examined in a real-time 3D format. This would give developers a clearer understanding of the impact of code changes on 3D VRI models.

The third activity is M&S and Analysis, which includes visual object analysis. One instance of visualization calls for 3D visualization and analysis, and another, calls for interaction and immersion. Visualization may involve multiple participants and multiple exchanges of data at varying levels of resolution. Information analysts would like to view and share control of 3D objects in cave-like facilities and in advanced telepresence systems. A typical use case for a C3 or C5 venue is to view a slice of a 30 foot reactor or to walk around inside a reactor to view the impacts of corrosion-related unidentified deposits or even to evaluate rod vibration simulating a loss-of-coolant accident.

The last activity evaluated was the lectern or speaker environment and research partnerships. The primary need for having guest speakers, whether for technical exchanges or educational outreach, precipitated the deployment at CASL of an augmented reality (AR) podium using a transparent image plane. This venue is ideal for speakers, teachers, or trainers. The interactive touch venue is also desirable for exchanging instructional information or sharing desktops for collaborative research. The AR podium represents a very advanced form of immersive telepresence. This use case has needs for a telepresence venue and touch environments to support various research workflow activities.

Generally, all users described a need for an ideation space where users can think or "ideate." Ideation is typically done is non-VR and 3D modeling environments. Tools used to ideate usually include the basic ability to sketch and draw, tools not inherent to VR environments or general visualization venues.

5 Challenges to Collaboration System Design

Without question the most difficult task in designing the VOCC system of systems was getting users to clearly identify and/or communicate their true collaboration needs. Most scientists deal with capability and thus technology requirements in a truly adhoc fashion. As a normal course of work execution scientists and researchers do not stop and take inventory of capability or even critically assess their approaches to innovation. It is clearly beneficial for collaborative systems designers to socialize their collaboration strategy, to teach users basic requirements definition and management, and to let users "share in the burden and risk" of choosing technology solutions. Getting users involved in all aspects of technology evaluation will result in their quicker adoption of the collaboration strategy [5].

Secondarily, the most difficult task to collaborative system selection is demonstrating and more importantly quantitatively characterizing the business value afforded by the collaboration venue ("collaboration metrics"). A 2010 Salire Partners (a technology value path provider) report states that 80% of companies see a positive return on investment (ROI) on collaboration technologies [6]. The average payback in months for government organizations is ~33 months. At the end of five years payback should exceed 100%. Frost & Sullivan, the industry research and growth consulting firm, looked beyond the more obvious return on investment (ROI) calculation a company may make before investing in these tools, such as the cost of operating the solution against savings derived from travel avoidance [7]. VOCC will be in the months ahead seriously focused not only on developing meaningful collaboration metrics to quantify typical cost savings, but also those associated with characterizing innovation ROI.

6 Summary

VOCC is a very unique system of systems collaboration platform that will facilitate and expedite desperately needed innovation in nuclear LWR design. Its physical one-roof will provide an optimal co-location environment for industry designers and researchers to more quickly solve complex energy problems. Its extensibility via the virtual one-roof will permit ORNL to share computational and visualization resources with industry in a manner that has never been done before. VOCC wants to democratize its capabilities in high performance computing, visualization, and modeling and simulation with its core partners via collaborative venues. Lastly co-location, whether physical or virtual, spurs collaboration resulting in a convergence of these two disparate yet complementary knowledge spaces.

References

1. Clancy, W.J.: Situated Cognition: On Human Knowledge and Computer Representations, p. 1. Cambridge University Press, Cambridge (1997)
2. http://www.verizonbusiness.com/resources/solutionbriefs/sb_lo oking-for-ways-to-improve-productivity-mobility-management-for-government_en_xg.pdf

3. IBM. An Agile approach to User Experience and Design, http://www-01.ibm.com/software/ucd/agileuxd.html

4. Zhao, Z., Johnson, S., Chen, X., Ren, H., Hsu, D.: Research on a Visual Collaborative Design System: A High Performance Solution Based on Web Service. In: Zhang, W., Tong, W., Chen, Z., Glowinski, R. (eds.) Current Trends in High Performance Computing and Its Applications, pp. 611–615. Springer, Heidelberg (2005)

5. Incorporated, A. S. Building toward a unified communication strategy (2009), http://www.adobe.com/products/acrobatconnectpro/webconferencing/pdfs/connectpro_unifiedcommwhitepaper_1_09.pdf

6. http://www.cisco.com/en/US/solutions/collateral/ns340/ns856/ns870/c11-59761300_return_collab_wp.pdf

7. Frost & Sullivan Develops New Metric, the Velocity of Collaboration, to More Accurately Demonstrate the Value of Collaboration Solutions, http://www.prnewswire.com/news-releases/frost-sullivan-develops-new-metric-the-velocity-of-collaboration-to-more-accurately-demonstrate-the-value-of-collaboration-solutions-102761559.html

One of Industrial Design Case to Share Tacit Knowledge

Hisashi Shima

Lenovo Japan Ltd.
Minatomirai Center Building 21F, 3-6-1 Minatomirai Nishi-ku Yokohamas-city
Kanagawa Japan
shima@lenovo.com

Abstract. The objective of this research is to study the sharing of tacit knowledge, especially, in industrial design development teams. Nowadays, in-house design teams have to be more productive and efficient than they were previously even if inexperienced designers are working on the project. An important challenge in industrial design is to determine a suitable solution or compromise when many factors are involved. We tried to list the important factors involved in finalizing a design and shared these factors with an industrial design team. Experienced and inexperienced designers were made to assign AHP (analytic hierarchy process) scores based on this list according to their personal understanding. First product was not enough same score of AHP, but three times more closed, it is assumed the more and more shared tacit knowledge, with this process.

Keywords: tacit design knowledge, product development, brand design, tacit dimension, empathy development, usability.

1 Introduction

As is the case for all other departments in today's companies, industrial design teams are facing the pressure of becoming more productive and efficient even if they employ young or inexperienced designers. In particular, the product development schedule is an important factor in product development. On the other hand, nowadays, design projects are more complicated than those previously undertaken as they must take into account human factors, marketing, legal, and engineering requirements etc. This paper focuses on quicker and more efficient sharing of tacit knowledge in industrial design teams by the application of AHP (analytic hierarchy process) to products. Generally, AHP is used for deciding the most suitable design. The present process is for understanding and visualizing each designer's tacit knowledge. It will especially help young designers by making it easy for them to recognize how to prioritize the design requirements and decide on the most suitable design during the development phase.

This paper presents a case study in which AHP was applied to the design development of the ThinkPad X-series Tablet Pen. First, AHP scores were assigned by all the designers, and then, these scores were analyzed and compared. It was found that the AHP scores assigned by experienced designer are in greater agreement with management and market expectations than those of inexperienced designers.

M. Kurosu (Ed.): Human Centered Design, HCII 2011, LNCS 6776, pp. 435–439, 2011.
© Springer-Verlag Berlin Heidelberg 2011

2 Methods and Procedure

2.1 Process Concept

To understand previous projects, we started by applying AHP to products that have already been announced and shipped to the market. Fig. 1 is a conceptual chart of the process discussed in this paper. In general, for a given project or product, a development team is formed initially. Sometimes less industrial design experience person have not good compromise point or solution. In our study, the AHP scores assigned by inexperienced designers to the first product differed substantially from those assigned by experienced designers. However, the scores assigned by these designers to the second product were much closer to those assigned by the experienced designers, and in the case of the third product, the scores assigned by both groups of designers were almost the same.

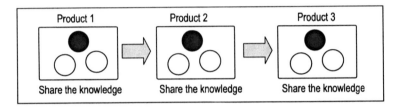

Fig. 1. Process concept

2.2 Process

We present an overview of the process followed by us, in which a team of designers assigned AHP scores.

In AHP, a score matrix is constructed according to a number-based scoring scheme, which is given as follows.

1: Column and row equally important
2: Column slightly more important than row
3: Column more important than row
5: Column significantly more important than row
7: Column exceedingly more important than row

Generally, such a scoring process requires many participants, but the scores presented in this paper were assigned by only three people because in this case, the objective is not "decision making" but "knowledge sharing."

Aesthetic view: Basically, the aesthetic appeal of form and shape; beauty and neatness design.

Usability view: Practical benefits of using the product.

Branding view: The concept of "branding" is not taken lightly in this scoring process. Designers choose shapes that are most suitable from branding perspective so as to conform to the brand strategy.

Affordance view: This factor is purely concerned with an outsider's perspective; it helps in enhancing the user-friendliness of the product design.

Table 1. Decision factor

	Aesthetic View	Usability	Brand View	Affordance View
Aesthetic View		1/5	1	1/3
Usability	5		3	1
Brand View	1	1/3		3
Affordance View	3	1	1/3	

Table 2. Weighing factors

	Weighted
Aesthetic View	0.125
Usability	0.485
Brand View	0.246
Affordance View	0.246

The following tables are example of the first project made the AHP score, the each table are not same as all product and project, this score made for Digitizer pen case. If the score is for notebook pc, the weight and score are different.

The following are the explanations for Figs. 2–5.

Fig. 2. This one is a pre-mass production sample. This shape did not receive wide acceptance from the management because it is not easy to recognize where to push and the feeling is not clear.

Fig. 3. The part of the pen around the click position is recessed.

Fig. 4. The click button is given a step shape; the higher surface indicates where to push.

Fig. 5. A dimple is added to the click button; the dimple is located where the button needs to be pushed.

From the usability viewpoint, the difference between the designs shown in Figs. 3 and 4 is negligible.

Fig. 2. (Case 0)

Fig. 3. (Case 1)

Fig. 4. (Case 2)

Fig. 5. (Case 2)

3 Results and Discussion

The division of labor in the team is as follows: Once a concrete idea is proposed, the design specialist depicts all the elements and factors in a single image, which is then presented to the management.

It was found that the decisions based on the AHP results in Table 5 are the same as the management decisions; in addition, these decisions are the same as those based on the AHP scores assigned by experienced designer. On the other hand, the marketing acceptance is good as well.

Table 3. Score for all cases from the aesthetic viewpoint

	Case 0	Case 1	Case 2	Case 3
Case 0		3	3	1
Case 1	1/3		1	1/3
Case 2	1/3	1		1/3
Case 3	1	3	3	

Table 4. Scores of all cases considering all factors

	Aesthetic View	Usability	Brand View	Affordance View
Case 0	0.375	0.375	0.375	0.137
Case 1	0.124	0.124	0.124	0.137
Case 2	0.124	0.124	0.124	0.313
Case 3	0.375	0.375	0.375	0.412

Table 5. Resulting AHP score

	Aesthetic View	Usability	Brand View	Affordance View	Total
Case 0	0.0469	0.0303	0.0533	0.0337	0.1614
Case 1	0.0155	0.0601	0.0176	0.0337	0.1270
Case 2	0.0155	0.0601	0.0176	0.0770	0.1702
Case 3	0.0469	0.1819	0.0533	0.1014	0.3834

4 Conclusion

The above results show that AHP helps the sharing of each designer's tacit knowledge in the manner depicted in Fig. 1. There was a substantial difference between the AHP scores of the designers for the first project, but by the third phase, the AHP scores of all the designers were almost the same.

However, this paper describes just one case study involving a small team, and the study duration was short. Hence, continuous research with the same team is required in addition to research with other designer teams.

Acknowledgment. I would like to thank our management at Lenovo for supporting this work and my team members for bringing it to a successful conclusion.

References

[1] Merholz, P., Schauer, B., Verba, D., Wilkens, T.: Subject to change. O'Reilly Media, Inc., Sebastopol (2008)
[2] Norman, D.: Emotional Desing: Why We Love (or Hate) Everyday Things. Basic Books, New York (2005)
[3] Norman, D.: The Design of Future Things. Basic Books, New York (2007)
[4] John, S.P., Adlin, T.: The Persona Lifecycle. In: Keeping People in Mind Throughout Product Design. Elsevier Inc., Amsterdam (2006)
[5] Grice, P.: Studies in the Way of Words. President and Fellows of Harvard College (1989)
[6] Utterback, J.M., Vedin, B.-A., Alvarez, E., Ekman, S.: Design-Inspired Innovation. World Scientific Publishing Co. Pte. Ltd., Singapore (2006)
[7] Atkinson, K., Wells, C.: Creative Therapies: A psycholodynamic approach within occupational therapy. Nelson Thornes Ltd. (2000)
[8] Polanyi, M.: The Tacit Demension. Routlege & Kegan Paul Ltd., London (1966)
[9] Simple AHP process web site, http://www.oidc.jp/mono/MonoMain/method5_1.htm (access March 01, 2011)

Task Analysis for Behavioral Factors Evaluation in Work System Design

Lingyan Wang and Henry Y.K. Lau

Department of Industrial and Manufacturing Systems Engineering,
The University of Hong Kong, Hong Kong, P. R. China
{lywang,hyklau}@hku.hk

Abstract. This paper deals with the application and development of a systematic methodology called Task Analysis which is based on the analytical investigation of the task allocation processes and bottlenecks in terms of work system goals, in order to evaluate synergy between worker's essential motions and mental activities of different functional levels which contributes to conduct worker's adaptive behavioral performances during the execution of production operation. A comprehensive consideration of adopting this approach to analyze some key behavioral factors in work system design is expanded to acquire consecutive work performance feedback, determine the instructional work goals, describe the detailed work flowchart, structure the clear interaction assessment, improve the standard procedures, and supply the useful criteria.

Keywords: Hierarchical Task Analysis, Cognitive Work Analysis, Behavioral Factors, Work System Design.

1 Introduction

Human is a key element in safe and reliable industrial production processes, effective and efficient behavioral function allocation of human operators in the work system that able to accommodate flexible production demands continues to be of interest to a substantial amount of automated manufacturing enterprises due to the dexterity and flexibility of human workers' manual handling operations in conjunction with the automation.

Task Analysis explicitly analyses how a predetermined work task is appropriately carried out to achieve by workers' physical activity and cognitive processes according to the inherent work system goal. It includes a detailed description of task definition, task structure, task allocation, task duration, task frequency, equipments, environmental conditions, and any other declarative or procedural factors required for workers to perform a given task. Then the structured behavioral functions operating procedures and rigorous characterized conditions flow to be included within the work system can be accurately specified and conducted. In other words, Task Analysis is the study of the way workers perform their work, it takes account of what they do, how they do, and why they do. Eventually, the capability of Task Analysis is extended to provide and develop a framework for the investigation and representation of workers' behavioral performances to facilitate the design of complex work system.

M. Kurosu (Ed.): Human Centered Design, HCII 2011, LNCS 6776, pp. 440–448, 2011.
© Springer-Verlag Berlin Heidelberg 2011

The entire development of this research conducive to implementation in work system design is illustrated in a modeling study of workers' behavioral factors evaluation, and the outline is arranged as follows: Section Background gives a brief overview of Task Analysis and sets the objective for effectuation; Section Method provides a clear and precise description of the procedures used in task analysis; Section Framework lays out the main features of this approach via its functional model; Section Conclusion discusses the merits of the proposed framework and presents the relevance of possible future work.

2 Background

2.1 Task Analysis

Task Analysis is studied in relation to the context in which it is performed to examine how specific work flows through understanding and assessing behavior of human, machine or its combination, it is a critical stage in work system design that involves task definition, task decomposition, task data collection, data analysis, and documented representation production of the analyzed task suitable for work system purpose [1]. The importance of Task Analysis and numerous methods have been well documented in the literature of various human factors and ergonomics studies, such as interface design, usability evaluation, man-machine collaboration, system control, error reduction, workload measurement, and so on.

Richardson et al. presented a reformulation of Hierarchical Task Analysis that focused upon the analysis of user goals rather than an existing task implementation to encourage novel and apt interface design based on the sub-goal template scheme which provided a notation for goal-oriented task analysis [2]. Norros and Savioja developed a new activity theory based approach named Core Task Analysis and an integrated evaluation method named Contextual Assessment of Systems Usability to analyze the appropriateness and acceptability of human conduct in complex work system with high usability and reliability requirements [3]. Tan et al. extended task analysis capability in hierarchical task analysis structure to model the collaboration between human and robot in cell production operation system for the sake of effective man-machine system [4]. Barbera et al. derived and organized a task-decomposition-oriented methodology to acquire and structure the complex real-time processing control system for the control tasks of autonomous vehicles [5]. Carstens et al. used task analysis to identify a web-based healthcare delivery system process flow affiliated with elder patients transitioning, conducting to reduce the likelihood of error and gain improvement within the system [6]. Dey and Mann performed a complete task analysis which was consisted of a structured written questionnaire and subsequent observation of experienced sprayer operators to measure the workload of operating an agricultural sprayer equipped with a navigation device [7].

These task analysis methods can be classified into two different categories as sequential approach and contextual approach [8]. The sequential model of task analysis like Hierarchical Task Analysis [9] or Goal Operator Method Selection [10] is considered as a goal oriented procedure of hierarchical sub-goals achievement via going through a sequence of formalized processing actions serially reached by the operator.

And the contextual model like Executive-Process Interactive Control [11] is formulated for modeling human multiple-task performance while the processor activates a wide range of cognitive capabilities to cope with work constraints in the scope of a known task without a predetermined restrictive sequence. Both of these two approaches have their own strengths and weaknesses, the sequential one is specific but lack of flexibility to unexpected emergencies, while the contextual one has the property of adaptability but hard to control with identical standard, it is only through integration that can give perfect records to the task analysis for behavioral performance.

2.2 Objective

This paper aims to create a new structured model which has the potential to bridge the gap between fixed and flexible working cell, and then provide a way to improve the implementation of behavioral performance measures involved in work system based on the theoretically rigorous approach initiated by Task Analysis. It is proposed that this integrated framework, by combining the sequential model Hierarchical Task Analysis together with the contextual model Cognitive Work Analysis [12], as well as incorporating the generic procedures of task analysis embedded in the work domain into it, will greatly enhance the understanding of behavioral factors situation, and then design a suitable work system.

3   Method

3.1 Hierarchical Task Analysis

Hierarchical Task Analysis (HTA) typically provides a model to analyze and represent the behavioral components like setting goals, defining tasks, identifying subtasks, planning operations, making diagnoses and decisions that essentially occur during a complex task executing in a wide context. It helps to graphically sketch the overall task, show the correct sequence and guide the formulation of constraints by using a practical structure chart to breaks tasks into subtasks and operations which interact through various inputs and outputs without focusing on too much detail. In the first instance the prospect for HTA as an adjunct tool to describe and analyze a complicated work system in terms of its goals which are expressed via some real operation units and objective criteria is quite capable of producing a goal-based systems analysis. And then the operation is broken down into subordinate operations which are defined by sub-goals with the purpose of producing an outline of the hierarchy that is concern with the adequacy of the hierarchical relationship description between the goals and sub-goals.

There are some basic heuristics and broad principles for proposing a framework within which HTA can be conducted to guiding the progressive goal description and adaptation other than a rigidly prescribed convention: define system purposes and boundaries, access multiple sources of system information, describe system goals and sub-goals, review the sub-goal groupings and triggering conditions, stop re-describing sub-goals at right point, and revise the final sub-goal hierarchy [13].

It is HTA that used in multifarious circumstances. Ainsworth and Marshall applied a universal task analysis technique HTA in the armed services and nuclear industries to allocate function, identify human error and assess systems by direct observation, questionnaire, and scenario modeling [14]. Shepherd devised a tabular format to illustrate the task taxonomy which analyzed constraints of the tasks and their associated sub-goals, for the purpose of investigating redesign opportunities in a batch control process [9]. Hellier et al. helped to uncover the complexities of HTA as a basis for predicting potential errors of a chemical sample analysis procedure by accomplishing observational studies and interviews with chemists [15]. Marsden and Kirby circumvented allocation problems of system function by focusing attention on the purposes of enabling impartial function allocation via sub-goal hierarchy in HTA [16]. Lane et al. demonstrated how HTA can be used as a systematic human error prediction and mitigation technique to prevent error or reduce the effects of error by means of modeling the process of drug administration [17].

3.2 Cognitive Work Analysis

Cognitive Work Analysis (CWA) is gaining momentum as a structured evaluation approach which emphasizes on the analysis, evaluation, and design of the complex and dynamic socio-technical systems for investigating and examining the significant effect of some key psychosocial factors on the performance of work system, thus to measure, implement, and improve the procedures for psychosocial personnel subsystem analysis in work system design. It leads to the consideration of the reason why the work system exists, the environment where the work takes place, the constraint that the system ability is performed, the domain that the activity is conducted within, the way how this activity is achieved, and the people who is performing it [18].

A theoretic foundation is extended the basic concepts of CWA by discussing the methodological guidelines as five defined phases: Work Domain Analysis, Control Task Analysis, Strategies Analysis, Social Organization and Cooperation Analysis, and Worker Competencies Analysis. It contributes to the general understanding of specific work-related psychosocial factors which are grouped into five categories as work demand, work relationship, work perception, work autonomy, and work reward. Comprehensive cognitive task analysis data and questionnaire data of workload, performance feedback, work content, support, conflict, stress, benefit, morale, communication, and union can be specified at each stage of the five CWA phases through the theoretical and empirical work, in order to estimate the structural equation modeling of the work system performance.

CWA was originally developed at the Risø National Laboratory in Denmark [19], and has been developed and applied in a wide range of socio-technical domains. Higgins explored the supervisory control of discrete manufacture by extending the suitable usage of CWA in particular Work Domain Analysis [20]. Naikar and Sanderson proposed a complementary framework for evaluating design proposals of a new Airborne Early Warning and Control System by describing its unique characteristics based on developing the first phase of CWA [21]. Miller presented a recursive diagnostic framework which conformed to the broader aspirations of CWA to represent an appropriate foundation for patient system information display [22]. Ahlstrom described how the CWA modeling tools could help to extract the development of weather display concepts and set up a high-fidelity simulation environment, thus to provide possible improvements in aircraft efficiency and safety of terminal operations

[23]. Jenkins et al. evaluated the advantages of exploring the Social and Organizational Analysis in the military domain through introducing a constraint-based description which is focused on the transfer of information and optimum working practices between workers within the system [18].

3.3 Procedures of Task Analysis

Task Definition. In general, task is a combination of goal and operation, the goal is a desired state of affairs while the operation is an activity for attaining this goal. Therefore, the basic idea of task refers to the purposeful work activities that user is attempting to accomplish for the overall system goal. Task definition involves a description of what the user has to do (the mission), what the user needs to know (the domain), and how the system supposes to work (the way) in the context of the whole system consisting of integrant parameters like machines, humans, skills and knowledge.

Task Decomposition. Task decomposition is the way how a task is split into subtasks to identify what goals, plans, and operations are involved in the overall task. Developing initial task decomposition can help to break the whole task situation down into sub-parts for further analysis of how the functions are allocated and how the operations are arranged. Function allocation represents the causal and sequential information-flow relationships of the functions performed in a rigorous measure. Operation is a generic term emphasizing on the associated series of actions that users carry out in the prescribed sequence, and each operation in turn could lead to a hierarchical structure.

Task Data Collection. It is necessary to resort to collecting information about task parameters and components which is accessible to compose the user's situation and activities while performing the task. Here are some useful methods for task data collection: sampling, observation, questionnaire and interview. To explain these in detail, sampling means the thorough process of selecting a random sample, observation refers to the documented description of observing actual user's behavior with little interference, questionnaire is a fixed set of some simple questions about the task information, and interview involves talking to users with a series of systematic questions about all kinds of task comments.

Task Data Analysis. Once the task data is collected, the next step is relevant to analyze it on the practical aims of producing a qualitative and quantitative synthesis of the detailed characteristics of a given task that is propitious to specify the design of the work system. This analysis goes beyond subjective intuitions or conjectures, and it concentrates on the interpretation of the data with classic statistical criteria in schematic forms which are derived from task behavior assessment methods like Event Trees Analysis [24], Hazard and Operability Study [25], Failure Modes and Effects Analysis [26], and Influence Diagrams [27]. The most important thing revealed by data analysis is that the analysis result is directly serving as reference material of great benefit to assessing and identifying critical bottlenecks in the task flow, specifically low-value activities, excessive workloads, inconsistent procedures, and so on.

Task Representation. Traditionally, the key function of representing the task is to provide a blueprint of visible and apparent task structure that supports the user to understand what aspects of the task are predominant, how much detail they are represented, and what sequences they are obeyed after collecting information into systematic format to sustain task execution. A logical representation that states and documents exactly what the task involves on the basis of its goal is structured by using some deductive approaches which are derived from graphical flowcharts or diagrams like Function Performance Diagrams, Input-output Diagrams, Critical Process Charts, Information Flow Charts, and Signal Flow Graphs [28].

4 Framework

4.1 Theoretical Framework

A theoretical framework which is founded on the aforementioned Hierarchical Task Analysis in tandem with Cognitive Work Analysis, namely Generalized Task Analysis Framework is presented in Fig. 1 to offer a heuristic procedure for capturing both the behavioral and cognitive aspects of work activities according to the main work aims in complex and dynamic industrial systems. The description of this framework is generally generated as decomposing a set of goals and states with constraints into their constituent sub-goals hierarchy in successive stages, as well as executing the sequential performance by which the final sub-goal hierarchy with plans are achieved to carry out the actions intervened in different cognitive representation situations that workers are likely to meet while operating the task.

It is therefore the central feature of GTAF in an overall sense is that it consists of two complementary unit named Sequential Processing Unit and Contextual Processing Unit for structuring and adjusting process of simple work actions which are the lowest level of task decomposition and have no further structure. The Sequential Processing Unit focuses on representing the mechanisms of work system through the breakdown of prototypical task into an extensive hierarchy charting of goals and operations, meanwhile, the Contextual Processing Unit is explicitly concerned with measuring behavioral human performance in conjunction with human cognitive variation in context of task data analysis execution stage, after the initial task definition with a high proportion of determining the instructional goals and also the constraints that affect the ultimate goals but cannot be changed.

To summarize, this structural modeling depicts the tangible way in which an exhaustive guide of task analysis is implicated for analysis, verification, implementation and development at each step within the process to determine more detailed tasks goals, resources, constraints, priorities, functions and processes. For clarity, an intensive theoretical layered description of Generalized Task Analysis procedures is organized as follows: defining and describing the task goals and domain in detail; refining task goals and splitting task into object based hierarchical subtasks until the basic task at the lowest level is simple enough to act on, then prioritizing and sequencing these sub-tasks; observing worker behavior and collecting empirical data; assessing and extracting the behavioral data to clearly manifest endogenous human variables feedback as reference; arranging and modulating a coherent task procedure to meet the primary task goal and fills out the principal sequence of steps in accordance with the actual behavior.

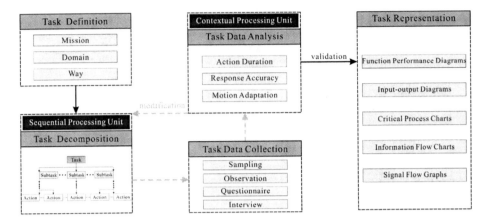

Fig. 1. Generalized Task Analysis Framework

4.2 Mathematical Framework

In addition to the previously mentioned qualitative interpretation, a quantitative consideration is proposed to examine the significant effect of three key behavioral factors on the work system performance via a composite measurement of the empirical performance data Action Duration, the observational cognition data Response Accuracy, and the available questionnaire data Motion Adaptation. A mathematical framework is defined to takes into account these synergy variables for quantifying the generalized behavioral performance output value, and the function is given by:

$$G = \sum_{i=1}^{n} (\sigma_d d_i + \sigma_a a_i + \sigma_m m_i) \tag{1}$$

Where σ is a parameter which determines the priority weight for each variable, d_i corresponds to the action duration, a_i is a rate of response accuracy, m_i is a degree of motion adaptation, n is the number of subtasks at the lowest hierarchical stage.

5 Conclusion

To conclude, this proposed conceptual framework incorporates Hierarchical Task Analysis as well as Cognitive Work Analysis into work system design, and thereby it offers a number of potential benefits to the behavioral factors evaluation aspect of the highly complex and dynamic production system which consists of extensive interactions between human, machine, and such relevant work factors. It raises work efficiency, work accuracy, and worker adaptation by adjusting work goals, assignments and procedures in accordance with consecutive worker performance variation feedback. It also could be used to ensure work safety through investigating work hazard, and increase work system availability via assessing work reliability.

The possible development trend of Task Analysis in work system design is Green Task Analysis because low carbon production is generally recognized as a valuable aid to harmonious society from an environmental tactical perspective. Consequently, the next stages of this research will focus on enhancing awareness of green inception in task analysis process through developing a green model for appropriate work performance evaluation, since minimizing waste is a universal goal for workers regardless of their specific objectives. This Green Task Analysis could be leveraged to energy saving and emission reduction, and eventually, in order to accommodate the great strategy of sustainable development.

References

1. Diaper, D., Stanton, N.: The Handbook of Task Analysis for Human-Computer Interaction. Lawrence Erlbaum Associates, Mahwah (2004)
2. Richardson, J., Ormerod, T.C., Shepherd, A.: The Role of Task Analysis in Capturing Requirements for Interface Design. Interact. Comput. 9, 367–384 (1998)
3. Norros, L.L., Savioja, P.J.: Towards a Theory and Method for Usability Evaluation of Complex Human-technology Systems. @ctivités 4, 143–150 (2007)
4. Tan, J.T.C., Duan, F., Zhang, Y., Kato, R., Arai, T.: Task Modeling Approach to Enhance Man-Machine Collaboration in Cell Production. In: The 2009 IEEE International Conference on Robotics and Automation, Kobe, Japan, pp. 152–157 (2009)
5. Barbera, T., Albus, J., Messina, E., Schlenoff, C., Horst, J.: How Task Analysis Can Be Used to Derive and Organize the Knowledge for the Control of Autonomous Vehicles. Robot. Auton. Syst. 49, 67–78 (2004)
6. Carstens, D., Patterson, P., Laird, R., Preston, P.: Task Analysis of Healthcare Delivery: A Case Study. J. Eng. Technol. Manage. 26, 15–27 (2009)
7. Dey, A.K., Mann, D.D.: A Complete Task Analysis to Measure the Workload Associated with Operating an Agricultural Sprayer Equipped with a Navigation Device. Appl. Ergon. 41, 146–149 (2010)
8. Baindbridge, L.: The Change in Concepts needed to Account for Human Behaviour in Complex Dynamic Tasks. IEEE T. Syst. Man. Cy. A. 27, 351–359 (1997)
9. Shepherd, A.: Hierarchical Task Analysis. Taylor & Francis, New York (2001)
10. Kieras, D.: GOMS Models for Task Analysis. In: Diaper, D., Stanton, N. (eds.) The Handbook of Task Analysis for Human-computer Interaction, pp. 83–116. Lawrence Erlbaum Associates, Mahwah (2003)
11. Rubinstein, J.S., Meyer, D.E., Evans, J.E.: Executive Control of Cognitive Processes in Task Switching. J. Exp. Psychol. Human 27, 763–797 (2001)
12. Vicente, K.J.: Cognitive Work Analysis: Toward Safe, Productive, and Healthy Computer-based Work. Lawrence Erlbaum, Mahwah (1999)
13. Stanton, N.A.: Hierarchical Task Analysis: Developments, Applications, and Extensions. Appl. Ergon. 37, 55–79 (2006)
14. Ainsworth, L.K., Marshall, E.: Issues of Quality and Practicability in Task Analysis: Preliminary Results from Two Surveys. Ergon. 41, 1607–1617 (1998)
15. Hellier, E., Edworthy, J., Lee, A.: An Analysis of Human Error in the Analytical Measurement Task in Chemistry. Int. J. Cogn. Ergon. 5, 445–458 (2001)
16. Marsden, P., Kirby, M.: Allocation of Functions. In: Stanton, N.A., Hedge, A., Brookhuis, K., Salas, E., Hendrick, H. (eds.) Handbook of Human Factors and Ergonomics Methods, pp. 34-1–34-8. Taylor & Francis, London (2005)
17. Lane, R., Stanton, N.A., Harrison, D.: Applying Hierarchical Task Analysis to Medication Administration Errors. Appl. Ergon. 37, 669–679 (2006)

18. Jenkins, D.P., Stanton, N.A., Salmon, P.M., Walker, G.H., Young, M.S.: Using Cognitive Work Analysis to Explore Activity Allocation within Military Domains. Ergon. 51, 798–815 (2008)
19. Rasmussen, J., Pejtersen, A.M., Goodstein, L.P.: Cognitive Systems Engineering. Wiley Interscience, New York (1994)
20. Higgins, P.G.: Extending Cognitive Work Analysis to Manufacturing Scheduling. In: Proceedings of OzCHI 1998, pp. 236–243. IEEE Computer Society, Adelaide (1998)
21. Naikar, N., Sanderson, P.M.: Evaluating Design Proposals for Complex Systems with Work Domain Analysis. Hum. Factors 43, 529–542 (2001)
22. Miller, A.: A Work Domain Analysis Framework for Modelling Intensive Care Unit Patients. Cogn. Tech. Work 6, 207–222 (2004)
23. Ahlstrom, U.: Work Domain Analysis for Air Traffic Controller Weather Displays. J. Saf. Res. 36, 159–169 (2005)
24. Ferdous, R., Khan, F., Sadiq, R., Amyotte, P., Veitch, B.: Handling Data Uncertainties in Event Tree Analysis. Process. Saf. Environ. 87, 283–292 (2009)
25. Dunjó, J., Fthenakis, V., Vílchez, J.A., Arnaldos, J.: Hazard and Operability (HAZOP) Analysis. A Literature Review. J. Hazard. Mater. 173, 19–32 (2010)
26. Urbanic, R.J., ElMaraghy, W.H.: Using a Modified Failure Modes and Effects Analysis within the Structured Design Recovery Framework. J. Mech. Design 131, 111005-1–111005-13 (2009)
27. Howard, R.A., Matheson, J.E.: Influence Diagrams. Decis. Anal. 2, 127–143 (2005)
28. Stanton, N.A., Salmon, P.M., Walker, G.H., Baber, C., Jenkins, D.P.: Human Factors Methods: A Practical Guide for Engineering and Design. Ashgate Publishing Limited, Burlington (2005)

Understanding the Business Realities: An Interview Technique Which Can Visualize the Job Problems

Ayako Yajima[1,2], Yuji Shiino[2], and Toshiki Yamaoka[1]

[1] Department of Design and Information Science Faculty of Systems Engineering
Wakayama University,
Sakaedani 930, Wakayama city, Japan, 6408510 Wakayama Japan
[2] FI Techninal Center Field Innovation Unit, Fujitsu Ltd.
yajima.ayako@jp.fujitsu.com

Abstract. We have developed the Customer Satisfaction (CS) Gap interview and analysis method. This method is based on ethno-cognitive interview and analysis method which is a method to grasp the business reality. For using this method, we tried that it was applicable in visualization in gap and an analysis of consciousness of customer satisfaction between receiving services and offering services. We interviewed six people and the time required about per 1 person for 1.5hours.We carried out it both Service recipient side that felt law customer satisfaction and service provided (that is ourselves). As a result there is a clear difference in the value of the CS, we are able to catch CS gap structurally.

Keywords: Customer satisfaction, qualitative and quantitative method, CS gap, framework, Cognitive psychology, ethnography, ethno-cognitive interview.

1 Introduction

When the system development is analyzed about failure factor in one project, a service provider is said that it is important to catch the real business field, action, their consideration. However, the target domain that becomes IT is complex in recent years. Therefore, the gap has extended about their wants between the management layer, user section, and a system section. There is a risk to rely on the only requirement from customer's counterpart. It is not considered to how to get communication and hearing, the method of the interview. In this study, we have developed qualitative and quantitative method that could capture customer satisfaction gap interview and analysis technique structurally.

1.1 The Survey of Customer Satisfaction

The satisfaction investigation for the service provider is carried out regularly by a specialized military unit in and out of the office. In Japan, JCSI bring the evaluation index that a company comparison is possible and defines the excellent company using the index. In addition, they carry out the calculation of the benchmark, and the index and result helps customer understanding regardless of industry. Even if we execute the survey of customer satisfaction, the result is interpreted in each section and the

M. Kurosu (Ed.): Human Centered Design, HCII 2011, LNCS 6776, pp. 449–457, 2011.

post and is used as an improvement index only in the best shape in the part. In a lot of enterprises, though the result of the survey of customer satisfaction is important in corporation principles, an appropriate question cannot be asked for the reasons that the definition of CS cannot be clarified. And they are not able to catch the customer's evaluation correctly.

The evaluation of the satisfaction rating is difficult. Even if the service provider has given the value and satisfied for customers, the customer doesn't feel value. Even if we investigate and analyze mean of customer satisfaction, service provider can't offer the value for service recipient. To catch the gap of the satisfaction rating, It is necessary to understand the expectation and the perception gap to both potential during the organization and the organization. Therefore, it is necessary for CS gap investigation technique to be able to structurally catch the gap.

1.2 Problem of Survey of Customer Satisfaction

There are two problems in the survey of Customer Satisfaction.

1) Problems in search procedure

We should consider the participant cooperates in satisfaction rating investigation. It is difficult whether we are doing an appropriate question to an appropriate participant to the investigation. Therefore, it is necessary that the investigator have a new framework that can make the expectation and the perception gap of the satisfaction rating visible.

2) Quality of information that can be acquired for investigation and problem in the interpretation.

It is difficult even if the service recipient side expresses dissatisfaction, the demand, and the expectation to the service provider and it is difficult how extremely precise it to be. In the company, the result of the survey of customer satisfaction doesn't seem to be related to the achievement for the short term.

1.3 Improvement of Technique and Analysis Technique of Survey of Customer Satisfaction

About general technique for investigating customer satisfaction measurement, A lot of company are executing the questionnaire and hearing according to the scale, though it differs depending on a type of business and a corporate policy.For a general investigation process, it is as follows.

①Project and design: Decide the survey technique according to the research subject.
②Making of assorted traits: Try to set the question sentence and the question order.
③Investigation: It investigates by using the made questionnaire.
④Date collection and analysis : Using the Framework that we originally set
⑤Making of result report material: Make it to visible like understanding the gap

Before we investigate, it is decided for us to customer's needs, requirement, and range of investigation. Niger et al (2010) said that it is important to show the relation between the level and the goal of the offered satisfaction rating. In the marketing area, the survey design is done by becoming not only the senior customer segment but also a senior customer and using the model by whom the people who get it are assumed.in the project design. Moreover, it tends to be decided that the surveying technique is done by either questionnaire survey or group interview, it is necessary for survey of CS to execute both the qualitative research and the fixed quantity investigation, mixed methodology.

2 Surveying Technique - For Proposing Technique

2.1 Idea of Investigation

To use the survey of customer satisfaction effectively, it proposes the following methods,

1) Supplementation by interview that uses theory of cognitive psychology
Information that can learn the background and the essence that relates satisfactorily about the service offer side and the service acceptance side cannot be taken only by the questionnaire.

The gap in customer satisfaction measurement is an expectation and a gap between organizations that the service offer sides and are each service acceptance sides. To know the gap is to trace it to the source of the gap. Not only the questionnaire investigation but also interviews are combined and the investigation is executed for that. Using the interview that puts element of service acknowledgment psychology and matrix framework that can structurally catch content, An investigator can catch not only respondent's opinion but also the section that surrounds them and their opinion. It is most important for the service provider to have to accept negative comment and candid advice.

2) Visualization of CS gap -Meaning of structural arrangement of gap
In text information written in what the participant spoke and the questionnaire, a lot of significant content exists. It is talked about as an episode. It is important that we show those qualitative information in shape that the member who is not participating in investigation understands the context that the gap is caused as much as possible.

3) Feedback of survey of customer satisfaction
The service provider shares the result with the service recipient.The service offer side can construct mutual trust with the service recipient side by executing the above-mentioned three points. The service offer side accumulates the method of canceling the gap (fit) as knowledge of the organization, and the profit is used.

3 Finding Out of CS Gap by Interviewing and Analysis

This survey method is a mixture investigation method that combines the quantitative survey with a qualitative investigation. The quantitative survey is questionnaire that hears own evaluation and service recipient's evaluation expectation. We investigate it both service recipient side and service provider side. Concretely we use the interview method based on the theory of ethnography and cognitive psychology interview and the structurizing interview that makes key word in process along time series starting point.

3.1 Interview Technique Based on Cognitive Psychology

The method is an interview technique for combining theories of the ethnography and the cognitive psychology. This is a technique to which a current action and the awareness of the issues of "Person" can be acquired by a multipronged cut by the use of the viewpoint and the word of the person who talks as much as possible. Ethnographic interview that Spradley(1997) advocates classifies "Open-ended Question" into three.

i) Descriptive question: Question to talk about specific scene and the realities in own word
ii) Structural question: Question that expresses caught viewpoint how the interview participant recognize their information structurally and their specific scene.
iii) Comparison question: Question to know difference when two events or more are arranged, mainly the difference that the person for the interview thinks about.

On the other hand, the interview based on the cognitive psychology is applied to the investigation interview method that the police in Britain developed.
This technique defines four aspects.

①A story not related to the content to be heard of is done.
②It hears it by the time series not only for order but for reversely.
③Aspect change method
④The action is expressed by the aspect of the artifacts used when they work and move that area.

The interview technique for building in these four aspects is developed, and the part is used for the CS gap interview and the analysis. Even the person who did not know the theory was made to be able to make useful seast where these aspects were built in, and to do a question necessary to take information necessary for the CS gap.

3.2 CS Gap Interview and Analysis Method

The CS gap interview uses and executes result of the questionnaire and the question work-sheet executed beforehand.

It is an interview method of flexibly advancing the question according to other party's talking while presenting the question work-sheet between the interviewer and the person for the interview.

1) Question work sheet
As for this sheet, six kinds of standardized forms are prepared. For the interview participant's business and consideration, we can find out to hear their mind and ordinal behavior by using the sheets that including two more aspect such as the interpersonal relationship axis, the time axes, and the space axes.
2) Questionnaire item
The questionnaire item of each process related with the customer on the service provider side is set. Concretely, there are setting about 4 phases, Project proposal /analysis design /development introduction/ operation maintenance. 11~15 items are set to each phase. Moreover, common cooperation between the knowledge legend and the section to all the above-mentioned phases etc. set about 15 items as a common denominator. The phase that relates to the CS gap to which it wants to evaluate the service provider side is selected and executed based on the set item.

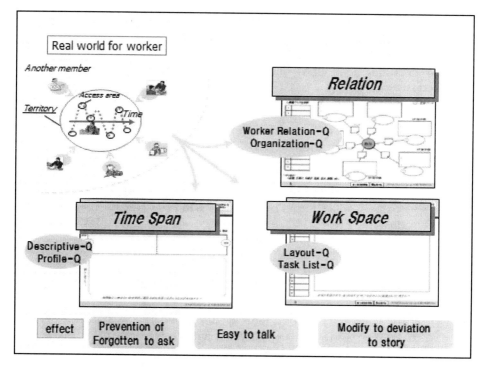

Fig. 1. Main of original question worksheet. six kinds of standardized form. For CS Gap, we uses "Relation" sheet. Especially, when you interview the CS Gap, The seat of the relation axis is customized.

The answer column is constructed with two elements. At first, It numbers it in order of the item attaching importance. The answer of each item of the question by the following aspects is selected from the measure for evaluation. About each item are

   · Self-evaluation on service provider (recipient) side
   · Forecast of evaluation of service recipient side (provider) that service provider (recipient) thinks about.

3) CS gap matrix
This matrix sheet is used after Questionnaire survey and interview survey when analyzing them. We analyze the text data which was written in a free answer on questionnaire and the episode that had been talked by the interview. The content is caught by the flow of the time series with the qualitative data as "Mass of the discourse". And, the interpretation of the content is added and the label of short sentences is added.

The CS gap matrix consists of five rows. A gap that expects of the service offer side (acceptance side) and is current is expressed by five stages.

The mapping does the text data that corresponds to each gap.

Five stages of the gap are five of the following.
·Talking that service acceptance side (offer side) expects of service offer side (acceptance side)
·Talking concerning going of service offer side to expectation from service acceptance side actually
·Talking to hope for correspondence to service offer side compared with expectation from service acceptance side
·It talks including the demand and the opinion on the service acceptance side that the service offer side one-sidedly thinks about though there is no expectation from the service acceptance side.
·Talking including service offer side's having gone to demand of service acceptance side now

It is personally unrelated to CS, and fills it in on the remarks column when there is an important talking.

3.3 Execution Procedure to Achieve Investigation

The flow of the investigation is as follows.

①Prior answer to questionnaire
The questionnaire survey executes it from the interview to the service provider side and the service recipient side. The same questionnaire is investigated.
②Execution of interview that uses question work-sheet
An original question work-sheet is used. It talks about person's relation and the event along the time series and it talks about the episode, the awareness of the issues, and the desire of the current state to the starting point.
③Structurizing interview using questionnaire item
The points in evaluation is applied to the service recipient side and both service provider side for a certain key word (for instance, organizational strength and technology, etc.). The difference between the gap and the gap is caught overall.
④In-depth interview for ②,③
The episode that arrives at the background of the talked content and the applied points in evaluation is deeply heard.
⑤The mapping does prior result of the questionnaire based on all investigations. Moreover, the content talked about by using the CS gap matrix is arranged and analyzed
⑥Document creation

The result is brought together in shape that the service recipient side and both service provider side can be shared.

4 Trial: Case Study in IT Vender –Improvement of Customer Satisfaction

4.1 Objectives

We try to investigate the Big customer which were with remarkably low satisfaction rating to enterprise on service provider when the system is introduced.

In that case, we applied the developing this technique for this scene and verified the effect.

4.2   Areas of Interview, Participant and Procedure

There are 6 subject in total as for the service provider side and the service recipient side. The service provider side are 3 people, and section chief in the systems engineer and business section at the financial department. The service acceptance sides are three people, and section chief classes in the system department. In this case, the mixed investigation method was executed, which is a combination of the questionnaire survey with the interview that applies the cognitive psychology.

The questionnaire and the interview were executed to three in-house people who were the service offer sides in the beginning. Next, the questionnaire and the interview were executed to three of the customer system departments that were the service acceptance sides. We interviewed it one by one and had the time for 1.5 hour/man interview.

4.3   Areas of Analysis and Procedure

The investigator interprets the questionnaire and the content of the interview based on the framework of the CS gap both customer and IT vender in house. We do the mapping to the CS gap matrix as a result of the interpretation. The talked content can able to acquire the remark of which part of the gap matrix, and check the tendency.

4.4   Results

1) CS gap from questionnaire
The points in evaluation of the questionnaire was quantified and the mapping was done to four sections quantitatively as shown in Fig.2. We knew 3 important findings.

i) There is a clear gap in the evaluation of the customer and the IT vender. And, the perception gap is being pointed out also by the customer.
ii) The IT vender side also recognizes customer's dissatisfaction. However, the recognition is not transmitted to the customer.
iii) There is no perception gap. However, it is dissatisfied with the IT vender and both customer. They want to clear dissatisfaction, however they do not understand why it to be good.

2) CS gap from CS gap matrix
From the content talked about by the interview, it analyzed and expressed what the cause of the CS gap was. As a result, it has been understood that the basis of the CS gap is in the difference of the recognition that originates in the restrictions of an organization each other and the system, etc. The information system construction is one of the its own management means for the customer.

However, for IT vender, that really is a business. The idea concerning these key words (Suggestion skills/Developmental skills/organizational strength/project management skills/communications skills) has the gap between the IT vender and the customer. So the customer keeps feeling odd compared with the IT vender while advancing the matter.

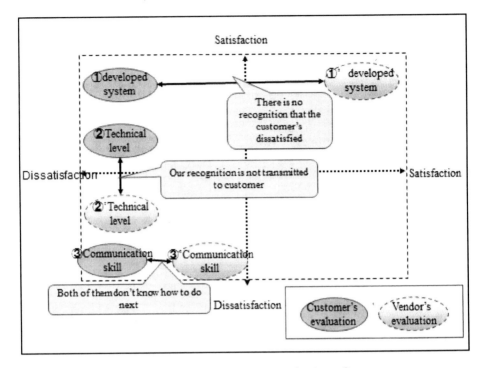

Fig. 2. How is the customer evaluating us?

4.5 Consideration

When you see the CS gap by using the development technique, they found the difference of recognition according to restriction of cultural thought and organization system that is existed behind customer and IT vender. It existed about two point from its difference as follow,

· Gap by comparison with vender of the other companies
· Gap caused by sticking to of both parties done by specific episode

When comparing it in shape to see both of the consideration of the IT vender with the customer, it is understood that the customer evaluation has already fallen even if the IT vender range of the responsibility is exceeded. Customer satisfaction measurement doesn't go up even if the vender advances the project well within the range of the trust.

5 Conclusion

We develop the survey technique for the customer satisfaction gap interview and analysis technique. The technique is constructed with the questionnaire and the interview based on the theory of the cognitive psychology and the social science. The interview tool and matrix frame is used with a new approach. As for the effect to use the CS gap matrix and the framework, it has been understood whether the interview is structurally caught, and where it is the essence of dissatisfaction. It became possible to be searchable as customer satisfaction measurement was structurally caught by trying the proper move method. The effect to use the framework like the CS gap matrix etc. gripped the interview log and the tendency to the talked content was able to be gripped by doing the mapping as it was in the customer starting point.

The accuracy of the technique will be improved in the future.

And, it keeps verifying covering and the accuracy of the way of hearing by using the text mining technology.

References

1. Japan Information Technology Services Industry Association, http://www.jisa.or.jp/e/
2. Inagaki, I., et al.: Engagement management strategy. Nihon Keizai Shimbun publisher, Japan (2010)
3. Spradey, J.P.: The Ethnographic Interview. Sage, Thousand Oaks (1997)
4. Hilgard, E.R., Loftus, E.F.: Effective interrogation of the eyewitness. International Journal of Clinical and Experimental Hypnosis 27, 342–357 (1979)
5. Miline, R., et al.: Investigative Interview-Psychology and Practice (Japanese translation) Kitaoji syobou (2004)
6. Yajima, A.: Attempt of visualization in IT industry by fieldwork. Journal of the Society of Instrument and Controle Engineers 48(5), 411–416
7. King, N., Horrocks, C.: Interviews in Qualitative Research. Sage, Thousand Oaks (2010)

Part V

Applications of Human Centered Design

Nonspeech Sound Design for a Hierarchical Information System

Rafa Absar and Catherine Guastavino

School of Information Studies, Centre for Interdisciplinary Research on Music Media and Technology, McGill University, 3661 Peel, Montreal, QC, H3A 1X1, Canada
{Rafa.Absar,Catherine.Guastavino}@mail.mcgill.ca

Abstract. This research describes a human-centered design methodology for creating nonspeech sounds to enhance navigation in a visual user interface. This paper describes how the sound design methodology proposed in [10][11] was extended to sonify a novel 3D-visualized information system for sighted users navigating a hierarchical structure. The method ensures that the sounds designed are not based on personal or ad hoc choices, and instead exploits the creativity of a user group as an application of participatory design in sound. Recommendations are derived from this case study on how to design auditory cues for familiar or novel user interfaces to convey structural information in an informative and intuitive way.

1 Introduction

Effective navigation is a process of high interest in recent times due to the explosion of information available. However information navigation is not always an easy task when navigating large complex information systems. The effectiveness of navigation depends on the structure of the information presented and the design of the structural cues [12][14]. If the structural layout and cues do not allow users to create an accurate mental model of the system, it can lead to cognitive overload [3]. Furthermore, the addition of only visual cues would increase the visual workload and may lead to higher cognitive and visual overload. Replacing some of the cues traditionally presented in the visual modality with auditory cues may reduce some of this overload. Hence, a proposed solution is the addition of nonspeech auditory feedback to complement or reinforce the visual structural cues (see [1] for a review).

This research describes the process of designing auditory feedback for a 3D-visualized hierarchical information system [5], which presents a challenging case of navigation. We hypothesize that conveying some of the information through the auditory modality can aid in navigation tasks by enhancing the formation of a mental model of the structure. Using a secondary modality may also improve users' affective reactions and consequently enhance the overall user experience.

Auditory interface design is often based on designers' personal preferences or available technology, rather than the formal structured methodologies found abundantly in other fields, such as graphical interface design. However, the design approach used here is based on a semiotic approach to sound design, proposed and evaluated by Murphy et al [10][11], to convey information about the spatial layout of

M. Kurosu (Ed.): Human Centered Design, HCII 2011, LNCS 6776, pp. 461–470, 2011.

webpages to visually impaired users. The focal point of the design method is the "rich use scenario" presented to a panel of participants, similar to a story or radio play, which they use as a tool to trigger creative sound ideas. While the rich use scenario encodes the designer's vision, the user panel method ensures that the sounds designed will not be based on personal or ad hoc choices, and instead exploits the groups' creative ideas as an application of participatory design in sound [8].

The following sections describe the specific case study, the outcomes of the iterative design panel sessions and consequently, recommendations are derived on how to extend the design methodology for designing auditory cues for familiar or novel user interfaces to effectively convey structural information.

2   The Design Methodology

The focal point of the methodology is defined as a rich use scenario describing a unique character and his surrounding environment [10]. Use scenarios (formal descriptions of the usage of an application) have been used in the field of HCI where they are presented to groups of users or designers [2]. The guideline in [11] suggests using panels of four or five members, who do not need to be experts in sound design, nor familiar with the application or its usage. It is suggested to use the panelists to describe and evaluate non-speech sound functions for different task descriptions, in three iterative design sessions. The methodology has been followed here to create a rich use scenario, describing a person in a unique situation using the application. Gaps occur at appropriate points in the story, where the panelists are asked to suggest sound ideas. A brief outline of the steps for the design method described in [11] with some modifications in this case are given below:

1. A task description for the sound functions of the application is prepared.
2. A user description based on a vision of a plausible user is prepared.
3. Based on the previous steps, a short story in which the interaction among the character and application plays an important role is written, with the perspective of the character. In the story, blanks are left for the sounds to be designed.
4. A design panel session with 4 or 5 panelists is organized. The session is started by reading the use scenario, keeping a brief pause in the place of each sound. Having read the story, it is discussed. Then the story is read again with the sentence that includes a blank space for a sound effect. The panelists are asked to try and describe what kind of sound would be appropriate. This is repeated for each sound. The entire session is recorded.
5. The panel's ideas of the appropriate sounds are implemented.
6. A second session with different people is organized. The implemented sounds are used when reading the story. Screenshots of the system are shown to the panel after an initial reading of the use scenario to help them acquire an idea about what the system looks like as described in the use scenario. All other steps remain the same as the first session.
7. The reactions and new ideas of the second session are analyzed. The original sounds are modified and new ones are created as suggested by the second panel.
8. A third session with a different set of participants is organized. Using sounds from stage 7, the use scenario is presented to the third panel and participants are

asked to choose sounds at relevant points in the story. After the initial reading, a demonstration of the system is presented, illustrating the task descriptions and animations at relevant points in the story.

A description of the system, the sound functions and the design process follows.

3 The System and Sound Functions

The 3D visualization application used, based on the McGill University Library online catalogue databases and their Library of Congress Subject Heading (LCSH) organizations, was designed and implemented by Julien [5]. The system has also been tested and evaluated on users by comparing with a traditional text-based system and found to give good performance results [6]. The interactive interface integrates the searching and browsing of large category hierarchies with their associated documents using a visual representation of the semantic hierarchy (Figure 1). This point-to-move application tool allows users to search and explore a semantic hierarchy by using the metaphor of a physical 3D space. Users can visually inspect subjects on labels hovering over circular areas and can traverse the hierarchy by following branches linked to each subject area.

The interface includes 3D animations to represent the depth and width of the tree. Each node of the hierarchical tree represents an area or subject. The animations allow the user to "zoom" into a desired node, or fly around the tree to see different perspectives and help users acquire a mental model of the semantic relationships between the contents of the tree. Adding auditory feedback that complements and reinforces the visual information may further aid in navigating the structure. Descriptions of the various sound functions designed follow.

- *Sound to distinguish the difference between leaves and internal nodes*: This is important information for information searchers since it signifies where to stop searching in a specific subject area and move to another. Visually this is represented by a slight difference in color; but this is often difficult to distinguish in a large category hierarchy. To enhance this difference with auditory cues, a distinctive sound is required to differentiate leaves from internal nodes.
- *An overview sound to display the density of information of a node*: Each node has a number of subjects or text collections under it, which can be an important marker for searchers if discerned quickly. A sound that gives an overview of the density of information under that node would be useful in this way. Shneiderman's information seeking theory [13] gave rise to the idea of generating an overview before navigation. Several studies have supported the use of nonspeech sounds to display overview information [7], [9].
- *Sound to indicate the hierarchy level number or depth*: Since the hierarchy is so large and spread out, it is often difficult to keep track of the level of depth traversed by the user. An auditory cue can indicate the relative depth level of a certain node. Brewster [4] has shown that nonspeech audio earcons provide an effective way of representing depth in hierarchical menu structures.
- *Sound to display the selection of a node*: This would be an auditory reinforcement of the visual selection of a node to show the contents of that node.
- *Sound to reinforce "flying" to a node*: This is again a supplementary sound to reinforce the visual animation of zooming or flying into a subject area.

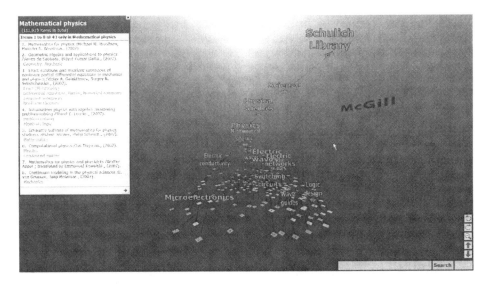

Fig. 1. A screenshot of the hierarchical information system

4   The Experiment: Design Panels and Results

4.1   Rich Use Scenario

The use scenario for this study was written from the perspective of a university student preparing for an assignment using an information visualization system. The scenario described a young university student, his mood and his surroundings.

Sam had an assignment due in two days and he still hadn't touched it, as usual. He ignored the inviting sounds coming from outside his window of people enjoying a sunny day, and decided to procrastinate no further. He logged on to his favorite information search system – he preferred to use a visualization system rather than an online library catalogue for his assignments.

The scenario also briefly described the technology or system he was using. The design methodology [11] suggested that too detailed descriptions of the technical aspects not be given, so as not to hamper the creative process.

The system looked like a top-down tree, in which each of the leaves of the tree were a subject area. He could browse the tree by subject to find areas of interest or even search for any item he wanted.

It went on to describe the activities he carried out on the system, with gaps or spaces in the regions of the story that required sound effects.

He started browsing through the tree, by moving the mouse over the labels of the tree leaves to find anything that caught his interest. When he found something he wanted to look at, he clicked on it [sound 1]. There was an animation that showed the zooming in or flying into that part of the tree accompanied by [sound 2]. From this subject, he went deeper and deeper into the tree by clicking on subjects branching out from above ones. The deeper he went, he still could keep track of his depth, as there was a sound that showed him his relative depth level [sound 3]. And once he reached the end of a branch (no more subjects under that) he would hear a specific sound [sound 4] that clearly told him he had come to the end of that particular branch. He also liked the option of being able to right-click on a subject area and hearing [sound 5] that told him the density of information present under that area (such as how many books and papers are available under the subject).

4.2 Participants

Thirteen participants were recruited in total for the three design panels, with each panel consisting of four or five participants. They were all University students, aged between 18 to 35 years. The criteria required for all participants were that they had no known form of hearing impairments and were fluent English speakers.

Table 1. User panel 1 reactions

Task description	Sound description suggestions
Clicking on a label	Generic clicking sound
Zooming or flying into a part of the tree	A whooshing sound
Indication of current depth level in tree	– Ticking sound, quicker with increased depth (e.g. a Geiger counter). However they realized it would be annoying. – Change in pitch: deeper levels should have lower frequency. – Ambient sounds of leaves rustling, more intense the deeper the level. – Opposite idea: Leaves rustling with less intensity for increased depth coupled with birds chirping, with louder chirping for more depth.
Indication of reaching a leaf (end of branch)	– Generic error sounds – Sound of a lock, a thud or a door closing – A distinctive bird chirp
Overview of the density of information of a subject	– An applauding crowd, louder with more information (again they realized it would be annoying) – A falling thud: heavier thuds for denser information – Page flipping: small number of pages for less dense information, flipping through a book for more information.

4.3 User Panel 1

The first panel session was made up of four graduate students, two of who had experience in sound editing or music, two without and one from information science. They were presented with the use scenario, with spaces at relevant points in the story where they had to input their sound ideas. In the use scenario, the hierarchical system was described as a top-down "tree", with labels on "leaves", leading to the groups' identification of forest sounds as a metaphor for auditory cues.

Based on the sound ideas described by the panel in Table 1, sounds were selected from the McGill Multimodal Interaction Lab (MIL) audio resources and online open-source sound resources [15], all recorded at 44.1 kHz with a bit-depth of 16 or 24 bits.

- Sound 1: 4 clicking sounds were selected: 2 single clicks and 2 double-clicks.
- Sound 2: 2 whooshing sounds were selected and edited to different lengths to match animation times by changing the speed or tempo.
- Sound 3: 3 options were selected: 1) 3 choices for birds at 3 depth levels, achieved by choosing parts of the sound files where the birds sounded more intense for deeper levels. The difference in average intensity level between each depth level was scaled to 6 dB. 2) The same procedure was followed for 2 choices in wind or leaves rustling. 3) 2 choices were given for the combined sounds of birds and leaves rustling. As the depth level increased, the sound of leaves rustling was reduced by 6 dB while that of the birds were increased by 6 dB in each step.
- Sound 4: 2 options were selected and scaled to the same RMS level: 1) 5 short distinctive bird chirps 2) 2 choices of the sound of a door or lock.
- Sound 5: 2 options were selected: 1) The sound of a book dropping was scaled up in 3 incremental files with 6 dB steps. 2) The sound of page flipping was edited to create 5 levels: the first level was one page flipping, the second of two pages, and the fifth of several pages flipping rapidly.

4.4   User Panel 2

Panel 2 consisted of five graduate students, two with formal musical training, three with none. However two had experience in information system design. The implemented sound ideas from Panel 1 were presented to Panel 2 at the relevant parts of the use scenario. They discussed these sounds, their preferences, or came up with new sound ideas. Panel 2 was also presented with system screenshots, to allow them to visualize the application in the scenario.

Based on their sound suggestions in Table 2, the original sounds from Panel 1 were modified and additional sounds selected from the MIL audio resources and online resources [15], all recorded at 44.1 kHz with a bit-depth of 16 or 24 bits.

- Sound 1: 3 clicks were selected, by eliminating one sound the panel did not like.
- Sound 2: One whooshing sound remained. A new swishing sound and two new "watery" sounds were selected.
- Sound 3: The combined sounds of birds and leaves rustling remained. Piano notes, acquired from the IOWA music database [16] were added (pitch change of notes from C4 to B4, or F1 to F7 represent depth levels).
- Sound 4: The sound of a book dropping or thud was selected.
- Sound 5: The page flipping option was edited so that the length of each file was almost the same by increasing the speed of the page flips for the longer sounds.

Table 2. User panel 2 reactions

Task description	Sound description suggestions
Clicking on a label	2 of 4 variations of single or double clicks preferred by all panelists.
Zooming or flying into a part of the tree	– Most thought the whooshing sounds were too abrasive; less rough whooshes would be preferred, especially for the longer sounds. – Suggested more pleasant sounds like a running brook might work.
Indication of current depth level in tree	– Just leaves or wind: sounds too harsh or stressful at deeper levels. They would not want it to sound stormy when deep in their search. – Just birds: preferred it more than just the leaves rustling. – Combination: all preferred the combined sounds more; they agreed it was easier to tell the difference in cues. – One panelist said she would prefer a tonal sound that changes in pitch to show level changes (higher pitches for higher levels)
Indication of reaching a leaf (end of branch)	– Most thought the chirps would not be easy to distinguish if paired with bird sounds in Sound 3. – They agreed a thud would work better, and liked the sound of a book dropping as a cue here.
Overview of the density of information of a subject	– They did not like book dropping volume changes as a cue. – They liked the page-flipping cue, but mentioned the sounds should be the same length, with faster flipping for more pages.

4.5 User Panel 3

Panel 3 consisted of four graduate students, two of who had musical training, while two had none; one of them was from Information Science. The third panel was presented with the modified sounds from Panel 2 and they discussed their preferences for these sound options. Panel 3 was also presented with a demonstration of the system and animations mentioned in the use scenario, as opposed to screenshots. This allowed them to match the sounds with the task descriptions more effectively.

From these reactions, Sound 1 (subject selection) was selected to be the single-click. The new swishing sounds of variable lengths were selected for Sound 2 (the zooming function). The combined birds and leaves rustling, with increased sound of birds and decreased sound of wind for increased depth, was selected for Sound 3 (hierarchical depth level indication). The sound of a book dropping was selected for Sound 4 (end of branch distinction). And the sound of pages flipping was selected as the final overview sound for Sound 5 (information density sonic overview).

Summary of observations and findings:

- The user panels help in group confirmation of the designer's vision. The group together successfully explored creative ideas with discussion.
- The use scenario helps to trigger the creative sound ideas. A story that is easy to connect to for the panel members helps in discussion initiation.
- If the application in the use scenario is not fairly familiar to the panel members, displays or demonstrations of the application should be shown.

Table 3. User panel 3 reactions

Task description	Sound description suggestions
Clicking on a label	General consensus: single click preferred over the double-click.
Zooming or flying into a part of the tree	– Three panelists liked the new swishing sounds. – Watery sounds were not appropriate in the context, swishing sounds match better with the visuals.
Indication of current depth level in tree	– Woods sounds preferred over to the piano (easier to distinguish and match with the system visuals). One panelist with piano training preferred the piano notes. – All panelists agreed that for the piano, the more easily distinguishable notes were the F1 to F7 notes, rather than C4 to B4.
Indication of reaching a leaf (end of branch)	– They all agreed the book dropping sound worked well as a cue here.
Overview of the density of information of a subject	– They all liked the page flipping cue, and agreed they would easily be able to distinguish the changes in information density with this cue.

- Words in the use scenario to describe the application have to be carefully selected so that the description does not lead the panel members to visualize an inaccurate version of the application and hence result in inappropriate sound choices in the required context.
- The iterative sessions with different panel members help to identify problems throughout the sound design process and lead to more creative input to be processed for each sound.
- Panelists have an almost immediate negative response to sounds that are harsh, abrasive, highly reverberatory, loud, long or busy (attention demanding).
- Auditory icons (environmental or real-life sounds) seem to be generally preferred than earcons (abstract or musical sounds). An exception arises when the panelist is musically trained in the specific musical instrument used in the earcons.

5   Discussion and Conclusion

It is recommended in [11] that application details not be described in too much detail to the panel, since it could hamper the creative process. The use scenario is meant to "generate creative input rather that focus discussion on the details of the system". In keeping with this suggestion, in the first panel session, the only description of the system given was that in the use scenario; no examples of the application were shown. However, since they were entirely unfamiliar with the novel system, this led to some confusion during the session, as panelists had a difficult time trying to visualize the system, task descriptions and animations described in the use scenario. Before being able to generate creative input, some time had to be spent in answering panelists' questions and trying to elaborate on the task descriptions. Hence, instead of "less technical details hampering the creative process", it almost had the opposite effect.

Thus, in the second panel session, screenshots of the application were shown after the use scenario was introduced. This led to better understanding of the system, but questions still remained on the animations and the "3D" view of the system. Consequently for the third panel, a demonstration of the application and the described animations in the task descriptions was shown. This yielded much better results in the session, consensus was reached much faster, and the group was able to match sounds with the animations much more effectively.

Therefore, while it is feasible to not delve too deeply into the details of the application in environments familiar to users, such as webpage-browsing, as was investigated in [10][11], this is not the case in novel systems. In cases where the application is entirely new and unfamiliar, it is necessary to tailor the rich use scenario to allow the panel to be able to visualize the application, while still triggering creative sound ideas by including inspiring details about everyday life.

Another observation was that the way the use scenario is written can influence the type of feedback or sound ideas the panel comes up with. For example, in the use scenario, the words "trees" and "leaves" were used to describe the hierarchical system and this may have influenced the panel to lean towards "forest-like" sounds of leaves rustling and birds chirping. However, when the second panel was shown the screenshots of the hierarchical system, they commented that it did not look like a tree or forest, and does not match the metaphor. The third panel however thought the animations matched the forest feel, when shown the demonstration of the system, and hence favored the forest metaphor. Hence, the way an application is described, or the specific words chosen should be selected carefully, so that it does not lead the panelists to have an inaccurate mental vision of the application and therefore, result in inappropriate sound choices or metaphor identification in the required context.

Thus, depending on the type of application in which auditory feedback is being designed, it is suggested that the use scenario be tailored so as to give the panel an overall feel of the system, while still keeping the essence of a story or a radio play. This is so that, when the scenario is presented, the members of the design panel can use the story to generate creative sound ideas, while not being concerned with parts of the application that they do not understand. Hence, for fairly familiar applications such as web browsers or file systems, it may not be necessary to show them any instances of the application. For 2D applications that have functions that may be difficult to explain in the use scenario, a few screenshots of the system can be shown after the use scenario has been presented and initially discussed. This illustration may bring about new sound ideas from the panel. However for completely new or multimodal applications, with 3D graphics or animations, a demonstration of the system starting from the first panel session is suggested. This would reduce any confusion or questions the panelists may have regarding the task descriptions in the use scenario and should help facilitate the creative process.

Natural and synthesized sounds will be used and interpolated to create multiple levels of increasing complexity to convey structural information in the hierarchy based on the results of the design process. Future work involves integrating the sounds into the hierarchical system and conducting a controlled evaluation of the system on users. The evaluation of performance effects and affective reactions will help to further verify the effectiveness of the sound design methodology in the current context of hierarchical information systems.

References

[1] Absar, R., Guastavino, C.: Usability of non-speech sounds in user interfaces. In: Proceedings of the 14th Inter. Conference on Auditory Display (ICAD 2008), June 24-27 (2008)

[2] Bodker, S.: Scenarios in user-centered design – setting the stage for reflection and action. In: Proceedings of the 1999 Hawaii International Conference on System Sciences, Maui, Hawaii, United States, January 5-8, pp. 3053–3063 (1999)

[3] Brewster, S.: Nonspeech auditory output. In: The Human-Computer Interaction Handbook: Fundamentals, Evolving Technologies and Emerging Applications, pp. 220–239 (2003)

[4] Brewster, S.A., Raty, V.P., Kortekangas, A.: Earcons as a method of providing navigational cues in a menu hierarchy. In: HCI 1996: Proceedings of HCI on People and Computers XI, pp. 169–183. Springer, London (1996)

[5] Julien, C.-A., Guastavino, C., Bouthillier, F., Leide, J.E.: Subject Explorer 3D: A Virtual Reality Collection Browsing and Searching Tool. In: Proceedings of the 38th Annual Canadian Association for Information Science Conference, CAIS 2010 (2010)

[6] Julien, C.-A.: SE-3D: a Controlled Comparative Usability Study of a Virtual Reality Semantic Hierarchy Explorer. Unpublished Doctoral Thesis, McGill University, Montreal, QC, Canada (2010)

[7] Kildal, J., Brewster, S.A.: Exploratory strategies and procedures to obtain non-visual overviews using TableVis. International Journal of Disability and Human Development 5(3), 285–294 (2006)

[8] Muller, M.J.: Participatory design: the third space in HCI. Journal of Human-Computer Interaction: Development Process, 165–197 (2009)

[9] Murphy, E.: Designing auditory cues for a multimodal web interface: A semiotic approach. Unpublished Doctoral Thesis, Queen's University, Belfast, Ireland (2007)

[10] Murphy, E., Pirhonen, A., McAllister, G., Yu, W.: A semiotic approach to the design of non-speech sounds. In: Proceedings of the 2006 International Workshop on Haptic and Audio Interaction Design, pp. 121–132 (2006)

[11] Pirhonen, A., Murphy, E.: Designing for the Unexpected: The role of creative group work for emerging interaction design paradigms. Journal of Visual Communication Special Issue on Wearable Technology 7(3), 331–344 (2008)

[12] Shneiderman, B.: Designing the user interface: Strategies for effective human computer interaction, 3rd edn. Addison-Wesley, Reading (1998)

[13] Shneiderman, B.: The eyes have it: A task by data type taxonomy for information visualizations. In: Proc. of the 1996 IEEE Symposium on Visual Languages, pp. 336–343 (1996)

[14] Webster, J., Ahuja, J.S.: Enhancing the Design of Web Navigation Systems: The Influence of User Disorientation on Engagement and Performance. MIS Quarterly 30(3) (2006)

[15] The Freesound Project, http://www.freesound.org

[16] The University of Iowa music instrument samples database, http://theremin.music.uiowa.edu/

Social Networking Applications: Smarter Product Design for Complex Human Behaviour Modeling

Tareq Ahram and Waldemar Karwowski

Institute for Advanced Systems Engineering,
Department of Industrial Engineering and Management Systems
University of Central Florida
Orlando, FL 32816, USA
tahram@ucf.edu

Abstract. The advent and adoption of internet-based social networking has significantly altered our daily lives. The educational community has taken notice of the positive aspects of social networking such as creation of blogs and to support groups of system designers going through the same challenges and difficulties. This paper introduces a social networking framework for collaborative education, design and modeling of the next generation of smarter products and services. Human behaviour modeling in social networking application aims to ensure that human considerations for learners and designers have a prominent place in the integrated design and development of sustainable, smarter products throughout the total system lifecycle. Social networks blend self-directed learning and prescribed, existing information. The self-directed element creates interest within a learner and the ability to access existing information facilitates its transfer, and eventual retention of knowledge acquired.

Keywords: Smart Products; service systems, Systems Engineering; Social Networking.

1 Introduction

The concept of smartness of consumer products and services has been investigated by several authors. This section presents a synthesis and summary of the most innovative work that influenced research in this field. Allmendinger and Lombreglia [1] highlighted smartness in a product from a business perspective. They regard "smartness" as the product's capability to predict business errors and faults, thus *"removing unpleasant surprises from [the users'] lives."* Ambient Intelligence (AMI) group [2] describes a vision where distributed services, mobile computing, or embedded devices in almost any type of environment (e.g., homes, offices, cars), all integrate seamlessly with one another using information and intelligence to enhance user experiences [3,4,5]. The advent and adoption of internet-based social networking sites such as MySpace [TM] and Facebook [TM] has significantly altered social interactions of their users. Users of social networking sites vary their activities; some may be very active sharing their daily life experiences with comments and pictures, while others simply use the sites as a personal directory service. The educational community has taken notice of the following positive aspects of social networking:

M. Kurosu (Ed.): Human Centered Design, HCII 2011, LNCS 6776, pp. 471–480, 2011.
© Springer-Verlag Berlin Heidelberg 2011

- Peer feedback, increasingly fast response times for scientific discovery, collaborative design and research
- Creation of blogs and support groups of individuals going through the same or similar difficulties
- Providing a social context in line with the university, company, design group, or field of study
- A venue with links not directly related to a given educational alignment or resource.

Social networking applications support the development of a methodology to better assess and predict imprecision and variability in user behaviour by applying advanced mathematical and soft computing techniques to aid in studying human social, cultural and behavioral aspects. Application of soft computing techniques helps identify erroneous, problematic activities and issues that might otherwise go undetected for their obscurity, complexity, or elaborate inter-relationships. In addition to these above, social networks themselves are highly adaptable, flexible, and mobile. For example, the "blogging" paradigm became the "micro-blogging" concept known as Twitter™, which now is integrated with Facebook's "status updates." Arguably, social networking provides an effective method of satisfying the primal human desire of communication.

Rapid technological advancements and agile manufacturing created what is called today smart environments. Definitions of smart environments may be taken into account as a first reference point, since smart products have to be considered in the context of their environment. For example, Das and Cook [6] define a smart environment as the one that is able to acquire and apply knowledge about an environment and adapt to its inhabitants in order to improve their experience in that environment. It is noticed that the knowledge aspect has been recognized as a key issue in this definition. Mühlhäuser [2] refers to smart product characteristics that are attributed to future smart environments: i.e., "integrated interwoven sensors and computational systems seamlessly embedded in everyday systems and tools of our lives, connected through a continuous network." In this respect, smarter products can be viewed as those products that facilitate daily tasks and augment everyday objects. In 2007, AMI identified two motivating goals for building smart products [7]:

1. Increased need for simplicity in using everyday products, as their functionalities become ever more complex. Simplicity is desirable during the entire life-cycle of the product to support manufacturing, repair, or use.
2. Increased number, sophistication, and diversity of product components (for example, in the aerospace industry), as well as the tendency of the suppliers and manufacturers to become increasingly independent of each other which requires a considerable level of openness on the product side.

Mühlhäuser [2] observed that these product characteristics can now be developed due to recent advances in information technology as well as ubiquitous computing that provides a "real world awareness" in these systems through the use of sensors, smart labels, and wearable, embedded computers. According to Mühlhäuser [2], product simplicity can be achieved with improved product to user interaction (p2u). Furthermore, openness of a product requires an optimal product to product interaction (p2p). Knowledge intensive techniques enable better p2p interaction through self-organization within a product or a group of products. Indeed, recent research on semantic web service description, discovery, and composition may enable self-organization within a group of products and, therefore, reduce the need for top-down constructed smart environments [8]. Smart products also require some level of internal organization by making use of planning and diagnosis algorithms as stated by [2]:

> "A Smart Product is an entity (tangible object, software, or service) de-signed and made for self-organized embedding into different (smart) envi-ronments in the course of its lifecycle, providing improved simplicity and openness through improved p2u and p2p interaction by means of context-awareness, semantic self- description, proactive behavior, multimodal natural interfaces, AI planning, and machine learning."

Major characteristics of smart products are illustrated by comparing their essential features. For example, [9] define six major characteristics for smart products illustrated in Table 1 below. Table 2 provides a comparative presentation of the main characteristics of smart products. These characteristics include the following:

- Context-awareness - the ability to sense context
- Proactivity - the ability to make use of this context and other information in order to proactively approach users and peers
- Self-organization - the ability to form and join networks with other products.

In addition to the above characteristics, Mühlhäuser [2] and SPC emphasize the fact that smart products should support their entire life-cycle. In addition, special care should be devoted to offering multimodal interaction with the potential users, in order to increase the simplicity characteristics of the products.

Table 1. Smart Products Characteristics [9]

Characteristic	Description
Personalization	Customization of products according to buyer's and consumer's needs.
Business-awareness	Consideration of business and legal constraints.
Situatedness	Recognition of situational and community contexts.
Adaptiveness	Change product behavior according to buyer's and consumer's responses to tasks.
Network ability	Ability to communicate and bundle with other products.
Pro-activity	Anticipation of user's plans and intentions.

Table 2. A Comparison of Smart Product's Characteristics [7]

Maass and Varshney [9]	Mühlhäuser [2]	Smart Products consortium [4]
Situatedness	Context-aware	Situation- and context-aware
Pro-activity	Proactive Behavior	Proactively approach the user
Network ability	Self-organized embedding	Self-organized embedding in smart product environments
	Support the entire life-cycle Multimodal Natural Interfaces	Support the user throughout whole life-cycle Multimodal interaction
Personalization		
Business-awareness		
Adaptiveness		
		Autonomy Support procedural knowledge Emerging knowledge Distributed storage of knowledge

1.1 Social Networking for Smarter Products and Services Design

Communication of ideas, as a core for effective education and collaborative design, is the basis of distance and virtual learning. Social networks blend self-directed learning and prescribed, existing information. The self-directed element creates interest within a learner and the ability to access existing information facilitates its transfer, and eventual retention of knowledge acquired. There may also be a competitive element for educators to explore, since design activities are transparent in social networking. Ziegler [10] observed that social networking sites may radically change the educational system, since they offer the *"capacity to motivate students as engaged learners"*, *rather than what he considers the usual "passive observers of the educational process."* However, there are also conflicting views in the literature regarding the usefulness of social networks in education and design. In today's interconnected world, social networking provides a great source of information and knowledge sharing that has not yet been fully explored to support collaborative products design and education.

Selwyn [11] performed an observational study of a group of students' online interactions with Facebook™ in the UK. Though the author cited many limitations of the study, some interesting findings included an observation that the social network site did not serve a meaningful role in making new partnerships. Rather, it maintained strong links already established in an emotionally close-knit group of people. Social networks share many functional elements with blog, a term coined recently as a shortened form of "web log," describing a page that is frequently updated with comments, links, images, and other media pertaining to a given subject. The blog makes a statement and offers a space below for readers to comment and respond. Social networks have taken the blog concept and applied it to a directory concept. People who are "linked" together can receive updates from others micro-blog inputs. Some educators (need references here!) claim that these links and updates can be used in a variety of educational ways. The concept of social networking can be extended to collaborative design and modeling as means of facilitating team work and sharing product design experience in order to enhance team learning process, including collaborative online discussions, idea generation, peer review activities, and even debate [12,13].

1.2 Social Networking in Education

The proliferation of broadband-enabled interactive devices, such as smart cellular phones and media players, with social networking gadgets and application allows social communication and collaborative education activities to occur outside of lecture times. Another offshoot of social networking and blogging sites is that of wiki articles and their massive compilation, Wikipedia. A traditional understanding of an academic resource that "anyone can edit" seems unreasonable. The seemingly micro-managed and endlessly peer reviewed "live" nature of the document made Wikipedia a compelling new way to create, store, and integrate vast stores of knowledge. The pull of social networking technologies cannot be ignored, as they have attracted millions of users in a short amount of time since their introduction. Shortcomings such as the necessity of pre-existing offline relationships and, as of yet, unexploited educational and design opportunities may be addressed by serious initiatives and the integration of such technologies into modern educational and design methods and practices [14]. The accessibility of these networks is more pervasive now than ever, thanks to gaming consoles and mobile devices. Prensky [15] claims that today's students "think and process information differently" from their pre-digital world counterparts. People born after the mid 1980s are part of a group of "digital natives" who take information technology and its use for granted. Today, data can be created anywhere and on a great variety of computing platforms. The ability to create and view data anywhere can translate to new learning opportunities. Several universities have already turned towards the web and outlets previously used only to sell music and video as a way to disseminate lecture materials. Apple Computer's iTunes ™ software dominates the digital media player market. It has recently launched "iTunes University," a subset of its online media store devoted to distributing lectures and presentations from various academic institutions. All of these novel technologies and media distribution platforms offer unimagined learning opportunities. As of yet, the educational elements are largely unused compared to their strictly entertainment-related digital media. Many opportunities exist for providing students with this media but dissemination is not enough alone. Serious educational games, educator's involvement, and classroom activities sent to these services offering *interaction* rather than "passive observation" would be valuable aids to the learning process. There is no doubt that today's traditional students consume more media and games than previous generations. They need only be given some structure and appropriate interactive learning media to augment their already media-enriched lives.

2   A Social Networking Systems Engineering Approach to Study Complex Human Behaviour

The contemporary systems engineering process is an iterative, hierarchical, top down decomposition of system requirements [16]. The hierarchical decomposition includes Functional Analysis, Allocation, and Synthesis. The iterative process begins with a system-level decomposition and then proceeds through the functional subsystem level, all the way to the assembly and program level. The activities of functional analysis, requirements allocation, and synthesis will be completed before proceeding to the next lower level. SysML is a general-purpose visual modeling language for specifying, analyzing, designing, and verifying complex systems which may include hardware, software, information, personnel, procedures, and facilities (OMG SysML: http://www.omgsysml.org). SysML provides visual semantic representations for modeling system requirements, behavior, structure, and parametrics, which is used to integrate with other engineering analysis models [17].

Traditional machine learning techniques have some limitations for modeling human behavior, mainly the lack of any reference to the inherent uncertainty that human decision-making has. This problem can be partially solved with the introduction of Soft Computing (SC) to model human behaviour via social networking applications. SC is an innovative approach to building computationally intelligent systems that differs from conventional (hard) computing in that it is tolerant of imprecision, uncertainty and partial truth. The guiding principle of soft computing is to exploit the tolerance for imprecision, uncertainty and partial truth to achieve tractability, robustness and low solution cost. SC consists of several computing approaches, including neural networks, fuzzy set theory, approximate reasoning, and search methods, such as genetic and evolutionary algorithms. SC technologies provide an approximate solutions to an ill-defined problems encountered in social networking application and can help creating human behavioral models in an environment, such as during conflicts , in which users are not willing to give feedback on their actions and/or not able to fully define all possible interactions due to social and cultural barriers. Different techniques provide different capabilities to support the development of smarter products and services. For example, Fuzzy Logic provides a mechanism to mimic human decision-making that can be used to infer goals and plans; Neural Networks a flexible mechanism for the representation of common characteristics of a user and the definition of complex stereotypes; Fuzzy Clustering a mechanism in which a user can be part of more than one stereotype at the same time; and Neuro-Fuzzy systems a mechanism to capture and tune expert knowledge which can be used to obtain assumptions about the user.

Systems engineering teams along with product and service designers are responsible for verifying that the developed products and services meet all requirements defined in the system specification documents. The following procedures outline the relevant systems engineering process steps [18, 21]:

- **Requirements analysis** - review and analyze the impact of operational characteristics, environmental factors, functional requirements and develops measures suitable for ranking alternative designs in a consistent, objective manner. Each requirement should be re-examined for consistency, desirability, applicability, and potential for improved return on investment [19]. This analysis verifies that the requirements are appropriate or develops new requirements for the smart product operation.

- **Functional analysis** - systems engineers and product designers use the input of performance requirements to identify and analyze system functions in order to create alternatives to meet system requirements. Systems engineering then establishes performance requirements for each function and sub-function identified.
- **Performance and functionality** - systems engineering allocates design requirements and performance to each system function. These requirements are stated in appropriate detail to permit allocation to software, systems components, or personnel. Performance and functionality allocation process identifies any special personnel skills or design requirements.
- **Design Synthesis** - designers and other appropriate engineering specialties develop a system architecture design to specify the performance and design requirements which are allocated in the detailed design. The design of the system architecture is performed simultaneously with the allocation of requirements and analysis of system functions. The design is supported with block and flow diagrams. Such diagrams support:

  o Identifying the internal and external interfaces
  o Permitting traceability to source requirements
  o Portraying the allocation of items that make up the design
  o Identifying system elements along with techniques for its test and operation
  o Providing a means for comprehensive change control management

- **Documentation** - the primary source for developing, updating, and completing the system and subsystem specifications. Smart product requirements and drawings should be established and maintained.
- **Specifications** – to transfer information from the smart product systems requirements analysis, system architecture design, and system design tasks. The specifications should assure that the requirements are testable and are stated at the appropriate specification level.
- **Specialty engineering functions** - participate in the systems engineering process in all phases. They are responsible for system maintainability, testability, producibility, human factors, safety, design-to-cost, and performance analysis to assure the design requirements are met.
- **Requirements verification** - systems engineering and test engineering verify the completed system design to assure that all the requirements contained in the requirements specifications have been met.

Model-based interactive human system approaches for design and modeling of smart systems and products differentiate between human performance and effectiveness criteria. These criteria determine a total system mission performance level and acceptability that is directly attributable to specific actions allocated to human performance metrics. These are indicators measure which performance effectiveness criteria are met [20, 21].

Currently there are few applications to facilitate human behaviour modeling in social networking applications. One of the applications that support a full Human

Systems Integration (HSI) within a systems engineering process is DOORS^{TM} by Rational. DOORS or Dynamic Object Oriented Requirements System specifically tracks requirements for product or software design (see Figure 1). Since the requirements process has many shared elements to knowledge management, DOORS facilitates requirements entry, organization into hierarchies, and display. Users make changes and link any requirement to sub requirements and related requirements. DOORS require individual users to have accounts. Each account can be restricted to elements of the database and given read-only or administrative-level rights. Changes made are tracked by user, allowing managers to trace changes down to the individual user level. Figure 1 shows a typical DOORS session. Requirements are shown in the left pane in hierarchical order and detail views are shown on the right.

The little triangles on the right of the detail view in Figure 1 indicate that the specific requirement is linked to another element, usually a related requirement. Test plans and verification methods are also linkable. DOORS provide a structured framework for adding, viewing, and changing requirements.

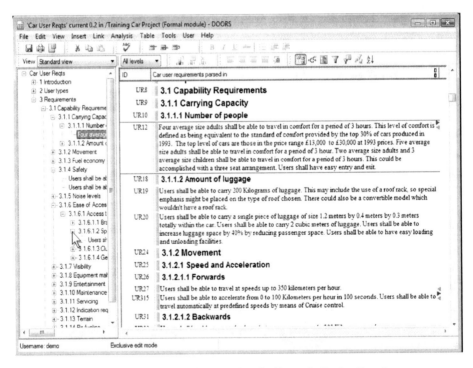

Fig. 1. IBM Rational DOORS Screenshot for Example Design Requirements

3 Conclusions

As an introductory contribution to the application of social networking and systems engineering process for the design and development of smarter products and services, this paper provides a motivation and quest for integrated social networking approach to systems engineering and to study complex human behaviour. While a large number of disciplines and research fields must be integrated towards development and widespread use of smarter products, considerable advancements achieved in these fields in recent years indicate that the adaptation of these results can lead to highly sophisticated yet widely useable collaborative social networking applications for smart products. It is believed that the application of systems engineering and social-networking to design and modeling of smarter products and services should prove useful in supporting and facilitating the understanding of complex human behaviour and to better identify crucial user needs.

References

1. Allmendinger, G., Lombreglia, R.: Four Strategies for the Age of Smart Services. Harvard Business Review 83(10), 131–145 (2005)
2. Mühlhäuser, M.: Smart Products: An Introduction. In: Constructing Ambient Intelligence - AmI 2007 Workshop, pp. 154–164 (2008)
3. Weiser, M.: The computer of the 21st century. Scientific American 265(3), 66–75 (1991)
4. Ahola, J.: Ambient Intelligence. ERICM News (47) (2001)
5. Arts, E., de Ruyter, B.: New research perspectives on ambient intelligence. Journal of Ambient Intelligence and Smart Environments 1, 5–14 (2009)
6. Das, S., Cook, D.: Designing Smart Environments: A Paradigm Based on Learning and Prediction. In: Shorey, R., Ananda, A., Chan, M.C., Ooi, W.T. (eds.) Mobile, Wireless, and Sensor Networks: Technology, Applications, and Future Directions, pp. 337–358. Wiley, Chichester (2006)
7. Sabou, M., Kantorovitch, J., Nikolov, A., Tokmakoff, A., Zhou, X., Motta, E.: Position Paper on Realizing Smart Products: Challenges for Semantic Web Technologies, Report by Knowledge Media Institute (2009),
   http://people.kmi.open.ac.uk/marta/papers/ssn2009.pdf
8. Chandrasekharan, S.: The Semantic Web: Knowledge representation and affordance. In: Gorayska, B., Mey, J.L. (eds.) Cognition and Technology: Co-existence, Convergence, and Co-evolution, pp. 153–174. Benjamins, Amsterdam (2004)
9. Maass, W., Varshney, U.: Preface to the Focus Theme Section: 'Smart Products'. Electronic Markets 18(3), 211–215 (2008)
10. Ziegler, S.: The (mis)education of Generation M' Learning. Media and Technology 32(1), 69–81 (2007)
11. Selwyn, N.: Screw blackboard..do it on Facebook!: An investigation of students' educational use of Facebook. Paper Presented to the Pole 1.0 – Facebook Social Research Symposium, November 15, at University of London (2007),
    http://www.scribd.com/doc/513958/Facebook-seminar-paper-Selwyn (accessed October, 2009)
12. Davies, J., Merchant, G.: Looking from the inside out: academic blogging as new literacy. In: Lankshear, C., Knobel, M. (eds.) A New Literacies Sampler. Peter Lang, New York (2007)
13. Dubet, F.: Dimensions and representations on student experience in mass university. Revue Francaise de Sociologie 35(4), 511–532 (2004)
14. Madge, C., Meek, J., Wellens, J., Hooley, T.: Facebook, social integration and informal learning at university: It is more for socialising and talking to friends about work than for actually doing work. In: Learning, Media and Technology. Routledge, New York (2009)

15. Prensky, M.: Digital Game-Based Learning. McGraw-Hill, New York (2001) ISBN 0-07-136344-0
16. Hitchins, D.K.: Systems Engineering: A 21st Century Systems Methodology. John Wiley & Sons, Chichester (2007)
17. Friedenthal, S., Moore, A., Steiner, R.: A Practical Guide to SysML: The Systems Modeling Language. Morgan Kaufmann, Elsevier Science (2008)
18. Defense Acquisition University (DAU) Guidebook, Chapter 4: Systems Engineering (2004)
19. Ahram, T.Z., Karwowski, W.: Measuring Human Systems Integration Return on Investment. In: The International Council on Systems Engineering – INCOSE Spring 2009 Conference: Virginia Modeling, Analysis and Simulation Center (VMASC), Suffolk, VA. USA (2009)
20. Ahram, T.Z., Karwowski, W., Amaba, B., Obeid, P.: Human Systems Integration: Development Based on SysML and the Rational Systems Platform. In: Proceedings of the 2009 Industrial Engineering Research Conference, Miami, FL, USA (2009)
21. Systems Engineering Fundamentals. Defense Acquisition University Press (2001)
22. Karwowski, W., Ahram, T.Z.: Interactive Management of Human Factors Knowledge For Human Systems Integration Using Systems Modeling Language. Special Issue for Information Systems Management. Journal of Information Systems Management, Taylor and Francis (2009)

Usability Tests for Improvement of 3D Navigation in Multiscale Environments

Tathiane Mendonça Andrade, Daniel Ribeiro Trindade, Eduardo Ribeiro Silva,
Alberto Barbosa Raposo, and Simone Diniz Junqueira Barbosa

Department of Informatics, Pontifical Catholic University of Rio de Janeiro
Caixa Postal 38.097, 22.453-900, Rio de Janeiro, RJ, Brazil
tandrade@inf.puc-rio.br, danielrt@tecgraf.puc-rio.br,
eribeiro@tecgraf.puc-rio.br,
{abraposo,simone}@inf.puc-rio.br

Abstract. The interest in virtual 3D environments has increased in the past years due to the popularization of the technology and the huge human ability to visually convey and grasp information. However, unlike the real world, 3D navigation, especially in multiscale environments, is no longer natural to humans, becoming confusing and resulting in unpleasant experiences. To improve the quality of the users' navigation in and across multiscale 3D environments, three techniques were developed, based on a structure called cubemap. The 3D application chosen to apply these techniques was the Petrobras 3D System for Integrated Visualization in Exploration and Production (SiVIEP). This paper describes and reports the evaluation of these three techniques, using usability tests, which were performed to validate the more adaptable solution, ensuring the efficiency of the proposed techniques in assisting and facilitating the task of 3D navigation.

Keywords: 3D Navigation, Multiscale Environments, Usability, Cubemap.

1 Introduction

The popularization of 3D systems has brought new means of manipulating three-dimensional objects available to a wide audience of users, from basic to expert, depending on the system. However, 3D systems may require more effort from users to understand the models presented on the screen [1]. Therefore, we need to carefully design the interaction between the user and the 3D system to have a high quality of use.

Besides users' lack of experience at navigating in 3D environments, we find poorly designed navigation tools and also the problem known as desert fog [2], condition where the surrounding environment does not provide sufficient information to enable users to make a decision.

This paper briefly describes and reports the evaluation of three techniques developed to improve the quality of the users' navigation in and across multiscale 3D environments. A *multiscale environment* allows to users to view objects at different scales, together with information at different levels of detail, from a single screw to an oil field spanning dozens of miles [3, 4]. The techniques are based on a structure

M. Kurosu (Ed.): Human Centered Design, HCII 2011, LNCS 6776, pp. 481–490, 2011.

called *Cubemap* [5], which has the purpose of providing information about the virtual environment at any given moment, like creating a representation of the environment in relation to the observer's position. Based on cubemap, the three techniques proposed in [6] and investigated in our usability tests are: *Fly with automatic speed adjustment, Collision detection and treatment*, and *Examine with automatic pivot point*.

The evaluation technique chosen was the usability test. The 3D application chosen to apply these techniques was the Petrobras 3D System for Integrated Visualization in Exploration and Production (SiVIEP), developed by the Computer Graphics Technology Group (TecGraf) / PUC-Rio. SiVIEP displays the objects of a petroleum field, integrating all of the related tasks, and allows changes in the field during the 3D visualization.

The goal of this study is to assess the efficiency of the navigation techniques, based on the cubemap structure, through usability tests with different user profiles and for different tasks.

This paper is organized as follows: In the next section, the cubemap concept is presented and the Navigation techniques are described. Section 3 presents SiVIEP and its features. Section 4 describes the creation, preparation and execution of the tests with the users. Finally, conclusions are discussed in Section 5.

2   Cubemap

According to McCrae et al. [5], cubemap aims to provide information about the virtual environment at any given moment. The method consists of rendering six images in six different directions from the viewer's position, each one corresponding to a side of the cube. Cubemap uses a field of view of 90° in camera perspective, allowing the combination of the resulting *frustums* to cover the entire environment located between the clipping planes.

Whenever the camera changes position or rotation, the cubemap is updated. The distances from the generated fragments to the viewer are calculated. The computed distance values are normalized in relation to the near and far planes, and stored in the alpha channel of the positions related to the fragments. This way, the cubemap provides a depth representation of the environment at every given moment without the need for any kind of preprocessing, which is a desirable feature in the case of dynamic scenes.

The cubemap update process requires six additional rendering steps, which adds considerable cost and can result in loss of performance. To prevent this from occurring we use a lower resolution rendering of the faces, since only an estimate needs to be obtained. A resolution of 64×64 was enough to reach the level of precision the users needed.

3   Navigation Techniques

This section presents the three navigation cubemap-based techniques developed to solve the aforementioned 3D navigation problems.

3.1 Fly with Automatic Speed Adjustment

In multiscale environments, the user is free to view the information from a simple screw up to oil fields with hundreds of miles. In this case, the navigation speed is directly related to the scale of the environment. When you navigate across an oil reservoir, for example, the speed should be faster than the one used to analyze the interior of a platform.

In certain environments, the scale does not change too much and is well known, allowing pre-computation of the values for the navigation's speed. However, these values can not always be estimated, either by incomplete models and units errors or by large scale variations.

To address this issue, the proposed method uses the distance values stored in cubemap to determine the scale and to adjust the speed of navigation. This allows us to estimate the appropriate speed and make smooth transitions between different scales.

3.2 Collision Detection and Treatment

Not allowing collisions can be a determining factor in virtual environments. In immersive environments, a collision with an object can cause the loss of immersion, causing discomfort to the user, especially in interaction that uses stereoscopy effects.

In other cases the lack of treatment of collision can cause the effect known as *desert fog* [2], especially for non-advanced users. For example, crossing walls in a closed environment or cutting through the ocean floor during navigation may cause the user to lose the necessary information to make further navigation decisions.

The information stored in cubemap determines the closest objects in the path taken by the observer. The distances are used to calculate a repulsive force that smoothly deflects the camera from obstacles.

3.3 Examine with Automatic Pivot Point

To inspect an object, two actions are required: 1) determine the object as the pivot point and 2) use the *examine* tool. This latter action allows the camera to rotate and zoom around the object of interest.

The location of the center of rotation is essential for the proper functioning of the *examine* tool. Among the problems with this interaction technique, there is the choice of a pivot point which is away from the user's field of view (Fig. 1a). Another factor is the center of rotation located in front or behind the object to be examined (Fig. 1b). In either case the result is not the one the user expected.

We observe that some users, even advanced ones, make mistakes when using these tools. The most common mistake is to forget to establish a new center of rotation. Even users who have not forgotten to accomplish this task reported that they felt uncomfortable about having to explicitly do it.

The proposed solution was to automatically determine the pivot point to be examined. For this, the point located in the center of the user's vision, information stored in the cubemap, is used as pivot point wherever the *examine* tool is activated. It only requires the user to point the center of his vision at the object of interest.

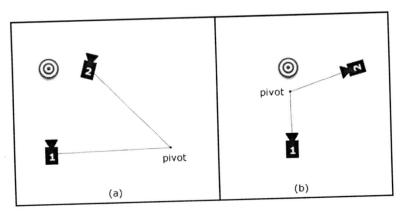

Fig. 1. Problems with the choice of inadequate pivot points. In both cases, when the camera moves from position 1 to position 2, the user loses sight the object being examined.

4   SiVIEP

The SiVIEP application was developed due to a need from Petrobras, a Brazilian Oil & Gas Company. The system interface is characterized by the presentation of geoscience and engineering models, which are represented by 3D objects under different scales and levels of detail (Fig. 2).

4.1   System Features

The SiVIEP application has a high complexity of requirements given to its composition of 3D objects, developed and designed with high-quality and close to reality.

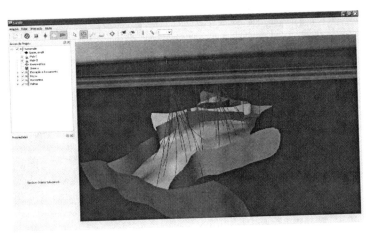

Fig. 2. SiVIEP interface example with a view of a petroleum field

SiVIEP users can navigate through the objects using tools like: *Fly* and *Examine*. The *fly* tool allows the camera to fly over the scene, at a specific speed. This tool can be controlled by the user by pressing the arrow keys of the keyboard to move the camera while guiding the direction of the motion with mouse movements. The *examine* tool allows the user to examine a certain object through drag movements using the left mouse button, which makes the camera rotate around the pivot point. The pivot point can be fixed by another tool of the system: *Change pivot point*.

5 Evaluation

The evaluation was planned according to the DECIDE *framework* [7], which presents a model to elaborate the evaluation of user–system interactions. After the preparation, the tests with the users were conducted to assess the following usability factors: ease of learning, efficiency, and user satisfaction with the system.

The results of the evaluation were analyzed and solutions were designed to solve the problems we had found.

5.1 Evaluation Planning

The evaluation was planned according to the guidelines of the DECIDE framework. The purpose of this framework is to ensure that all necessary steps in conducting an evaluation have been correctly applied correctly.

The test required selected users to have two different profiles, to demonstrate the different views and preferences between them. The two groups of users defined were: *advanced* users, with experience in the use of 3D visualization and 3D modeling applications, i.e., who use this type of software at least once a month; and *non-advanced* users, with little experience with 3D visualization applications, except for games, and who do not use 3D applications frequently. Some authors [8, 9] came to interesting conclusions about the behavior of non-advanced users. For example, they do not understand how a tool works or have a wrong understanding of it, causing them to expect a different behavior when using it. In some cases, they try to accomplish all tasks with a single instrument, even when they need a different set of tools for the task.

From the 16 participants of the tests, 6 were considered advanced, 8 non-advanced, 2 were used for the pilot tests, one from each group. They were between 19 and 30 years old, and 9 of them were male and 7 were females.

Before the interaction, we conducted a semi-structured interview with the users in order to detect the profile to which they belonged. After the interview, a short demonstration was presented to the user so that he/she could get an idea of how to navigate in SiVIEP. In some occasions, where individuals from the non-advanced group did not understand how to manipulate the controls and use the tools available, it was necessary to allow the user to quickly interact with the system, which did not interfere with the test results. Such situations had low occurrence in this test.

After the demonstration, we explained to the users how the test would be conducted and asked them to sign an informed consent form. Then, we applied the tasks in random order, to minimize the effects of system learning from one task to the other.

Each task was applied in the modes with and without the features of Multiscale navigation and inspection we were investigating. The users read the task description and were able to clarify any doubts about it. Because we wanted to test each cube-map-based navigation technique, we defined 3 tasks (A, B and C) for users to perform, each one involving one or more techniques.

The technique *Fly with Automatic Speed Adjustment* was to be used in all tasks because the user needs the *fly* tool to navigate through the objects to reach the location required by the tasks. Task A tested only the *Fly with Automatic Speed Adjustment* technique by establishing a situation where the user had to move from a specific side of the oil platform, traversing through a corridor on the same side and finally getting to the final point, which was the other platform located within his/her field of view. Task B primarily tested the *Examine with Automatic Pivot Point* technique: the user had to inspect four different points inside a specific oil platform, and to facilitate the task, the user could enable the *Change Pivot Point* tool followed by the *Examine* tool each time he/she passed by an inspection point. Finally, task C primarily tested the *Collision Detection and Treatment* technique: the user had to traverse an open and closed corridor inside the oil platform, and on this path there were platform objects as obstacles the user could or could not pass through during the interaction.

5.2 Test Execution

The user–system interaction was recorded in video and the users' comments in audio, both during the interviews and the test execution. The tests did not have a time limit. The observers took notes to record information that could not be captured by the recording equipment.

To execute each task, two different versions of SiVIEP were tested by each user: *automated*, in which the techniques to test were turned on so that the user did not have to worry about some navigation aspects; and *manual*, in which the techniques mentioned were turned off and some navigation parameters had to be explicitly set by the users. The versions were applied in a random order to reduce the learning effect of using the first version over the second. Data collected from the observation and the interviews were triangulated.

Primarily, pilot tests with two users were executed, one from the advanced group and the other from the non-advanced group. This pilot test was necessary and important because it revealed the material needed some changes in order to focus even more on the navigation techniques problem.

After executing each task, some questions were asked to the user, aiming to explore his/her opinion on the interaction with the system in each version (manual or automated).

At the end of the test execution, the user answered a questionnaire which consisted of the following statements about the interaction with SiVIEP, with a response scale ranging from 1 to 5, here identified as Q1, Q2, Q3, Q4 and Q5:

Q1: Using SiVIEP was (Hard .. Easy)
Q2: The system is similar to some other program you have used (Little .. A lot)
Q3: You performed the tasks successfully (None of them .. All of them)
Q4: What did you think of the manipulation and navigation of the model objects? (Hard .. easy)
Q5: In general, what do you think of your interaction with SiVIEP? (Poor experience .. Great experience)

5.3 Evaluation Results

This section describes the results of the usability tests and the problems found during the user–system interaction.

First, we asked the users' opinion about the beginning of the interaction. The users from the advanced group thought the beginning was easy, but it was even easier when testing the first task in the second version, because of the learning effect, and especially when the first version was the automated one. The users from the non-advanced group thought the beginning was difficult, but this situation got better when the first version tested was the automated one. According to most users from both groups, the automated version made the navigation simpler and avoided more errors.

For better comprehension of the results, they are organized in table 1, which contains the number of users, of each group, who preferred a specific version after doing each task.

In the advanced group, there was a difference in the users' preferences. In tasks A and C, the users preferred the automated version slightly: in the first case, because the automated version avoided some navigation and direction errors, and in the second case because it simulated the realism and provided a better sense of direction. The users who preferred the manual version, in these tasks, justified their choice by saying that they liked the automatic adjustment, but preferred to control the navigation speed. And, about the collision treatment, they thought that it was easier make task C without this technique due to the high speed this version can achieve when the camera is near the objects. Regarding task B, the users' opinions were divided. Those who chose the automated version justified their choice with comments like "this version helps me avoid some mistakes by distraction" since, in the manual version, the user can forget to select the center of rotation and get lost into the navigation field. On the other hand, the users who chose the manual version justified their choice by saying that they lost the vision of the center of rotation more easily, since the automatic adjustment set the pivot point automatically. But, they also felt lost during the navigation when forgot to set the pivot point manually and in their words "the screen jumps from a place to another into the navigation field".

Almost all users from the non-advanced group preferred the automated version of the navigation in tasks A and B given the facility to control and the precision of the manipulation. In task C, the users preferred the manual version given the speed with which they completed the task, since in this version there is not a collision treatment, and the speed is faster than the other version when the camera approaches the objects. However, four of the six non-advanced users complained about how difficult it was to control the camera in this version.

Table 1. Preference versions according to feedback from users of both groups

Task	Advanced Group		Non-advanced Group	
	Automated version	Manual version	Automated version	Manual version
A (fly)	4	2	7	1
B (examine)	3	3	7	1
C (collision)	4	2	2	6

Tables 2 and 3 present the questionnaire results of the advanced and non-advanced users' opinion about the navigation in SiVIEP, respectively. Individuals in the advanced group are herein called UA1, UA2, UA3, UA4, UA5 and UA6, whereas the individuals of the non-advanced group are identified as UN1, UN2, UN3, UN4, UN5, UN6, UN7 and UN8. To facilitate the reading, we are repeating the questions below:

Q1: Using SiVIEP was (Hard .. Easy)

Q2: The system is similar to some other program you have used (Little .. A lot)

Q3: You performed the tasks successfully (None of them .. All of them)

Q4: What did you think of the manipulation and navigation of the model objects? (Hard .. easy)

Q5: In general, what do you think of your interaction with SiVIEP? (Poor experience .. Great experience)

Table 2. Results of the questionnaire as applied to the users in the advanced group, where 1 is the worst response value and 5 is the best

	UA1	UA2	UA3	UA4	UA5	UA6	Avg
Q1	4	4	3	4	4	4	3,8
Q2	4	3	5	2	3	3	3,3
Q3	3	5	4	5	5	4	4,3
Q4	4	4	5	4	4	4	4,2
Q5	4	4	4	3	3	4	3,7

Table 3. Results of the questionnaire as applied to the users in the non-advanced group

	UN1	UN2	UN3	UN4	UN5	UN6	UN7	UN8	Avg
Q1	3	4	4	3	4	2	4	3	3,3
Q2	1	4	3	1	2	4	5	2	2,8
Q3	3	5	5	3	4	5	5	4	4,3
Q4	3	4	4	3	3	3	4	4	3,5
Q5	3	5	4	4	5	4	4	4	4,2

Some responses were not different than expected due to the users, such as the values obtained for questions Q2 and Q4, in both groups. The questions that had unexpected answers were Q1 and Q5: even though interaction with a 3D system is something new to the non-advanced users, they judged the interaction as easy and had a very good experience with it.

After answering the questionnaire, users could make suggestions for improvements of SiVIEP through written responses. The suggestions were very important to justify some answers and behaviors during the test. Two users from the advanced group and two users from the non-advanced group suggested an insertion of keyboard shortcuts during the interaction with the *fly* tool. Another suggestion was to momentarily increase the speed in the automated version, by means of the scroll wheel, thus customizing the behavior in this version. One user, from the non-advanced group, said that it was difficult to locate himself in the model, arguing thus for the definition of a target symbol to replace the mouse pointer, indicating the direction of the user navigation. Another user, from the non-advanced group, suggested the insertion of control tools for zooming.

6 Conclusions

This work presented a usability test of three navigation cubemap-based techniques developed to improve the *examine* tool, by providing a way to automatically determine the pivot point, and the *fly* tool, in which collision support and automatic speed adjustment in relation to the scale were implemented.

The usability tests were performed to assess the ease of learning and user satisfaction with the proposed techniques. The test consisted of three tasks, in which each one tested a technique proposed. They were performed by two user groups with different profiles.

The results indicate that the techniques presented here may improve the users' navigation experience. Usability tests confirmed that, from the 6 individuals of the advanced group and 8 of the non-advanced group, respectively, about 70% and 90% preferred the version of the application that included the proposed techniques. In face of their experience with 3D applications, users from the advanced group preferred the manual controls, most of the time, but acknowledged that the automatic adjustment minimized the mistakes that could happen and found the collision treatment and detection very important to the interaction. Users from the non-advanced group were surprised to see that they were able to interact well, most of the time, with a 3D system like SiVIEP, because that was their greatest concern before the interaction had begun. They preferred the automated version, most of the time, given the facility that this version offered for navigation and manipulation of objects.

The 3D navigation in multiscale environments is becoming more common, especially due to the popularization of technology and the need to develop 3D systems with increasing complexity in areas such as engineering, geoscience, games, and maps, to name a few. This popularization brings further problems to be solved in terms of navigation and user interaction, providing a lot of challenges to HCI researchers and practitioners. We believe the usability tests we have conducted shed some light in the problems different users experience during interaction and navigation in these environments.

Acknowledgments. All authors thank TecGraf, PUC-Rio for its support to their research work. Simone Barbosa also thanks CNPq for its support though grant #313031/2009-6. Alberto Raposo thanks FAPERJ for its support though grant #E-26/102.273/2009.

References

1. Andrade, T.M., Seixas, M.L.A., Raposo, A.B.: Uso de Técnicas da Engenharia Semiótica para Avaliação de um Sistema 3D na Área de Exploração e Produção de Petróleo. In: XXXVII Seminário Integrado de Software e Hardware – SEMISH 2010, pp. 425–439. SBC, Belo Horizonte (2010)
2. Jul, S., Furnas, G.W.: Critical zones in desert fog: aids to multiscale navigation. In: UIST 1998: Proceedings of the 11th Annual ACM Symposium on User Interface Software and Technology, pp. 97–106. ACM, New York (1998)

3. Perlin, K., Fox, D.: Pad: an alternative approach to the computer interface. In: SIGGRAPH 1993: Proceedings of the 20th Annual Conference on Computer Graphics and Interactive Techniques, pp. 57–64. ACM, New York (1993)
4. Bederson, B.B., Stead, L., Hollan, J.D.: Pad++: a zooming graphical interface for exploring alternate interface physics. In: UIST 1994: Proceedings of the 7th Annual ACM Symposium on User Interface Software and Technology, pp. 17–26. ACM, New York (1994)
5. Mccrae, J., Mordatch, I., Glueck, M., And Khan, A.: Multiscale 3d navigation. In: I3D 2009: Proceedings of the 2009 Symposium on Interactive 3D Graphics and Games, pp. 7–14. ACM, New York (2009)
6. Trindade, D.: Técnicas de Navegação 3D Usando o Cubo de Distâncias. Tese de mestrado. Departamento de Informática. PUC-Rio, Rio de Janeiro (2010)
7. Preece, J., Rogers, Y., Sharp, E.: Interaction Design: Beyond Human-computer Interaction, 1st edn. John Wiley & Sons, New York (2002)
8. Fitzmaurice, G., Matejka, J., Mordatch, I., Khan, A., Kurtenbach, G.: Safe 3d navigation. In: I3D 2008: Proceedings of the 2008 Symposium on Interactive 3D Graphics and Games, pp. 7–15. ACM, New York (2008)
9. Ware, C., Osborne, S.: Exploration and virtual camera control in virtual three dimensional environments. In: SIGGRAPH 1990: Proceedings of the 17th Annual Conference on Computer Graphics and Interactive Techniques, pp. 175–183. ACM, New York (1990)

SemaZoom: Semantics Exploration by Using a Layer-Based Focus and Context Metaphor

Dirk Burkhardt, Kawa Nazemi, Matthias Breyer, Christian Stab, and Arjan Kuijper

Fraunhofer Institute for Computer Graphics Research, Fraunhoferstraße 5,
64283 Darmstadt, Germany
{dirk.burkhardt,kawa.nazemi,matthias.breyer,christian.stab,
arjan.kuijper}@igd.fraunhofer.de

Abstract. The Semantic Web is a powerful technology for organizing the data in our information based society. The collection and organization of information is an important step for showing important information to interested people. But the usage of such semantic-based data sources depends on effective and efficient information visualizations. Currently different kinds of visualizations in general and visualization metaphors do exist. Many of them are also applied for semantic data source, but often they are designed for semantic web experts and neglecting the normal user and his perception of an easy useable visualization. This kind of user needs less information, but rather a reduced qualitative view on the data. These two aspects of large amount of existing data and one for normal users easy to understand visualization is often not reconcilable. In this paper we create a concept for a visualization to show a bigger set of information to such normal users without overstraining them, because of layer-based data visualization, next to an integration of a Focus and Context metaphor.

Keywords: Human-Computer-Interfaces, Semantics Visualization, Information Visualization, Semantic Web.

1 Introduction

With the development of the Semantic Web huge knowledge bases were created to use the features of storing data in a structured and linked form. Typically a knowledge base consists of an ontology in the format of RDF or OWL. Nowadays adequate visualizations are available for visualizing the schema and its instances. The most common visualization type for semantic data is a graph-based visualization, because it is one of the most known and understandable visualization metaphor for normal users. An actual and important challenge is the visualization of massive data in adequate form, because today knowledge bases containing more than 100.000 instances and sometimes also more than 2 million instances. Of course, the visualization of all the available instances is not appropriate, because it will overstrain the user by visualizing too many entities. But by exploring and searching within huge knowledge bases the resulting number of instances is sometimes more than 50 instances and it is hard to automatically reduce it to a lower value. So in these cases a visualization or a visualization metaphor is needed for visualizing a higher numbers of entities.

M. Kurosu (Ed.): Human Centered Design, HCII 2011, LNCS 6776, pp. 491–499, 2011.

In the research area of information visualization some approaches exists for providing the possibilities to present also higher numbers of entities in an acceptable way. Especially interaction metaphors like Focus and Context are established forms for visualization large numbers of nodes. A typical and easy to use representative is the Fisheye View, where during the interaction time the entire context is visible, next to a parallel zoomed area in which a detail view on the data is given. The interaction metaphor has as central approach the use of a magnifier, that zooms the area of interest e.g. around the mouse. The more an entity is away from the area of interest, the smaller it is displayed, so for instance entities on the borders of the visualizations are very small. Over all the time, the entire context of the presented data is visible. So also the relations between the elements are visible all the time. With such an interaction metaphor, the user is able to explore also larger subsets of a dataset by regarding the context of the entities.

In this paper we are going to present a concept of a Focus and Context visualization named *SemaZoom*. With *SemaZoom* a graph of instances from an underlying ontology will be visualized and can be explored. It uses the Fisheye Effect for providing the possibility to read all entities and regarding that the context is visible all the time. *SemaZoom* contains also another new approach, for presenting concepts and instances in a different way. Normally a semantic structure is visualized in a single graph, presents the concepts and instances equally. This approach has as disadvantage that many nodes (including the concepts) will be presented, but often only the instances are relevant. So the user sees also objects that are not relevant for him. In fact there are also graphs, which only present instances. But the corresponding concept often contains additional and relevant information, which can be important for the user. For instance, if a user searches for golf, the concept can indicate if the instance will be the sport activity or the automotive. For this reason we extended the graph by using fisheye effect with a second layer that will show the concept to a presented instance. So the *SemaZoom* presents the semantic data from an ontology in a graph of instances by using the fisheye effect and in a second layer with drawn colored background shapes surround the instances from the first layer. These shapes address concrete concepts. The names of the concept can be written directly on the shapes or in a legend next to the shapes. This approach guaranties that also additional information from the schema level are available for the user, during his exploration through the instance data. Furthermore the space for presenting information will be exhaustively used in efficient way, thus reducing the cognitive overload of the user on a low level.

2   Related Works

For visualizing of large amount of data, different kinds of established visualization approaches are available. Especially by visualizing semantics data are taxonomy-based and graph-based layout techniques appropriate. Furthermore, different kinds of interaction techniques and metaphors are available for realizing the interaction and exploration through the data. One of the most trivial form of visualizing data are the presentation of all available nodes, but in particular by massive data this approach is not adequate, because the users will get overstrained. Today different approaches exist to avoid an overstraining of the users, for instance the reducing of the data on relevant nodes or efficient visualizing strategies to use the space on a screen optimal. The approach for reducing data on relevant data is a critical once, because for this a method is required, which is able to determine relevant and irrelevant nodes and to avoid that the systems hides nodes, which are necessary for the user. So in this paper we are concerning on strategies, that will not hide nodes, but rather allow to define a focus on the data. This is only possible with zoom and filter techniques basing on the visualizing level.

One visual technique for providing a filter technique is Panning. First systems provide this feature by scrolling, panning or paging [5]. The underlying idea bases on moving a viewport over a document. The most known representative is scrolling, which is realized in application by scrollbars. In general panning can be realized with 4 different main interaction forms [6]: with scrollbars, by using a dragging metaphor, by using the map metaphor and by a combination of map and pointer.

By large numbers of items or nodes panning will be no adequate metaphor for interacting and exploring through the data, because the scrolling etc. to another position in a document takes a long time. If zoom functionality will be added, the browsing through a document or in general through the dataset is much faster, because of the smaller size of nodes that represents information over which has to be browsed [5]. For a general use, the zoom function allows the user to get an overview of the existing data and when he needs more information, he can take a "deeper look" into the data by zoom-in to the data elements of interest.

A disadvantage by providing a zooming functionality is the loose of the context, when the user zooms into a data area of interest. An approach that compensates this aspect is the use of Multiple Views. By this approach the user has at least two views on the data. On for the main interaction, often this is a bigger view area, to work within the data. Next to it, there are often one or more other views, which provide an overview to the entire data set. This metaphor is commonly used for geographical information e.g. GoogleMaps, where a user has und the bottom right side a small overview window from the map e.g. from a country or continent.

Next to the general metaphors for working with large datasets, there are also solutions which combining some of these basic metaphors. A result of such a combination are so called Focus and Context metaphors or also known as Overview and Detail metaphors. An overview to a number of established techniques for Focus and Context is given in Cockburn et al. 2008 [1].

A common known representative for Focus and Detail is the Fisheye view effect. One of the first implementation on a graph-based visualization is described in Sarkar et. al. [3]. The fisheye effect makes nodes bigger within a focused area than nodes which are out of this area. In general the nodes are becoming smaller, the more they are distanced from the focused area. In fact all elements of the graph are visible during the whole time of interaction, which regards the visibility of the context. Additionally the user is able to see and read a node from the graph, if the node is within the focus are e.g. by moving the mouse over it.

An example implementation from fisheye view on graph-based visualization for showing semantics data is presented in Zhang et. al. [2]. In this implementation the schemas and instances are handled equally, so that the benefit of a structured information source will not be full supported, but with this kind of implementation the user gets an easy form to interact through a bigger network of semantically information.

3 Concept of the SemaZoom Visualization

Graph-based visualizations are most common used visualization type for visualizing semantics data. They provide an easy view on the data and furthermore they allow an easy and understandable navigation through the concepts and instances. But semantics data-sources are often very voluminous so that a useful navigation through this massive data is impossible.

This aspect is the challenge, which the *SemaZoom* navigations try to solve, next to an effective presentation of the different objects that exist in semantic data structures for concepts and instances.

3.1 Differentiation between Concepts and Instances

In the scenario of an information search, the required results are often concrete instances, which describing explicitly specific information. Concepts are commonly used for filtering instances from a given result set. The reason is that concepts in the context of semantic structures defining general characteristics for concrete occurrences in form of instances, which can be classified by properties, but of course the values can be differently for each instance. In our visualization framework we are using for filtering the result set by selecting or deselecting concepts other kinds of visualization like the *SeMap* [7]. As a consequence for supporting the simultaneous use of visualization for providing different possibilities and regarding different aspects, with our SemaVis-framework it is possible to composite a personal so called "Knowledge-Cockpit" [4] by coupling preferred visualization (see also the screenshot in Fig. 3.).

The focus of this visualization lays on the visualization of the instances and its relations. Because of the fact that instances are linked with each by semantically relations, graph-layout algorithms are suitable. In comparison to graph-based visualization, which are showing concepts and instances in a same way (e.g. ZoomRDF [2]), we only visualize instances in the graph, which allows us to present more instances without overstraining the user. This graph consisting of semantically instances and relation is presented in one layer.

An assignment of instances to their concept provides an orientation for the user, because the concepts implicitly containing additional information to an instance. This additional information to an instance is not required during most of the time by interacting or exploring through the data, so that the concepts do not have to be presented conspicuously. For achieving such a restrained presentation, we show conceptual information like the information of altitudes values are presented on maps. We realize it by coloring the background in dependence of the assigned concept. This coloring will be done in a second visual layer, which is underneath the Instance and Relation Layer (see Fig. 1.). So the background of every instance is colored, by an existing legend it is possible to identify the concrete concept, where an instance is assigned to. Semantically concept relations cannot be visualized with this kind of visualization. The semantically concepts in form of background shapes are drawn into the second layer, which is placed under the instance layer. Shapes will be updated or redrawn automatically, if the positions of the instances are changing.

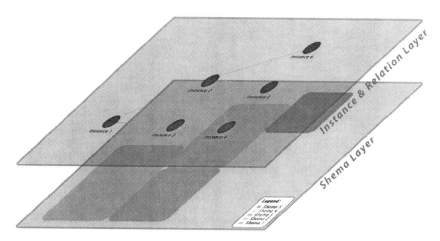

Fig. 1. Visual concept for *SemaZoom* to visualize semantic instances and concepts

3.2   Usage of Focus and Context Metaphor

By the presentation of semantics data, e.g. from an ontology management system, are often very large data sets are to visualize. For supporting an adequate visual presentation of such large data sets within graph-based visualization some implementations do exist by using a Focus and Context view. A well-known and with often positive results evaluated metaphor is the Fisheye-view metaphor.

We use the fisheye effect within the *SemaZoom* visualization for providing an interactive data presentation. So the user can navigate through big dataset to search explore for information. Because of the use of the Fisheye view, the user is able to see focused information in a normal size so that they are good readable, but he also has permanently an overview about the context, how an instance is related to another once. Because of the changing size and position of every instance by the use of such a dynamic zoom-metaphor, the background shapes that represents the concepts, have dynamically be updated, to ensure that these information will be shown correctly during the entire time of interaction.

For the generation of the fisheye view effect, we are using a similar calculation, like it is used in Sarkar et. al. [3], we only changed some parameters, which are providing a better interactivity for a web-based usage.

3.3   Implementation and Feature Summary

As a result we implemented the concept prototypically as a visualization module that can be used in our developed SemaVis-framework. An impression of the implemented visualization is given in Fig. 2.

This kind of visualization supports the presentation of instances, instance relations and concepts as semantic objects. For the navigation the visualization supports multiple paths, where a user does not have to go back for changing to another instance. In the current version *SemaZoom* is only less configurable, which makes it not recommendable for the use in adaptive systems. Actually only the color of the background shapes, which represents the concepts, and the color of the instance relations can be changed. A summary to the supported features is given in Table 1.

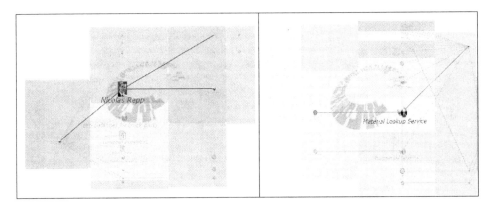

Fig. 2. Screenshots of the *SemaZoom* with different configurations

Table 1. Summary of supported semantics data features by this type of visualization

	Semantics					Navigation		Adaptation			
	Concepts	Individuals	Relations	Multiple Inheritance	Properties	Single path	Multiple path	Size	Color	Order	Predictions
SemaZo	✔	✔	✔	-	-	✔	✔	-	(✔)	-	-

4   Case Study

The described *SemaZoom* visualization was developed as a part of the Core-Technology-Cluster (CTC) of the THESEUS Program (Theseus 09), a 60-month program partially funded by the German Federal Ministry of Economics and Technology. The partners in the THESEUS Program research under the device "New Technologies for the Internet of Services" heterogeneous technologies for gathering and offering semantic information on web. The program itself consists of twelve projects, divided in THESEUS Use Cases and the THESEUS Core-Technology-Clusters. Where the six Core-Technology-Clusters are led by research institutions and focus on fundamental research areas, the THESEUS Use Cases are led by enterprise institutions and bridge the gap between fundamental research and industrial dissemination. Different enterprise partners focus on their usage scenario of the different areas of information processing. For example the Siemens Corporation investigates the processing of medical-related information. In this THESEUS Use Case (Medico) different usage scenarios identify different user groups: There are medical doctors, who use the information of the patient's clinical and medical records to find similar cases and provide the adequate care for them. On the other hand you have the patients themselves, who should be able to understand about their disease and find for example groups or a community with similar ailments.

Another example for a THESEUS Use Case is THESEUS Texo, which is conducted by SAP Research Germany and investigates different models and techniques for providing, engineering, disseminating and using web-based services. The different user roles e.g. service-engineers, domain-experts or service-consumers handle the same information in different ways. Complex information structures defined as ontologies makes it necessary to provide here a best fit of visualizing the information structures for the different user-groups and their precognitions.

A third example for a THESEUS Use Case is Contentus, led by the German National Library. Here you find the same heterogeneity of users. There are domain-experts, who have the required knowledge in a specific scientific domain, e.g. experts for German Literature, but are not experts in using and processing complex ontology-based information-systems. Of Course you find in Contentus the average user too, who just explores knowledge domains and expects a very simple to use visualization and user interface.

Beyond the THESEUS Use Cases there are six THESEUS Core-Technology-Clusters (CTCs) investigating different fundamental research questions regarding semantic information processing. The CTC are mainly conducted by research institutes. The CTC for Ontology Management, led by "Forschungszentrum Informatik", investigates for example managing, reasoning, editing and inferencing ontologies. The CTC Situation Aware Dialog Shell investigates different questions regarding context-aware information processing.

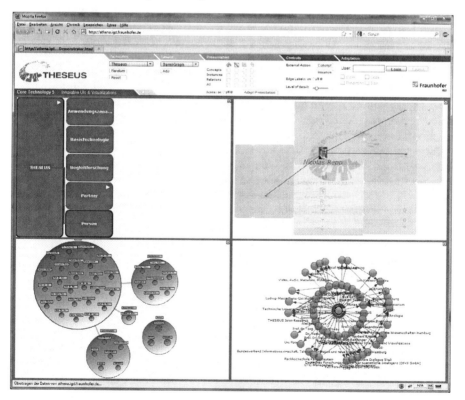

Fig. 3. *SemaZoom* integrated into the SemaVis Knowledge-Cockpit

The *SemaZoom* visualization was developed as a CTC-component of the THE-SEUS Core-Technology-Cluster Innovative User Interfaces and Visualizations and is used in different THESEUS Use Cases for exploring knowledge domains within the specific scenarios of the Use Cases. *SemaZoom* was tested in different Use Cases in their specific semantic knowledge domain. The first tests in real scenarios, where users were able visualize medical documents and information in the UseCase Medico. Because of the high number of related items, with *SemaZoom* it was possible to explore coherence documents and reports to a patient. We also integrated the *SemaZoom* visualization into our demonstrator with which a user can generate is own personal "Knowledge Cockpit" (see Fig. 3.).

5 Discussion

Interacting through semantics data is an up growing challenge, because the currently existing approaches and implementations are designed for specific aspects, often they are focusing on technical details, for instance to highlight ontology-based features for organizing the data in an ontology management system. Normal users, respectively experts of foreign domains, need other kinds of visualizations. With this *SemaZoom* visualization we tried to provide an easy to user interface for semantics data, but during the conceptualization and also during the implementation we get some challenges, where further work is required and useful. A general aspect is the classifications of the user profiles, particularly depending to their IT skills. Depending to those user profiles a mapping is necessary, which kinds of visualization will be "easy to use". A very easy visualization metaphor can be a helpful tool for computer novices, but for advanced users or experts it can be an annoying tool, because it provides to less information.

This is also a necessary question for the *SemaZoom* to define which information is useful and should be presented to a user during his interaction with the visualization. Currently we provide information only to the name of an instance or concept, but no further details to or surround this object. Also the names of the relations between instances are currently hidden, because it can be confusing if in general too many text labels will be presented, especially if a large data set is to visualize.

A specific challenge for this concrete visualization is the form, how the semantic schema with its hierarchy of concepts can be presented and also their relation to each other. In this version we show only the concrete concept to an instance, but if another instance from a higher or lower concept level is presented next to it, it can be confusing for the user. An approach to avoid this can be the use of different brightness levels as an indicator for the level within the schema hierarchy, but for this the entire schema must be known by the visualization, which generates a high overhead especially by large semantics databases, because all the information must be known by the visualization. This circumstance makes it hard for the use of client-server based search technologies.

6 Conclusion

In this paper we presented a concept for a layer-based Focus and Context metaphor for the usage on semantics data and for supporting a semantics exploration. The concept consists of two steps. First, we generate the presentation of the data with a graph-based visualization that shows the instances and instance relations in one layer and concepts in a second layer. The concepts contain additional information to the presented instances, like the type of it e.g. an automotive, a sport activity etc.

We also integrated a Focus and Context metaphor, which allows the presentation of a large number of instances without overstraining the user, because these kinds of metaphors provide the feature to show an overview of the data, next to details to viewed nodes of interests. In our visualization we are using the Fisheye View metaphor as one representative of Focus and Context. As a result of this Fisheye View effect, the user can explore parts of the visualized data by focusing it and the elements within the focus area will be shown in a normal well-readable size. Objects out of the focus area are presented with a very small size, but they are visible all the time, so that the context – especially the instance relations - is always observable for a user.

By combining both aspects, we created a new form for a semantics visualization with an interactive style. The *SemaZoom* visualization provides an easy understandable presentation of semantics, which also allows the possibility to explore bigger datasets. In particular by the visualization of semantics data, this kind of visualization provides a well overview on the data and parallel the user is able to look in detail on instances of interests.

Acknowledgements. This work was supported in part by the German Federal Ministry of Economics and Technology as part of the THESEUS Research Program. For more information please see http://www.semavis.com.

References

1. Cockburn, A., Karlson, A., Bederson, B.B.: A review of overview+detail, zooming, and focus+context interfaces. ACM Comput. Surv. 41(1), Article 2 (2009)
2. Zhang, K., Wang, H., Tran, D.T., Yu, Y.: ZoomRDF: semantic fisheye zooming on RDF data. In: Proceedings of the 19th International Conference on World Wide Web (WWW 2010), pp. 1329–1332. ACM, New York (2010)
3. Sarkar, M., Brown, M.H.: Graphical fisheye views of graphs. In: Bauersfeld, P., Bennett, J., Lynch, G. (eds.) Proceedings of the SIGCHI Conference on Human Factors in Computing Systems (CHI 1992), pp. 83–91. ACM, New York (1992)
4. Nazemi, K., Burkhardt, D., Breyer, M., Stab, C., Fellner, D.W.: Semantic Visualization Cockpit: Adaptable Composition of Semantics-Visualization Techniques for Knowledge-Exploration. In: International Association of Online Engineering (IAOE): International Conference Interactive Computer Aided Learning 2010 (ICL 2010), pp. 163–173. University Press, Kassel (2010)
5. Gutwin, C., Fedak, C.: Interacting with big interfaces on small screens: a comparison of fisheye, zoom, and panning techniques. In: Proceedings of Graphics Interface 2004. ACM International Conference Proceeding Series, vol. 62, pp. 145–152. Canadian Human-Computer Communications Society, School of Computer Science, University of Waterloo, Waterloo, Ontario (2004)
6. Kaptelinin, V.: A comparison of four navigation techniques in a 2D browsing task. In: Conference Companion on Human Factors in Computing Systems, CHI 1995, pp. 282–283. ACM, New York (1995)
7. Nazemi, K., Breyer, M., Hornung, C.: SeMap: A Concept for the Visualization of Semantics as Maps. In: Proceedings of the 5th International Conference on Universal Access in Human-Computer Interaction (HCII 2009), pp. 83–91. Springer, Heidelberg (2009)

Scientometric Analysis of Research in Smart Clothing: State of the Art and Future Direction

Kyeyoun Choi[1], Huiju Park[1], Eui-Seob Jeong[3], and Semra Peksoz[1,2]

[1] Institute for Protective Apparel Research and Technology,
Oklahoma State University, 74074 USA
[2] Department of Design, Housing and Merchandising,
Oklahoma State University, 74078 USA
[3] Korea Institute of Science and Technology Information, 443-270, South Korea
{kyeyoun.choi,hui.park,semra.peksoz}@okstate.edu

Abstract. The purposes of this study were to investigate research trends on smart textile and clothing and to suggest future research directions on smart textile and clothing by using scientometrics approach. The research of smart clothing was divided into five categories: technology, human factors, application, manufacturing, and consumer demands and retailing. Technology emerged as the dominant category suggesting technological development of smart materials and wearable input devices have been intensively studied and have provided a solid foundation for smart clothing research. The number of research on output devices and data and power transportation showed a gradually increasing trend since 2000. Analysis on technical collaboration among each research field showed a high correlation between input technology and the three main categories: smart materials, functional application and, manufacturing. Material sciences, electronic engineering and computer sciences were shown to be major research disciplines to lead smart clothing research based on quantity of publications.

Keywords: Smart Clothing, Scientometrics, Technology, Research.

1   Introduction

Smart clothing that has sensing, actuating and responding capabilities to a stimulus from the immediate environment [1] has evolved to be highly sophisticated since the time when a wearable computer was initially introduced as an "existential computer" in the late 1970s and early 1980s [2]. Smart clothing should satisfy a garment's aesthetic aspect and be appropriated for an end-user with smart functionality, enhanced by embedding technology [3]. The technologies applied to smart clothing are quite extensive in accordance with their applications. Because of this, the research on smart clothing are not only actively performed by clothing and textiles scientists alone, but done collaboratively by research groups consisting of scientists with different backgrounds (interdisciplinary research). Advances in textile technology, computer engineering, and materials science are promoting a new breed of functional fabrics. In addition, fashion designers are adding wires, circuits, and optical fibers to traditional

M. Kurosu (Ed.): Human Centered Design, HCII 2011, LNCS 6776, pp. 500–508, 2011.

textiles and creating garments that glow in the dark or change their shapes, responding stimuli, such as sound, light, and heat etc. [4]. Meanwhile, electronics engineers are using a fabric, woven or knitted, with conductive yarns and embedding sensors into body suits that map users' whereabouts or vital signs [5]. However, up to now, many of the developed technologies have merely confirmed their possibilities rather than realizing functionality on a specific clothing system because researchers with different backgrounds rarely shared their knowledge in interdisciplinary research. In addition, quantitative and systematic research about drawing a comprehensive map to understand current research performed on smart clothing has been rarely explored in relevant research fields. Therefore, scientometric or bibliometric approach to understanding and forecasting emerging and most useful techniques in collaborative disciplines for designing smart clothing should be performed.

Scientometrics is the science of measuring and analyzing science and it is often performed using bibliometrics which is a measurement of the impact of (scientific) publications. Scientometric data classified by years, countries, fields of science, etc. help in evaluating the scope of scientific research and trends in investigations and their levels [6]. Scientometrics aims to provide a point of criticism to research evaluation using only numerical factors and to describe scientific activity quantitatively. Even though the index or data through scientometric approach does not represent whole scientific activities in a field, it has been used because it clearly reflects the main part of the target research and identifies relevant areas. The comprehensive and reliable approach is mainly used in science fields, such as biology, engineering, sociology, etc. However, this methodology is barely used in clothing and textile field. The scientometric approach is expected to provide practical implications to researchers for understanding current research trends, as well as future research demands and directions for the development of smart clothing. Therefore, the purposes of this study were to investigate research trends and to identify future research for textile and clothing based systems using scientometrics.

2 Research Method

The research of smart clothing was divided into five categories based on the literature review [7]: (1) technology, (2) human factors, (3) application, (4) manufacturing, and (5) consumer demands and retailing. These categories were further subdivided into 16 subcategories (Table 1). Research papers from over 11,000 journals published by Web of Science of Thomson Reuters since 1990 were used as the database. The source was generated by using Boolean operators which resulted in 7,858 relevant articles: the main research keywords, such as "smart, intelligent, wearable" and "textile, fabric, clothing, garment" were entered into the search fields and combined with the Boolean logic, such as "AND" and "OR"[8]. Each identified article was independently reviewed and screened by two researchers with expertise in the subject area to eliminate non-pertinent papers. Finally, 842 publications were left after the reviewing titles, keywords, authors, and abstracts of papers in each subcategory. The filtered papers were analyzed based on country of origin, publication year, number of articles, and research fields.

Table 1. Categories on smart clothing

Main Category	Sub Category	Abb.	Description
	Smart Material	SM	Materials
	E-textile	ET	
Technology	Input	IP	
	Output	OP	
	Communication	CM	Clothing + Device
	Data Management	DM	
	Energy	EN	
Human Factors	Durability	DU	
	Wearability	WE	Evaluation, Testing
	Aesthetic:Testing	AE	
Application	Functionality	FU	
	Entertainment	ENT	Prototype Suggestion and Development
	Accessories	ACC	
Manufacturing	Manufacturing Process	MP	Manufacturing Solution, Process, Production
Consumer Demands and Retailing	Concept Design	C-design	Market needs, Future Demand
	Clothing Demand	CD	

3   Result and Discussions

3.1   Nationwide Research Trends on Smart Clothing

As shown in Figure 1, of the 46 countries involved in smart clothing research, EU (37.6%), U.S. (24.1%), Japan (12.5%), South Korea (9.8%), China (9.0%), India (2.3%), and Canada (1.8%) emerged as leading countries. EU showed the highest number of publications with England (6.1%), Italy (4.9%), Germany (4.7%) and Switzerland (4.4%) being the four major countries performing smart clothing research. Examining individual countries revealed that approximately 62% of the studies published were from five countries: U.S., Japan, South Korea, China, and England. While the U.S. and Japan took the lead in smart clothing research since the 1990's, England and China sharply increased their number of publications since 2006.

Material sciences, electronic engineering and computer sciences were shown to be major research disciplines to lead multidisciplinary research on smart clothing based on quantity of publications (Table 2). Five leading countries also showed greater number of publications in material science, engineering and computer science. EU and U.S. show more focus on research about engineering and computer science, while Japan, South Korea, and China focus more on material science and engineering. The number of publications on smart clothing has sharply increased for the last two decades as shown in Figure 2. The majority of which were publications on smart materials, input and output devices in functional application.

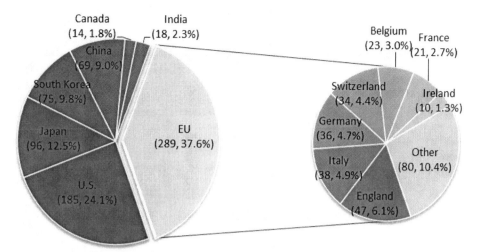

Parenthesis: number of research papers, percentage

Fig. 1. Research papers of top ranked countries (between 1991 and 2010)

Table 2. Research field of leading countries on smart clothing

	EU	U.S.	Japan	South Korea	China
Materials Science	77	64	42	37	44
Engineering	114	80	54	35	16
Computer Science	113	81	22	20	15
Chemistry	35	16	6	10	12
Information	16	6	0	1	3
Physics	25	12	9	6	10
Communication	29	16	4	3	0
Total	**409**	**275**	**137**	**112**	**100**

By 2000, research on smart clothing was solely focused on technology while other research fields were rarely studied. Overall, the total number of publications on smart clothing increased during the last decade. 716 of 842 research articles (85%) were published between 2001 and 2010 (about 5.6 times more than the number of articles between 1991 and 2000). Having the highest number of publications, *technology* emerged as the dominant category suggesting technological development of smart materials and wearable input devices have been extensively studied and have provided a solid foundation for smart clothing research. The number of research on output devices, and data and power transportation in technology showed a gradually increasing trend since 2000. After 2001, research topics such as *applications* and *manufacturing* have been actively studied. The number of publications on these two topics has sharply grown in the last decade, while the number of publications on *human factors* and *consumer demands* have shown gradually increasing or stagnant trends. As the overall research interest in smart clothing grew, the number of papers providing general overview also increased.

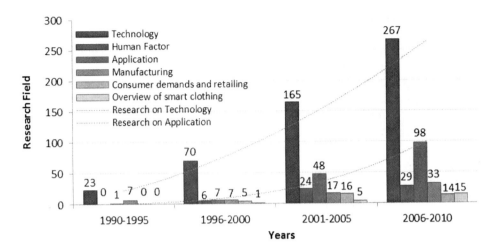

Fig. 2. Chronicle research trend by research field since 1990

3.2 Direction of Research and Technology Development in Each Category

The partitioning of research on smart clothing and their key technologies are presented in figure 3 and table 3. 'Technology' (525 publications, 62.4%) was shown as the most dominant research fields and included *smart materials, input devices, output devices, communication, data management and energy.* 205 publications were focused on minimized or wearable *'input devices'* for smart clothing. Examples of the input devices included biomedical sensors embedded to clothing, wireless sensing, hand gesture input device, ubiquitous input device, flexible substrates, flexible textile antenna, and radio frequency tag. A second most prominent topic was *'smart materials* (198 publications)' responding to environmental stimuli. Temperature sensitive smart materials such as phase change materials, shape memory alloys and polymers showed the largest number of publications. Other smart materials included conductive polymer used for electric heating and data transmission and power supply, multifunctional nano-fiber, carbon fiber, and optic fiber for bio-sensing and other functionalities. *'Output devices',* (18 publications) such as actuator, visualization, network, dynamic information display system, and immersive projection technology, were among the subjects most commonly studied. *'Communication'* (26 publications) included studies about bluetooth technology, wideband communication, intra-body communication, and wireless communication network. *'Data management'* (15 publications) includes smart transistors, circuit design and testing, event recognition system, information transmission system and annotation database management. *'Energy'* (10 publications) included Piezoelectric energy harvesting technology for increasing energy efficiency in smart clothing.

'Human factors' (59 publications, 7.0%) included *durability, wearability,* and *aesthetic evaluation* of smart clothing. Research on *'durability'* (14 publications) was focused on long-lasting desired performance and quality which were associated with physical and mechanical properties of material for smart clothing. *'Wearability'* is operationally defined as an index of suitability for smart clothing including usability, functionality, comfort, wearer's psychological comfort and satisfaction. Analysis revealed that research on wearability (44 publications) centers mainly on context awareness, convenient operational interface, ubiquitous computing environment, thermal protection and physiological impacts of smart clothing on human body. On the other hand, *aesthetic* aspect of smart clothing was shown to be rarely studied (1 publication).

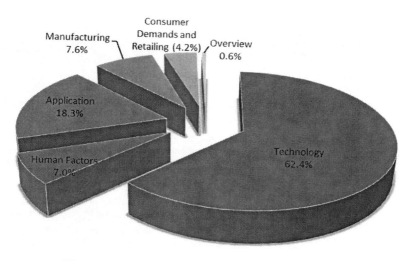

Fig. 3. Overall partitioning of research on smart clothing

'Application' (154 publications, 18.3%) included *functional* and *recreational* application, *accessories* and *prototype* development. More than 90% of *'functional applications'* (140 publications) were studied to meet needs such as providing remote health monitoring, safety and protection for specific target markets such as the elderly, hospital patients, soldiers, emergency personnel and firefighters. In particular, smart clothing for remote health monitoring system showed the largest number of applications using wearable sensors detecting vital signs and telecommunication with hospitals. Applications for *'entertainment'* (3 publications) included studies about a wearable computing system for game and color-responsive textiles. Publications in *'Accessories'* (11 articles) category was related to sensor glove, multi-functional wrist watches, and smart-shoes which can measure in-shoe plantar pressure.

'Manufacturing' (67 publications, 7.6%) included *smart clothing assembly, e-textile and smart material production, functional finishing* and *challenging issues for mass production*. More research was conducted on engineering of smart materials than practical solutions to smart clothing assembly and standardized industrial production.

'Consumer demands and retailing' (34 publications, 4.2%) included *existing and emerging market needs*, which included enhanced protection, context-awareness and ubiquitous computing.

Table 3. Key technologies on smart clothing

Categories		Key technologies
Technology	Smart Material	Responding to environmental stimuli Temperature sensitive smart material (Phase change materials, shape memory alloy and polymer, conductive polymer for electric heating
	E-textile	Smart nano-textile, optical fibers, Strain-sensing fabric, Electrical characteristics of conductive yarns, Modeling and simulating e-textile, e-textile craft, conductive yarns by coating or bulk treatment
	Input	Device: minimize, wearable biomedical sensors embedded to clothing, wireless sensing, hand gesture or ubiquitous input device flexible substrates, flexible antenna, radio frequency tag
	Output	Actuator, visualization, network, dynamic information display system, immersive projection technology
	Communication	Bluetooth technology, wideband, intra body, and wireless communication network
	Data Management	Smart transistors, circuits design and testing, event recognition system, Information transmission system, and annotation database management.
	Energy	Piezoelectric energy harvesting technology
Human Factors	Durability	Long-lasting performance and quality control
	Wearability	Suitability, usability, functionality, wearer's physical comfort, satisfaction
	Aesthetic: Testing	Rarely studied
Application	Functionality	Remote health monitoring, safety and protection for specific target user (elderly, hospital patients, soldiers, emergency personnel, firefighter, etc.) Detecting vital sigh, telecommunication with hospitals
	Entertainment	Game and color-responsive textile
	Accessories	Sensor glove, multi-functional wrist watches, smart shoes
Manufacturing	Manufacturing Process	Smart clothing assembly, e-textile and smart material production, functional finishing, challenging issues for mass production
Consumer Demands and Retailing	Concept Design	Prototype design
	Clothing Demand	Existing and emerging market needs, enhanced protection, context awareness and ubiquitous computing
	Overview	Review for smart clothing, smart materials

3.3   Collaborative Research Trends across Disciples

Seven key disciplines were found as leading research fields: material science (MS), engineering, computer science (CS), chemistry (CH), information technology (IT), physics (PH) and communication (CM). In particular, engineering, material science and computer science were shown to be dominant disciplines which have produced the majority of publications on smart clothing. In the field of engineering, electrical and electronic elements of smart clothing have been actively studied. Other research issues included nano technology and instrumentation of smart clothing. In the field of material science, active multidisciplinary research was found. Polymer science, composites and textiles have been also studied. In the field of computer science, information systems were mainly studied. Other research issues included software engineering, artificial intelligence, hardware and architecture, interdisciplinary application, computational biology and theory. Table 4 shows correlation coefficients among research fields, which shows the extent of interdisciplinary effort in smart clothing research. Analysis on technical collaboration among each research field showed a high correlation between input technology and the three main categories: smart materials, functional application and, manufacturing. Higher correlations were found in collaborative research for smart material and durability (.47), E-textile and manufacturing process (.42), e-textile and functionality (.33), input device and manufacturing process (.36), functionality and accessories (.37), and functionality and manufacturing process (.32).

Table 4. Correlation coefficients among categories

Working Category	SM	ET	IP	OP	CM	DM	EN	DU	WE	FU	AC	MP
ET	0.08											
IP	0.20	0.24										
OP	0.29	0.09	0.14									
CM	0.02	0.08	0.17	0.11								
DM	0.02	0.03	0.14	0.13	0.12							
EN	0.12	0.29	0.14	0.05	0.03	0.17						
DU	0.47	0.02	0.05	0.12	0.00	0.00	0.04					
WE	0.13	0.24	0.20	0.04	0.02	0.02	0.09	0.12				
FU	0.09	0.33	0.26	0.03	0.03	0.02	0.24	0.03	0.37			
AC	0.02	0.09	0.12	0.01	0.01	0.01	0.07	0.00	0.10	0.37		
MP	0.22	0.42	0.36	0.07	0.04	0.03	0.20	0.07	0.21	0.32	0.08	
C-design	0.04	0.15	0.29	0.03	0.04	0.03	0.06	0.10	0.13	0.15	0.05	0.29
CD	0.02	0.16	0.11	0.01	0.02	0.01	0.06	0.01	0.07	0.20	0.06	0.16

SM: Smart Material, ET: E-textile, IP: Input, OP: Output, CM: Communication, DM: Data Management, EN: Energy, DU: Durability, WE: Wearability, FU: Functionality, AC: Accessories, MP: Manufacturing Process, CD: Clothing Demand

4   Conclusion

Scientometric approach in this study provided an overview of research trends on smart clothing and identified the mainstream research fields. The focus of most researches has been on technology, which has provided a solid base for practical application suggestions and further research development. Relevant research on smart clothing have been mainly focused on composition and application of smart material, development of interface and e-textile for realizing smart function, input and output devices, development and evaluation of smart clothing prototype in mechanical engineering, computer science, and chemistry. In other words, current research on smart clothing was more focused on development and evaluation of the smart function with little consideration on human comfort, aesthetic and wearability of the clothing.

As evidenced by quantitative analysis, there is a scarcity of studies that address human factor and consumer demands, which requires active collaboration between apparel and electronic industries to provide practical solutions to industrial manufacturing process for mass productions. Therefore, more studies on human factors, consumer demands and multidisciplinary collaboration are expected to make a synergistic effect on the development of smart clothing and commercialization. Considering that smart clothing is the mixture of multidisciplinary collaboration, the unique nature of smart clothing is expected to require unconventional manufacturing processes, product distribution, retailing and maintenance. Studies on these issues will provide a solid infrastructure for commercialization to make smart clothing more practical and beneficial to mass market. Since the data base for this study was limited to journal papers searched by a specific search engine, this study does not reflect all existing publications in smart clothing. Future studies should include patents of smart clothing and relevant technologies, which will provide a more comprehensive view of overall research and development trends of smart clothing.

References

1. Tao, X.: Smart Fibres, Fabrics and Clothing. Woodhead Publishing Ltd., Cambridge (2001)
2. Mann, S.: Smart Clothing: The Wearable Computer and Wearcam. Personal and Ubiq. Com. 1(1), 21–27 (1997)
3. McCann, J., Hurford, R., Martin, A.: A design process for the development of innovative smart clothing that addresses end-user needs from technical, functional, aesthetic and cultural view points. In: Ninth IEEE International Symposium, Wearable Comptuters, pp. 70–77 (2005)
4. Fibers 2 Fashion, http://www.fibre2fashion.com/industry-article/18/1738/development-of-shape-memory-fabrics-garments1.asp
5. Gould, P.: Textiles gain intelligence. Materialstoday, 38–43 (2003)
6. Zolotov, Y.: Scientometric Studies, vol. 58, pp. 903–904 (2006)
7. Cho, G.: Smart Clothing Technology and Application. CRC Press, Florida (2010)
8. Brusilovsky, B.: Partial and System Forecasts in Scientometrics. Technological Forecasting and Social Change 12, 193–200 (1978)

An Experimental Study of Home Page Design on Green Electronic Products Web Site

Fei-Hui Huang

Department of Marketing and Distribution Management,
Oriental Institute of Technology,
Pan-Chiao, Taiwan 22061 R.O.C.
fn009@mail.oit.edu.tw

Abstract. The objective of this study is to understand users' electronic commerce needs and expectations in order to elicit the design requirements of a useful Web home page interface centered on green electronic products (GEP). In this study, an experiment was conducted to investigate the user needs captured by their external and mental patterns in order to apply them to the user-oriented Web home page design. The importance of Web site and home page design including gender differences have been found in this experiment and from the experimental results. Finally, consideration for designing a useful home page for a Web site are summarized as follows: (1) the home page design should be easy-to-use, easy-to-understand, and easy-to-digest and clearly show the Web structure and site purpose to users; (2) the home page should be streamlined to show information more efficiently; (3) gender differences should be considered in Web site and page design to meet different types of users' needs.

Keywords: User-centered interface design, green electronic products, user-Web interaction, Web pages design.

1 Introduction

Web services have become an essential part of daily life for many people. They use the Web browser to connect to the internet for searching information, communicating, shopping, banking, and other commercial and non-commercial activities. More and more companies and government organizations, therefore, are developing Web sites to expand their market potential and to convey green information in establishing an international image. Regarding the green electronic products (GEP), Huang et al. (2009) pointed out that there is a problem of market penetration and a substantial area for improvement in providing consumers environmental information as shown by the large difference between actual and willing buyers. Here, an idea for developing a Web site is put forth to promote or deliver green and GEP information to reinforce internet users' environmental values. However, the World Wide Web (WWW) has grown to 234 million Web sites in 2009 (Netcraft, 2009). This makes the information from the Web rather huge and a Web site providing users with nice interactive experiences is of increasing importance. Designing web pages according to users' mental models influences user interactions on the Web site (Bargas-Avila et al., 2007). The

M. Kurosu (Ed.): Human Centered Design, HCII 2011, LNCS 6776, pp. 509–518, 2011.

home page of a Web site may represent, moreover, an electronic vision of the reception lobby where potential customers can visit a company or it may merely be composed of attractive graphics to capture attention (Liu et al., 1997). In order to achieve such benefits and a good user-Web interaction, the Web pages / home page needs to be designed consistent with the theme of the site and Internet user's cognition.

The aims of this study are: (1) to acquire the users' preferences towards web services; (2) to study user-Web interaction, searching performance, and user satisfaction in using the proposed Web sites; and (3) to elicit the design requirements of a user-oriented home page interface for the B2C (Business to Customer) Web service in dealing with usability problems.

2   Relevant Literature

The WWW may provide services with low cost and great convenience, and is a global information resource for the consumers to obtain specific product/service information in anticipation of a purchase or general information about a brand, product, or service category (Peterson and Merino, 2003). As such, an idea for developing a Web site is put forth to disseminate global consumer electronic products and environmentally friendly information to online users, because the internet has already become an important information source and a powerful tool nowadays for people to retrieve information.

2.1   Web Site and Page Design

Web sites can be classified into 4 categories by their use purpose, including entertainment, information, communication, and commerce (Lindgaard & Dudek, 2003). Here, the GEP Web site belongs to the category of information-services. The critical areas of successful Web site design include structure, layout, navigation, and orientation (Katerattanakul and Siau, 1999). Also, online users' information seeking processes and needs on the site have to be considered. Web site design includes information architecture design, e.g., the organization, navigation, labeling, and search systems that offer accessibility to the end user, readability design, search design, and page design. The main focus of this research is the home page design on the GEP Web site presenting freely available information in traditional Chinese. The home page is the foundation of a well designed Web site and must be developed as the central starting point to which users can rely on to reorient themselves in their information search. The home page interface refers to the first-level web page linked to the site of its respective organization. In order to create a dedicated GEP home page with rich and well-rounded interface, its design should include building complete and easy-to-use web functions, positioning a rational interface framework, displaying the contents orderly, and providing complete, transparent, and accountable information. Therefore, the content of a homepage is important for users to understand the scope of the site and to get more information. Presentation of information on a Web home page must take into consideration graphics, colors, the amount of information displayed, and the way that the information is organized (Zhang et al., 2000).

2.2 User-Web Interaction and Usability

The usability or Human-Computer Interaction (HCI) criteria are important in making the customer's interaction with the Web site a satisfying one through the Web-based interface. The goal of the HCI is to meet user needs and expectations as much as possible, and then to improve the usability of software system (Dix et al., 2004). User interaction itself amounts to the interplay of cognition and information processing (Spillers, 2004). Developers need to collect quantitative data on the thoughts and feelings from user-Web interactions in addition to physical movements to identify anticipated needs, requirements and expectations from the Web site and to ground the system in a comprehensive understanding of the information-foraging process in context (Garg-Janardan and Salvendy, 1986) to build effective and efficient human-centered electronic information systems. For continued improvement of user-Web interactions, the knowledge of users' behaviour when interacting with the interface will need to be taken into account more (Rose and Levinson, 2004). It may correlate with users' cognitive needs, expand users' knowledge, minimize the cost of interaction and provide less information load. It is important to understand on-line consumer behaviour by looking at the user's interaction with the Web site both as a store and as a system (Koufaris, 2002). A Web page interface is a complex mix of text, links, graphic elements, formatting, and other aspects that affect the Web site's usability. Usability is user-friendly and user-centered in interface and functional aspects. Usability as "the extent to which a product can be used by specified users to achieve specified goals with effectiveness, efficiency, and satisfaction in a specified context of use" (ISO 9241-11, 1998) and as "the quality of an interactive computer system with respect to ease of learning, ease-of-use, and user satisfaction" (Rosson & Carroll, 2002). If the user is not satisfied with the Website usability at the first visit, 40% of users will not visit the site again (Nielsen, & Loranger, 2006). Therefore, it is important to know user's satisfaction in the home page design. The satisfaction as "the sum of one's feelings or attitudes toward a variety of factors affecting that situation" (Bailey & Pearson, 1983).

Usable Web design usually follows common design practices, and thus an understanding of current Web site design practices may aid designers in creating usable designs (Jones & DeGrow, 2011). Before developing the GEP home page interface, 8 Web sites have been selected as references. Six of them are environmental sites including environmental protection administration (http://www.epa.gov.tw/), greenliving information platform (http://greenliving.epa.gov.tw/GreenLife/), asus corporate social responsibility (http://csr.asus.com/chinese/), redesigning society's relationship (http://www.greenelectronicscouncil.org/), the electronic product environmental assessment tool (EPEAT) (http://www.epeat.net/), and United States environmental protection agency (http://www.epa.gov/). The other 2 sites are Yahoo.com and Google.com.

3 Method

In this research, three GEP Web sites, which have the same information and logical structure, with different home page configurations and Web pages design have been developed. Within each Web site, the home page and other pages have a consistent design style presented to the users.

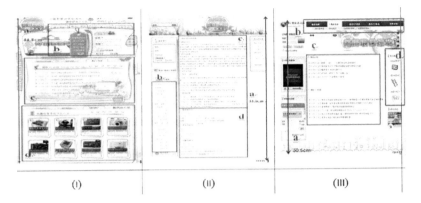

(I)   (II)   (III)

Fig. 1. Screen captures for the three home pages

3.1   Home Page Interfaces Design

The screen captures of the three home page interfaces are showed on Figure 1. Each site provided the same information, e.g., about Us, environmental news, GEP news, consumer electronic products, and GEP information, and the same Web services, including a forum, membership system, keyword search, online user services and the site activities, e.g., voting event. The difference among the three home page interface design are: (a) vertical page length, (b) menu design, (c) news and key information design, (d) consumer electronic products presentation, and (e) color, image, symbol character, visual, and layout design.

3.2   Experimental Design and Research Procedure

The independent variable of the study is the configuration of home page on the site. The dependent variables were pre-experiment questionnaire, searching performance, user-Web interaction, searching time, and post-experiment questionnaire.

In all home page conditions, each participant had to finish three tasks, whose orders were randomized to minimize learning effects, consisting of: (1) to search for a consumer electronics with the highest rating by consumers and the most environmentally-friendly product; (2) to find information relating to Earth Day and the reasons why companies provide GEPs; and (3) to seek a GEP with a water resource saving symbol. At the beginning of the experiment, an experimenter introduced the experiment and the tasks. After filling out a pre-experiment questionnaire, each participant took around 10 minutes to become familiar with the assigned Web site. After a 5 minute break, he/she was instructed to finish a series of tasks as accurately as possible with no time constraint. After the participant completed the tasks and took 5 minutes break, he/she had to fill out a post-experiment questionnaire. Participants took approximately 1.5 hours to complete the experiment.

3.3   Participants

Thirty-six undergraduate students, 18 males and 18 females, were paid to participate in the experiment. The males' mean age was 21.72 (SD 1.81) years, interaction with

WWW resources averaged 8.94 (SD 2.24) years, and online purchasing came out to 3.61 (SD 2.57) years. The females' mean age was 22.11 (SD 1.75) years, interaction with WWW averaged 9.47 (SD 2.67) years, and online purchasing came out to 3.42 (SD 1.42) years. Participants were assigned to use one of the experimental Web sites for finishing assigned tasks. Each site collected 6 males and 6 females' experimental data. The experiment was carried out in a lab with one participant at a time.

4   Results

4.1   Analysis of Pre-experiment Questionnaire

To secure the validity of the survey, the variable generated from the instrument were tested for internal consistency reliability before the data analyses. The internal consistency coefficient was 0.805 for the variable of importance of the Web services. The results of pre-experiment questionnaire survey indicated that the averages of the importance for the 25 questions are from 4.81(SD 0.4) to 3.36(SD 0.8). The most important Web services for Internet users is ease of operation to find wanted information, followed by operating effectiveness for finding the wanted information within a reasonable time, online customer service, search tools, interface layout, image information, and so on. The mean of importance of the back to home page function is 3.97 (SD 1.03). Also, the results of t-test indicated that there were significant differences in the question items of customer service [$t34=2.153$; $p=0.038 < 0.05$] and presenting site purpose [$t28.854=2.432$; $p=0.023 < 0.05$] between males and females. This reveals that the Web site's customer service for males [mean 4.89 (SD 0.32)] is more important than for females [mean 4.44 (SD 0.74)], and a clear display of the site's main purpose for females [mean 4.3 (SD 0.59)] is more important than for males [mean 3.8 (SD 0.79)]. Finally, the results of Web search tools ranking showed that the most important tool for participants is keyword search. The next most important tool for male students are advanced search followed by images as links to retrieve information; as for female students the other highest ranked tools are images as links and text hyperlink design.

4.2   Analysis of User-Web Interaction and Searching Performance

All participants were able to accomplish the experimental tasks successfully. The results of T, Mann-Whitney U, and Median test showed that perceived task load, operation time, the number of clicked action on the site, or the number of errors between different genders or among 3 different configuration home pages were of no significant difference ($p>0.05$). The average of TLX is 49.28 (SD 11.91). The user's perceived average of performance is 35.56 (SD 24.95). The mean of the number of clicked actions in home page I, II and III are 32.6 (SD 8.1), 35.3 (SD 8.9), and 31.8 (SD 9.4) times respectively. The mean operation times in home page I, II and III are 1466.2 (SD 409.2), 1286.8 (SD 259.4), and 1331.1 (SD 373.1) seconds respectively. Finally, the major action for the users is surfing related information in product Web pages, followed by back to home page, errors (e.g., clicking wrong button/link leading to incorrect page/information), products ranking, and back to previous page.

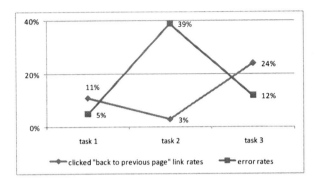

Fig. 2. The error and clicked link rates in different tasks

The significant difference were among experimental tasks in the number of errors (Median=17; p=0.043 < 0.05). Moreover, the results of correlation analysis revealed that between the number of errors and the number of "back to previous page" clicks [r(9) = -0.734, p =0.024] has a negative correlation. The results are shown in Figure 2. The users performed with an error rate of 5%, 39%, and 12% in task 1, task 2, and task 3 respectively. In addition, they clicked the "back to previous page" link with rate of 11%, 3%, and 24% in task 1, 2, and 3 respectively.

4.3   Analysis of Post-experiment Questionnaire

The experimental raw data were analyzed by ANOVA. The main effects on users' satisfaction with Web menu, color, image, and symbol character design among the different configuration home pages were significant. Multiple comparisons of Tukey's method indicated that there were significant differences between home page I and home page III on the user satisfaction with color (p=0.000), image (p=0.029), and symbol (p=0.013) design. Also, the differences of the user satisfaction with Web menu (p=0.01) and color design (p=0.032) between home page II and home page III were significant. The results of t-test indicated that there was a significant difference between males and females on their satisfaction with the layout of home page I [t10=-3.796; p=0.004 < 0.05]. Also, the differences in home page ☐ between the different genders on their satisfaction with the layout [t10=-2.86; p=0.017 < 0.05], comprehensive information coverage [t10=-2.27; p=0.049 < 0.05], and menu design [t10=-2.39; p=0.038 < 0.05] were significant.

The results from home page design ranking revealed that the most important design is site purpose, followed by layout, Web functions, search function, visual design, and comprehensive information. Regarding the Web menu design, the results of menu layout design ranking showed that a horizontal drop-down list is preferred by the participants. Finally, the results from the open-ended questions for the home page design preferences revealed that 48% of participants suggested a home page with good information classification, 45% of them prefer a design using the environmental colors, images, logos, green marks, or a combination to present the site purpose on home page, 43% of them liked combining static and dynamic information with a simple layout design and the main information, e.g., news and product ranking showing in the center of the screen, 25% them suggested a simple and clear menu design, and 24% them did not like a long or business like home page design.

5   Discussion

Based on the results of the pre-experiment questionnaire, the Web site service can be broken down to 3 fundamental parts. First, the designer has to provide internet users with an easy-to-use Web site (e.g., easy operation, operating effectiveness, customer service, such as helping on-line users solve their operation problems, and search functions). For searching Web information, the most used function for users is keyword search. Secondly, the Web pages have to be easy to understand by the users; for example, user-centered layout design, providing image information, attention to visual design, and displaying the site's main service. Finally, the Web page has to provide text information, information tables, user-to-user recommendations, and hot products ranking for assisting users to easily digest wanted, accurate, and complete information. In addition, none of the users encounter the situation of disorientation problems. The correlation between the number of errors and the number of times the user clicked back to previous page reveals that the users are able to obtain desired information in fewer steps if they clicked the "back to previous page" link to restructure their searching approach. It also shows the importance of the Web page in presenting the Web logical structure. In order to reduce errors and search actions, the issues of search tools and intuitive navigation have to be taken seriously and improved.

Regarding the home page design, the results from the pre-experiment questionnaire and experiment revealed that the most users do not think the home page design is important, and different home page interface design do not affect their searching performance. That is because users care most about retrieving the wanted information successfully in the information search process. However, the importance of the home page for users has been shown by their interaction with the Web site and their satisfaction with the interface design. Based on the study results, several suggestions for the development of a GEP home page interface that is easy-to-use, easy-to-understated, and easy-to-digest have been discussed, and are described as follows.

- Applying appropriate color, image, and symbol character design to present the site purpose and to satisfy users: The results revealed that the primary color to present the environmental image is green for most users. Also, matching the appropriate colors in the Web page design is better then using just white and green colors for users. Regarding the image and symbol character design, home page I has a higher satisfaction than home page III due to home page I 's use of images and metaphors consistent with the real world, e.g., images of trees, grass, leaves, and plants and symbols of GEP logo and green earth, for presenting more environmental elements to users. These results are also consist with users' opinion from the open-ended question, 45% of participants prefer the designer using the environmental colors, images, logo, green marks, or a combination of all designs to present the purpose of GEP Web site on home page. The results also revealed that several users do not like the green Web site being commercialized.
- Arranging the main information and the Web functions grouped logically in an appropriate location and page space to avoid overlapping with other Web objects within a suitable Web page length: The second most important component of home page design for users is Layout. 24% of participants do not like a long home page design. They preferred the length of home page be designed to fit their screen size

or a little longer. For effective presentation, they also suggest that the main information, e.g., news and product ranking, should be located centrally on the screen and in an appropriate page space. Regarding the Web functions, several users suggested that the membership system should be located on the top right corner of the screen. In addition, users have higher satisfaction in using the horizontal menu with a horizontal drop-down list of home page III than using the horizontal menu with a vertical drop-down list of home page I. Even though the mouse design assists in vertical screen movement, users still preferred the horizontal menu with a horizontal drop-down list design to have a larger space for displaying major information in the central location. The other reason users do not like the Web menu design on home page I is that the menus have been designed in separate areas and some of the drop-drown lists of menus overlap with other Web objects or functions. This is consistent with the study showing 25% of participants prefer a simple and clear menu design within a suitable page space and location.

- Providing complete key information with a static and dynamic visual design: The visual design and providing comprehensive information are important for home page design as well. The Web site provides a lot of information and a range of navigation features to allow users to employ multiple approaches to support their search. The visual design should be considered in the development of the home page to present the integral and whole information of the Web site and to facilitate the users in understanding the relation between the Web site component or items and the navigation function or Web site structure. Also, 43% of users like combining static and dynamic information with a simple layout design on the home page allowing them to pay more attention to their desired information. They preferred that the important content be more eye catching, but too much multimedia animation should be avoided. Moreover, 48% of participants prefer the home page to provide classified and complete information effectively. For displaying such information in the limited screen to satisfy user's needs and expectation, the headline or key information design, e.g., tab, text hyperlink, image as a link, and marquee, and integration of information with text, images, photos, colors, or symbol design has been suggested.

Finally, gender differences in preference to navigation services and content structure within the information services Web site has been found. The male users prefer the Web site providing keyword and advanced search function on the home page. The female users like the keyword search and search function of an image as a link. After several years of experience, all users learned to search information by using the keyword search. However, the gender difference in preferred search logic has been found on their other preferred ways of searching which may also imply how users develop their search strategies. The females using the Web site pay more attention on their main service and tend to retrieve the wanted information by follow the site's navigation logical structure. But the males like to do the active search by using the keyword search and advanced functions; therefore, they tend to prefer more customer services, e.g., on-line help with operating problems, on the home page. In addition, males prefer more animated visual design, such as flash animation, to catch their attention to the main information and tend to read more text-based information than females, but do not like too much text information on the home page. The females pay attention on the home page design presenting site purpose and Web page's visual design based on images more than males did.

6   Conclusions

The GEP Web site is constructed to create an environmental image and to provide information of global consumer electronic products to Web users. In order to disseminate the green and the products information to online users effectively, the issue of home page design for GEP Web site has been investigated in this initial research stage. This study is aimed towards researching searching performance, user-Web interaction, and user satisfaction to obtain users' needs and expectations on the proposed home page interfaces through the experimentation for improving the usability. The importance of Web site design and home page design and gender differences have been found in this study. The summary for designing a useful home page in this study are: (1) adhering to the ideas of easy-to-use, easy-to-understand, and easy-to-digest and clearly showing the Web structure and site purpose to users; (2) streamline the home page to show information more efficiently via space saving techniques, avoiding a long lengthy Web pages with too much text-based information, placing menus in the same area and using a horizontal drop-down list design, presenting the main information on the main area, and designing each function on the Web page, such as keyword search, flash information, etc., with independence of presentation to avoid overlapping; and (3) gender differences should be considered in Web site design to support different types of users' needs.

Acknowledgements. The authors would like to express their gratitude to National Science Council of Taiwan for the funding under the grant number NSC-98-2221-E-161-004.

References

1. Bailey, J., Pearson, W.: Developing of tool for measuring and analyzing computer satisfaction. Manag. Sci. 29(5), 530–545 (1983)
2. Bargas-Avila, J., et al.: Usable error message presentation in the World Wide Web: do not show errors right away. Interacting with Computers 19(3), 330–341 (2007)
3. Dix, A., Finlay, J., Abowd, G.D., Beale, R.: Human-Computer interaction, 3rd edn. (2004)
4. Garg-Janardan, C., et al.: The contribution of cognitive engineering to the effective deasign and use of information systems. Inform. Services Use 6(5/6), 235–252 (1986)
5. Huang, F.H., et al.: 2009/7/19~24. E-Shopping Behavior and User-Web Interaction for Developing a Useful Green Website. In: Proceedings of 12th International Conference on Human-Computer Interaction. USA, San Diego
6. ISO 9241-11:1998, Ergonomic requirements for office work with visual display terminals (VDTs) – Part 11: Guidance on usability
7. Jones, S.T., DeGrow, D.: Fortune 500 Homepages Design Trends. IEEE Transactions of professional communication scheduled 54(1), 1–15 (2011)
8. Katerattanakul, P., et al.: Measuring Information Quality of Web Sites: Development of an Instrument. In: Proceedings of the Twentieth International Conference on Information Systems, Charlotte, NC, pp. 279–285 (1999)
9. Koufaris, M.: Applying the technology acceptance model and flow theory to online consumer behavior. Information Systems Research 13, 205–223 (2002)

10. Lindgaard, G., Dudek, C.: What is this evasive beast we call user satisfaction? Interacting with Computers 15(3), 429–452 (2003)
11. Liu, C., et al.: Web sites of the Fortune 500 companies: facing customers through home pages. Information & Management 31(6), 335–345 (1997)
12. Netcraft "Web server survey" on December 24 (2009), http://news.netcraft.com/archives/2009/12/24/december_2009_web_server_survey.html (access date: June 15, 2010)
13. Nielsen, J., Loranger, H.: Prioritizing Web Usability. New Riders Press, Berkeley CA (2006)
14. Peterson, R.A., et al.: Consumer information search behavior and the internet. Psychology & Marketing 20(2), 99–121 (2003)
15. Rose, D.E., et al.: Understanding user goals in Web search. In: Proceedings of the 13th International Conference on World Wide Web, New York, NY, USA, May 17-20 (2004)
16. Rosson, M.B., Carroll, J.M.: Usability engineering: scenario-based development of human-computer interaction. Morgan Kaufmann, San Francisco (2002)
17. Spillers, F.: Task Analysis Through Cognitive Archeology. In: Diaper, D., Stanton, N. (eds.) The Handbook of Task Analysis for Human-Computer Interaction, Laurence Erlbaum Associates, New Jersey (2004)
18. Zhang, X., et al.: Information quality of Commercial website home pages: An explorative analysis. In: Proceedings of ICIS, Brisbane, Australia (2000)

Attribute Description Service for Large-Scale Networks

Donald Kline and John Quan

University of Alaska Fairbanks, 202 Chapman Building, Fairbanks AK 99775-6670
{dpkline,jquan2}@alaska.edu

Abstract. An analysis of requesting resources from large-scale networks reveals a fundamental challenge. As the network grows, more and more resources become available, and so finding resources that fit experimental test criteria becomes difficult and time consuming. For example, the National Science Foundation sponsors GENI—an experimental network with a goal to gain enough resources to model the Internet at scale. Currently, GENI contains relatively few contributed resources donated from businesses and academia, and so matching resources to tests is rather simple. However, experimenters plan to conduct network experiments that are very complex and difficult to accurately model by using the vast numbers of resources expected in GENI. When GENI reaches its final state, finding the right resources that fit experimental test criteria out of many thousands of donated resources may be as difficult as conducting the experiment itself. This dilemma underscores the importance of establishing an attribute description service that promotes a standardized language for all interactions between the end users and the large-scale network.

Keywords: Database, Data dictionary, Design, Human Factors, Standardization, Languages, Attribute, control framework, component, GENI, Large-Scale Network, resource, classification.

1 Introduction

Large-scale network (LSN) experimentation may have unprecedented impacts on the future Internet, from meeting consumer needs to improving network topologies. In fact, several countries are forming intra-and inter-national LSN infrastructures to increase effectiveness in socio-economic partnerships, education, and experimentation. By connecting a myriad of resources into collaborative pools from which users can share information, these countries intend to test new network architectures and services, and research better ways to make our current Internet structure more trustworthy.

In the socio-economic arena, Cisco and Korea partnered in a $2 billion multi-year project called the Inchon Free Economic Zone (IFEZ). "The strategic collaboration with IFEZ aims to ... provide unprecedented solutions for citizens to take advantage of a technology-enabled lifestyle across areas encompassing public services, commerce, health care, education and safety ..." [1]. In addition, the US, the EU, and Japan are funding research LSNs for exploring future Internets at scale. This paper focuses on the US sponsored Global Environment for Network Innovation (GENI).

M. Kurosu (Ed.): Human Centered Design, HCII 2011, LNCS 6776, pp. 519–528, 2011.
© Springer-Verlag Berlin Heidelberg 2011

1.1 GENI

Four control frameworks (CF) form the basis of GENI: PlanetLab, the Open Resource Control Architecture (ORCA), ProtoGENI, and the Open Access Research Testbed for Next-Generation Wireless Networks (ORBIT) [2]. They apply unique rules and specifications to solve similar problems, such as connection, authorization, and security. Each CF is a sovereign entity that orchestrates its own infrastructure, with Slice-based Federation Architecture (SFA) defining the basic unit of exchange. Version 2.0 of SFA (2010) specifies components and slices as integral concepts in GENI. Components are the foundational elements, such as computers, routers, and programmable access points, which hold resources, such as a CPU, RAM, and bandwidth. Slices contain components, or an aggregate of components, and a researcher then runs an experiment or service within the slice. The SFA establishes basic rules concerning instantiating, scheduling, authorizing, and registering a slice [3]. Currently, it does not clearly define the resource specification (RSpec) of a component, and so each CF provides an RSpec with only limited device descriptions. However, each researcher may need several, hundreds, or even thousands of contributed components for large-scale experimentation, and the resource descriptions are so sparse that the researcher cannot efficiently determine the test suitability of the components. Therefore, the time to determine component qualifications and the need for attribute descriptions grow with the number of resources required. Figure 1 exemplifies a PlanetLab RSpec that provides a bandwidth limitation but contains little other component data [4].

```
<site id="s15">
  <name>CarnegieMellon</name>
  <node id="n40">
    <hostname>planetlab-1.cmcl.cs.cmu.edu</hostname>
    <bw_limit units="kbps">5000</bw_limit>
  </node>
  <node id="n41">
    <hostname>planetlab-2.cmcl.cs.cmu.edu</hostname>
    <bw_limit units="kbps">5000</bw_limit>
  </node>
  <node id="n42">
    <hostname>planetlab-3.cmcl.cs.cmu.edu</hostname>
    <bw_limit units="kbps">5000</bw_limit>
  </node>
</site>
```

Fig. 1. Sample RSpec from PlanetLab

Furthermore, each CF complicates the matter by maintaining their own definitions outside of their RSpecs for the nodes and slices they control. These definitions vary greatly even though the devices available on each network are all very similar. The following node definitions are from different CFs, and the RSpecs do not incorporate these resource definitions. Given the amount of variation just between these definitions, it is plain to see that if every CF were to maintain their own high-level resource definitions there would be limited correlation between descriptions from different CFs. These differences would increase the complexity of stitching resources together and providing a seamless user interface.

For example, the PlanetLab User's Guide (2007) states "Node. A node is a dedicated server that runs components of PlanetLab services. Slice. A slice is a set of allocated resources distributed across PlanetLab. To most users, a slice means UNIX shell access to a number of PlanetLab nodes" [5]. These definitions may work well for "most users," but may not provide enough detail to enable experimental repeatability because one cannot exactly match component attributes to the original experiment.

The draft ProtoGENI Control Framework Overview (2009) describes nodes as "...raw ProtoGENI nodes are currently just that, raw. None of the Emulab setup is done on those nodes, like experimental interfaces initializing interfaces with IP addresses, building accounts for other project members, starting programs automatically with the event system, etc. This will eventually be supported in the 'cooked' interface" (p. 19) [6]. This definition implies that many attributes will be included for nodes in the future, but other CFs may find the data unusable in the Emulab format.

ORBIT is designed to provide precise, repeatable wireless experiments. It specifically defines its nodes in the ORBIT tutorial (2010) as "a PC with a 1 GHz VIA C3 processor, 512 MB of RAM, 20 GB of local disk, two 100BaseT Ethernet ports, two 802.11 a/b/g cards and a Chassis Manager to control the node, ... The Chassis Manager has a 10BaseT Ethernet port. The two 100BaseT Ethernet ports are for Data and Control. The Data ports are available to the experimenter. The Control port is used to load and control the ORBIT node and collect measurements" [7]. This is an excellent resource description, probably because ORBIT experimenters need this level of detail to document their experiments. Unfortunately, other CFs cannot use their own RSpec format to display this information.

2 Attrbute Description Service

An analysis of requesting resources from LSNs reveals a fundamental challenge. As the network grows, more and more resources become available, and so finding resources that fit experimental test criteria becomes difficult and time consuming. The Attribute Description Service (ADS) attempts to meet this challenge to improve the human-computer interactions between LSNs and their users, and to manage the large number of resources that each CF will ultimately control. It is based on a description standardization that still gives each CF the ability to manage their version of ADS as they see fit. ADS accomplishes this in a two-pronged approach: it extends the current CF RSpec by providing structured descriptors kept in a queriable database, and it develops a data dictionary to globalize these descriptions so that each CF can easily determine the resources of the others.

2.1 Extending the Description Language

The first goal of ADS is to extend a framework's description language to provide detailed descriptions of its components so users can easily determine the suitability of test resources. One way to accomplish this is by modifying the current node registration process to include a resource description sub-process in which the resource owner lists the resource's attributes. This might use an automated method, such as the Windows systeminfo command or Macintosh system_profiler command, a manual entry method, or a mix of both. The CF may then store this information in a method of its own choosing. However, storing it as a separate database within the CF allows for separation between its connection, authorization, and security protocols, but ensures other CFs within GENI have access to the attribute data. The database the CF chooses can take many forms, such as a relational database (including object-oriented), or a native XML database, and each has its advantages and disadvantages.

Ultimately, the CF must decide which database solution is most appropriate for its needs. This allows the CF to track the types and attributes of its resources, and increases the ease with which experimenters gather experimental resources. It allows users to extend the primary question, "What resources are available?" into "What resources are available that fit my criteria?" (Figure 2).

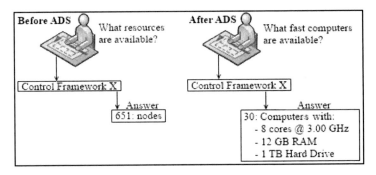

Fig. 2. Attribute Description Service creates useful descriptions

2.2 Creating a Data Dictionary

The second goal of ADS is to standardize the descriptions across all networks to codify existing classifications and form the basis for new ones, thereby creating a language of valid descriptions. By standardizing this language in a data dictionary across multiple networks, the contents of each network become clear to the user. As noted by the GMOC: GENI Concept of Operations (2010), "A GENI experimenter who has a slice that spans multiple aggregates will benefit from a single view of their slice which depicts data such as operational status, network link utilization, etc" [8]. Therefore, a GENI experimenter will find enormous benefit from a single view of their slice that can span multiple CFs or LSNs, and that uses established resource attribute descriptions. Standardizing resource descriptions enables richer depictions because it facilitates resource quality indexing based on classifications and attribute values, and services are easier to incorporate because they do not require a layer of indirection to translate their language or create resource descriptions.

Elmasri, et al. (2011) describes how numerous organizations currently use data dictionaries, which are databases that store meta-data about another database. Typically, data dictionaries define both the high and low level details about one database (p. 306) [9]. The ADS data dictionary uses an adaptation of these ideas. Instead of attempting to define low-level designs of CFs, the ADS data dictionary focuses on standardizing high-level aspects of numerous databases across the world. Specifically, ADS intends to provide the following metadata:

- Definitions of components, resources and attributes
- Descriptions of the relationships between resources and attributes
- Tailored user descriptions

There are challenges associated with standardizing descriptions across multiple systems. These include finding methods for all systems to incorporate the new standard easily, even though each of their systems is unique. Section 2.1 mentioned multiple variations that allow for different types of CFs to incorporate resource descriptions into their own system. However, that alone is not enough to provide a uniform language across an entire LSN. CF boundaries will not be seamless if each CF creates its own resource descriptions and standards, and this would cause confusion for the users. Another problem with having each CF create their own language is the duplication of work required to provide meaningful descriptions to the end users. Each system would have to make their own standard, and then form their own methods for translating between their system and all other frameworks. ADS provides a solution that avoids these issues through a standardized language.

For ADS to incorporate a standard across a LSN, it must easily translate between every CF. Since the actual words in the language are consistent across variations, the different CFs may store data in different ways and only the data structure needs to be translated (Figure 3).

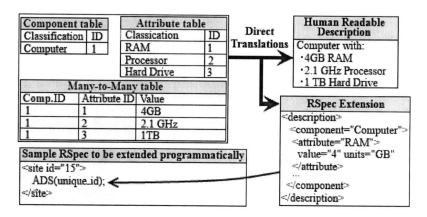

Fig. 3. Attribute Description Service extending RSpec with DBMS

For ease of use, ADS should be a remotely accessible service that provides metadata from the data dictionary. This data dictionary defines which attributes and relationships are valid, and then share that information with each CF. Each CF is free to choose its own implementation for storing resource descriptions. Then it can incorporate the ADS data dictionary service to control the language used, and thus follow the standard.

Definitions of Components, Resources and Attributes. The GENI System Overview (2008) defines components as the smallest building blocks of LSNs. "A component encapsulates a collection of resources, including physical resources (e.g., CPU, memory, disk, bandwidth), logical resources (e.g., file descriptors, port numbers), and synthetic resources (e.g., packet forwarding fast paths)." It goes on to define resources, as "... abstractions of the sharable features of a component that are allocated by a component manager and described by an RSpec. Resources are divided into computation, communication, measurement, and storage. Resources can be contained in a single physical device or distributed across a set of devices, depending on the nature of the component" [10].

Today, LSNs consist of numerous types of devices including physical and virtual machines, optical sensor networks, wireless networks, programmable switches and routers, commodity internet, National Lambda Rail (NLR) and Internet2. In addition, the types and variations of resources federating into LSNs are constantly growing. Alaska alone is looking at supercomputers, satellites, and SCADA systems as potential federators with GENI [11]. In order for ADS to be applicable to LSNs, it must be capable of fully describing every device while still managing to group similar types of resources together.

Figure 3 shows a simplified relational database prototype where resources and attributes are separate tables. This design decision accomplishes two things. First, data is inherently stored in normalized form which Elmasri, et al. (2011) explains is "...to achieve the desirable properties of (1) minimizing redundancy and (2) minimizing the insertion, deletion, and update anomalies" (p. 517) [9]. By describing component and resource classifications separately from attributes, ADS creates a way to group similar resources together. Furthermore, assigning a classification to a component or resource makes no demands on the attributes that define that device. This allows valid descriptions to range the full spectrum of granularity without losing the most basic grouping element, its classification.

Uniquely defining and naming each classification creates an understandable language that makes it clear to the user what types of items are in each classification, and to which classification each resource belongs. When one encounters a device that does not match any current classification, one simply submits a new classification definition to the governing body of the data dictionary. This governing body should rely heavily upon CF input and the same online dictionaries and wikis that real users have available when defining each classification.

Attributes will have associated values. For example, an attribute "% free space" could have the value 0.50. The value associated with each attribute must be defined to display the value's contents correctly. For instance, some attributes will have a numeric value while others will require a string of characters Moreover, some attributes (such as an IP address) will require a specific data structure. Yet another attribute may store the scheduled downtime of a device over the next month.

Descriptions of Relationships. Figure 4 shows the logical layout of attributes, components, and resources in the ADS dictionary. The matching of resources and components is the list of all valid pairs between the two sets of classifications. For example, the component "access point" matches resources like "antenna" and "network port." However, it does not match the resources "graphics card" or "GPS."

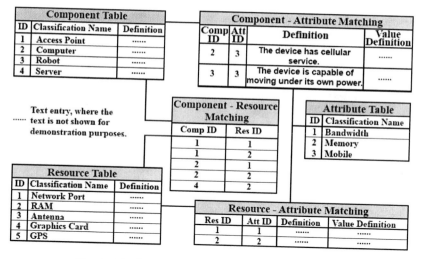

Fig. 4. A logical layout of the data dictionary metadata

These matches would then define that a valid ADS description of an access point may have antenna and network port resources, but not those resources that are not in the matching. CFs would then use this metadata to validate their resource descriptions by comparing their descriptions to these matches. In addition, each attribute description will need to define the classification of the attribute and the meaning of the attribute's associated value. The method of defining the classification and the value definition of the attribute is similar to the process of defining the classifications for components and resources, except that it is necessary to redefine these terms for each component and resource that the attribute may reference. In other words, the meaning of an attribute has the potential to change depending on the device with which it is associated. As shown in figure 4 for instance, an attribute called 'mobile' could describe a computer as able to receive cellular service, or it could describe a robot capable of moving under its own power, and these definitions are stored in the matching of components and attributes.

Tailored User Descriptions. Different users of LSNs also will have different demands. These users need a way to personalize their experience with the language so that it provides the correct amount of detail to all users. The amount of detail a user requires will vary depending on the knowledge and background of the user, and their intended use of the resources (e.g., an experienced computer scientist performing mathematical analysis on the weather will have more stringent demands than a college student completing a homework assignment).

ADS can use attribute description "presets" to automate resource searches for a particular user class. By assigning detail levels to the attributes, ADS can provide a meaningful description for every type of user. The detail level would serve as a marker to describe how important that attribute might be to a particular user class. For example, the storage capacity a computer has available is important, and would likely have a detail level of one. Whereas, the size of a hard drive's cache would have a higher detail level because the likelihood of that attribute being important is low. When a user that has their detail preference set at 'one' views a resource description they only see the attributes that have a detail level less than or equal to their preference (e.g., in this example the user would see the storage capacity of the computer, but not the cache size).

Figure 5 displays the maximum amount of detail stored by ADS and some examples of "User Group" buttons to provide a simple means to focus a resource search. For instance, many LSN researchers may only require basic information about the components that make up their slice, such as the connection speed between components, and the RAM and CPU cycles used. However, a networking experiment may require in-depth information about the switches, Network Interface Cards, repeaters, bridges, routers, hubs, and firewalls that make up his or her experimental slice.

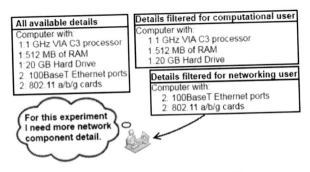

Fig. 5. Tailored Descriptions for User Groups

3 Synergies

The initial design of ADS as a service guides the creation of uniform device descriptions so that real users can read, understand, and search through the growing number of resources federating into LSNs. Thus, the primary incentive to use ADS is to improve the human computer interactions associated with LSNs. ADS enhances the interface for real users by providing a common language that does not change as experimenters, donors, and service providers traverse the numerous CFs that make up LSNs.

ADS provides a layer of abstraction so that a user can request suitable resources from any CF within a LSN as if it were built on only one infrastructure. If the front-end application knows the connection, authentication, and data structure details, the user receives CF resource descriptions in a universal format. Only the data structure of the description will need translated, because the values, classifications, and definitions associated with the description remain the same. By eliminating the need for CFs to define complex translations between each other's resource descriptions, ADS drastically reduces the amount of work required to federate CFs and LSNs together.

Furthermore, ADS enables quality indexing for all types of devices. Such a service would be similar to the windows experience index, which provides a numerical measure of computer quality based on the quantitative attributes of the operating system and computer hardware [12].

LSN experimenters need instrumentation and measurement services to gauge the effectiveness of their experiments in the same way that a physicist needs these services for his or her experiment. Naturally, if CFs can leverage ADS to find resources within its own LSN and within any other LSN, services can do the same. Some services that might be desirable to LSN experimenters and donors are logging, content management, distributed hash tables, and storage.

Of course, ADS does not only apply to experimental LSNs. If ADS facilitates linking humans with resources across experimental CFs and LSNs, it may also be useful in other LSNs such as defense, communications, or public networks such as IFEZ. In addition, maintaining a universal data dictionary may have commercial applications because software and hardware designers can hone their products to better meet market demands or create products to compete in emerging markets.

4 Conclusion

In conclusion, LSNs are sprouting up all over the world, but with no standard method of classifying the vast amounts of resources that their underlying CFs will eventually contain. This may make the task of finding experimental resources as difficult as performing the experiment itself, unless some means to delineate components is implemented

ADS provides a possible solution to this problem by using a two-pronged approach: to develop a global data dictionary to codify existing terms, and to universally define new ones. A CF may then extend its current resource database using these descriptions in a format of its own choice. Extending its sparse RSpec using ADS enables physical, qualitative, and time dependent attribute descriptions that allow a researcher to ask the question, "What resources are available that fit my criteria?" Moreover, maintaining a data dictionary of quantitative and qualitative attribute definitions across an expanding infrastructure enables researchers to find relevant resources to fine-tune experiments. Thus, ADS promotes a standardized language for all interactions between the end users and the large-scale network.

Furthermore, universities that develop popular experimental services for its own CF may share their service with other CFs. Commercial businesses might use ADS to market needed services and develop new customer bases. Of course, if experimental networks can use ADS, other LSNs can take advantage of it to manage resources, as well.

In addition, it provides a global change management function that is currently beyond the capabilities of any LSN. Standardizing and centralizing a description service provides modification traceability and a means to roll back deleterious changes. Finally, all users benefit by applying quality indexing as a reference to compare components from different sources.

References

1. Incheon Metropolitan City and Cisco in Strategic Collaboration to Accelerate the Future of City; Cisco Global Center for Smart+Connected Communities to Serve As a Springboard For Network-enabled Services and Solutions in Incheon M2PressWire (March 30, 2010), http://web.ebscohost.com
2. GENI Spiral 2 BBN Technologies (2010), http://groups.geni.net/geni/wiki/SpiralTwo
3. Peterson, L., Ricci, R., Falk, A., Chase, J.: Slice-based Federation Architecture. V2, 4-8 (July 2010), http://groups.geni.net/geni/attachment/wiki/SliceFedArch/SFA2.0.pdf
4. PlanetLab RSpec Planet Lab (February 2010), http://svn.planet-lab.org/browser/sfa/trunk/sfa/managers/pl/pl.xml
5. User's Guide PlanetLab (2007), http://www.planet-lab.org/doc/guides/user#slice
6. ProtoGENI Control Framework Overview GENI Project Office (2009), http://groups.geni.net/geni/attachment/wiki/ProtoGeniControl FrameworkOverview/022709%20%20GENI-SE-CF-ProtoGENIOver-01.4.doc
7. Tutorial/Testbed ORBIT (2010), http://www.orbit-lab.org/wiki/Tutorial/Testbed
8. GMOC Concept of Operations GENI Meta-Operations Center (2010), http://gmoc.grnoc.iu.edu/uploads/8i/Gu/8iGu80-LqQB37VU4ZE1i5g/GENI_Concept_of_Operations-final.pdf
9. Elmasri, R., Navathe, S.: Fundamentals of Database Systems, 6th edn. Pearson Education, Inc., Boston (2011)
10. GENI System Overview BBN Technologies (September 29, 2008), http://groups.geni.net/geni/wiki/GeniSysOvrvw
11. Model Federation Framework VMI-FED (2010), http://assert.uaf.edu/geni/model_fed_framework.html
12. Windows Experience Index Microsoft (2010), http://www.microsoft.com/windows/windows-7/features/windows-experience-index.aspx

Study of Honest Signal: Bringing Unconscious Channel of Communication into Awareness through Interactive Prototype

YounJung Kwak[1], Tiia Suomalainen[2], and Jussi Mikkonen[3]

[1] 599 Gwanak-ro Gwanak-gu Seoul/ 151-748, Seoul National University #49-101
[2] Espoo / PB 11000, FIN-00076 Aalto TKK
[3] Hämeentie 135 C, Helsinki / PB 31000, FIN-00076 Aalto TAIK
lemonloaded@hotmail.com, tiia.suomalainen@gmail.com,
jussi.mikkonen@aalto.fi

Abstract. Efforts are made to understand people in the context of their social network; especially in unconsciously carried communication channel which Alex S. Pentland coined as 'honest signal'. This project explored 'optimizing honest signals to lessen gap between intended expression and received impression' through designed technology device. Experiment setting was controlled into 'presentation-speech situation', developed in three phases. Phase 1 was basic research, testing impressions given by postures and finding significant body-part for honest signal. In phase 2 'where and how person will be given feedback through designed device to aware his/her unconscious movement' was progressed. In Phase 3, finalized prototype- headset and shoes-evaluation test was made to check if prototype helped user to control honest signal during presentation by notifying such movement. This study has tested that if people have more awareness to honest signals they are sending, they are capable of enhancing control over signals. It will enable people to optimize signals, collecting more of wanted impression than not. In communication aspect, it offers new potentials interactive device or interaction can do for people, by making what was not cognitive into realizable signals or by making certain messages stronger.

Keywords: Communication channel, honest signal, unconscious movement, interactive prototype, psychology reflexive, behavior feedback, behavior control.

1 Introduction

Advancement in technology and new realm of exploration has opened up a new possibility of collecting detailed human behavior in a level that was previously not possible. Efforts are made to understand people in the context of their social network; especially in unconsciously carried psychology-reflexive-bio-mechanism communication channel which Alex S. Pentland coined as 'honest signal'. With more honest signals detectable or shared, usage of the signal for communication in social network is an attractive exploration subject. Though question arises; although 'honest

M. Kurosu (Ed.): Human Centered Design, HCII 2011, LNCS 6776, pp. 529–536, 2011.
© Springer-Verlag Berlin Heidelberg 2011

signal' is natural, is delivered message is as meant or not. After all, if honest signal is delivering wrong or unwilling message, carefully planned consideration needs to be made in using honest signal device for better communication.

Therefore, project centered on validity of 'optimizing signals to lessen gap between intended expression and received impression through design technology device'. Next question inseparably followed; would more awareness to honest signals help people to enhance the control over signals. Like toys help children allay stress, interactive device may work as medium to control signal by affecting emotion or behavior. During development of the study, the scope of the study limitation was set basing on discussion over 'when does one want to hide honest signals and instead show what s/he wants to impress'. Experiment setting was controlled into 'presentation-speech situation'. Other influential factors for such setting were given time and feasibility of experiment.

This project was schemed in three phases.

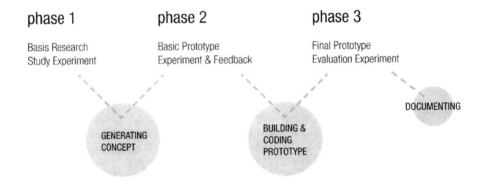

Phase 1 was basic research, testing impressions given by postures and finding significant body-part for honest signal. The theory was [body postures create signals. If then, helping to fix postures will help people generate wanted impression- i.e. confidence]. The object of this phase was to find hint to body posture for getting specific signal across.

Phase 2 was conducted after revising theory as [unconscious movement is important honest signal. If people can have more awareness to such movement, they may be able to reflect and control it]. After assuring the feasibility of the theory, the second phase progressed on to 'where and how person will be given feedback through designed device to aware his/her unconscious movement'. Tests were made with rough prototypes.

Phase 3 was evaluation testing. With tests from phase 2, devices that can measure and interact movement was built. Usage and where it would be put-on was concreted. The prototype consists of two devices- headset and shoes. It will be used in training for or in real presentation at home, school, and offices. The feedback is given only to the user when unnecessary movement continues for excessive time. The aim of the evaluation was to check if prototype helped user to control honest signal during presentation by notifying such movement.

2 Experiments

2.1 Phase 1. Basic Research

The objective of phase 1 was to find impressions given by postures and to find significant body-part for honest signal. With the theory [body postures create signals. If then, helping to fix postures will help people generate wanted impression- i.e. confidence], the experiment started with confidence posture signal. Experiment was set to explore what people think of signals given by postures, and to observe what people actually do when they are or are not confident. Confidence was chosen for its adequateness in 'what is known by people', 'clarity of signal', and 'controlled situation'.

In this phase, a posture-photo survey and a set-up scenario experiment was conducted. The survey was finding specific attitude impression people were receiving from various postures, and the scenario experiment was for finding significant signal factor that photo survey has missed. There is possibility that there is significant communicational body posture which may be keys to finding new scope of signal exploration.

Questionnaire's posture photos was composed by breaking gestures into separate body part movement- head, shoulder, arms, legs, hands- and mixed body movement. The photos were given with attitude vocabularies to match with received impressions. 15 people answered online anonymous survey.

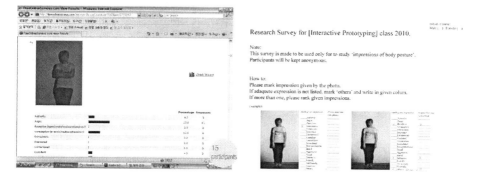

Fig. 1. Photo survey image (l) Online view (r) Survey draft

When looking at the photo survey result, all body movement was undoubtedly creating impressions to people. The level of context deliverance- clarity in chosen vocabulary- however showed differences. Certainly hand, shoulder, arm, leg, leaning and head movements, head and leg postures created most significant impressions. The photo survey was not showing compelling result. Another problem found was, survey photos centered on upper body posture and was too partial to conclude the most important body part for creating signal.

After the limit of photo questionnaire, scenario experiment was devised to observe how people express or react to uneasy atmosphere. The scenario experiment consisted with 4 participants with roles. Audiences were given roles to create specific atmosphere, and presenters were unaware of this. Presenters were measured with accelerometer on their head to track spinal movement. The whole process was video recorded. The presenter and the audiences were both subjects of observation. Soon after the first experiment, we have found our theory needed to be fixed. Body posture was important, however, movement was stronger factor to impressions.

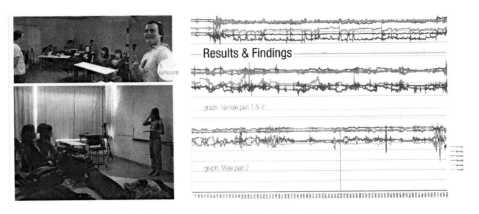

Fig. 2. Scenario experiment (l) photo (r) graphs from sensor

Video recorded scenario presentation workshop showed that movement created strong impression, especially in real life where people are always in motion. Certainly leaning or having certain body parts open was significant in displaying manners, and audiences consciously used these factors as well as verbal means to play given roles. The presenters who did not know roles audiences were playing, however, displayed more of constant unconscious fidgeting (movement). Nervousness was displayed by fidgeting with feet. Especially the presenter with uncomfortable atmosphere was very active with fidgeting the feet. Also continuous staring at down at the notes gave nervous and unprofessional signals.

When observing further into video and after looking at data collected from accelerometer, uneasy feelings were displayed more through lower body movement and head movement than hand and shoulders.

2.2   Phase 2. Testing Theory

In phase 2, research has started with changing hypothesis from posture to motion center; [unconscious movement is important honest signal. If people can have more awareness to such movement, they may be able to reflect and control it]. Another idea discussed was that if attitude affects behavior change, then attitude might as well be influenced with behavior change. New hypothesis started with new grounds of stating what helps to change attitude (or state of mind) as well as what makes one look more convincing in presentations. It was this point that designed technology device could be used to give movement feedback to users and bringing awareness of their movements to them.   Thereby enabling users to change impressions they were creating: helping users communicate content, give confidence, and optimizing performance.

After assuring the feasibility of the theory, the second phase progressed on to 'where and how person will be given feedback through designed device to aware his/her unconscious movement'. Three rough prototypes- headset, hand-held device and vibrating-shoes- were made to test where in body will feedback device be placed. Also aptness of sound-based feedback and vibration-based feedback was tested. 3 participants gave presentation with one of devices on, 4 with each device on. Feedback system was operated according to test conductor's judgment, and participants gave review.

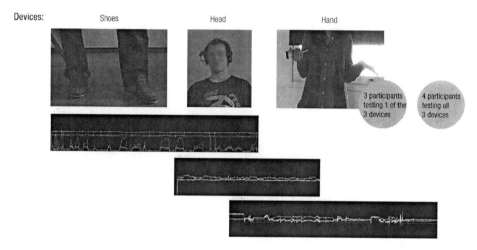

Fig. 3. Tested Devices: prototypes with sensor, vibration & sound control

To given feedback, all participants showed significant decrease in movements. To placement of devices, hand was out-casted for its naturalness of motions. Additionally, participants liked that the feedback was given to the place that was doing the unwanted movement.

With vibration given at unnecessary movements, all participants clearly showed significant decrease in movements. They have answered knowing that there will be buzzing to unnecessary movement had helped to gain awareness, and for some participants gave stress to stand more still even when vibrator did not buzz.

To placement of devices, vibration in hand was the most unrecognized. Most participants answered feedback through hand was familiar, but also added hand movement was natural. Comments were added with notion that certain movements helps to make presentation more active/ interesting. Vibration in shoes was new, though participants answered positively in experiential feeling. Strong reason to supporting feedback to leg movement was that upper body was used to create positive gestures in presentation whereas lower body is kept stable most of the times. To vibration in head, experiential responses differed; some felt it too strong whereas some felt completely at ease with it. Participants especially liked that the feedback was given to the place that was doing the unwanted movement.

Though all participants answered that being aware of movement and knowing to what movement the device is giving feedback would be helpful in enhancing presentation manner. This was incorporated later by giving vibration directly where movement are made by placing buzzer in body parts where movement is measured- ex. right leg movement to the right shoes. It was also added that such feedback may be more useful when practicing than in real presentation situation.

2.3   Phase 3. Evaluation Test

With the collected data, the usage and where it would be put-on was concreted. The prototype consists of two devices- headset and shoes. It will be used in training for or in real presentation at home, school, and offices. The feedback is given only to the user when unnecessary movement continues for excessive time.

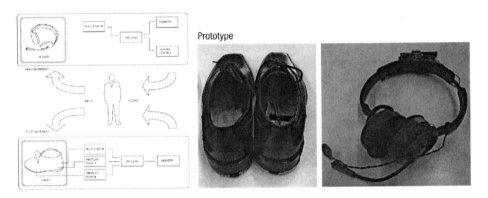

Fig. 4. The final prototype: (l) simple flow chart of devices (r) final prototypes: shoes with auto sensor and vibration feedback, headset with auto sensor and vibration & sound feedback (selectable)

For final evaluation, 3 people participated. The aim of the evaluation was to check if prototype helped user to control honest signal during presentation by notifying such movement. Evaluation procedure consisted of six steps. First, the participants were asked to fill short questionnaire mainly describing preconceptions they have of themselves as presenters. Second, after filling the questionnaire participants were asked to give short 2-5-minute presentation. Presentation was given after practicing, but without using the designed device. Third, Participants were asked practice the presentation with the designed devices- using both devices either simultaneously or separately. For precise evaluation and usability testing, participants were first-time users of devices and were not given any usage instruction. They were asked to answer review questionnaire. Four, participants gave 5-minute presentation the following day. Five, the final presentation was given and taped. Six, after the presentation short interview was conducted.

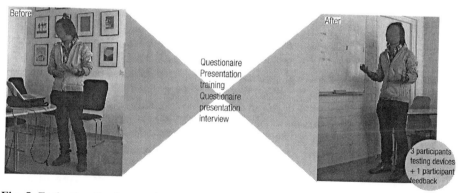

Fig. 5. Evaluation Testing: participant K replied after using the device, he noticed himself too still thus boring. In second presentation, he tried to make action for better presentation.

All participants reported to have gained knowledge of themselves as presenters and to have changed/tried to change their behavior because of the prototype. They also confirmed that being more aware of their movements helped them to control them. Practicing with prototypes also made them carefully use the movements made during presentation much more.

Two out of three participants reported headset having affected their performance by reminding them to look up. "When I practiced, I noticed that I look at my notes too much, but headset reminded me to look up from my notes".

Two out of three also reported that shoes had helped them to improve their performance. One reported that practicing with shoes made him realize that he moved too little and he begun trying to move more for emphasis. According to videos of his two presentations he was right about not moving at first and succeeded in trying to move more in the second presentation. Other participant realized that he was moving too much, he was surprised to have got so many vibrations while practicing.

Comparing presentation videos and questionnaires proved that not everyone is well aware of the movements they are making while presenting. One participant answered in first questionnaire not being very well aware of his movements, but guessed that most of his movements are in hands. However his hands did not move significantly, but his entire body moved very notably. Neither had other participant previously known about being too still.

3   Conclusion

Though question arises; although 'honest signal' is natural, is delivered message is as meant or not. After all, if honest signal is delivering wrong or unwilling message, carefully planned consideration needs to be made in using honest signal device for better communication.

To the question 'is delivered honest signal is as meant (intended) or not', sometimes signals are delivered wrong or unwillingly. The photo survey showed even with same postures, impressions people received had differences. During the video recorded scenario experiment, type of signal was changed from posture based to movement base for more accurate and significant impression. One was to control unconsciously created signals, and second was to create wanted signals. People were collecting stronger communicational message in movements than postures. In communication aspect, it offers new potentials interactive device or interaction can do for people, by making what was not cognitive into realizable signals or by making certain messages stronger.

To the question 'would more awareness to honest signals help people to enhance the control over signals'; this study has tested that if people have more awareness to honest signals they are sending, they are capable of enhancing control over signals. It will enable people to optimize signals, collecting more of wanted impression than not. 3 people participated for evaluation testing, but 7 people participated for rough prototyping testing (phase 2). The 7 has also given presentation and review of usage. Certainly there was limit to the study and case numbers as well as difference to degree and place of movement made by various participants. However, overall evaluation went well and valuable data was collected throughout the study. Most responses were given to active foot movement in this experimental project. Though even participants who are not foot-active reported to have gained more knowledge of themselves as presenters and felt to have improved as presenters. In longer usage period results would most likely have been more significant. What specially enlightened us was that by making user more aware of the movements meant also shoe prototype made user more aware of the movements that were not made.

Designing of an Effective Monitor Partitioning System with Adjustable Virtual Bezel

Sangsu Lee, Hyunjeong Kim, Yong-ki Lee, Minseok Sim, and Kun-pyo Lee

Dept. of Industrial Design, KAIST
335 Gwahangno Yusung-Gu,
Daejeon, Republic of Korea
{sangsu.lee,hyunjkim73,ykl77,kplee}@kaist.ac.kr,
simminseok@gmail.com

Abstract. We suggested a new monitor environment with an adjustable virtual partition in order to incorporate advantages from both the multiple monitor and single monitor. We conducted a user study by making a prototype. Results showed that the prototypes enhanced the user work performance while it reduced the temporal demand. We believe that our design suggestion and the user study results can make a contribution to future single large monitor distributions from the user's need of a bigger screen which provides a more immersive experience, as well as to a new computing environment such as laptops and tabletop computing that does not allow multiple monitor establishments.

Keywords: Window management, Interface design, Multiple monitor, large monitor, partitioning.

1 Introduction

The use of monitors on computers is continuously researched as an important topic in the HCI field. As people want more effective work environments, many workplaces nowadays are equipped with multiple monitors, and according to previous researches, those setups actually enhance work efficiency [4, 10].

However, there is a contemporary tendency to find strength of a single monitor environment. Foremost reason is the falling price for large displays. In addition, the immersive experience in terms of consuming movies and games has become more important, and those activities are better supported by a single large monitor [2]. Thirdly, people may better concentrate on one work, as there are studies that conclude that a sufficiently large screen brings about higher productivity than dual monitors [1, 2].

Yet, while many studies were conducted to understand advantages of different screen setups, there are little attempts to combine the strength of both single and multiple monitor environments. As actual use cases vary, so do the needs for a flexible monitor environment. Thereby, the monitor bezel plays a key role. For example, in the case of exploring a large city map on a dual monitor setup, the bezels cutting through the map is disadvantageous [3]. Yet, the same setup might prove to be beneficial when working on tasks that require switching between multiple documents or

M. Kurosu (Ed.): Human Centered Design, HCII 2011, LNCS 6776, pp. 537–546, 2011.

programs. In this case the bezel helps users to organize their work into different activities that are partitioned physically by the different monitors [4].

In this paper we have thought of a new environment, in which the strength of both multiple and single monitor setup can be combined. A dual-monitor, on the one hand, might bring productivity advantages by physically separating the workspace into two parts, yet the partition is a static one that is defined by the physical size of the displays. A virtual partition on a large single monitor, on the other hand, allows changing the proportions by moving the partition. It even allows removing the partition to use the display as one large monitor. Through this, a user can maximize the positive aspects of a dual-monitor environment by having a partition in the wanted location, while still maintaining the merits of a single monitor.

Based on this idea, we made a prototype as a new monitor environment with a flexible partition that users can adjust according to their needs. To verify the efficiency, we conducted a user-study with 32 participants.

2 Related Works

2.1 Concept of Space

Grudin [4] presented some of the first work describing everyday multiple-monitor users where he points out the benefits of an arbitrary division of space, claiming that a single large space has disadvantages, as it limits the possibility to park objects out in a defined periphery. This claim implied the necessity of a virtual partition.

When considering the idea of partitions in the virtual world it comes naturally to draw analogies to the real space and architecture. Architecture is primarily about the designing of space, but also the designing of constructional entities that demarcate space. The difference is concerned in the way in which space is experienced. In one respect there are spaces in which one can dwell and act, in another there are spaces where there is only room for the roaming eye. So, there are spaces for use, as opposed to spaces for merely looking at. The English 'space' and the French 'espace', both derived from the Latin 'spatium', are primarily concerned with extensiveness and the distance between objects. The German translation 'Raum', on the other hand, etymologically akin to the English word 'room' considers space as both that which is enclosed as well as the enclosure itself. This means that a German-speaker will think of 'Raum' as a small separated portion of the limitless space. With regard to this dual meaning of 'space' and 'Raum', our present-day concept of space for the digital world ought to be redefined, according to the notion of continuity and distance on the one hand, and the idea of limitation and enclosure on the other.

2.2 Partitioning Real Space

In the example of partitioning the space of a house, people generally value more rooms over room sizes. A house with one large bedroom is not the same as a house with two bedrooms of moderate size. In the two-bedroom house, the second room is used for different purposes, e.g. as a guest room or office. Different tasks are optimal in one or the other room. The wall makes a difference by creating a space with a dedicated purpose and an exclusive accessibility. Here we already can see that reasons for partitioning space in the real world are diverse. They range from aims to order and structure the space, to separating functional units from each other, to providing visual separations, and to enable privacy as well as democracy and equality.

2.3 Partitioning Virtual Space

Based on the concept of partitioning space in the real world, previous researches have approached the question of why and how people partition digital worlds. Henderson et al. proposed Rooms [6], an comprehensive virtual desktop system, allowing users to arrange sets of windows to correspond with tasks and switch among the sets with very simple input actions. Grudin [4] examined how to partition digital worlds effectively and how people with a lot of display space arrange information. He identified a Focal and Peripheral Awareness in Multiple Monitor Use and concluded that having to open and close windows, obstructs the train of thought, as finding features buried in menus is more time-consuming than visually scanning the full set of functions on a second monitor. Based on Grudin, Czerwinski et al. studied the productivity and performance benefits of very large displays [2]. The study resulted in design guidelines for enhancing user interaction across large display surfaces. Furthermore, Hutchings proposed various window management techniques for a more efficient and enjoyable multiple-monitor experience [7,8,9], and Kang and Stasko conducted a comparative study to quantitatively examine how people perform common tasks with one or two monitors [10]. Both, Hutching's and Kang's studies emphasized the faster performance and the reduced workload of participants with two monitors, and also stressed the subjective preference for multiple monitors. Finally, Bi and Balakrishnan [1] discovered benefits from using a wallsize large high-resolution display by observing users' behaviors when using such a display for daily computing. Those were the facilitation of multi-window and rich information tasks, the enhancement of users' awareness of peripheral applications, and the opportunity for a more immersive experience.

The two distinct usage patterns in partitioning screen real estate and managing windows on a large display were (1) the reduction of window minimizing and maximizing, and the increase of window moving and resizing, as well as (2) the division of space into task zones, where the center part represents the focal region, and the remaining space the peripheral region.

Related to these studies, this research identifies ways that can help people make better use of the increased screen area through the application of a flexible virtual partition. Furthermore, it reports on a preliminary study of people performing specific tasks on a computer while using a flexible virtual partition to divide a single display.

According to Robertson et al. [3], the bezels in a multiple display present both opportunities and problems. There are several practical solutions to address this issue, such as *WinSplit Revolution* [11], software to provide the ability to move a window to a specific position on the screen. However, current solutions are just focusing on managing windows, not on the adjustable 'bezel' concept. This study thus will introduce and explore the effectiveness of a virtual bezel for single display that remains adjustable by user at any time.

3 System Design and Implementations

Our design goal is to combine both strength of a dual and a single (large) monitor through adjustable virtual partitioning of display (Figure 1). Our basic design concept is allowing the partition (bezel) between the two monitors in the existing dual monitor environment to move organically so that it can take in advantages from both sides. This will allow expanding a monitor's width for certain situations as well as widening one side into whole size, providing full space for a single task according to different needs.

The largest consideration in terms of interface was how to adjust this virtual variable partition. First, having the single monitor idea as the most direct and initial idea, we had a tangible stick on the monitor to move the partition physically. However, according to observing the monitor environment of the laboratory personnel, we found out that it was difficult to tangibly control the monitor directly in many situations. In the end we concluded that adjustment using the mouse, which is most frequently used in the PC environment, is the optimum choice both for the performance and for user convenience.

The final prototype was made so that the Logitech MX500 mouse's middle button shortcut can make adjustments. Program was made as a C++, residing in the tray in the Window environment and makes a single monitor into a virtual dual monitor that has left and right parts. The operated program is initially designated to the left part but can be put on to the right part by moving the window. The user can move the mouse while holding on to the button shortcut on the mouse to move the partition to a desired position.

Fig. 1. Dual and Single large monitor setup (left and middle) and our Virtual Bezel concept (right)

4 User Study

We conducted a user study in order evaluate our suggested prototype. Study measured the task completion time for task efficiency and task workload at the time. Also, a qualitative study was conducted on behalf of the user experience aspect in order to see if our expected user scenario was actually useful.

4.1 Expected User Scenario and Benefits

When we design and conduct user study with our virtual bezel concept, our expected user scenario and benefits are as follows.

First, when users are working simultaneously with different programs, adjusting the screen proportion might increases the work efficiency. Through the prototype, it might possible to extend the work efficiency and convenience by a flexible adjustment of the proportion between the work screen and the source screen. When the source screen has much information, expanding the source area to find the wanted page will be beneficial. Also, when focusing on the content production, extending the contents area will help to focus on the work.

Secondly, while users are working in full-screen mode, checking the parallel task might be more effective. One of the largest benefits of a single display is being able to concentrate on a task while feeling a sense of immersion by having one program in a full-screen mode. It is even greater for visual work such as image editing, watching a movie, or studying at a map. In case of a dual-monitor, these tasks might be inefficient, since the remaining screen is useless to the full-screen mode. However, a dual-monitor has the advantage of displaying peripheral information through gadgets, messengers, or weather applications on the remaining monitor, while being in full-screen mode. We expect that users might take both benefits of them by using our prototype environment.

Task switches do not cut the flow of perception and support a fast processing.

Also, when users are executing two tasks in turns, under the 'Windows' environment, switching between two windows usually is done by using ALT-TAB. However, there is confusion in switching windows in a multiple monitor environment, as the focus on a layer can be on either monitor. In our prototype, a user might able to arrange two windows on each screen area onto the same monitor. By having the partition located on the left end or the right end, it is possible to maximize the two windows individually. Under such a condition, by moving the partition from the right end to the left end, it is possible to clearly identify the focus on a layer. By being able to mentally grasp that each program naturally is located in each part, it might be possible to switch a task without ending the flow of perception.

We will test our expected user scenario during the user study as well as other quantitative and qualitative analysis.

4.2 Study Setup

We recruited 32 users from design students (avg. age=26.8, SD age=4.26): 16 who used a single monitor, and 16 who used a dual monitor configuration for their daily computing. To evaluate the effect of the virtual bezel on both dual and single monitor users we tested 8 of the 16 dual monitor users in the dual monitor setting (G1) while 8 of them conducted tasks using the virtual bezel (G2). Similarly, 16 of the single monitor users were divided so that 8 of them were tested in a single monitor setting (G4) while 8 conducted tasks on our prototype (G3) (Figure 2).

To control every variable except for the bezel settings, every study was conducted in a same lab using the same environment: 24 inch screen size, 16:9 ratio, 1920x1080 resolution, LCD monitor on a PC with Window 7 installed. The users were to conduct a virtual task of producing a school seminar poster (Figure 3). Information and design requirements for the poster were listed on a MS PowerPoint file in a text format and

> **G1** (8 people): Use _dual-monitor_ in day work / Use _dual-monitor_ setup for this study
> **G2** (8 people): Use _dual-monitor_ in day work / Use _single monitor with virtual bezel_ setup for this study
> **G3** (8 people): Use _single monitor_ in day work / Use _single monitor with virtual bezel_ setup for this study
> **G4** (8 people): Use _single monitor_ in day work / Use _single monitor_ for this study

Fig. 2. User groups for this study

they were to design an A1 size poster using Adobe Illustrator with the given information. In regard to the familiarity of the task, using a program with another program taken as a source is a quite common activity in office environments and thus is a task conducted frequently in existing studies that tested the dual monitor environment's efficiency.

Fig. 3. Users can adjust virtual partition in any position according to their needs using the prototype we made

We made the users to think aloud while the task operations were video-taped. After the test we conducted a NASA-TLX [5] survey, a workload analysis method, and listened to the post-task comments through interviews. NASA-TLX, a subjective workload assessment tool, is separated into six items, i.e. Mental Demand(MD), Physical Demand(PD), Temporal Demand(TD), Performance(PR), Effort(EF) and Frustration-level(FR) to enumerate the degree of the demands quantified by comparison of each item. Through this process we made the users evaluate the prototype workload by themselves in a subjective manner in order to recognize the prototype's advantageous and shortcomings when compared to the existing counterpart.

On the other hand, after a short break after the task test we had all 32 participants experience utilization situations for a short period: 1) Checking the weather gadget on the wallpaper while watching a movie and expanding the movie to full size 2) Watching two different web pages alternately by moving partitions. Each set up situation was to be experienced for approximately 2 minutes, and then we asked questions on their subjective ideas on whether the prototype could be utilized effectively in real life situations.

4.3 Study Result

For task completion time, there was a significant difference between conventional and dual displays (G1: for group information, see Figure 2) and conventional single display (G4) setting: dual displays setting took 16% less than the task completion time in a single monitor setting. When introducing the virtual adjustable bezel, there was no significant time improvement for conventional dual-monitor users (G1 shows 28 minutes, G2 shows 30 minutes) However, participants using a single monitor in their daily work showed a significant improvement from 33 minutes (G4) to 27 minutes (G3) when applying the virtual bezel (t=3.230, df=14, p=0.006). (Figure 4)

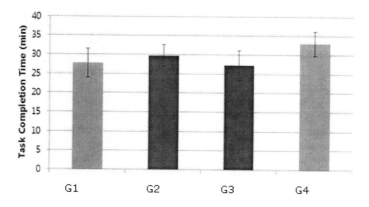

Fig. 4. Task completion time between user groups

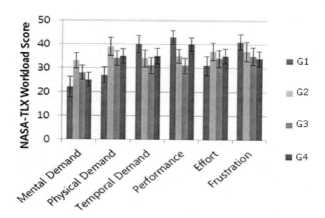

Fig. 5. Task workload between user group

Results of the workload analysis using NASA-TLX (Figure 5) shows that the virtual adjustable bezel brings the biggest improvement in performance. Everyday dual monitor users showed a decrease in their performance load from 43 (G1) to 35 (G2) (less is better), and everyday single monitor users also showed a decrease from

40 (G4) to 31 (G3). Both differences were statistically significant (t=4.287, df=14, p=0.001; t=5.286, df=14, p=0.000). For the temporal demand, which subjectively measures how much time pressure is felt during the task, dual monitor users showed a 15% decrease when using the virtual bezel (G1 and G2), and single monitor users showed an approximate 12% decrease in their temporal demand (G4 and G3). Both were statistically meaningful reductions (t=4.837, df=14, p=0.000; t=-4.185, df=14, p=0.001). However, everyday dual monitor users showed a huge increase in both their mental and physical demand when they used the virtual bezel. Other quantitative results are shown in Figure 5.

From the observation and interview, we could see the ways of users manipulating the adjustable virtual bezel during the given task. When working simultaneously with two different programs, users were trying to adjust the screen proportion between the work screen and the source screen. Since the source screen has much information, expanding the source area to find the wanted information was beneficial. Also, when focusing on the content production, extending the contents area helped to focus on the work. Also, users could switch tasks quickly and naturally, although not expected from our prototype. When executing two tasks in turns, under the 'Windows' environment, switching between two windows usually is done by using ALT-TAB. However, there is confusion in switching windows in a multiple monitor environment, as the focus on a layer can be on either monitor. In our prototype, a user was able to arrange two windows on each screen area onto the same monitor. By having the partition located on the left end or the right end, it is possible to maximize the two windows individually. Under such a condition, by moving the partition from the right end to the left end, it is possible to clearly identify the focus on a layer. By being able to mentally grasp that each program naturally is located in each part, it is possible to switch a task without ending the flow of perception.

6   Discussions

The results of this study show that the *virtual bezel* reduces the task completion time of a normal single display setting. It is, however, less efficient than a dual display setting. Since we use the same monitor type for every study setting, a dual-monitor setting (G1) with two monitors has twice as much screen space, compared to the single-monitor setting that applies the *virtual bezel* (G2, G3, G4). Thus, G1 can see more information than the other user groups. We need further studies that compare conventional dual-monitor settings with single-monitor virtual bezel settings, under the condition that in both setups the total screen size remains the same. With the same screen size in a single-monitor setting, the results show that applying a virtual bezel brings improvements in performance (G3 vs. G4). Most participants (G4, 8 out of 9) use ALT-Tab frequently for switching programs, while only one out of nine in a virtual bezel setting (G3) uses ALT-Tab for switching.

The NASA-TLX survey result shows that there is a significant improvement of the *performance load* and *temporal demand. Performance load index* shows a great improvement effect indicated by how much the users subjectively perceive to have achieved. This implies that users can feel more satisfaction with the *virtual bezel* for both the dual and single monitor setting. This is the level of their recognized performance being directly linked to the level of satisfaction. Thus, great improvement on this item is very encouraging.

On the other hand, the increase in *mental* and *physical demands* needs to be discussed. The main reason seems to be that users initially are not familiar with the virtual bezel concept and the interface using a mouse's customized button. Many participants (7 out of 16) mentioned that the interface is simple yet confusing. Although it might improve as they get used to the interface, a more intuitive interface, such as a gesture interface should be considered. Also, some participants (2 out of 16) pointed out that the partitioning of the screen using a virtual bezel lacked the 'dividing-strength' of a physical bezel. Thus, we may consider a real physical bezel for an adjustable partitioning in further studies.

As an interesting finding, some users (4 out of 16) commented that the virtual bezel concept may be useful in a mobile laptop environment where dual monitor setup is generally not possible. Thus, it will be promising to discover ways to use a small single screen more efficiently.

7   Conclusions and Future Works

As increasingly larger monitors become more available to users, it is appropriate to devise new interfaces that help carry out activities on computers more efficiently. In this paper, we introduced the idea of a virtual bezel that allows users to split a single monitor. We conducted an initial exploratory user study to understand and explore the effectiveness of an adjustable virtual bezel. The study showed that an adjustable virtual bezel can bring about productivity advantages by splitting the workspace, compared to a conventional single monitor of the same size.

This study was not able to find out how much effect long term utilization could bring when operations are made familiar. Future observations in the actual context with long term use will provide a deeper insight regarding a more efficient and convenient desktop work environment.

Acknowledgments. This research was supported by WCU(World Class University) program through the National Research Foundation of Korea funded by the Ministry of Education, Science and Technology (R33-2008-000-10033-0).

References

1. Bi, X., Balakrishnan, R.: Comparing usage of a large high-resolution display to single or dual desktop displays for daily work. In: Proc. CHI 2009, pp. 1005–1014. ACM Press, New York (2009)
2. Czerwinski, M., Smith, G., Regan, T., Meyers, B., Roberstson, G., Starkweather, G.: Toward Characterizing the Productivity Benefits of Very Large Displays. In: Proc. of INTERACT 2003 (2003)
3. Robertson, G., Czerwinski, M., Baudisch, P., Meyers, B., Robbins, D., Smith, G., Tan, D.: The Large-Display User Experience. IEEE Computer Graphics and Applications 25(4), 44–51 (2005)
4. Grudin, J.: Partitioning digital worlds: Focal and peripheral awareness in multiple monitor use. In: Proc. of CHI 2002, pp. 458–465. ACM Press, New York (2002)

5. Hart, S.G.: Nasa-Task Load Index; 20 years later. NASA-Ames research center, Moffett Field (2006)
6. Henderson Jr., D.A., Card, S.K.: Rooms: the use of multiple virtual workspaces to reduce space contention in a window-based graphical user interface. ACM Trans. on Graphics 5(3), 211–243 (1986)
7. Hutchings, D.R., Stasko, J.: mudibo: multiple dialog boxes for multiple monitors. In: Ext. Abstracts CHI 2008, pp. 1471–1474. ACM Press, New York (2008)
8. Hutchings, D.R., Stasko, J.: Quantifying the performance effect of window snipping in multiple-monitor environments. In: Baranauskas, C., Abascal, J., Barbosa, S.D.J. (eds.) INTERACT 2007. LNCS, vol. 4662, pp. 461–474. Springer, Heidelberg (2007)
9. Hutchings, D.R., Stasko, J.: QuickSpace: new operations for the desktop metaphor. In: Ext. Abstracts CHI 2002, pp. 802–803. ACM Press, New York (2002)
10. Kang, Y., Stasko, J.: Lightweight taskapplication performance using single versus multiple monitors_a comparative study. In: Proc. of GI 2008, pp. 17–12. Canadian Information Processing Society (2008)
11. WinSplit Revolution, http://www.winsplit-revolution.com/ (last accessed on January 13, 2011)

Bridging the Social Media Usage Gap from Old to New: An Elderly Media Interpersonal and Social Research in Taiwan

Shih-Hsun Lin and Wen Huei Chou

National Yunlin University of Science and Technology
Department of Digital Media Design and Graduate School of Computational Design
123 University Road, Section 3, 64002 Douliou, Yunlin, Taiwan
{Furanke,cristance}@gmail.com

Abstract. Understanding the media usage and interpersonal communication that the elderly have been familiar with is valuable for designing social media for the elderly. We conducted interviews for acquiring the data about attitudes and behaviors of the elderly, and then analyzed the transcripts to discover the patterns of the elderly in media usage and social life. The findings show that in media usage our subjects prefer watching TV and contacting people with phone/mobile phone in their leisure time. Also they prefer the habitually daily routine of watching TV in the living room although they appreciate the flexible selectivity of using a computer, but have difficulties with these new media. Activities such as weddings and funerals remain the important chances to retrieve relatives, and physical contact still remains the primary interaction for elders. Fitting in with the elderly habits in daily life to design is discussed.

Keywords: elderly, social media, media usage, media communication, interpersonal communication.

1 Introduction

Social media is becoming a more and more popular and influential medium of communication, especially among the younger generation. However, the elderly are accustomed to their familiar media usage to get information and communication devices to contact people, which have formed the way of the elderly media usage and interpersonal interactions in their social life. For encouraging the elderly to more easily participate in social issues, express opinions through social media, and then having their opinions to be respected, a social media that can fit into the lifestyle of those elderly is significant and crucial.

In this paper, we regarded social media as the media through which people can be connected and share information; with these uses, they can maintain relationships with friends and relatives, or even to build new relationships; those social media include social networks sites (SNS), the services implementing information and communication technologies (ICT), and computer-mediated communication (CMC) applications.

M. Kurosu (Ed.): Human Centered Design, HCII 2011, LNCS 6776, pp. 547–555, 2011.
© Springer-Verlag Berlin Heidelberg 2011

The reformation and evolution in communications and media through information technologies and the Internet has been developing into the information society, which influences the structure of society. Technologies allow information to be distributed more easily and faster; its complexity and costs might intensify existing social inequalities, causing large groups of misfits such as people who do not fit in with the information society [1]. The elderly are always going to find it difficult to fit into new-tech easily and smoothly; they are digital immigrants learning a new language of the digital age [2], this learning process is usually laborious to them.

Learning to accommodate the existing digital social media, this can facilitate the elderly to use social media, exchange opinions and social support via the Internet. How different CMC approaches used among the elderly were being explored [3]. Indeed, the elderly can benefit from ICTs through learning to overcome their anxiety on using the computer, and then to gradually accommodate this new communication approach. Not every elderly person can go through this learning process, especially those with lower education and speaking dialect primarily.

The social media nowadays are designed by so-called digital natives and the elderly attributes about media usage and communications are almost never being taken into account. As digital social media has become an influential new media [4], though social media connects people via the Internet, this is different from previous media usage and interpersonal interaction that the elderly are familiar with. For improving the elderly interpersonal and social relationships, we have to understand the media usage and characteristics of the elderly social life.

The digital divide in the information society brings about the issues of the inequalities on using new communication approaches [5], those also faced by the elderly. Developing an adequate new media environment for the elderly can facilitate them to participate in social media. The elderly are willing to use social media to connect with people and contribute contents, such as by expressing their opinions, only if they have sufficient support on using these new media [6, 7]. The objective of this research is to investigate the differences between the previous media and digital social media usage, for developing an adaptive social media which can fit the elderly's needs, close to their custom, and be more acceptable to the elderly. The social media is designed positively for the elderly so that it can support the elderly current practices and understanding rather than forcing them to adopt new and alien ways of keeping in touch [8].

2   Literature Review

Being old is not just a physical degradation, but also a social status that the elderly seek to manage, sometimes resisting, sometimes accepting [9]; as the attitudes to families and friendships are concerned, and the role in those relationships, the elderly seem to have their own perspectives.

The elderly have unique design requirements that are probably stronger than the potential for their experiencing physical or cognitive decline [9]. Thus, for giving the elderly the option to participate in mediated social interaction online through technology, developing the social media for the elderly should take their life situations, habits and attitudes into account [10]. For having social media used and accepted, it must not only be usable but also support current patterns of communication wherever possible and fit into the environment in which the elderly live, rather than introducing a completely new usage pattern to the elderly [8]. In the following sections, we first review the related researches in media usage among the elderly, and then the attitudes of the elderly on communications and relationships.

2.1 Media Usage among the Elderly

Those media the elderly are familiar with play the important roles in their lives; the elderly rely on watching TV which can provide information and entertainment, and the telephone as a communication media which serves them to contact people. In addition, there are researches that claimed the elderly would feel anxious on using new-tech products [11-13].

Watching TV is almost an important activity among the elderly in their daily lives; the reason they always watch TV is having no other adequate leisure activity to do; also the TV was very familiar to them when they were young. There are four reasons why the elderly are used to watching TV: 1) for relaxation, entertainment and as a companionship; 2) for getting information that can provide materials to talk; 3) watching TV can be a way to escape temporarily and give a chance to interact with friends or families by talking about the TV programs; 4) Passing time by watching TV had been formed as a habit in life [14].

The telephone is the preferred way of keeping in touch among the elderly, because it is seen as easy to use, and allows normal behavior, like chatting; and personal telephone calls are not recorded, or ever heard by another; the telephone also allows richer communication by allowing the user to hear someone's mood, or health, in their voice. Its immediacy also gives benefits by offering security in case help is needed quickly [8].

Other existing researches explored different devices that the elderly were interacting with, practically focused on the adaption of new functions and human interface [12, 15-17]. For example, the touch-based user interface would be easy to learn and adopt among the elderly, and they would be able to successfully use it regardless of the physical or cognitive weaknesses of the elderly [12].

The problems faced by the elderly in using and engaging with interactive media are currently less associated with physical and cognitive factors, more related to their previous experiences and attitudes [18]. There are main obstacles that the elderly face in contributing content by adopting user-generated content (UGC) , these are: 1) with respect to affect, personal integration, affiliation, release of tension, and creation, they did not perceive what UGC can offer them; 2) fear that the systems are too hard to learn, and concomitant low expectations towards support that can help them; 3) lack of social influences, the elderly did not perceive that someone important to them thought they should use the application; 4) pre-established negative attitudes towards digital technologies, including computer anxiety and concerns about privacy [6].

2.2 Attitudes of the Elderly to Communications and Relationships

Previous researches did not explore the needs of the elderly as a group suffering cognitive or physical decline, instead, focusing on the elderly attitudes to keeping in touch [10, 15]. For example, the elderly want to feel that real contact with someone, contacts reaching a level of intimacy, and that something representing for them can be put into the communication, such as their voices or handwriting [15]; whereas inputting text with the computer can produce neater and consistent text that is more readable, helping the elderly lift the worry of someone else's confusion with their handwriting [13].

Concerning the time spent on relationships, the elderly are more motivated to spend time on relationships that are emotionally rewarding and of significance to them, such as their families and close friends, and are less motivated to acquire new knowledge about the social world by meeting new people [15]. In the family relationship, the assumptions about symmetry being inherent in family ties can be misleading; that asymmetry is often a better way of accounting for the attitudes of the elderly to their families; asymmetry also offers a better basis for design [9].

In terms of social media usage, the factors most hindering the elderly to use SNSs would be: 1) The elderly fear their personal information being used by malicious third parties, because not only confusion on how to adjust the SNS privacy settings, also their lack of confidence in their personal computer skills; 2) fear of accidental social blunders in mediated social interaction also due to not understanding privacy settings and privacy management; 3) incompatibility of their perceptions of social relationships with their preconceptions and assumptions about SNSs ; 4) the extension of the interaction habits formed through very-long-term relationships to a new interaction environment was regarded as costly [10]. Those concerns should be taken into account of designing the social media for the elderly.

From the elderly viewpoint there is the differentiation between the time for communication and the time not to communicate; they are careful not to intrude on others, and would like to be treated by others in the same way [15]. There are elderly people who hardly understand that communications among the younger generation using CMC are always continuous but transient. However, the elderly who can communicate by e-mail think that they can send informal messages by e-mail to breach the restriction of time zone, and also allow them to compose an e-mail at their leisure [8].

3   Method

A ground theory approach is applied in this research. We firstly carry out in-depth interviews of five 55-60-year-old subjects for exploring the patterns of the elderly on media usage and social connection in daily life. This semi-structured interview included several approaches of media usage investigations: 1) what they have and use? 2) how they connect with others? 3) what they do? 4) when and where they use them? And in the second part protocol analysis, we use NVivo to code the transcripts and sort them into six main categories.

3.1   The Subjects and the Interview Circumstances

There were five elders to be invited to become the subjects in this research. These subjects interviewed in Taiwan have the following attributes: 1) their ages are 55-60 year-old, because of our research focusing on those who are about to retire or have retired; 2) they are not familiar with using the personal computer for communication, especially inputting traditional Chinese characters by keyboard is a challenge; 3) they are used to using Taiwanese to communicate with people, that is a dialect pervasive in Taiwan among the elderly in this age.

The circumstances in which the interviews were carried out are familiar to the subjects. Two subjects (one male, one female) were interviewed respectively, and three subjects (one male, two females) were interviewed as a focus group; in the context of the focus group there also were other three people whose age was under 55 year-old giving some opinions to serve to facilitate these three subjects in expressing their perspectives much more. These sessions were voice-recorded for later review and analysis.

3.2 Questions for Interview

The questions designed for interview are in order to realize the characteristics of media usage and social life among the elderly (Fig.1 shows the question framework); how they use media and communication devices as well as interact with their friends and relatives in daily life. Fundamentally, these questions are based on two dimensions, media usage and social life. Semi-structured interviews were carried out by asking five subjects these prepared questions and the extended questions according to the subjects' answers. After interviews voice-data was acquired and this was transcribed and analyzed in the next stage.

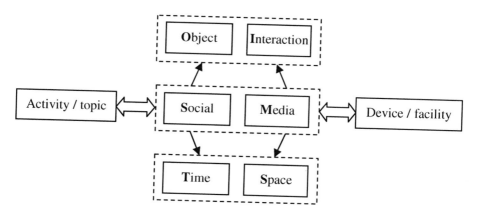

Fig. 1. The question framework: the core including two main nodes - Social and Media; Object for people contacted like friends; Interaction for the way to interact with people or media; Time for the questions related to time and Space for questions related to space

3.3 Analysis

In the process of analysis there are four steps: 1) convert the voice-recorded data of interviews to transcripts; 2) preliminary free coding; 3) organize the free codes to tree codes; 4) find the elderly patterns through examining these coded data back and forth.

Firstly all conversations in the interviews were converted to transcripts using a word processor and then imported to NVivo, the software used to code and analyze data. Secondly the transcripts were read and coded to be free codes. Thirdly the free codes were organized into tree codes, meaning to organize codes in hierarchy, some of these codes were combined and renamed, and then classified these codes under the six main categories according to the framework of the designed questions. These categories are: 1) social, 2) media, 3) time, 4) space, 5) object and 6) interaction. Finally these coded data were examined to find the patterns of the elderly media usage and social life.

4 Findings

Some trends appeared from the analysis. As the interview questions were designed for investigating the patterns of the elderly in media usage and social life, these findings are divided into two sections:

4.1 Media Usage of the Elderly

In the categories of media and time, the subjects expressed they spend much time watching TV, at least 3 hours a day. They always turn on the TV when having nothing else to do. The TV is their main media device in getting information and for entertainment in daily life. This result is in accordance with some previous research, they believe the elderly majorly acquire information and pass time by watching TV [14]. Our subjects usually watch TV news at first, and other programs would be tour and food programs, variety shows and drama series. Non-native language programs are less popular, because it made our subjects strenuous that watching these programs they need to read the subtitles on the TV screen.

Although watching TV news is for knowing what happened outside, they consider it annoying that the same TV news content televised is repeated many times in one day; also because of this, they would not be concerned about missing TV news that will be televised again soon. Due to their patterns of elderly daily life, the time they watch TV is usually fixed; they seldom change their living schedule for watching a TV program.

Other media such as radio, magazines or newspapers are used in some context by some of our subjects; they might listen to radio while driving, otherwise read newspapers or magazines while waiting for services at a store. Our subjects are not subscribing to newspapers or magazines. And because of presbyopia, the subjects expressed inconvenience that when reading newspapers or magazines; they need to wear reading glasses.

The subjects indicated that if not watching TV, they would not know how to pass time in daily life; while watching TV they can focus on the TV programs. If they cannot watch TV, they probably want to be accompanied. So the TV can act as a companion for the elderly who have much spare time in retirement.

In the categories of space and interaction, two subjects using a computer to get information on the Internet, they appreciated the selectivity of this new media that can help them to find the content that they are interested in, but it made their eyes feel tired easily when using the computer. One female subject expressed it would be better that the flexible selectivity of using a computer could be adapted to the TV in the living room. Compared to staring at the computer monitor, watching TV is more comfortable for them; also the living room environment is cozier and more suitable for a small group of people.

4.2 Communications and Relationships in the Elderly Social Life

In the categories of social and interaction it reveals our subjects mainly use phone/mobile phone for connecting with people. Not only when using the phone/mobile phone can freely express in oral, also their contacts mainly use phone/mobile phone as their communication device. Three subjects expressed that they would use text messages in some situations; they usually use text messages to inform the person who could not answer the phone, or to inform many friends to attend an activity or a party. However, typing text messages on the mobile phone by Chinese phonetic input method in Bopomofo[1] is not easy to use for them; and the subjects indicated the characters on the mobile phone buttons are too small to see clearly, they needed to wear reading glasses while typing text messages.

In the categories of interaction and object, it was found that having no interaction with friends and no way to learn friends' recent situation would cause losing contact with friends. The only way to maintain the relationships with friends are through the phone/mobile phone; but some of our subjects deemed that if they cannot meet friends regularly, their relationships would be estranged gradually. Our subjects believe touch in person via face-to-face interaction is important. However, the restriction of distance made them not able to meet each other regularly, with only one contacting the other by phone, which would result in the relationship fading away, and possibly, to disappear someday.

In the subjects' social networks the people who are relatives would be gathered together at weddings or funerals, which remain the important chances of interactions. Probably they would have not contacted one another for a long time, but due to these chances, their connections are hardly broken off; not like the connections with friends could be weakened or disappear, because of losing contact. One female subject expressed that when meeting classmates in the class reunion, she felt comfortable and hardly had a strange feeling; the relationships with classmates built in student life seems to easily be reconnected.

Although some of our subjects have an e-mail account, they rarely communicate with friends via e-mail because of their friends not using e-mail. The e-mails they received were almost always advertisement e-mails, as a result of leaving their e-mail address in the service applications such as bank services.

The subjects who use a computer to communicate are mainly using voice devices to have conversations, the experiences which are similar to using phone/mobile phone; sometimes, if having a webcam, they would use it to see each other.

5 Discussions and Conclusion

Our subjects prefer information in streaming media format, such as video and radio broadcasting, and non-native language programs are less popular. Watching TV and controlling it by a remote controller are very familiar for our subjects; they have no difficulties in selecting the program or channel that they want to watch. In terms of

---

[1] Bopomofo consisting of 37 characters and 4 tone marks is widely used as an educational tool and Chinese computer input method in Taiwan
(http://en.wikipedia.org/wiki/Bopomofo).

that they still participate in call-in TV and radio programs, and turn on the TV or radio until they go to bed. For creating an experience similar to watching TV, referring to interactive TV would be valuable [19, 16].

People share opinions and daily lives to develop relationships. In relationship connection, people maintain and develop what they care about and are cared for. The research shows that the elderly prefer passive ways of gathering information; they seldom grasp or attempt information actively. Therefore they feel new media needs consistent attention or manipulation in strenuous circumstances. Also the elderly social connections are actually based on substantial relations already existed. As their existing substantial relations are not all fancy on new technologies, a big ratio have difficulty with special input devices and system of traditional Chinese.

Our subjects almost did not write letters, they were almost not used to expressing thoughts by writing. Oral expression is the straightforward way for communication to them, and they can express themselves in their familiar dialect, regardless of the frustrated feeling in accommodating the Chinese input method with a keyboard. Therefore, voice input cooperated with manuscript input is more acceptable to the elderly. Regarding the presentation of information in text, for the elderly the readability needs to be considered, text-to-speech or voice messages are also a feasible way.

For facilitating or encouraging the elderly to participate in social media, it would make uses of social media more acceptable to the elderly, the way of using social media much closer to their habits in daily life. The influences of Physical and cognitive decline, the attitudes of keeping in touch, and the ways to contact people, which all conditioned the elderly participation in social media in the digital age.

As most digital social media are formed and followed by the younger generation, to understand the media usage and behavior of previous generations is essential for bridging the inconsistent way of accommodating the accelerating artificial world, and improving the dignity of elderly life. In future studies researchers can invite the elderly to participate in a design process of social media, and then to produce a prototype to validate the requirements for this group of elderly in Taiwan.

References

1. Dijk, J.V.: The network society: social aspects of new media. Sage, Thousand Oaks (2006)
2. Prensky, M.: Digital natives, digital immigrants. On the Horizon 9, 1–6 (2001)
3. Xie, B.: Multimodal Computer-Mediated Communication and Social Support among Older Chinese Internet Users. Journal of Computer-Mediated Communication 13, 728–750 (2008)
4. Kaplan, A.M., Haenlein, M.: Users of the world, unite! The challenges and opportunities of Social Media. Business Horizons 53, 59–68 (2010)
5. Frissen, V.: The myth of the digital divide. In: E-merging media: Communication and the Media Economy of the Future, pp. 271–284. Springer, Heidelberg (2005)
6. Karahasanovic, A., Brandtzæg, P.B., Heim, J., Lüders, M., Vermeir, L., Pierson, J., Lievens, B., Vanattenhoven, J., Jans, G.: Co-creation and user-generated content-elderly people's user requirements. Computers in Human Behavior 25, 655–678 (2009)
7. Ryu, M., Kim, S., Lee, E.: Understanding the factors affecting online elderly user's participation in video UCC services. Computers in Human Behavior 25, 619–632 (2009)

8. Dickinson, A., Hill, R.L.: Keeping in Touch: Talking to Older People about Computers and Communication. Educational Gerontology 33, 613–630 (2007)
9. Lindley, S.E., Harper, R., Sellen, A.: Designing for elders: exploring the complexity of relationships in later life. In: Proceedings of the 22nd British HCI Group Annual Conference on People and Computers: Culture, Creativity, Interaction, vol. 1, pp. 77–86. British Computer Society, Liverpool (2008)
10. Lehtinen, V., Näsänen, J., Sarvas, R.: A little silly and empty-headed: older adults' understandings of social networking sites. In: Proceedings of the 23rd British HCI Group Annual Conference on People and Computers: Celebrating People and Technology, pp. 45–54. British Computer Society, Cambridge (2009)
11. Collins, D.: A comparison of Web browser interactions by older users using a keyboard or a mouse as an input method (2008)
12. Häikiö, J., Wallin, A., Isomursu, M., Ailisto, H., Matinmikko, T., Huomo, T.: Touch-based user interface for elderly users. In: Proceedings of the 9th International Conference on Human Computer Interaction with Mobile Devices and Services, pp. 289–296. ACM, Singapore (2007)
13. Sayago, S., Blat, J.: About the relevance of accessibility barriers in the everyday interactions of older people with the web. In: Proceedings of the 2009 International Cross-Disciplinary Conference on Web Accessibililty (W4A), pp. 104–113. ACM, Madrid (2009)
14. Koçak, A., Terkan, B.: Media use behaviours of elderly: A Uses and Gratifications Study on Television Viewing Behaviors and Motivations. GeroBilim Journal, Journal on Social and Psychological Gerontology 1, 70–86 (2009)
15. Lindley, S.E., Harper, R., Sellen, A.: Desiring to be in touch in a changing communications landscape: attitudes of older adults. In: Proceedings of the 27th International Conference on Human Factors in Computing Systems, pp. 1693–1702. ACM, Boston (2009)
16. Mitchell, V., Nicolle, C.A., Maguire, M., Boyle, H., et al.: Web-based interactive TV services for older users. Gerontechnology 6, 20–32 (2007)
17. Müller, C., Neufeldt, C., Schröer, L.: Designing a large social display for an old people's home (2010)
18. Turner, P., Turner, S., Van De Walle, G.: How older people account for their experiences with interactive technology. Behaviour & Information Technology 26, 287–296 (2007)
19. Mitchell, K., Jones, A., Ishmael, J., Race, N.J.: Social TV: toward content navigation using social awareness. In: Proceedings of the 8th International Interactive Conference on Interactive TV & Video, pp. 283–292. ACM, Tampere (2010)

Research in the Use of Product Semantics to Communicate Product Design Information

Chung-Hung Lin

No.1, Lingtung Road, Department of Technological Product Design, Lingtung University,
Taichung, Taiwan
chunghung@teamail.ltu.edu.tw

Abstract. In the past twenty years, following the transformation of industrial technology and structure, needs created by modern life, issues regarding environment protection, and simply new ways of thinking have revolutionized the core value of product design. There is more to be considered than just the convenience that a product can bring. Consumers nowadays also want to learn how to use a new product in the most efficient way, and many industrial designers see that as one of their goals. The use of Product Semantics in product design is a proper means to communicate the information that consumers need in order to improve their overall experience using the product. Recent research focuses on the application of product design semantics, in order to discover how designers convey messages through the use of semantics, i.e. styles, colors, functions and textures, etc. It seeks to determine how the best way for the designer to communicate all that information to the consumer. A designer want the user to be able to operate his product under "zero obstacle" conditions, to understand the message the product carries, and to enhance the consumer's pleasure and comfort during its operation. From the perspective of industrial design, there are two summarized main points:

1. Product design is now defined as a system to communicate product information based mainly on product use circumstances.
2. The message that a product carries has become the most important factor throughout the design process.

Keywords: Product design, product semantics, message communication, designer.

1 Preface

In today's society, our living environment has become focused on "people" instead of "objects". Therefore the main focus of design is humanism, and the style of the product should be focused on communication with "people". The purpose of design is to express a designer's idea, which means giving the designed object ways to communicate feelings or meanings. In conclusion, the supreme goal of product design is for the operator to understand and complete the operation. Under this principle, a product carries a mission of "translation" and is the bridge between a user and a designer. Therefore, product semantics can be used to discuss psychology and changes in our society or culture. The foundation of product design is based on "people", and also

M. Kurosu (Ed.): Human Centered Design, HCII 2011, LNCS 6776, pp. 556–565, 2011.
© Springer-Verlag Berlin Heidelberg 2011

needs to include factors coming from different groups, history and cultural backgrounds. Shortening the distance between designer and consumers is the first goal of product design. Creating useful, functional products is the primary task of the modern industrial designer.

Nowadays professional designers are widely recognized for their outstanding contributions which is important in product function. On the other hand, when it comes to product design, modern society has come to understand that the focus should definitely be on people (users). The need for a human-oriented approach to product design has given rise to the use of "semantics" as a factor in good product development. One of the most widely discussed issues in product design is how to convey "the meaning" of the product. According to Dr. Krippendorff, product designers need to consider differences in perception, culture and history, along with different symbols. These topics are meant to encourage designers to focus on "people" in product design.

This also gives industrial designers another direction for product design. Since the display of a product affects a person's feelings, Product Semantics can be useful for discovering consumers' affections, sentience and symbols. Furthermore, it discusses how to think to make a proper message communication while using a product [1]. While encouraging an increased psychological need for a product, in addition to its core function, the product should also communicate culture essence, meaning and circumstances of usage, and various symbols.

From the foregoing statements, one can conclude that Product Semantics is the study of existing symbols and their meaning under regular circumstances. The designer also needs to consider individual users' varying physical and physiological functions, as well as various mental, social and cultural phenomena. Recent research has sought to determine whether "Product Semantics" is a suitable criterion for product design. The answer should be positive; not a stiff rule but a flexible principle under the consideration of "humanness".

2 Review of Literature

2.1 The Origin of Product Semantics

In 1983, Prof. Klaus Krippendorf and Prof. R. Butter first launched the concept of "Product Semantics" which they defined as "Research on the symbols of usage under different styles of people's creation, and use the idea in design." In 1984 Krippendorf further defined Product Semantics as an awakening from the old way. Products, he said, should represent more than just physical functions. Products themselves should show or imply to consumers how they are to be operated. In addition, products should carry symbolic meanings and should be able to blend into people's lives.

The theory of Product Semantics comes from language symbol system. Semiotics is the practice of using symbols in scientific and phenomenological research. Saussure, the father of Semiotics, said "Language is a semiotic system to express ideas." He believes each symbol has at least two meanings. A symbol is a "Signifier"-a symbolic model of an object. In addition, a symbol is also "Signified", communicating the meaning behind the symbol, which can be an idea, cultural essence or other symbolism. Professor M. McCoy at the Cranbrook Academy of Art in America considers the following five features of product semantic design when evaluating product styling:

1. Environmental Context: Forms, sizes, material and colors of a product should be in symmetry with its natural environment and social circumstances.
2. Memory: When a new product is launched, it doesn't necessarily need a brand new modeling language, but should seek to use memories people share in old images instead. Product consonance can be created by the use of familiar sequences and messages.
3. Operation: In order to properly guide consumers through standard operation of the product, operating instructions should be clear and easy to understand. They should thoroughly describe the control, display, appearance, material, and color of the product, and should explain the function and relationship of every control key, switch, and operating device.
4. Process: Semantic design is not about making products mysterious. On the contrary, the invisible inner operation needs to be proclaimed through outward display. Even though there are many things the realms of new technology that one cannot witness directly, designers need to interpret the inner operation of a product for the consumer. Not only must the designer communicate the technical function of a product, but he should also encourage the consumer to use his heart or intuition to envision its operation.
5. Ritual of Use: Simple styles can satisfy the needs of simplicity and efficiency in our daily lives. On a ceremonial occasion, for example, a product can be given an alternative design interpretation. A product can be made solemn, warm, cold or rational in order to complete a scenario design and foster the desired psychological interaction between the object and its user.

2.2 The Definition of Product Semantics

The origin of Product Semantics in industrial design can be traced to the 1960s. Semantics is literally the study of the meaning of words. Semantics is the science of exploring and researching meanings of languages. "Product Semantics" is a concept borrowed from linguistics; coming originally from semiotics. However there is a social, historical, and philosophical background behind its existence.

Product Semantics is the study of the images and meanings of a product. A product's style, color, texture, function or context form the meaning and concept that the product seeks to convey. Just as semantics is the study of the meaning of language in general, Product Semantics is the study of Product Language. The structure of Product Semantics theory comes originally from "the use of symbols " from Hochschule für Gestaltung, in Ulm, Germany. To trace its history back even farther, it is related to the "theory of marks" from Charles and Morris at the New Bauhaus School of Design in Chicago. The Product Semantics concept was launched in 1983 by K. Krippendorf in America and R. Butter in Germany, and was defined by Cranbrook Academy of Art in America in a seminar of Product Semantics held by IDSA as follows: "Product Semantics is research into the interaction between the operation and the message that is carried by a man-made object (Figure 1), and how we use that knowledge in industrial design ."[2]

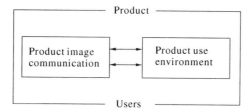

Fig. 1. The interactive relationship product messages and product use

2.3 Marks and Symbols of Product Semantics

Marks and symbols are signs for things, or are devices that can be used to distinguish status. There are numerous ways to show them including words, pictures, colors, images and actions, and psychological characteristics are created by images in people's minds, through a concept or impression. Signs and symbols are built through sensory perception, consciousness or knowledge, or through a sequence of psychological activities. By reflecting on an object, an idea or a thought can be associated with one of the object's features.

The theory of "Semantics" teaches us that people's knowledge or feelings are values or impressions on their minds created by sight, hearing, touching, memories or imagination which are received by them as signs and symbols in the first place. Product semantics is based on "signs". A product's appearance -its shape, color, texture, and materials-convey information to a user that enables him to see it and touch it accurately. Additionally, when it comes to the style of a product, designers should take into consideration the differences in cultures, customs or behaviors in order to find the true meaning of a sign (Figure 2).

Generally speaking, judging from inward and outward aspects of a product, there are a sequence of design considerations including functions, structures, shapes, materials, colors, interfaces, human factors... that must coalesce to form a product. The ultimate objective of a product is not only to satisfy its functional requirements, but also to take the user's psychological circumstances into account through the use of Product Semantic Signs and Symbols which communicate the meaning of the design by integrating semantic forms, colors, and textures.

Fig. 2. The basic component of Product Semantics is "signs"

3 Semantic Forms

There are two main concepts in the theory of Product Semantic Forms: One is the meaning of the form and the other is the message of the form. The analysis of the structure of semantics belongs to expressions of meanings. Messages communicated through Semantics help a designer understand the affects of a product's form on a user, and also help deliver the messages accurately. In product design, semantics is necessary for communicating symbolic messages. According to Pierce's theory, there are three circumstances in which the use of Product Semantics is appropriate as a means to convey a product's symbolic messages:

1. An accurate display of the meaning of a product can enhance the quality of the interface for a user.
2. Signs or symbols are used in communication between designers, and are a very helpful coordination element for users as well.
3. The ways to display product signs or symbols are not necessarily alike, but depend on the specific society, culture, and customs in which the product is marketed.

Product Semantics involves two main concepts: "The appearance of the meaning of a form" and "The conveyance of the message of a form." Analysis of the structure of semantics belongs to expressions of meanings. Messages should be conveyed through Semantics so the designer can understand the effects of a product's form on a user and can deliver the product's messages accurately. As illustrated in the following diagram, Product Semantics has three principal aspects, "product signs", "the appearance of a meaning" and "the conveyance of a message". (Figure 3)

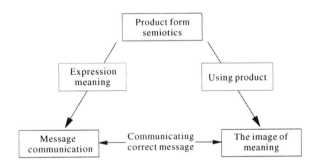

Fig. 3. Relations among signs, meanings and messages

In light of the foregoing discussion, the focus of our research is "the communication of product information". From there we extend the meaning and symbol of product forms, and are thereby able to distinguish the meaning of signs and symbols in practice. Moreover, according to the description of signs and symbols, we can also determine every model (forms, textures, lines, etc) of a sign in order to establish the existence of "Semantic Product Forms".[3]

4 Semantic Colors

The expression of color is the most abstract part of Product Semantics. Color is a bridge between the meaning of a product and the consumer's emotions. People have strong, direct feelings and impressions toward colors. In product design, colors are not only beautiful and decorative, but they also carry symbolic meanings. As the primary visual factor for appreciation of beauty, colors deeply affect our visual experience and emotions. Colors play an important role in forming a conception. Colors and styles that are in symmetry in a single design bring meaning and life to a product. Semantic colors in product forms come from people's visual and physical reactions to colors. This process allows users to connect their own experience with a product, and to discover meanings that a product carries.

Colors in a product design are expressed by Hue, Value and Saturation and their relationships. Different colors and combinations can arouse different emotions: red symbolizes passion and vitality, blue represents distance, calm and tranquility, purple means mystery, white represents simplicity, black shows solemnity, nobility and old age, grey is modest, etc. Every color carries different emotions and represents different symbolic meanings.

The careful use of colors in product design can not only display the meanings of functions, and ways of operation, but can also make users feel more comfortable. For example, a red button can indicate an emergency switch, a green button generally means safe for operation, and a yellow button can be used to signal precaution. Shapes and colors of a product carry meanings as well. For example, black and silver surfaces look edgy and solemn. That is why many 3C products are in silver-gray or black, although some are combined with yellow, red, or green to add a sense of fashion. If products such as purses or shoes are rendered in red, pink, or other warm colors, they will carry a graceful female taste. White is a very classic color. The Macintosh computer from Apple is white, which carries a vivid emotional and physical character and has become the image for Mac. MUJI, on the other hand, is based on a simple, natural design.

5 Semantic Materials

Semantic Materials is a branch of Product Semantics that deals with material properties, textures and tissues... compounds that deliver messages. In addition to styles, signs or symbols, lines, etc, the meaning of a product is received through being seen and touched by a user. When designers choose materials, the first consideration is function; for example, intensity, wear resistance, and process ability, as well as possible user-defined requirements. Textures and tissues of materials give the user a visual and physical experience with a product that also generates emotions and certain symbolic meanings. Therefore, designers need to take people into consideration when choosing materials. Textures and tissues are forms of art, and the right material enhances emotional feelings and shortens the distance between products and people.

Different textures bring out different emotions. Glass and steel, for example, are edgy and modern, as in a SONY laptop computer, where the surface is made of clear compound metal to give a sense of transforming technology. By contrast, wood and bamboo, are natural, relaxing and down-to-earth. Wraps used by MUJI are in a wood pattern, giving a sense of nature and seeming more environment-friendly. Patterns (lines, arrangements and setups) are also signs and symbols that can help users understand the meaning of product functions. For example, several thin lines are used to prevent slipping, a rough surface is designed for holding, and a dented part invites the user to press down. Texture of materials and performance characteristics of tissues affect the ultimate visual result of a product.

6 The Display of Signs in Semantic Product Forms

An analysis based on Semantic theories from Peirce, Sanssure and Morris will still need to be proven in practice. It's based on a designer's understanding of Semantics that shows the meaning a product carries (Figure 4). Therefore, we need to know how to apply Semantics to explain values of styles of products, so designers should view a consumer's questions as information.

Product semantics is about communication which means every message that a product can deliver to a user.[4] The message here is an application of signs. Generally consumers express their inner thoughts into outward behaviors, then eventually an understanding. For example, the button on a toaster is often a long, vertical gap which people associate with the idea of pressing down. For example, if users can tell a button is for turning on and off from its shape, texture, color, then guide a user how to operate by its special pattern, and through signs or colors let a user make various adjustments such as turning the volume up, down or changing the . A user can get used to all of the operation through this kind of message convey. As for the product which but a symbolic meaning, it's a practical simulation, a clear display. If we apply icon, index and symbol three signs from Peirce's Semantics into product design, it will create a sequent of solution as in Figure 5.

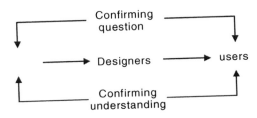

Fig. 4. The transformation of positive product meaning

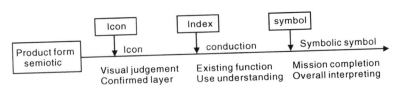

Fig. 5. Application of Perice's Semantics (adapted by the author)

Table 1. The meaning of product semiotics (adapted by the author)

Form	Meaning	Material	Meaning	Color	Meaning
Circle	Rotatable	Trace	Preventing slides	Red	Warming
Square	Can be pushed	Coarse	Technology	Black	High quality
Spiral	Adjustable	Glossy	Fashion	Gray	Male
Concavity	Can be pressed	Tiny holes	Exhausted	Green	Environmental protection

Now we can find out that product signs are based on people's transition from cognitive behaviors into the message of the need of operation, the efficient convey is no more than "interpretation of meanings." As for an appearance of a product, it gives people some visual signs, theses signs are special "languages" that a product carries, and these languages affect people's opinions of the product. Symmetrical Cartesian geometries shows the solemnity of a structure and creates a peaceful, graceful and cheerful atmosphere. Solemn but lively circles have a sense of inclusion which is good for creating a satisfactory and harmonious atmosphere. Curves represent movements and vitality and can be used to create a enthusiastic, free and friendly atmosphere. Free curves are close to the nature and full of life and can be used to create a simple, natural and environment friendly atmosphere. These signs are attached to a product whether a designer put them there intentionally or not and affect a user to feel positive or negative.[5]There are three most common product signs which are shapes, textures and colors. The meanings are categorized as the following Table 1.

7   The Study of Cases of Product Design

The research is to study various signs and see if the messages that a product carries can guide a user to operate successfully and allow a user to achieve the goal of efficient product operation. And how do we apply Product semantics into product design to show the meaning of a product or its symbolic meaning is the key of making a product alive. Therefore, in order to show that the theory of sign and semantic product theory are helpful to making a product easier to understand and closer to consumers' needs. In the research, the concept is a flash light design along with the theory of sign and Product semantics. Case 1: The object is a tall triangular prism, there is a small cylinder and light on the top. On the edge, there is a small triangle switch facing up. This kind of design implies a force pushing towards the tip and is able to switch the light on. This design is a kind of expression of symbol as long as how we find out the meaning of an arrow (Figure 6).

Case 2: The object is a cone, the head is big and the end is small, the light bulb is at the big end, the light at the end is expending in a cone shape, and the handle is triangle which implies pressing down with a palm to turn on the light. A design like this is in a geometry form which has a convergent effect and the handle can spread up (Figure 7).

Fig. 6. Design of a triangular prism flash light

Fig. 7. Design of a corn shape flash light

Fig. 8. Design of a triangular prism shape flash light

Case 3: In the middle place, a small circle attached to a square and a triangle on the both sides, then from the square and triangle extends a big round line, a circle means being able to spin around this small object, and also able to adjust the relative position of the square and the triangle. And about the shapes, a square means a bottom, a triangle means gathering and releasing the light, and using yellow to mark where the light is (Figure 8).

After the discussion of the three cases above, we can see whether shapes, colors or signs all have to be placed at proper parts of a product and each sign has to be easy to distinguish and recognize by a user.

8 Conclusion

Overall the develop of the theory of product design in this research is based on three rules. One is a design of outward appearances according to the needs of functions and also for covering inner components. Another one is about ergonomics for operation emphasizes on body sizes and a design which is based on operation needs. The last one is based on the technology today and its manufacturing skills or needs of material properties; it's a tendency of quantity and modeling focused. All the above come with limits, they are product but people orientated. In today's society, our living environment has become focused on "people" instead of "objects". Therefore the main focus of design is humanism[6], and the style of the product should be focused on communication with "people".

Through the study in the article we can experience various product designs with mental knowledge. Can we use a mark to form a product? And being able to present a clear image of a product's spirit. Signs can guide a user through operation smoothly, and that is the goal designers all share. Hopefully the result of this research will help to clarify "the meaning of product signs" and its meaning and value, and being able to use it in designs of product forms, colors and textures. Therefore, when it comes to product design, we should use semantic signs to the best according to a user's habits and experiences, that is the goal of product design. The main focus of this study is how to use signs and symbols in product design in a proper manner.

References

1. Lin, C.H.: Design Theory and Value. Garden City Publish, Taipei (2001)
2. Marx, P., Wartburg, Elizabeth, Hsu (Translate): The Design Application of the Mass Semiotics. China Social Science Publish, Beijing (1992)
3. The semiotics theory is judging "the form" to express the significance, The semiotics theory is using the symbol to display the design significance. such an expression way belongs to the function and the spirit
4. Languish, J., Lin, S.H.: Objects and Images, Product Semantics any use?, pp. 132–133. University of Industrial Arts, Helsinki (1992)
5. Ho, C.-Y., Shih, R.: On the Relation Between Product Semantics and Semiotics, p. 6. University of Technology, Taipei (1997)
6. Kuang, C.L.: Contrasting Designing, p. 23. The Artist Publish, Taipei (1995)

Adapting User Interfaces by Analyzing Data Characteristics for Determining Adequate Visualizations

Kawa Nazemi, Dirk Burkhardt, Alexander Praetorius,
Matthias Breyer, and Arjan Kuijper

Fraunhofer Institute for Computer Graphics Research, Fraunhoferstraße 5,
64283 Darmstadt, Germany
{kawa.nazemi,dirk.burkhardt,alexander.praetorius,
matthias.breyer,arjan.kuijper}@igd.fraunhofer.de

Abstract. Today the information visualization takes in an important position, because it is required in nearly every context where large databases have to be visualized. For this challenge new approaches are needed to allow the user an adequate access to these data. Static visualizations are only able to show the data without any support to the users, which is the reason for the accomplished researches to adaptive user-interfaces, in particular for adaptive visualizations. By these approaches the visualizations were adapted to the users' behavior, so that graphical primitives were change to support a user e.g. by highlighting user-specific entities, which seems relevant for a user. This approach is commonly used, but it is limited on changes for just a single visualization. Modern heterogeneous data providing different kinds of aspects, which modern visualizations try to regard, but therefore a user often needs more than a single visualization for making an information retrieval. In this paper we describe a concept for adapting the user-interface by selecting visualizations in dependence to automatically generated data characteristics. So visualizations will be chosen, which are fitting well to the generated characteristics. Finally the user gets an aquatically arranged set of visualizations as initial point of his interaction through the data.

Keywords: Adaptive Visualizations, Human-Centered Interfaces, Human-Computer-Interfaces, Semantics Visualization.

1 Introduction

Nowadays data are stored in large and complex knowledge bases like digital libraries. Besides public accessible digital libraries, companies store information of project results, employee contracts, development and research results, etc. in knowledge bases. Often these databases and libraries consist of semantic data sources, so that data entries are stored in structured form and are related with each other. This form of structured and linked data allows new approaches for exploring and searching information. In the past the focus in the research community lays on defining adequate structures or schemas and in visualizing this information mostly in form of graphs.

M. Kurosu (Ed.): Human Centered Design, HCII 2011, LNCS 6776, pp. 566–575, 2011.

From the research field of information visualization also other kinds of visualization are known e.g. Treemap, Sunburst visualizations [12] or Timelines [11]. For supporting data exploration for users, different visualizations are given and selectable, so that a user is able to use his preferred visualization technique. So a user can choose these visualizations, with which he can finish his tasks efficiently. Such a self-defined set of visualizations which are linked with each other were introduced in our previous works as "Knowledge cockpit" [10]. A Knowledge Cockpit provides users many advantages in exploring and searching the underlying data. The visualizations are often aspect-oriented and useful for special operations like navigating through hierarchies or giving an overview about complex link structures between nodes.

For supporting the user, adaptive user interfaces are an alternative way to adapt visualizations to the user's behavior (like the presented system in [10]. The adaption bases on analyzing the previous user interaction or the analysis of summarized interactions of a user group, for instance by using group-based recommender systems. These adaptive systems are designed for giving a direct support to the user like highlighting relevant nodes or sorting elements by relevance. Another approach of adapting information visualizations is the analysis of the underlying dataset. But currently merely limited systems are available which supports the adaption on data structures. Often the data adaptation is only used for trivial adaptation within a single visualization, like turning relation labels on or off. The scientific visualization community focuses on adapting single visualizations to the user's needs and preferences or his task context and goals. But the analysis of the underlying data is an often neglected topic.

In this paper we describe a concept for a system that analyzes the data to extract specific characteristics, which can be utilized for optimizing the visualization. For this approach significant data characteristics were previously identified and specified which automatically can be detected by a data analysis system. It is important that specific characteristics can be automatically extracted, because the data that should be visualized can be vary in its characteristics, especially by using digital libraries or knowledge bases. Besides the data, also characteristics of visualizations will be identified visualizing various and possible data characteristics. Examples could be how the data schema with its structure can be visualized or how efficiently the layout-placement in a single visualization is using the given space. Based on these determined characteristics beside the underlying data and beside the visualization, we developed another system and a new algorithm to ascertain visualizations, which provides adequate information visualization for the given dataset. In a complete scenario the user is able to make a search request to a given semantic knowledge base, where the server generates a response. The adaption system analyses the resulting data for its characteristics and determines optimal visualizations techniques which fit to this data. So the user gets a pre-initialized knowledge cockpit with optimally selected visualization techniques. After pre-initializing the knowledge cockpit, the user is able to intervene and change the set of activated visualizations by closing or opening other visualizations – related to the goals and tasks of the user. For our system we are using different aspect oriented visualizations, which are intuitive and easy to use. We evaluated that a suitable determination of adequate visualizations and the combination of intuitive useable visualization at an initial interaction point will guaranty a general intuitive and easy to use data exploration.

As already mentioned in various publications [1][2][3], many different visualization methods have been proposed in the past. They significantly differ in several aspects, including the supported tasks, the target domain and the type of data which is being visualized by the methods [4].

This suggests that each visualization method may have different strength and weaknesses in context of different user tasks and different properties of a dataset, but also in context of different system configurations.

2   Related Works

There are not many works that are concerned with the approach of calculating the best possible visualization method from a given dataset alone. In fact, there are always many aspects of the given circumstances, which determine the best visualization method for a given dataset.

A good visualization system is based on a model that takes into account the above mentioned circumstances to figure out the best possible visualization method. Golemati et. al. [4] uses such a model, which is based on a feature profile for each visualization method and matches it against a computed property list in order to choose the best possible visualization method within a given information retrieval task context. This information retrieval task context consists of three different kinds of contexts that have been identified and which will determine the best visualization method for a given dataset. These contexts are a document collection context, a user context and a system context.

The user context consists of a dynamic and a static user profile, where the dynamic profile is updated after every information retrieval session and contains a user specific preference database with the likes and dislikes of single visualization methods. The static user profile contains user properties, such as the user's education, his information retrieval knowledge, his age, gender and profession, his abilities and stuff like that. The system context consists of properties about input devices, output devices and other hardware equipment. Finally there is the document collection context, which contains information about the origin of the dataset, whether it comes from a query or if it was selected from branches of a categorization. It also contains metadata, like the author and title of the document, given keywords or the department of issue and it contains information about the format of the document and whether it is full text, image, manuscript only or something else.

The feature profile of a visualization method consists of properties that give information about the numbers of visualization dimensions, the visualization metaphor used by the visualization method, whether interactive browsing is supported and for which kind of documents, if color coding is supported of term frequency can be shown.

Golemati et. al. [4] constructed a rule database containing rules with the following format: (context-property, vis-method-property, score), where context-property is a property from the information retrieval task context, that means the user, system or document collection context, vis-method-property is a property of a visualization method and score is a numeric metric in the range (-10,10) expressing how appropriate visualization methods having the specific vis-method-property are considered for contexts where the particular context-property holds.

While this is a valuable approach, we found that many custom rules are necessary to put this concept into practice. For every new visualization method that gets developed you not only need to define its visual properties, you also need a lot of new rules being created, so that the system can calculate whether the visualization might be appropriate or not.

Additionally, it is claimed by Ahlberg et. al. [5] that successful visualization systems do not depend on a single powerful visualization method that might be computed, but rather on a whole set of visualization methods, appropriate for various tasks and data types.

So we decided that it will be necessary, not only to compute a set of visualizations instead of a single best visualization, but it also might be possible to use context properties and visualization method properties alone to compute which set of visualization methods might be the best one. This would free visualization system designers from the need of creating a lot of custom rules.

3 Concept for an Adaptive User-Interface

In accordance to concept of Ahlberg et. al. [5], our procedure chooses a final set of visualization methods from all the visualization methods available within the visualization system.

In this section, we describe the procedure of selecting a set of appropriate visualization methods on the basis of what is described as collection context by Golemati et. al. [4]. They also describe a user context and a system context. In our paper, the system context is not taken into account, but our concept could easily be extended to do so. The user context is reflected in the use of priorities for each aspect of the collection context, which we will call semantic dataset. In order to describe the procedure of selecting the appropriate visualization methods and add them to a final set of visualization methods, we start with several definitions that will help us to do so.

3.1 The Definitions

We first define a visualization method (vm) as a set of visualization properties (vp). Second, we define a manual-visual-criteria-test (MVCT) as a tripel, which contains a brief description of data-properties (dp) a visualization method might be able to visualize and one or several examples of visualization methods, which are rated with a certain value using the MVCT, so it becomes possible to use the MVCT to rate any new visualization method in a comparable fashion to the examples stated within the examples of the MVCT. The output of an MVCT is called a visualization property (vp) of visualization method and its value is always within the range (0,1), where 0 means that the visualization method does not support the visualization of the corresponding data property and 1 means the visualization of the corresponding data property is perfectly well supported.

There are two different kinds of MVCT's. The first describes generic properties that are available within all semantic datasets to some degree (e.g. amount of instances). The second kind of MVCT's describes visualization method specific properties, which visualize data properties that are not part of every semantic dataset (e.g. aspect of time).

Next, we define data-criteria-tests (DCT's) which are used to gain information about a semantic data structure like rdf-based data. Each DCT contains a well-defined range of values and an algorithm that calculates a value of that range from the semantic data structure.

Now we define the already above mentioned data-property (dp) as a set of DCT's and a formula which combines the results of the DCT's to a value within the range of (0,1), where 0 means that the data property is available within the semantic data structure that is being analyzed and 1 means the data-property is perfectly reflected within the semantic data structure (e.g. a perfectly balanced tree). It now becomes possible to calculate a semantic dataset (sd) as a set of data-properties from the underlying semantic data structure. Note, that the brief description of a data property as used in MVCT's needs to explain the formula of the data property.

Finally we define a mapping function which consumes 3 parameters: a visualization method (vm), a semantic dataset (sd) and a priority vector (pv). It calculates the distance between the semantic dataset and the visualization method by applying the Euclidean norm.

$$distance\left(\vec{sd}, \vec{vm}, \vec{pv}\right) = \sqrt{(dp_1 - vp_1)^2 * p_1 + \ldots + (dp_n - vp_n)^2 * p_n}$$

The priority vector contains the importance of each data-property in the user context (UC). In the context of this paper, all the priorities contained in the priority vector are assumed to be equal, because additional research is needed to define a user context that allows the automated derivation of priorities for the use with the mapping function.

3.2 The Adaption Procedure

Now that we have made the definitions above, it becomes possible to describe the procedure of adapting user interfaces by analyzing data characteristics for determining adequate visualizations. A mandatory requirement is that all the visualization methods have already been rated by visualization method experts through the use of the MVCTs.

The above mentioned procedure starts by extracting the list of available visualization methods and for each visualization method the set of visualization properties. In the next step, all the DCT's are applied to the underlying semantic data structure. These results in a set of data properties, which form the semantic dataset that we need to continue the procedure with the appliance of the mapping function. We give the calculated semantic dataset to the mapping function and run it once for each available visualization method to calculate the Euclidean distance between each visualization method and the semantic dataset.

Afterwards, the visualization methods are sorted by increasing distance to the semantic dataset. The last step of the procedure is to repeat adding the visualization method with the lowest distance to the semantic dataset to the final set of visualization methods, until a predefined maximum of visualization methods has been added to the final set.

Note that using this procedure, two similar visualization methods with both having a short distance, might introduce redundant visualizations into the final set, while another good visualization method with a slightly greater distance might be eliminated from the final set. This might be avoidable by thinking of the visualization methods as vectors within a supported visual feature space. It then becomes possible to computing the angle between the visualization method that has been added recently to the final set and all the other visualization methods that are not yet within the final set. A visualization method with a small angle could be considered very similar to a method that might already be contained within the final set. A big angle lets us assume that the corresponding visualization method is quite different from the ones already added to the set. This is why the angles could be used to rate the distances of the remaining visualization methods in such a way, that the visualization methods with small angles get their distances increased and visualization methods with big angles get their distances decreased, because what really is important is not the distance between the semantic dataset and a visualization method, but the utility of it. If the final set would only consist of one visualization method, the visualization method with the smallest distance would be the one with the maximum utility, but because we compute a set of visualization methods, we need to take into account the supported features of the visualization methods that already have been added to the final set.

3.3 About Visualization Properties and Data Properties

Data-visual properties are the core of our concept. For each data property, there is a corresponding visual property. The visual aspect of the property is measured by a visualization expert through the application of an MVCT to a visualization method. The data aspect of the property is computed through the application of a set of DCT's on a semantic data structure and the combination of their results. We have identified several properties, but research is still in progress and we hope to identify more properties and to develop a taxonomy for these properties. To accomplish our task, we started by analyzing different kinds of existing visualizations and tried to describe in detail what their strength and weaknesses were and what kind of data they can display. We drew inspiration from Katifori et. al. [6] and Landesberger et. al. [7]. So with time, we hope to develop a list of data-visual properties that are displayable by the ever growing pool of visualization methods which have be analyzed and for which DCT's need to be developed.

Table 1 shows the current state of our research. To understand the table we explain some of the terms used in the table. An Element can be a Concept, Instance or Relation. The sum of all Elements represents the data structure. An Object can be an Instance or Concept. VM stands for Visualization Method. You also should know that there are general properties (marked green) which we measured within any semantic data structure, and there are aspect oriented properties (marked orange) that need a specific kind of concept within the semantic data structure, so that it can be measured. Neither the list of general nor the list of aspect oriented properties is a complete list.

The description of the data-visual properties contain a short description of the visual aspect preceded by (V) and a short description of the data aspect preceded by (D).

Table 1. Overview to data/visual properties for graphical user-interfaces

DATA/VISUAL PROPERTIES	Description
Overview of Objects	V: How many Objects at once are shown? D: Count amount of Objects and map them to "size classes" according to [6]
Overview of Instances	V: How many Instances at once are shown? D: Count amount of Instances and map them to "size classes" according to [6].
Overview of Concepts	How many Concepts at once are shown? D: Count amount of Concepts and map them to "size classes" according to [6].
Overview of Relations	V: How many Relations at once are shown? D: Count amount of Relations and map them to "size classes" according to [6].
Handles Relation Crossings	V: How good are relation crossings handled? D: Use heuristics for Crossing number from graph theory (crossing number 0 = planarity)
Handles Dense Graphs	V: Can dense graphs be visualized? D: Measure density of graph as mentioned in [7].
Shows Hierarchy	V: How well does the VM show a hierarchy? D: Root concept with sub-concept exists? How many cycles exist within the data structure?
Shows Depth of Hierarchy	V: How good is the depth of a hierarchy shown? D: Find longest path from root concept to a leaf concept
Shows Width of Hierarchy	V: How good is the width of a hierarchy shown? D: Count concepts on level with most concepts
Shows Balance of Hierarchy	V: Is the size of branches of hierarchy well shown? D: calculate ratio of concepts in branches (e.g. variance of incident relations)
Shows Level of Hierarchy	V: Is it easy to see all Objects of a hierarchy level? D: count levels of hierarchy and weight them with the amount of concepts within the levels
Shows Multiple Parents	V: How well can multiple parents ob Obj. be shown? D: Measure amount of objects with multiple parents
Shows Multiple Hierarchies	V: Can the VM show multiple hierarchies/roles? D: Multiple parents of objects form different hierarchy trees
Overview of Elements	V: How well are complex graphs be shown? D: Count amount of elements and weight it with the distinctiveness of hierarchies
Overview of Scale Free Network	V: Are scale free networks visualized well? D: Does semantic structure show a power law degree distribution? [8]
Overview of Small World Graphs	V: Are small world graphs visualized well? D: Is average path length $L = \log(objects)$?
Overview of Bipartite Graphs	V: Are bipartite graphs visualized well? D: Measure whether odd cycles exist
Overview of Concept Lattice	V: Are concept lattices visualized well? D: See [9]
Show Instance Clusters	V: How easy is it to detect clusters of instances? D: use cluster analysis
Show Class Clusters	V: How easy is it to detect clusters of classes? D: use cluster analysis
Show Class with Instance Clusters	V: How easy is it to detect classes with many inst.? D: Use cluster analysis (e.g. find Concepts with many Instances)
Show Instance with Classes Clusters	V: How easy is it to detect inst. with many classes? D: Use cluster analysis (e.g. find instances having many parent classes)

Table 1. (*continued*)

Show Properties of Elem./Obj./Inst./Conc./ Rel.	V: How good are properties shown in general? D: Count amount of existing properties
Show Labels of Elem./Obj./Inst./Conc./ Rel.	V: Is it possible to see labels? Are they readable? D: Count amount of labels and weight them with their length
Overview of similar Elem./Obj./Inst./Conc./ Rel.	V: How easy is it to detect similar elements? D: Measure how widespread objects with similar attributes are
Average showable amount of Properties in Elem.	V: How many dimensions are shown at once when many elements are visible? D: Whats the average amount of properties of an object
Max showable amount of Properties on Elem.	V: How many properties can be shown when one element is focused? D: How many properties has the object with the most?
Average showable Pathlengths	V: What's the length of paths that can be shown when many elements are visible? D: Calculate the average path length
Max showable Pathlengths	V: What's the longest visible path that can be seen when focused? D: Calculate the longest path in semantic data structure
Overview of Evolution of Elements	V: Is it possible to view the history of change of the semantic data structure? D: count the amount of elements with history information and weight it with the amount of history available
Show Time Aspect	V: Is it possible to show elements that have a time aspect in a meaningful way? (e.g. timeline) D: calculate ratio between elements with time aspect and total elements
Show Geographical Aspect	V: Is it possible to show elem. that have a geographic aspect in a meaningful way? (e.g. geographical map) D: calculate ratio between elements with geographic aspect and total elements

4 Case Study: Adaption in Search Context

A representative example for using an adaptive system, which bases on the idea of selecting visualizations by the presented data, is in context of searching the web. Typically a user is typing some keywords and finally he gets visualized the available results. In currently existing search engines, the results are presented mostly in lists, but there also some first implementations, which are using graphical forms of visualization, often in form of graphs.

For another form of searching information, we used our semantics visualization framework for providing a user interface for searching within the free available free-base[1] data source and from a partner the Alexandria data source. Both data sources consisting of a kind of semantics data structure, but it is not so strong defined like it is by ontologies like OWL. Furthermore, both data sources are different in structure and also from the contained data.

In this use case, the user enters keywords or a phrase to search for entities and our conceptualized and implemented adaption engine generates the adequate fitting visualization in dependence to the results and enables them to the user. Therefor the adaption engines generates characteristics from the result dataset and tries to select visualization that are fitting best to the determined result set of data entities (an example is presented in Fig. 1).

---

[1] http://www.freebase.com

Fig. 1. A performed search for "George Orwell" in different data sources with different data characteristics and in consequence with different automatically selected visualizations

In the further interaction the user has always the possibility for selecting and deselecting the visualization, to support the feature of orchestrating an individuated so called Knowledge Cockpit [10], which also allows a user-centered search interaction.

5 Conclusion

The paper introduced into a concept for adapting user interface by selecting appropriated visualization in dependence of a given result set consisting of data entities. In difference to existing approaches, we do not focus on adapting a single visualization, which is today the most often used approach for adapting graphical user interfaces. By using complex and large databases, the advantage of our approach is the ability of using different kinds of aspect-oriented visualizations, which are coupled with each other for supporting an interactive exploration approach. This also supports the general process of information retrieval.

To support a data-driven user-interface adaption, we also specify characteristics which can be found on the data side and also on the visualizations side. By a manually evaluation of every available visualization and by automatically evaluating of the data set, which should be visualized e.g. a result set from a search request, we also described an approach for an algorithm to determine possibly fitting visualizations. So the users get an initial set of visualizations that are showing the result set or in general the data in an optimal form. During the further interaction, the user is always able to change the visualization by preferred kinds of visualizations.

By using this concept for instance during a search process, the user will be supported by finding required information faster. By coupling this idea for adapting user-interface with adaption methods for adapting single visualizations, a complete new form of interaction can be achieved, where a user gets information very efficiently.

Acknowledgements. This work was supported in part by the German Federal Ministry of Economics and Technology as part of the THESEUS Research Program. For more information please see http://www.semavis.com

References

1. Card, S.K., Mackinlay, J.D., Shneiderman, B.: Readings in Information Visualization: Using Vision to Think. In: Readings in Information Visualization: Using Vision to Think, Morgan-Kaufmann, San Francisco (1999)
2. Chi, E.H.: A Taxonomy of Visualization Techniques using the Data State Reference Model. In: Proceedings of the IEEE Symposium on Information Visualization (2000)
3. Shneiderman, B.: The Eyes Have It: A Task by Data Type Taxonomy for Information Visualization. In: Proceedings Visual Languages (1996)
4. Golemati, M., Halatsis, C., Vasilakis, C., Katifori, A., Lepouras, G.: A Context-Based Adaptive Visualization Environment. In: Proceedings of the Information Visualization. IEEE Computer Society, Los Alamitos (2006)
5. Ahlberg, C., Wistrand, E.: IVEE: An Information Visualization & Exploration Environment. In: Proceedings of IEEE Viz (1999)
6. Katifori, A., Halatsis, C., Lepouras, G., Vassilakis, C., Giannopoulou, E.: Ontology Visualization Methods - A Survey. ACM Computing Surveys 39, Article 10 (2007)
7. von Landesberger, T., Kuijper, A., Schreck, T., Kohlhammer, J., Wijk, J.J., van Fekete, J.D., Fellner, D.W.: Visual Analysis of Large Graphs. In: Conference of EUROGRAPHICS (2010)
8. Barabasi, A.L., Bonabeau, E.: Scale-Free Networks. Journal of Scientific America (2003)
9. Yevtushenko, S.: Computing and Visualizing Concept Lattices. Thesis, TU Darmstadt, Darmstadt (2004)
10. Nazemi, K., Burkhardt, D., Breyer, M., Stab, C., Fellner, D.W.: Semantic Visualization Cockpit: Adaptable Composition of Semantics-Visualization Techniques for Knowledge-Exploration. In: International Conference Interactive Computer Aided Learning 2010 (ICL 2010), pp. 163–173. University Press, Kassel (2010)
11. Stab, C., Nazemi, K., Fellner, D.W.: SemaTime - Timeline Visualization of Time-Dependent Relations and Semantics. In: Bebis, G., Boyle, R., Parvin, B., Koracin, D., Chung, R., Hammound, R., Hussain, M., Kar-Han, T., Crawfis, R., Thalmann, D., Kao, D., Avila, L. (eds.) ISVC 2010. LNCS, vol. 6455, pp. 514–523. Springer, Heidelberg (2010)
12. Stab, C., Breyer, M., Nazemi, K., Burkhardt, D., Hofmann, C., Fellner, D.W.: SemaSun: Visualization of Semantic Knowledge based on an improved Sunburst Visualization Metaphor. In: Proceedings of ED-Media 2010: World Conference on Educational Multimedia, Hypermedia & Telecommunications, Toronto (2010)

User-Oriented Graph Visualization Taxonomy: A Data-Oriented Examination of Visual Features

Kawa Nazemi, Matthias Breyer, and Arjan Kuijper

Fraunhofer Institute for Computer Graphics Research,
Fraunhofer Str. 5, 64283 Darmstadt, Germany
{kawa.nazemi,matthias.breyer,arjan.kuijper}@igd.fraunhofer.de

Abstract. Presenting information in a user-oriented way has a significant impact on the success and comprehensibility of data visualizations. In order to correctly and comprehensibly visualize data in a user-oriented way data specific aspects have to be considered. Furthermore, user-oriented perception characteristics are decisive for the fast and proper interpretation of the visualized data. In this paper we present a taxonomy for graph visualization techniques. On the one hand it provides the user-oriented identification of applicable visual features for given data to be visualized. On the other hand the set of visualization techniques is enclosed which supports these identified visual features. Thus, the taxonomy supports the development of user-oriented visualizations by examination of data to obtain a beneficial association of data to visual features.

Keywords: graph visualization taxonomy, user-oriented visualization, visual features.

1 Introduction

Information Visualization aims to provide visualization techniques to present data in an efficient and effective way [11]. As main objective the comprehensive preprocessing of the data can be named, so users can immediately gather information and interact with it to identify relevant aspects. The evolved visualization techniques in the Information Visualization community utilize different visualization metaphors to present the data and impart implied data characteristics to the user. These metaphors, in turn, utilize different visual features, like color, order, size, shape, etc., to gain their metaphoric expressiveness.

The purposeful usage of visual features depends on two factors: On the one hand an *appropriate mapping* of data characteristics to visual features is crucial. Visual features vary in their suitability for presenting and imparting specific data, which depends on the kind of given data characteristics. On the other hand visual features are *interpreted* by users in different ways, thus the resulting mental map in mind of the user may strongly differ from the intention the visualization designer aspired. As a matter of fact different visualization techniques support diverse sets of visual features. Thus, the varying support for visual features and the varying suitability of visual features for specific data characteristics leads to the following, scarcely contradicted statement: different visualization techniques differ in their suitability for presenting different data characteristics.

M. Kurosu (Ed.): Human Centered Design, HCII 2011, LNCS 6776, pp. 576–585, 2011.

In this paper we present a new user-oriented taxonomy of graph visualization techniques based on the data-oriented analysis of supported visual features. The taxonomy classifies the visualization techniques according to commonly supported *visual features*. Furthermore, the taxonomy implies a selection guide based on the given *data characteristics* for an appropriate mapping of data to visual features. The selection guide offers a sequence of the compatible visual feature according to their clearness and directness for the user. So the taxonomy can be used to identify appropriate graph visualization techniques depending on the data characteristics to be visualized.

The graph visualization taxonomy is the result of a three-parted process. In Section 2 the first process step dealing with identifying the visual variables is described. In the second process in Section 3 a survey of graph visualization techniques is presented which summarizes and arranges the visualization techniques into user oriented clusters. The outcomes of steps one and two are the input for process step three, described in Section 4. Here user oriented graph visualization taxonomy based on the visual features will be constructed. The last Section demonstrates how this taxonomy can be utilized to bridge the gap between data characteristics and human perception characteristics.

2 Fundamental Visual Features

To display information in an intuitive manner and thus impart knowledge to users visual features (or visual variables) are utilized. These visual features act like graphical metaphors which point out the information itself in a way which structures the visual communication. Each visual feature reflects perception characteristics which include regularities for cognitive processing [18]. To obtain an overview of visual features some related work is described here.

In the research field of Information Visualization visual features had been examined for a long time. The most prominent and fundamental works are these of Bertin [3] and Mackinlay [16]. In their observation they found out these visual features radically effect humans' perception of the displayed information. So they identified visual features and chunked them into reasonable groups.

2.1 Visual Variables

Bertin brought up the theory of visual variables in his work semiology of graphics [3]. This theory claims that these variables fundamentally impacts the way of perceiving graphically displayed information. Therefore he enlists the following visual variables: position, shape, orientation, color, motion, texture, value, and size where the variable value can be interpreted as luminance or grey tone to distinguish it from the variable color. These variables had been differentiated into the four categories associative, selective, ordered/ordering, and quantitative visual variables. In Fig. 1 these visual variables are graphically presented.

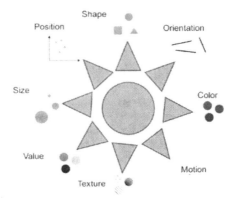

Fig. 1. Visual variables by Bertin [3], taken from [16]

2.2 Ranking of Perceptual Tasks for Visual Variables

Based on the work described in Section 2.1 Mackinlay extended these visual variables by length, angle, volume, connection, and containment [15]. Furthermore, he identified a ranking of perceptual tasks for the sets of quantitative, ordinal and nominal visual variables. These rankings are displayed in Fig. 2.

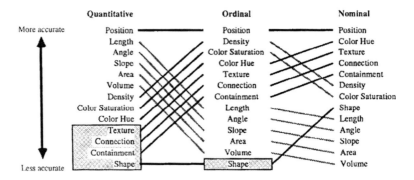

Fig. 2. Ranking of perceptual tasks for visual variables [15]

3 Graph Layout Techniques

For the hierarchy of visualization techniques two-dimensional graph-based layouts are examined. These are the most adequate visualization techniques for the standard user: he can easily interact with them and most accurately interpret the displayed data.

The identification of supplied visual variables in a graph visualization technique are examined according to these variables. Most existing graph technique surveys differentiate according to mathematical, behavioral, or perceptual (like visual clutter) aspects. For the approach presented in this paper a user-specific differentiation according to aspects of the visual result is required.

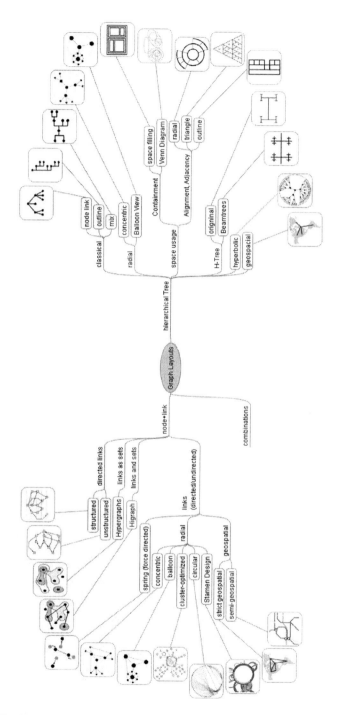

Fig. 3. User-oriented structure of graph layout visualization techniques

Therefore, existing surveys [4,5,8,10,12,17,19] were examined and restructured. The taxonomical differentiation is based on Keim [10,12] and is transformed to differentiate the visual result the user perceives [7,13]. To obtain a most complete picture of visualization techniques also some important works examining specialized techniques are regarded [1,2,6,9,14].

As a results of the visualization technique survey, the taxonomy (shown in Fig. 3) differentiates on the first level in i) tree layouts, ii) network layouts, and iii) combinations, which mainly depends on the given data structure to visualize. The second level differentiates predominantly according to the visual result the user perceives, which correlates with the set of supported visual variables.

4   User-Oriented Taxonomy of Graph Visualizations

In this Section the graph layout techniques (see Section 3) are analyzed according their support of visual features (Section 2.). The techniques are evaluated according the visual variables they support.

4.1   Support for Visual Features in Visualization Techniques

The evaluation of the support of specific visual features in the visualization techniques is presented in Fig. 4. If a feature is available in the visualization it is marked as 'X', if not it is marked as '–'. If a variable is included but not designated for displaying information but could principally be utilized for this it is marked as '(X)', if a variable is not available in a graph layout but could be integrated it is marked as '(-)'.

visual variable	hierarchical Tree								node+link								
	classical	radial	containment	alignment angle	alignment outline	H-Tree	hyperbolic	geospacial	directed links	links as sets	links and sets	spring	radial central node	radial cluster	radial circular	radial stamen	geospacial
Position	(x)	-	-	(x)	(x)	(x)	(x)	*-	x	x	x	(x)	-	x	-	-	*-
Length	x	-	-	-	-	x	x	*-	x	-	(x)	(x)	(-)	(-)	(-)	(-)	*-
Angle	(x)	(x)	-	x	-	-	(x)	-	x	-	(x)	(-)	(x)	(x)	(x)	(x)	-
Orientation	(-)	-	(x)	-	-	-	(x)	-	(x)	-	(x)	(-)	-	-	-	-	-
Size	x	x	x	x	x	x	x	x	x	x	x	x	x	x	x	x	x
Saturation	x	x	x	x	x	x	x	x	x	x	x	x	x	x	x	x	x
Color	x	x	x	x	x	x	x	x	x	x	x	x	x	x	x	x	x
Texture	-	-	x	x	x	x	-	(x)	-	x	x	-	-	-	-	x	(x)
Relations	-	-	-	-	-	-	-	-	x	x	x	x	x	x	x	x	x
Containment	-	-	x	-	-	-	-	(x)	-	x	x	-	-	-	-	-	(x)
Shape	x	x	-	-	-	x	x	x	x	x	x	x	x	x	x	x	x
Motion	-	(-)	-	-	-	-	(-)	-	-	(-)	(-)	x	(-)	(-)	(-)	(-)	-
Hierarchy	x	x	x	x	x	x	x	(x)	x	(-)	(x)	x	x	-	-	-	(x)

Fig. 4. Evaluation of support for visual features in the graph layout visualization techniques

4.2 Procedure for Construction of Graph Visualization Taxonomy

Based on the evaluation of support for visual features in the graph layouts in Fig. 4 and the feature classes quantitative, ordinal, and nominal by Mackinlay [16] (Section 2.2), the user-oriented taxonomy is build up. In this paper for each of these classes a single taxonomy is build up. Therefore, the following procedure is applied:

The taxonomy starts with the element *all*. Visual features which are supported by all techniques are assigned to this element. Afterwards, the features are to arrange according to the ranking (for quantitative, ordinal, or nominal, Section 2.2). The subsequent procedure has to be iterated until all layout techniques are included in the taxonomy, respectively each taxon at the end of this differentiation contains exactly one visualization technique. Techniques are grouped as taxon in the taxonomy which supports the first visual variable. The same holds for the techniques which do not support this visual feature. For both taxa those visual features are assigned to the taxa which are supported or not supported by all techniques included in the taxa.

4.3 The Resulting Oser-Oriented Graph Visualization Taxonomy

Applying the procedure for constructing, described in Section 4.2, the result is a taxonomy of graph visualization techniques. Due to the data-oriented ordering of the visual features in the procedure according to user-oriented perception of information displayed with the visual features, this resulting taxonomy is user-oriented based on the data-oriented examination of these features.

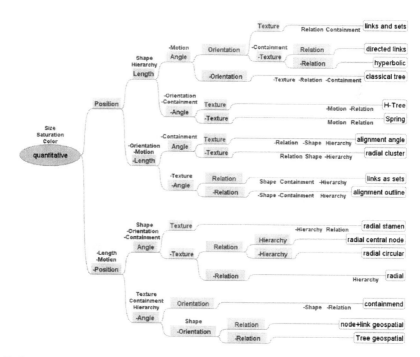

Fig. 5. Graph visualization taxonomy according to visual variables applicable for quantitative data

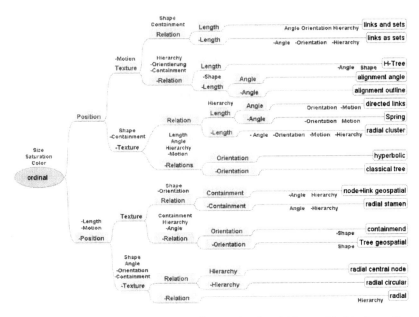

Fig. 6. Graph visualization taxonomy according to visual variables applicable for ordinal data

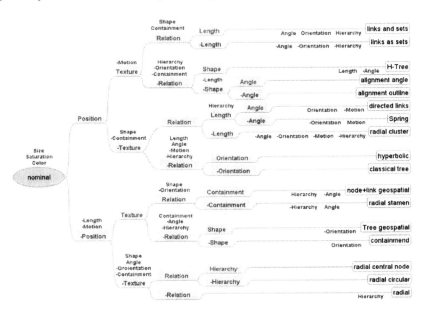

Fig. 7. Graph visualization taxonomy according to visual variables applicable for nominal data

The user-oriented taxonomies for presenting quantitative, ordinal, and nominal data with adequate visual features will be presented in the Figs. 5, 6 and 7. In these figures green colored features are available, red colored (and signed) features are not available for the specific taxon. Subsequent a guide is given how these taxonomies can be interpreted and be utilized to identify adequate visualization techniques for given kinds of data.

5   Application Scenario Using the Taxonomy

The taxonomy implies a correlation between data types and user-adequate visualization features. Thus, when the task is to present a specific data type the taxonomy depicts this visualization feature which should be utilized to visualize the data. At the same time the taxonomy presents the set of graph layout techniques which embody selected variables, if interpreted as decision tree. Thus it can be used to identify those layout techniques which can be applied to display an amount of data types.

Furthermore, on the right side the taxonomy demonstrates a sequence for the visualization techniques according to their expressiveness in terms of the number of available visual features for a given kind of data type. Thus, at the upper side this layout technique is presented which embodies most visual features. Walking down the graph layout techniques their expressiveness according to the embodied visual features decreases.

For example, when three data elements are to visualize which are of the type quantitative: According to the quantitative taxonomy when selecting the 'positive' paths in the tree the data should be visualized by utilizing position, length and angle. Furthermore, the set of graph layout techniques embodying these features are beneath the taxon 'angle'('links and sets', 'directed links', 'hyperbolic', and 'classical tree'). One of these techniques can be used to visualize the data in a user-oriented way utilizing those visual features. The final selection of the specific technique is a design decision and depends on the usage scenario and specific user group. Thus, the taxonomy preselects applicable layout techniques and provides the allocation of beneficial visual features for presenting the data that way the user can intuitively and easy percept and interact with.

6   Conclusion

In this paper a novel taxonomy is developed which offers the identification of adequate visual features for presenting data in a user-oriented way. To achieve this, existing visualization techniques are summarized and re-organized to a user-driven categorization. For the purpose of visualizing data for the standard user those visualization techniques were considered which are graph-based and two-dimensional. The visualization techniques were evaluated for their support of visual features. Based on the evaluation result the taxonomy had been constructed for quantitative, ordinal and nominal data types.

Using the taxonomy adequate visual features are depicted which should be utilized to visualize a given data element according to the elements data type. Furthermore, the set of visualization techniques is enclosed which support the identified visual features, which can thus be applied to display an amount of given data in a user-oriented way. In addition, the taxonomy presents a sequence for the visualization techniques according to their expressiveness in term of the number of supported features.

Due to the data type-specific ordering of the visual features according to their perception characteristics for users, thus forming a user-oriented sequence, the result is a user-oriented taxonomy for graph visualization techniques based on a data-oriented examination of visual features in these visualization techniques.

Acknowledgments. The here described work was developed as a part of the Core-Technology-Cluster for Innovative User Interfaces and Visualizations of the THESEUS Program, a 60-month program partially funded by the German Federal Ministry of Economics and Technology.

References

1. Daniel, A., Munzner, T., Auber, D.: Topo-Layout: Multilevel Graph Layout by Topological Features. IEEE Transactions on Visualization and Computer Graphics 13(2), 305–317 (2007)
2. Wilhelm, B., Jünger, M., Mutzel, P.: Simple and Efficient Bilayer Cross Counting. In: Goodrich, M.T., Kobourov, S.G. (eds.) GD 2002. LNCS, vol. 2528, pp. 130–141. Springer, Heidelberg (2002)
3. Bertin, J.: Semiology of graphics. University of Wisconsin Press (1983)
4. Brodbeck, D., Mazza, R., Lalanne, D.: Interactive Visualization - A Survey. In: Lalanne, D., Kohlas, J. (eds.) Human Machine Interaction. LNCS, vol. 5440, pp. 27–46. Springer, Heidelberg (2009)
5. Collins, C.: CSC 2524: Graph Visualizations, 2006. Department of Computer Science, University of Toronto (2006)
6. Consens, M.P., Mendelzon, A.O.: Hy+: A Hygraphbased Query and Visualization System. In: Visualization System (Video Demonstration). Proc. ACM SIGMOD 1993, pp. 511–516 (1993)
7. Harel, D.: On Visual Formalisms. Communications of the ACM 31 (1988)
8. Herman, I., Melancon, G., Marshall, M.S.: Graph visualization and navigation in information visualization: A survey. IEEE Transactions on Visualization and Computer Graphics 6(1), 24–43 (2000)
9. Holten, D.: Hierarchical Edge Bundles: Visualization of Adjacency Relations in Hierarchical Data. IEEE Transactions on Visualization and Computer Graphics 12(5), 741–748 (2006)
10. Keim, D.A.: Databases and Visualization. In: Jagadish, H.V., Mumick, I.S. (eds.) SIGMOD Conference, p. 543. ACM Press, New York (1996)
11. Keim, D.A.: Information Visualization and Data Mining. IEEE Transactions on Visualization and Computer Graphics 7(1) (2002)
12. Keim, D.A.: Visual techniques for exploring databases. Invited Tutorial. In: Int. Conference on Knowledge Discovery in Databases, KDD 1997 (1997)
13. Kosara, R., Hauser, H., Gresh, D.: An Interaction View on Information Visualization. In: Proceedings of Eurographics (2003)
14. Leissler, M.: A Generic Framework for the Development of 3D Information Visualisation Applications. Doctor thesis, Technical University Darmstadt, Germany (2004)
15. Mackinlay, J.: Automating the Design of Graphical Presentations of Relational Information. ACM Transactions on Graphics 5, 110–141 (1986)
16. Qeli, E.: Information Visualization Techniques for Metabolic Engineering. Doctor thesis, Philipps-University Marburg, Germany (2007)

17. Shi, K., Irani, P., Li, B.: An Evaluation of Content Browsing Techniques for Hierarchical Space-Filling Visualizations. In: INFOVIS 2005: Proceedings of the 2005 IEEE Symposium on Information Visualization, p. 11. IEEE Computer Society, Washington, DC (2005)
18. Tversky, A.: Features of Similarity. Psychological Review 84(2), 327–352 (1977)
19. Yee, K.-P., Fisher, D., Dhamija, R., Hearst, M.: Animated Exploration of Dynamic Graphs with Radial Layout. In: INFOVIS 2001: Proceedings of the IEEE Symposium on Information Visualization 2001 (INFOVIS 2001), p. 43. IEEE Computer Society, Washington, DC (2001)

Towards Compositional Design and Evaluation of Preference Elicitation Interfaces

Alina Pommeranz, Pascal Wiggers, and Catholijn M. Jonker

Section Man-Machine Interaction, Delft University of Technology, Mekelweg 4,
2628 CD Delft, The Netherlands
{a.pommeranz,p.wiggers,c.m.jonker}@tudelft.nl

Abstract. Creating user preference models has become an important endeavor for HCI. Forming a preference profile is a constructive process in the user's mind depending on use context as well as a user's thinking and information processing style. We believe a one-style-fits-all approach to the design of these interfaces is not sufficient in supporting users in constructing an accurate profile. We present work towards a compositional design approach that will lead designers in the creation of preference elicitation interfaces. The core of the approach is a set of elements created based on design principles and cognitive styles of the user. Given the use context of the preference elicitation suitable elements can be identified and strategically combined into interfaces. The interfaces will be evaluated in an iterative, compositional way by target users to reach a desired outcome interface.

Keywords: Compositional Design, Preference Elicitation, Interface Design.

1 Introduction

Knowing what a user likes and dislikes, i.e., his preferences, is important for intelligent systems in many domains. Preferences are part of an accurate user model needed to create system responses (e.g. recommendations or decision-theoretic advice) adapted to the user. Eliciting preferences from users is not a simple task. Traditional assumptions of economists that people have known, stable and coherent preferences are not correct [18]. People, confronted with a new decision task do not possess stable preferences, but have to construct them [19]. This construction process depends on the goal of the decision and the decision context (e.g., alternatives in an outcome set, how information is presented to the person).

Several techniques have been developed to elicit preferences, from implicit ones that learn preferences from the user's behavior [12], to explicit ones that ask users a long list of elicitation questions. However, only a few researchers [22] have explicitly put the focus on the user and the interaction between the user and the system and taken the constructive nature of preferences into account.

Given that the process of constructing preferences is important for people to arrive at an understanding of their own preferences and as the flow of process influences the outcome, we argue that more focus should be put on the design of preference elicitation interfaces. Apart from the constructive nature of preferences, one should also

M. Kurosu (Ed.): Human Centered Design, HCII 2011, LNCS 6776, pp. 586–596, 2011.

take into account the application goal and contextual factors, such as the importance of the decision, available time, number of alternatives and people involved as this influences the applicability of the elicitation technique. For example, a product recommender embedded in an online shopping environment has a different goal and offers a different environment than a system supporting an individual in taking a difficult medical decision. The interfaces will look completely different in terms of their interaction elements, but both can be equally successful in creating a useful preference profile.

In this paper we present a first outline of a compositional design approach for preference elicitation interfaces together with an example how we applied the approach. In addition, we provide design guidelines and discuss open issues offering directions for future research.

1.1 Research Hypothesis

As people construct their preferences when needed, a successful preference elicitation interface should be situated; i.e. based on a careful analysis of the purpose and the use context of the application. *H1: The purpose and use context of the application that elicits preferences determines the preference elicitation interface.*

The importance of information processing during preference construction by users, leads us to the next aspect: People's differences in information perception and processing leading to misinterpretations and wrong preferences. To be sure the user has a chance of understanding relevant information in an optimal way, an interface needs to support each user's cognitive style in an adaptive/-able way. Depending on the purpose of the application, (e.g. decision or negotiation support, tutoring system), different types of styles could be relevant (thinking styles, learning styles etc.). *H2: The use context determines the set of relevant cognitive styles to be considered for preference elicitation interfaces.*

Once the use context and the relevant styles are identified we can proceed with designing interfaces. There have been few attempts to create principles to guide the preference elicitation design [22]. Therefore, one of our contributions will be a set of design principles relevant for preference elicitation interfaces which has to be proven. *H3: Preference elicitation interfaces should satisfy the proposed set of design principles.*

The question remains how to design a successful preference elicitation interface, i.e. one creating an accurate profile, for a particular application in an efficient way? Do we have to start from scratch for every new task? As interfaces are composed of interaction elements, the question arises whether it is possible to reuse elements and combine them for a given use context, similar to component-based software engineering [14]. If the environments are sufficiently similar, it should be possible. Thus we propose a compositional approach in which new interface designs are built out of elements fitting particular user styles and contextual factors and iteratively improved. *H4: Preference Elicitation Interfaces can be designed and evaluated in a compositional manner.*

2   Background

2.1   Constructive Preferences

Studies in behavioral decision making have confirmed that preferences are not stable but constructive [19], i.e., people do not have well-defined preferences in most situations but construct them in the decision making context. Furthermore, people are not entirely rational, but also emotional and social beings. Therefore, besides aiming at maximizing the accuracy of their decisions they try to reduce cognitive effort and negative emotions while enhancing positive emotions and the ease of justifying a decision. While reducing cognitive effort, people might undergo several faults in the preference construction process [18], e.g., avoidance of trade-offs or focusing on too little information. There are different views on how people construct their preferences [15,26]. Simon and colleagues [27] found that while people processed the decision task, their preferences of attributes in the option that was chosen increased whereas those for attributes of rejected options decreased. Similar effects have been found in negotiation settings [7]. This is in line with one of the meta-goals named by Bettman and Luce [1], i.e. trying to maximize the ease of justifying a decision. Another aspect of constructing preferences has been brought forward by Fischer et al. [11] focusing on the goals of the decision task in relation to a so-called prominence effect. This effect occurs when people prefer an alternative that is superior only on the most important, attribute. In order to avoid unwanted effects we have to think carefully about the way we pose a preference elicitation task to the users. Payne and colleagues [18] have developed guidelines for measuring preferences taking people's behavior into account. Work focusing on the user side has been presented by Pu et al. [22].

2.2   Preference Elicitation Interfaces

Techniques commonly used in preference elicitation interfaces include knowledge-based find-me techniques [2], example critiquing and tweaking [8,24], active decisions and clustering or collaborative filtering [6]. The latter two are used mainly to create profiles for new users based on clusters of existing users and similarity [23]. There are also hybrid systems combining different approaches [3]. In knowledge-based systems, preferences are elicited by example-similarity; the user rates a given item and requests similar items. Tweaking can be used to limit similar items to only those satisfying the tweak. In example-critiquing approaches the user is presented with a set of candidates (e.g. products) to be critiqued. The user can either choose one of them or critique some of their attributes. In the Apt Decision Agent [24], e.g., people initially provide few criteria for an apartment to get a selection of sample apartments. Next, they can give feedback on any attribute of any apartment.

Not all techniques mentioned are relevant for decision support systems due to a lack of user-involvement. The user will be less likely to understand advice by a system, if the system has created a user profile implicitly [5]. A majority of the literature presenting these systems focuses on technical implementations and not the user. Few researchers proposed guidelines for user-involved preferences [18, 22].

3 Design Principles for Preference Elicitation Interfaces

The following design principles are derived from the diverse literature influential to the success of a preference elicitation interface.

(1) Support of human process of constructing preferences. The work of [22] provides a number of more detailed guidelines addressing this criterion: (1.1) show decision context, that also allows people to see the consequences of their decisions.(1.2) provide examples that can be critiqued by the users to refine their preferences. (1.3) give immediate visual feedback.

(2) Affective feedback. There is interplay between cognition and affect when people construct their preferences. Therefore, combining cognitive (e.g. choosing from a list of values) and affective (e.g. emoticons) elements in an interface might lead to more insights into the user's preferences.

(3) Value-Focused Preferences. In value-focused thinking [16] proposes a focus on fundamental values relevant for a decision before identifying possible alternatives and assessing their desirability. Generally, values are seen as more stable than preferences over attributes [25] preference elicitation interfaces [9, 29].

(4) Transparency. A major aspect influencing the success of decision support systems is the user's trust in the system [21]. System transparency is one aspect that can enhance users' trust [28]. By transparency we mean that the user understands what the system is doing, why it asks certain questions and how the current profile looks. Implicit elicitation methods often restrict the user from constructing preferences and suggestions based on the created profile are hard to understand [5].

(5) User-System Collaboration. Designing user interfaces means designing the interaction between the system and the user. For decision support it is important that the user and the system collaborate in establishing a good user profile. We define three criteria for the interaction: *(5.1) Natural Interaction.* Natural interaction refers to the usual way in which the users act in the physical world applied to computer systems. People use gestures, expressions, speech and movement to communicate. *(5.2) Real World Metaphors.* Part of designing the interaction with a system as natural as possible is using real-world metaphors that users can relate to. *(5.3) Mixed Initiative* [10] is a popular approach for collaborative problem solving (e.g. constructing a preference profile).

4 Proposed Approach

Use context is important in deciding what the interface should be like to support the user's construction of an accurate profile. We define use context in terms of the measurable aspects (to be extended) task goal, importance of the decision, available time, the number of alternatives and people involved (for an example in a negotiation support system see section 4.3).

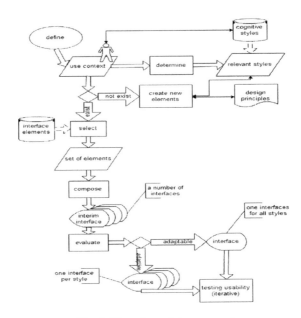

Fig. 1. Flowchart of the approach

4.1 Interface Elements

The core of our approach form so-called interface elements. Generally, an interface can be seen holistically or as a collection of elements. Simple elements (widgets) like buttons, list boxes, and comboboxes, fit all cognitive styles. These can be embedded in more complex elements, e.g. a spell checker in a word processor. Complex elements often represent a concept or an idea rather than just a simple way of entering data. Complex elements are meaningful in terms of a particular user goal, cognitive style and fit with a particular environment. We focus on combining complex elements. The ultimate goal is to develop and store many elements in a database with information on how each element relates to contextual factors, styles, and design principles. This database should contain only evaluated elements and be available to interface builders to speed up the design process.

4.2 The Approach Step-by-Step

Step 1: Defining the use context. The starting point of our compositional approach is the definition of the application's use context, i.e. aspects such as the general purpose or task goal, e.g. recommendation or personalized decision advice, the number of people involved, time constraints, the size of outcome space and the decision importance.

Step 2: Determining relevant type of styles. To assure optimal adaptation to the user's way of handling information the next step is to determine the type of style that is most relevant in the given use context. This knowledge could be asked from a database containing all types of styles and their relations to context factors. As this is not available currently the designer has to extract knowledge from literature.

Step 3: Defining target group's styles. Next, the designer needs to define whether the target user group consists of people with all styles in the chosen set of styles or whether to focus on a specific subgroup that shares one style. In the latter case it is enough to design one interface. Otherwise, it is possible to design one interface for each style or an adaptive interface covering all styles and can be adapted by the user.

Step 4: Selecting/Creating interface elements. Given a database with usable interface elements and their relationships to the use context factors, styles, and design principles, this step contains of asking the database to return all elements fitting the use context defined in step 1 and the type of style in step 2. In case the query does not return enough elements to cover all styles and design principles, the designer has to create new elements. In case these elements end-up in the final design due to positive evaluation these elements will be added to the database. The question is 'how to create these elements?' We propose a participatory design process. Since an element needs to fit a certain type of style and more specific different styles of that type, we suggest assessing each participant's style (e.g. by questionnaire) and grouping the participants per style. A leader of the design sessions needs to assure that the design principles are represented in the final designs. For the design sessions standard creative techniques can be applied (brainstorming, thinking hats, etc.). Another option is that the designer creates the elements herself using theories of the identified styles and following the design principles.

Step 5: Composing interim interfaces. The set of elements needs to be combined into complete interfaces. Ideally all possible combinations should be developed and evaluated by the target users to find the optimal one. However, with a high number of elements this is an impossible endeavor. Therefore, the designer should pick a subset of all possible interfaces that covers the design principles and styles. The concrete combination relation is still an open issue for research. One idea is to focus on combining elements addressing different styles in each interface allowing for creative, constructive feedback by users in the next step.

Step 6: Compositional evaluation. As the core of the approach is the compositionality of elements, we evaluate the interfaces also compositionally. That means the goal of the evaluation should not be to find the best interface in the designed set but to ask users to evaluate the different elements and offer ideas of how to combine them. (As elements may not be useful by themselves it is not possible to only evaluate elements by themselves.) In the compositional evaluation the participants interact with all interim interfaces. We suggest a formative evaluation (e.g. think aloud) to encourage discussion of ideas and constructive feedback.

Step 7: Composing stable interfaces. After participants interacted with the interfaces user feedback needs to be applied to the design of a new set of interfaces. This is done by combining the elements using a different composition relation. In case the designer chooses to let the system adapt its interface to its user based on a style, it is useful to create one interface per style. In case the designer creates one interface for all styles, the interface should be adaptable. Given the first case we suggest a participatory design approach in which the participants are grouped per style. Each group is asked to help in the design of an optimal preference interface. The outcome of this process is a set of stable interfaces, one per style. Given the second case (one interface) it is possible to group participants per style and let them negotiate a new interface or use mixed groups and combine the outcomes into a new interface. These are just suggestions. What way will lead to the best desired outcome(s), still needs to be shown by future research.

Step 8: Usability Testing and Optimization. The last step is an iteration of standard usability testing with target users to refine the end-design(s).

4.3   The Approach Applied: Preference Elicitation for the Job Domain

To clarify our approach we give an example of a current design case, i.e. a preference elicitation interface of an application that gives advice on job (contract) offers.

Step 1: The goal of the system is to support people deciding on a job offer that best matches their preferences. The importance of the final decision is high. Typically a job offer contains a number of attributes, e.g. salary, salary growth, vacation days, company car, and the combination of them to concrete offers leads to large number of alternatives. There are no time-constraints and the system is focused on a single user.

Step 2: For our example we selected the mind styles theory by Gregorc [15] which categorizes people based perceptual and ordering preference. Perceiving information can be abstract (based reason and intuition) and concrete (using senses). The order of information processing can be sequential or random. This leaves us with four types *concrete sequential, concrete random, abstract sequential* and a*bstract random.*

Step 3: Our application is directed at people with diverse mind styles.

Step 4: We designed elements and took some existing elements from the literature.

Virtual agent: The abstract random thinker likes to listen to and work with others and thinks best in personalized environments. This combined with principle 5.1 and 5.3 (mixed initiative) led to a virtual agent element asking the user about his preferences.

Tag Cloud is an interface element we used to reach system transparency (4) and for the abstract random thinker, because tagging as a social activity fits his preference for group activities.

Post-it notes serve as a real-world metaphor (5.2) for collecting information by writing it on the notes and structuring it by moving the notes around. This fits with concrete random thinkers, who like experimenting to find answers.

Outcome Cluster: Using design principle 1.1 and 1.3 we designed an element that gives an overview over how well job offers score given a preference profile. Using clusters instead of complete orderings serves the concrete random thinker, because it leaves room for interpretation.

Interest Profiling: Value-focused thinking (3) was the inspiration for this element. We use four profiles visualized by a set of images each reflecting a life-goal. Each profile represents a set of preferences. As value-focused thinking is a step-wise process from considering interests to choosing alternatives to compare, it fits sequential thinkers.

Decision Matrix: Decision or comparison matrixes are often used on product comparison websites (1.1). Preferences and offers are ordered by importance, from top to bottom and left to right respectively. Users can adjust preference values as well as the ordering and get visual feedback of the consequences (1.3).

ValueChart [4] gives an immediate visual feedback (1.3) on how well the five job offers in our system match the user's interests.

Affecttive Example Critiquing shows the details of a given job offer in form of a table with attribute-value pairs and allows users to critique (1.2) any pair with a set of smileys allowing affective feedback (2).

Preference Summary adds to system transparency (4). It fits the concrete sequential thinker as it represents the input in a factual way. It is basically a list of issues and their preferred values ordered by importance (like, want, dislike, do not want).

Step 5: Composing interim interfaces. We combined the elements into four proto-types each one serving one mind style. Together the prototypes cover all design principles. For a detailed description see [20]. (Design principles listed in brackets).

Abstract Random: Conversational Interface: In this interface we combined a virtual agent with a tag cloud because they fit the abstract random thinker and the tag cloud is a good representation of the internal "thinking" of the system. Putting the tag cloud into a thought bubble connects the elements. (4, 5.1, 5.3.)

Concrete Random: Post-it notes Interface: Both post-its and job offer clusters serve the concrete random thinkers preference for trial-and-error. Especially by combining the two elements in one interface we can offer the possibility to move post-its and immediately see the effects the preferences have on the job offers. The interface does not limit the user to any order or amount of input. (1.1, 1.3, 5.2)

Abstract Sequential: Offer Comparison: The decision matrix is a way to analyze the relationship between job offers and preferences. The downside of the matrix is that the preferences need to be filled in to a certain extend before the matrix gives any results. To shorten that process we added the interest profiling element. After choosing a profile the user gets pre-set preferences that can be adjusted. (1.1, 1.3, 3)

Concrete Sequential: From interest to issue: This interface supports a concrete, sequential style. It consists of four steps that the user is led through: selecting 3 most important interests, using ValueCharts to see how good job offers fit the interests, critiquing the attributes of job offers with smileys and getting a summary. (1,2,3,4)

Step 6: We did eight individual formative evaluation sessions and a group discussion with all participants. They had diverse backgrounds including IT, design disciplines and linguistics and by that different mind styles. In each individual session (1 hour) a participant interacted with all four prototypes (10 minutes each). We asked people to think aloud and conducted a semi-structured interview at the end of the session in order to get qualitative feedback on the prototypes. During the sessions and the interview we asked people to give constructive feedback on the different interface elements. We emphasized that it is not about finding the best interface, but informing the design process to compose new interfaces [25].

Step 7: We conducted a creative design session with the same participants. We asked two mixed styles groups of four people to compose a new interface. They were given a set of magnetic interface elements (the ones described and basic ones) and a magnetic board to assemble them on (inspired by PICTIVE [17]). In steps 6 and 7 we found that most participants preferred elements that explicitly let them construct preferences and explore results of those manipulations (e.g. post-its, job clusters). Both interfaces constructed in step 7 used multiple views, i.e. values, preferences, outcomes on the data which adapt as soon as one of the views is changed.

Step 8: We have combined the two interfaces and implemented the outcome in a prototype of our system. The user testing still needs to be completed.

5   Discussion and Future Work

We argued for the necessity of careful design of preference elicitation interfaces taking into account factors of the application's use context and different cognitive styles of the users. The background literature on people's construction of preferences and the diversity of techniques used in different domains supports our first hypothesis (H1). We extracted a set of design principles from the literature on human preferences in psychology, social sciences, economy and HCI. The usefulness of these design principles (H3) needs further empirical proof. Furthermore, we propose a first outline of a compositional approach for the design and evaluation of such interfaces. The core of the framework is a set of interface elements fitting the given use context and styles of the users. These elements can be composed into interfaces. With a handful of interfaces one can test a large number of elements that can be recombined in different ways for the next iteration. As interface elements can be reused across applications a database of evaluated elements can be built up to give guidance in future designs.

Our design of an interface eliciting job preferences showed that it was possible and efficient to combine several interface elements and evaluate them in a compositional way. Several research questions are still open. At this moment the optimal composition relation used for step 5 and 7 to combine the elements into interfaces is unknown. We have provided initial ideas of combining the elements. Further research is needed to find a ready-to-use way leading to consistent and usable interfaces. Concrete research questions are: (1) In case too many elements exist, which are the most suitable ones to combine? (2) Does a deliberate mixing of cognitive styles in the first composition step lead to more creative feedback in a formative evaluation? (3) Does the composition of elements into interfaces speed up the design time? We encourage readers to build on these first ideas in their own research.

Acknowledgments. The research is supported by STW, applied science division of NWO and the Technology Program of the Ministry of Economic Affairs. It is part of the Pocket Negotiator project with grant number VICI-project 08075.

References

1. Bettman, J.R., Luce, M.F., Payne, J.W.: Constructive consumer choice processes. Journal of Consumer Research 25(3), 187–217 (1998)
2. Burke, R.: Knowledge-based recommender systems. In: Encyclopedia of Library and Information Systems, vol. 69 (2000)

3. Burke, R.: Hybrid recommender systems: Survey and experiments. User Modeling and User-Adapted Interaction 12(4), 331–370 (2002)
4. Carenini, G., Loyd, J.: Valuecharts: analyzing linear models expressing preferences and evaluations. In: AVI 2004: Proceedings of the Working Conference on Advanced Visual Interfaces, pp. 150–157 (2004)
5. Carenini, G., Poole, D.: Constructed preferences and value-focused thinking: Implications for AI research on preference elicitation. Technical report (2002)
6. Chen, L., Pu, P.: Survey of preference elicitation methods. Technical report, Swiss Federal Institute of Technology, Lausanne (2004)
7. Curhan, J.R., Neale, M.A., Ross, L.D.: Dynamic Valuation: Preference Changes in the Context of Face-to-face Negotiation. Journal of Experimental Social Psychology 40, 142–151 (2004)
8. Faltings, P.P., Viappiani, P., Torrens, M.: Designing example-critiquing interaction. In: IUI 2004: Proceedings of the 9th International Conference on Intelligent User Interfaces, pp. 22–29 (2004)
9. Fano, A., Kurth, S.W.: Personal choice point: helping users visualize what it means to buy a BMW. In: IUI 2003, USA, pp. 46–52 (2003)
10. Ferguson, G., Allen, J.: Mixed-initiative systemsfor collaborative problem solving. AI magazine 28(2) (2006)
11. Fischer, G.W., Carmon, Z., Ariely, D., Zauberman, G.: Goal-Based Construction of Preferences: Task Goals and the Prominence Effect. Management Science 45(8), 1057–1075 (1999)
12. Goldberg, D., Nichols, D., Oki, B.M., Terry, D.: Using collaborative filtering to weave an information tapestry. Commun. ACM 35(12), 61–70 (1992)
13. Gregorc, A.F.: The Mind Styles Model: Theory, Principles, and Practice. AFG (2006)
14. Heineman, G.T., Councill, W.T.: Component based software engineering: putting the pieces together. ACM Press, New York (2001)
15. Johnson, E.J., Steffel, M., Goldstein, D.G.: Making better decisions: from measuring to constructing preferences. Health Psychology 24(8), 17–22 (2005)
16. Keeney, R.: Value-Focused Thinking: A Path to Creative Decision Making. Harvard University Press, Cambridge (1992)
17. Muller, M.J.: Pictive—an exploration in participatory design. In: CHI 1991, New York, NY, USA, pp. 225–231 (1991)
18. Payne, J.W., Bettman, J.R., Schkade, D.A.: Measuring constructed preferences: Towards a building code. Journal of Risk and Uncertainty 19(1-3), 243–270 (1999)
19. Payne, J.W., Bettman, J.R., Johnson, E.J.: The Adaptive Decision Maker. Cambridge University Press, Cambridge (1999)
20. Pommeranz, A., Wiggers, P., Jonker, C.: User-centered design of preference elicitation interfaces for decision support. In: Leitner, G., Hitz, M., Holzinger, A. (eds.) USAB 2010. LNCS, vol. 6389, pp. 14–33. Springer, Heidelberg (2010)
21. Pu, P., Chen, L.: Trust-inspiring explanation interfaces for recommender systems. Knowledge-Based Systems 20(6), 542–556 (2007)
22. Pu, P., Chen, L.: User-Involved Preference Elicitation for Product Search and Recommender Systems. AI Magazine 29(4), 93–103 (2008)
23. Rashid, A.M., Albert, I., Cosley, D., Lam, S.K., McNee, S.M., Konstan, J.A., Riedl, J.: Getting to know you: learning new user preferences in recommender systems. In: IUI 2002, USA, pp. 127–134 (2002)
24. Shearin, S., Lieberman, H.: Intelligent profiling by example. In: IUI 2001, New York, NY, USA, pp. 145–151 (2001)

25. Shiell, A., Hawe, P., Seymor, J.: Values and preferences are not necessarily the same. Health Economics 6(5), 515–518 (1997)
26. Shiv, B., Fedorikhin, A.: Heart and mind in conflict: the interplay of affect and cognition in consumer decision making. Journal of Consumer Research 26(3), 278–292 (1999)
27. Simon, D., Krawczyk, D.C., Holyoak, K.J.: Construction of Preferences by Constraint Satisfaction. Psychological Science 15(5), 331–336 (2004)
28. Sinha, R., Swearingen, K.: The role of transparency in recommender systems. In: CHI 2002 Extended Abstracts on Human Factors in Computing Systems, pp. 830–831 (2002)
29. Stolze, M., Strobel, M.: Dealing with learning in ecommerce product navigation and decision support: the teaching salesman problem. In: MCPC 2003 (2003)

Secure Online Game Play with Token: A Case Study in the Design of Multi-factor Authentication Device

Shinji R. Yamane

Aoyama Gakuin University eLPCO
Shibuya 4–4–25, Shibuya-ku, Tokyo, 1508366 Japan
s-yamane@computer.org

Abstract. Online game security is often discussed, but, in game development, security factors are tend to be an afterthought. It is helpful to consider the unique perspectives of the game designer, security engineer, and game player all together in the game development process. In this paper, we present a framework for a formal approach to understand the security interface. We also try to integrate different perspectives when analyzing cases which use a hardware security token for online games. This interface-level analysis of security attempts to achieve two goals: Firstly, to make the security in online gaming not merely an add-on feature but an integrated part of game development. Secondly, to bridge the gap between game design and security technology and allow the game designer and security engineer to collaborate toward their particular goals.

1 Introduction

Historically, gaming interfaces have brought new experiences to HCI. For example, once the classic computer game *SPACEWAR* had been developed, the analogue controller for a "spaceship" was also developed and integrated. Later, Nintendo and other companies developed and released various controllers to enhance the game experience, such as the "light gun," Power Glove, or EyeToy [1]. Konami's *Bemani* music video games brought a series of interface experiences to the arcade, then spread it to the home. Nintendo's Wii, Harmonix's *Rock Band*, and Microsoft's recent Kinnect feature some of the new generation of popular gaming interfaces.

As the new generation of gaming interfaces becomes popular, game design at the interface level provides new opportunities. Today, there are other game interfaces to be discussed. In the genre of online gaming, another game interface device is rising in popularity — the security interface devices. These hardware (or software) devices, sometimes wearable or embedded, that perform a user-side security function.

1.1 Security Challenges in Online Games

As online game markets have matured, security concerns also have arisen. Online gaming brought new security services and online interactions. As online games have evolved, so have the security devices that use to play them. Security for online games encompasses multiple research disciplines: software engineering, distributed

M. Kurosu (Ed.): Human Centered Design, HCII 2011, LNCS 6776, pp. 597–605, 2011.

computing, in-formation security management, human behavior analysis, psychology, legal enforcement, game design, and HCI.

In this wide range of security research, the current research on online game security mainly focuses on the two approaches. The first is the security engineering approach [2], including system development and network support. The second is security incident management including a kind of triage decision scheme to examine the incident would be in-system dependent or out-system dependent [3]. Both of these security approaches are based on a general-purpose methodology. However, this paper focuses on another part of security that makes online game security different from other applications: the user interface and device for gaming.

1.2 Multi-factor Authentication Requirements

Stronger authentication against the attack is one of the technological features required in the online games today. A "strong" passphrase for example has to protect player's personal information or virtual assets: it should release information only to authorized persons.

In the report on the issues of security and privacy in online games (especially in massively multi-player online games [MMOGs] or virtual worlds), ENISA (European Network and Information Security Agency) encourages service providers to require stronger authentication including the use of "second factor" when players perform any high-value or sensitive actions [4]. Multi-factor authentication has two categories: single-token authentication and multi-token authentication. Using a verified authentication process, users possess and control the *tokens* (typically a key or password) to be combined. For example, a PIN activated smartcard with a private key is a single-token, multi-factor authentication scheme. Another example, a combination of a password and a callback to a registered cell phone is a valid multi-token, multi-factor authentication scheme [5].

The various devices used in multi-factor authentication can be embedded or mixed by the interface designer or platform provider. From this viewpoint, we point out the uniqueness of the game user interface de-sign and what the online game interface design can provide to players.

2 Usability of Security in Online Gaming

As the attacks on online game accounts and in-game assets have be-come smarter, today's major online game service providers have implemented various security countermeasures more so than in other on-line commerce industries. At this point, online gaming provides a prototype to design the security interface device in ubiquitous computing. As in the design of most ubiquitous and embedded computers, the design of game interface focuses on cost factors, and security factors are often an afterthought. Considering the game security design at the interface level can bring new research opportunities for the interface tactics in this age of ubiquitous computing.

2.1 Token and Multiple Devices in Gaming

From the early days of CD keys or prepaid cards, the online games have implemented various tokens to use in authentication. In recent years, several kinds of token have been introduced and operated into the on-line games (Table 1).

Table 1. Devices for Digital Games

Type	Products and Services
Controller Device	Wii Remote Controller, *Rock Band*'s three Controllers, PS3 6-axis Controller, realworld-oriented interfaces
Programmable Device	gaming keypads, automouse, macro automation tools
Multi-purpose device	Smartphone
Security Device	CD keys, hardware token, software security device, *3D Secure* credit number, Smartcard

2.2 From Early Case to Ubiquitous Computing

This is not the first time in the short history of digital gaming that security devices have been implemented for gaming. However, their use has been limited to a particular platform. An early example of a combination of interface and token design is found in Japanese arcade game titles using the smartcard. From their early stage, Japanese arcade entertainment machines used the technology of sensors, mechanical interactions and many kind of displays [6].

Networked play with hardware security device began in arcade game such as *World Club Champion Football (WCCF)* (Sega, 2001–) [7], the hybrid of collectible cards and tabletop arcade games. The console has a flat panel that can read both the traditional trading cards(Player Card), and can read one smartcard(Club Card) that stores the gamer (professional club team manager in the game)'s data and skills[8]. The smartcard is called "Club Card" because the card represents the ID card of the player's football club, and the player cannot import his manager's skill data to the new Club Card until he completes his contract with the football club. By bringing one's own Club Card, the player can play the game continuously in any arcade game centers in Japan, nationwide.

As Ayatsuka and Rekimoto [9] pointed out that *WCCF* the early adaptation of a realworld-oriented interface in game control, we point out that *WCCF* also had been adopted a realworld-oriented user interface for authentication. In the following section, we present the framework, then turn to the current situation: MMOGs, provided by various publishers running on untrusted platforms.

3 Model and Application Area

In this section, we propose three interrelated dimensions for the analysis of embodied game security interface hardware devices. Fig. 1 illustrates a requirements for the framework of three components that represents each dimensions: security analysis (the traditional security aspect), interface idiom (the real-world-oriented interface aspect), and game play design (the game design/narrative aspect).

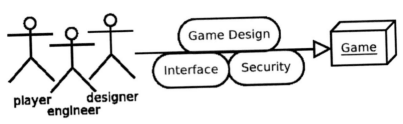

Fig. 1. Different Perspectives to Understand the Game

This framework of three components also establishes their actor counterparts: the security engineer, HCI researcher, and game designer. It is helpful to consider the different perspectives not in an ad-hoc and afterthought manner, but as a formal and integrated methodology. Later, we will demonstrate this framework to help explain how embodied interactions can be understood as the experience of authentication for online gaming.

Networked play with hardware security devices has connected players all over the world. One of the major tokens used in online game security (especially in MMOGs) is the hardware token which generates user's one-time password. In 2004, *Entropia Universe* (formerly *Project Entropia*), developed and operated by MindArk, started offering its "Gold Card Membership," which included a smartcard and reader for one-time password hardware token [10]. In 2008, Blizzard started to sell its "Blizzard (Battle.net) Authenticator", stand-alone one-time password generator for *World of Warcraft*. In 2009, Square-Enix also started to sell the "Square Enix authenticator," similar to Blizzard's, for *Final Fantasy XI* (Fig. 2). These game companies designed the game hardware devices for authentication in a different way: the membership card integrated with smartcard, or the general-purpose password generator hardware (Fig. 2). These may be the first such multiple and multi-level security devices to achieve widespread popularity.

Today's game security devices represent the range of new challenges in the game interface. To consider the effectiveness of the online game security interface, this paper proposes an analysis along some related dimensions: the traditional security system design, conceptual mapping of idiom, and narrativization.

Fig. 2. Hardware Tokens: Square Enix Security Token (left), Blizzard (Battle.net) Authenticators (middle), and MindArk's Gold Card Kit for *Entropia* (right)

4 Critical Analysis Using Framework

4.1 Security Analysis

Traditional security analysis pointed out the threat of a man-in-the-middle or Trojan attack to the multi-factor authentication mechanism [11]. Indeed, the technical support staff of Blizzard Entertainment Europe has openly acknowledged man-in-the-middle attack circumvented their one-time-password generator scheme [12] in 2010. Another security extension such as multifactor authentication with two-channel protection [13] will be examined in the near future.

Other than these security analyses, another important security analysis approach is "usability of security" which suggests the possibilities of security in domain-specific user interface design rather than the general-purpose user interface (see related works in section 6). This approach requires the security expert to collaborate with the user interface designer. This is suitable for our proposal, as our framework allows experts from different disciplines to work together during the game development cycle.

4.2 Interface and Its Idiom

To evaluate the usability of security, we focus on how to implement the function of authentication into the daily environment or game-related goods. In our context, the security device affords better authentication than existing tokens, such as the key or password. Adding new dimensions to the traditional security analysis, we demonstrate the evaluation of the mapping efficiency of idiom firstly (Fig. 3) The WCCF Club Card is a good example of effective interface design.

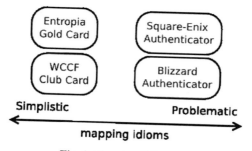

Fig. 3. Mapping Efficiency

The Club Card is simply a single-token, single-factor authentication scheme not combined with a password token: however, the physical design as members card supports player in mapping what he belongs to and what he can do directly.

A similar interface design can be found in *Entropia*'s Gold Card. Designed as a membership card, it generates a one-time password with the smartcard reader. On the other hand, the physical design of *WOW* and *FF XI*'s hardware token is not the card type but the key holder type. It does not require a smartcard reader, the user simply push the button to generate a one-time password. A well-implemented version of the security interface device could afford an even better idiom of security than the members' cards which exists today.

4.3 Narrative and Engagement

The final component is narrative. Narrative is a game design-dependent feature, and the game designer assumes the main role. In this paper's case, the arcade game and MMOGs do not feature experiences with a strong narrative such as in film and literature. Even the players' social relationship could bring the strong experience.

Nonetheless the well-designed narrative experience is important as the narrative within the larger game experience can guide the players to stronger security features. WCCF Club Card provides a good example of narrative again. As same as the trading cards, the Club Card has several designs with the color and graphics of several big football clubs. Pairing the social relationship or public memory of the big club with the authentication gesture interface, *WCCF*'s game design succeeded to enhance the player's passion of engagement.

Tanenbaum and Bizzocchi[14] demonstrated two ways to approach the interface narrativization with *Rock Band*. These are "iconic" narrativity (the look of the interface) and "functional" narrativity (the actual operation of the interface). Similar to the interface idioms in the last section, the "membership card" device represents a greater degree of functional narrativization than "push-button" security token. The "membership card with reader" combination also offers greater functional narrativization, if the card reader allows players to play and to purchase a member's authentication, as long as the current "push-button" devices of *WOW* and *FF XI* have no special narrativized interface design for secure play.

Comparing the one-time password generator device used for *WOW* and *FF XI*, the latter "Square Enix authenticator" is a simply general purpose device (See Fig. 2). Though the company logo on the device is the only sign of "iconic" narrativity, it offers little in terms of either authentication but also gaming. This narrative status is the end of the narrative scale (Fig. 4).

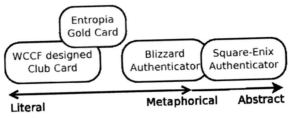

Fig. 4. Metaphorical & Literal Functional Narrativation

5 Discussion

In this paper, we have presented a three dimensional framework for the analysis of embodied or gestural interfaces for security in online gaming devices: the traditional security analysis, conceptual mapping of idiom, and narrativization. We do *not* claim that the vectors of these dimensions are isolated from the others. These dimensions are related, rather than isolated, and our approach attempts to bridge the roles of the security engineer, game artist and designer. We believe that these interface-level analyses, which focus to the security effort, redefine the security in online gaming not only add-on feature or afterthought nor system and network support matter, but as an integrated part of the next generation of game design and the development process.

We also note that we could get a comparison with the past arcade game which is different from current MMORPGs. The analysis on the gaming device is not limited to the newest home entertainment but opens new possibility of the past history of game interfaces.

6 Related Works

The formal approach to bridge the gap in understanding games has been proposed though the collaborations. *MDA framework* (standing for Mechanics, Dynamics, and Aesthetics) has been developed and taught as a part of Game Design and Tuning Workshops at the Game Developer Conference since 2001. This framework is also proposed as a tool to help both game designers and researchers, and for use in examining game AI improvements [15].

Canadian game researchers Tanenbaum and Bizzocchi [14] proposed the three dimensions of ludic, kinesthetic, and narrative experience and demonstrated the long analysis of *Rock Band* interfaces.

For a long time, usability was not an element of security [13], but an interface design issue. To provide effective and understandable security for users, the usability of security was addressed in the field of user-centered design. An initial case study focused on non-entertainment application, such as email security software [16]. In this early study, Whitten and others had suggested that "making security usable will require the development of domain-specific user interface design principles and techniques." Our paper extends this idea to online games and their domain-specific user interfaces.

7 Future Issues

From our point of view, studying the game security interface will be able to open up the possibilities in both in the entertainment and non-entertainment use of distributed computing by casual users among untrusted platforms. Toward the research goal, we have some future game research issues.

The usability analysis of software security device, typically the cellphone onetime-password application, is one of the biggest genre to consider. The online game industry has provided the new game design based on the unique feature of cellphone. For example, *Before Crisis —Final Fantasy VII—* (Square Enix, 2004–) was designed and depended deelply for Japanese cellphones at that time, and the game narrative even suggested that players to take a photography using their cellphone cameras in play [1]. This approach suggests to us that the new opportunity of game design is to send game users' unique information with embedded interface via another channel.

References

[1] Thulatimutte, T.: Controller mediation in human-computer play. Honors thesis, Stanford University (2005),
http://www.gamecareerguide.com/features/203/honors_thesis_
controller_.php

[2] McGraw, G., Chow, M.: Guest editors' introduction: Securing online games: Safeguarding the future of software security. IEEE Security and Privacy 7(3), 11–12 (2009)

[3] Bardzell, J., Jakobsson, M., Bardzell, S., Pace, T., Odom, W., Houssian, A.: Virtual worlds and fraud: Approaching cybersecurity in Massively Multiplayer Online Games. In: Baba, A. (ed.) Situated Play: Proceedings of the Digital Games Research Association (DiGRA)'s Third International Conference, pp. 742–751 (2007)

[4] Hogben, G. (ed.): Virtual Worlds, Real Money: Security and Privacy in Massively-Multiplayer Online Games and Social and Corporate Virtual Worlds. European Network and Information Security Agency, ENISA (2008), Position Paper,
http://www.enisa.europa.eu/act/it/oar/masively-multiplayer-
online-games-and-social-and-corporate-virtual-
worlds/security-and-privacy-in-virtual-worlds-and-gaming

[5] National Institute of Standards and Technology (NIST): Electronic authentication guideline. Special Publication (SP) 800–63, NIST (2006) Version 1.0.2 and Draft Revision 1 (December 2008),
http://csrc.nist.gov/publications/nistpubs/800-63/SP800-
63V1_0_2.pdf, http://csrc.nist.gov/publications/nistpubs/
800-63/SP800-63V1_0_2.pdf

[6] Sambe, Y.: Japan's arcade games and their technology. In: Natkin, S., Dupire, J. (eds.) ICEC 2009. LNCS, vol. 5709, p. 338. Springer, Heidelberg (2009)

[7] DigInfoNews: SEGA WCCF Intercontinental Clubs 2007–2008. Online news (2009),
http://www.youtube.com/watch?v=67Z4RdYrTG0

[8] Sega: World Club Champion Football. Official European Web Site (2004),
http://www.segawccf.com/

[9] Ayatsuka, Y., Rekimoto, J.: Active CyberCode: a directly controllable 2D code. In: CHI 2006 Extended Abstracts on Human Factors in Computing Systems, pp. 490–495. ACM, New York (2006)

[10] Project Entropia version update 5.6 (2004), Press release available online at
http://unionova.eu/old/history/vu/5.html

[11] Schneier, B.: Two-factor authentication: Too little, too late. Commun. ACM 48(4), 136 (2005)

[12] Ziebart, A.: Man in the middle attacks circumventing authenticators. WoW Insider (2010), Online article at http://www.wow.com/2010/02/28/man-in-the-middle-attacks-circumventing-authenticators/

[13] Anderson, R.J.: Security Engineering: A Guide to Building Dependable Distributed Systems, 2nd edn. Usability and Psychology, ch. 2. John Wiley & Sons, Chichester (2008)

[14] Tanenbaum, J., Bizzocchi, J.: Rock Band: a case study in the design of embodied interface experience. In: Proceedings of the 2009 ACM SIGGRAPH Symposium on Video Games, pp. 127–134 (2009)

[15] Hunicke, R., LeBlanc, M., Zubek, R.: MDA: A formal approach to game design and game research. In: Challenges in Game Artificial Intelligence: Papers from the AAAI Workshop. Technical Report WS-04-04, Association for the Advancement of Artificial Intelligence, pp. 1–5 (2004)

[16] Whitten, A., Tygar, J.D.: Why Johnny can't encrypt: A usability evaluation of PGP 5.0. In: Cranor, L.F., Garfinkel, S. (eds.) Security and Usability: Designing Secure Systems that People Can Use, pp. 679–702. O'Reilly Media, Sebastopol (2005); An earlier version was published in Proceedings of the 8th USENIX Security Symposium, pp. 169–183 (1999)

Author Index